Viruses of Fungi and Simple Eukaryotes

MYCOLOGY SERIES

Edited by

Paul A. Lemke

Department of Botany and Microbiology
Auburn University
Auburn, Alabama

Volume 1 Viruses and Plasmids in Fungi, *edited by Paul A. Lemke*

Volume 2 The Fungal Community: Its Organization and Role in the Ecosystem, *edited by Donald T. Wicklow and George C. Carroll*

Volume 3 Fungi Pathogenic for Humans and Animals (in three parts), *edited by Dexter H. Howard*

Volume 4 Fungal Differentiation: A Contemporary Synthesis, *edited by John E. Smith*

Volume 5 Secondary Metabolism and Differentiation in Fungi, *edited by Joan W. Bennett and Alex Ciegler*

Volume 6 Fungal Protoplasts, *edited by John F. Peberdy and Lajos Ferenczy*

Volume 7 Viruses of Fungi and Simple Eukaryotes, *edited by Yigal Koltin and Michael J. Leibowitz*

Other Volumes in Preparation

Viruses of Fungi and Simple Eukaryotes

edited by

Yigal Koltin
Tel Aviv University
Ramat Aviv, Israel

Michael J. Leibowitz
University of Medicine and Dentistry of New Jersey—
Robert Wood Johnson Medical School
Piscataway, New Jersey

MARCEL DEKKER, INC. New York and Basel

Library of Congress Cataloging-in-Publication Data

Viruses of fungi and simple eukaryotes.

(Mycology series ; v.7)
Includes index.
1. Fungal viruses. 2. Viral genetics. I. Koltin, Yigal [date]. II. Leibowitz, Michael J. [date]
III. Series.
QR343.V59 1988 576'.6483 88-397
ISBN 0-8247-7890-1

MARCEL DEKKER, INC.
270 Madison Avenue, New York, New York 10016

Current printing (last digit):
10 9 8 7 6 5 4 3 2 1

PRINTED IN THE UNITED STATES OF AMERICA

Rokuo Ushiyama (April 29, 1939-April 11, 1986)

We note with sadness the recent death of Dr. Rokuo Ushiyama. Dr. Ushiyama was born on April 29, 1939. After training in the Agriculture Departments of Gifu University and the University of Osaka Prefecture, he received the Ph.D. in Cell Biology from Oakland University in New Zealand in 1970. Subsequently he returned to Japan and in 1972 began work at the Tottori Mycological Institute, where he was promoted to Chief of the Department of Genetics and Physiology. He was also awarded a doctoral degree in Agriculture from the University of Osaka Prefecture in 1983.

Starting in 1961 he published a long series of papers on various aspects of plant pathology and mycology: These established him as a leader in his field. Although he took many approaches to virology, he was known especially for his skill as an electron microscopist. His morphological work with the viruses of *Lentinus edodes* was especially noteworthy and earned him the admiration of the many workers who have been frustrated by the difficulty of getting high-quality micrographs of fungal material. The search for molecular correlates of his morphological studies will occupy fungal virologists for years to come.

In addition to our memories of Dr. Ushiyama's scientific accomplishments, some of us were fortunate enough to know him personally. He played an active role as a speaker and organizer of many international meetings in microbiology and virology. At the Sixth International Congress of Virology, held in Sendai, Japan, in 1984, Dr. Ushiyama was co-chairman of the workshop on double-stranded RNA viruses of fungi. He took his obligation as our host in Japan very seriously and went out of his way to make all participants in that workshop feel welcome. With his typical concern for the feelings of others, he politely declined our invitation to speak at the EMBO Workshop on which this volume is based without letting us know that he was seriously ill. Dr. Rokuo Ushiyama died on April 11, 1986. This volume is dedicated to his memory.

Yigal Koltin
Michael J. Leibowitz

Series Introduction

Mycology is the study of fungi, that vast assemblage of microorganisms which includes such things as molds, yeasts, and mushrooms. All of us in one way or another are influenced by fungi. Think of it for a moment—the good life without penicillin or a fine wine. Consider further the importance of fungi in the decomposition of wastes and the potential hazards of fungi as pathogens to plants and to humans. Yes, fungi are ubiquitous and important.

Mycologists study fungi either in nature or in the laboratory and at different experimental levels ranging from descriptive to molecular and from basic to applied. Since there are so many fungi and so many ways to study them, mycologists often find it difficult to communicate their results even to other mycologists, much less to other scientists or to society in general.

This series establishes a niche for publication of works dealing with all aspects of mycology. It is not intended to set the fungi apart, but rather to emphasize the study of fungi and of fungal processes as they relate to mankind and to science in general. Such a series of books is long overdue. It is broadly conceived as to scope, and is to include textbooks and manuals as well as original and scholarly research works and monographs.

The scope of the series will be defined by, and hopefully will help define, progress in mycology.

Paul A. Lemke

Foreword

The mycoviruses are intracellular viruses found in both unicellular and filamentous fungi. The typical mycovirus is usually identified in the cytoplasm as an RNA (often double-stranded) encapsidated in a protein coat. Life would have been much easier for fungal molecular biologists and plant pathologists if the viruses of fungi infected and killed their hosts so that they could be studied with the simple plaque assay used in the analysis of other prokaryotic and eukaryotic viruses. Unfortunately, the mycoviruses do not kill their hosts; rather host and attendant viruses have reached a mutually satisfactory modus vivendi: The virus may be no more obtrusive or noticeable than a vacuole. The absence of an independent lytic cycle may explain why, for many years, the mycoviruses were known only as a fungal cellular component that induced interferon (Chapter 1).

The study of most of the mycoviruses began with the analysis of some property conferred upon the host by the presence of the virus. This route is most obvious in the analysis of the saccharomyces killer virus (ScV), where toxin production and immunity are conferred by the presence of the killer double-stranded RNAs. Detailed genetic and biochemical analysis led to the identification of host and viral genes required for toxin production, resistance to killing by the toxin, and maintenance of the virus (Chapters 5 and 8). It is now clear that M dsRNA encodes the preprotoxin and extremely likely that L encodes the viral capsid. Studies of mutants defective in toxin production have been important in unraveling the host secretory pathway, since the preprotoxin traverses the same pathway as host-encoded secretory proteins and is processed by many of the same proteases. Indeed, the genes for these proteases were first discovered as mutations that failed to produce killer toxin. Progress on the toxins of the *Ustilago maydis* killer system is following rapidly on the heels of the saccharomyces work (Chapter 10).

The ability to isolate intact virions, capsids containing either the L-A or M dsRNAs, permits a direct biochemical analysis of the killer virus. The components of the virions, the coding capacity of the dsRNAs, and the transcription and translation of the viral genomes can be studied by standard technologies (Chapters 5-9). The identification of the expected polymerases, viral capsid proteins, and mode of replication (conservative) has made the killer particles seem as if they had the accoutrements and habits associated with the more traditional reoviruses.

Perhaps the most powerful approach to unraveling the functions encoded by the killer genome has been the cloning of the cDNA for the gene encoding the M_1 toxin (Chapters 8 and 9). Fusion of the cDNA for the toxin gene to the regulatable phosphate promoter on a yeast plasmid has permitted a deeper understanding of the mechanism of toxin production and, most especially, immunity. The fact that a DNA plasmid can confer both toxin production and immunity is mirrored in the naturally occurring linear DNA killer plasmids found in *Kluyveromyces lactis* (Chapter 12). Remarkably, the kluveromyces killer DNAs function when transformed into saccharomyces although they are not co-immune with the saccharomyces killers and are incompatible with the presence of yeast mitochondrial DNA.

Surprisingly, yeast cells contain not only dsRNA particles but also retroviral particles. As was the case with the dsRNAs, the Ty retrovirus was not first recognized as a virus. The confusion surrounding the identity of the Ty element was compounded by the historical fact that Ty elements were first discovered as DNA transposable elements found at many locations in the genome. The elements were noticed because they caused mutations either by transposition into a gene (Chapter 2) or by homologous recombination events (Chapter 4) that resulted in deletions. The Ty element seemed like a legitimate yeast gene because many host genes were required for its transcription (Chapter 3). The correct identity of the Ty element was revealed only when an element was fused to the inducible galactose promoter and forced to transpose upon galactose induction (Chapter 2). These experiments showed that the Ty element transposed via an RNA intermediate and that viral particles containing the Ty RNA and a Ty-encoded reverse transcriptase were massively produced during the induction. Recent work on the structure of the RNA and the proteolytic processing has bolstered the connection between Ty elements and retroviruses. The presence of reverse transcriptase in yeast and the existence of naturally occurring DNA killers raise the possibility that the DNA killers are cDNAs made by reverse transcription of some ancestral dsRNA killer.

Many other fungi exhibit fascinating biological phenomena associated with dsRNAs, including the killer systems in pathogenic yeasts (Chapter 11) and hypovirulence in *Endothia parasitica* (Chapters 13, 14, and 17). The hypovirulence caused by *E. parasitica* has practical ramifications for the elimination of many fungal diseases. Strains of

E. parasitica containing dsRNA are avirulent, whereas those without the dsRNA cause chestnut blight. The treatment of a diseased chestnut tree with avirulent *E. parasitica* containing dsRNA viruses causes healing because the dsRNA passes from the avirulent strain to the virulent strain and confers avirulence upon the recipient (transmissible hypovirulence). Other pathogenic fungi, *Ophiostoma ulmi* (Chapter 15), *Helminthosporium victoriae* (Chapter 16), and *Rhizoctonia solani* (Chapter 18), are also infested with dsRNAs. For these fungi, the relationship between the presence of the dsRNA and the phenotype (i.e., hypovirulence of the fungus) is not firmly established. The ideal experiment would be the transfer of viruses to a fungal host free of viruses with the concomitant acquisition of hypovirulence. Remarkably extracellular transmission of the saccharomyces killer virus has now been achieved by infection of either spheroplasts or conjugating cells (Chapter 9). The message to plant pathologists is clear: Follow the path paved by yeast geneticists. Even the mycoviruses, once thought intractable, cannot resist the powerful combination of genetics and biochemistry.

Gerald R. Fink
Whitehead Institute/M.I.T.
Cambridge, Massachusetts

Preface

This volume in the Mycology series presents a summary of current research on the viruses of simple eukaryotes. Most of the best-studied hosts for these viruses are fungi, although the viruses of eukaryotic algae are also covered. The authors were participants in the European Molecular Biology Organization Workshop entitled "Viruses of Fungi and Lower Eukaryotes," held on August 24-29, 1986, in Jerusalem, Israel; however, this book is not a meeting proceedings volume. Instead, the authors have done their best to present their work to a general scientific audience, including students and those working in other disciplines. The chapters include literature reviews, as well as new data, in order to make this work useful to virologists, plant pathologists, medical mycologists, molecular biologists, geneticists, and students of the biological sciences.

Since the first volume in the Mycology series, *Viruses and Plasmids in Fungi*, edited by Paul A. Lemke (1979), research on viruses of simple eukaryotic organisms has progressed in many directions. Some mycoviruses, especially those of yeasts, are being studied as experimentally manipulable models in order to understand the mechanisms of virus replication and expression and host-virus interactions in eukaryotic cells. Thus, some scientists are attracted to these viruses and their hosts because they can be studied using genetic and molecular approaches that are more difficult or impossible in the viruses of higher plants and animals. As can be seen in this book, many mycoviruses bear striking resemblances to the viruses of higher organisms, especially to the retroviridae and reoviridae, and to the cryptic viruses of plants. Many of the viruses of simple eukaryotes infect hosts that are pathogens of agricultural and medical importance. The study of these viruses is not only of theoretical interest, but also of practical value. As is discussed in the text, a number of viruses infecting plant pathogenic fungi alter the virulence of these fungi for their plant hosts. In some cases, virus infection renders the host fungus hypovirulent, raising the prospect of using the virus to treat or prevent fungal infections. In other cases, viruses appear to act as virulence determinants in the fungal host. The relationship of viruses to host virulence has been demonstrated most conclusively in the case of the viruses infecting plant pathogenic fungi. However, viruses also infect fungi pathogenic for humans and other animals. The possible relationship of these viruses to fungal virulence remains an unknown but exciting possibility.

In addition to their importance as models for viruses of higher organisms and as determinants of host cell virulence, viruses of simple eukaryotes can be viewed as probes for the study of eukaryotic molecular and cell biology. As is true of all viruses, these viruses depend on multiple host cell gene products for their replication and expression. Therefore, the function of many cellular processes can be studied by means of their actions on viruses. Especially in the cases of the mycoviruses, mechanisms by which eukaryotic cells express genetic information are being studied. Significant progress has been made in using these viruses as probes to study transcriptional and translational mechanisms, protein transport, processing, secretion, and the mechanisms by which cells defend themselves against toxic proteins. The availability of these viruses adds another experimental approach to the study of their host cells.

Thus, some workers in the field are attracted to these viruses as general models for understanding eukaryotic cells and their viruses, while other are motivated by an interest in a specific virus, host, or disease process. Recent progress has narrowed the gap that may have once existed between these two approaches, since the techniques of molecular genetics are being successfully used to study these viruses both as experimental models and as modifiers of virulence of economically important pathogens. Therefore, we have each done our best to make this work accessible to our colleagues working in different systems and to the general scientific community.

These host-virus systems provide a diverse and exciting field of investigation. We view this work not as an encyclopedic compendium of work completed, but rather as a prospectus for many new experimental approaches. We hope that we have conveyed our enthusiasm along with the information.

Yigal Koltin
Michael J. Leibowitz

Contents

Series Introduction (Paul A. Lemke) *v*
Foreword (Gerald R. Fink) *vii*
Preface *xi*
Contributors *xv*

1. Discovery of a Mycovirus: An Indirect Route 1
Walter J. Kleinschmidt

2. Yeast Ty Elements as Retroviruses 15
Jef D. Boeke and David J. Garfinkel

3. Transcriptional Regulation of Ty Elements in *Saccharomyces cerevisiae* 41
Fred Winston

4. Recombination Associated with Yeast Retrotransposons 63
Susan W. Liebman and Susan Picologlou

5. Double-Stranded RNA Replication in *Saccharomyces cerevisiae*: Genetic Control and In Vitro Replication 91
Tsutomu Fujimura, Rosa Esteban, and Reed B. Wickner

6. Structure, Transcription, and Replication of Killer Virus dsRNAs 117
J. A. Bruenn, M. E. Nemeroff, M. Lee, D. F. Pietras, J. J. Dowhanick, and L. J. Field

7. Transcription and Translation of the Yeast Killer Virus Genome 133
Michael J. Leibowitz, Iffat Hussain, and Teresa L. Williams

8. Secretion and Action of Yeast K1 Killer Toxin 161
Howard Bussey, Charles Boone, Aleksandra Dmochowska, Deirdre Greene, Hong Zhu, Susan J. Lolle, Thierry Vernet, Daniel Dignard, and David Y. Thomas

9. Acquisition and Expression of the Killer Character in Yeast 179
S. L. Sturley, M. El-Sherbeini, S.-H. Kho, J. L. LeVitre, and K. A. Bostian

10. The Killer System of *Ustilago maydis*: Secreted Polypeptides Encoded by Viruses 209
Yigal Koltin

11. Killer Systems and Pathogenic Yeast 243
Judith S. Kandel

12. Kluyveromyces Linear DNA Plasmids 265
Norio Gunge

13. Structural Properties of dsRNA: Genetic Elements Associated with Plants and Plant Pathogenic Fungi 283
James Tartaglia and Donald L. Nuss

14. Detection and Interpretation of dsRNA 309
J. A. Dodds, R. A. Valverde, and D. M. Mathews

15. dsRNA and Disease Factors of the Aggressive Subgroup of *Ophiostoma ulmi* 327
H. J. Rogers, K. W. Buck, and C. M. Brasier

16. Viruses of *Helminthosporium victoriae* 353
Said A. Ghabrial

17. Viruses of *Endothia parasitica* 371
Neal K. Van Alfen

18. dsRNA Virus of *Rhizoctonia solani* 387
Aliza Finkler, Bat-Sheva Ben-Zvi, and Yigal Koltin

19. Viruses of Eukaryotic Chlorella-like Algae 411
James L. Van Etten, Anne M. Schuster, and Russel H. Meints

Index *429*

Contributors

Bat-Sheva Ben-Zvi, M.Sc. Tel Aviv University, Ramat Aviv, Israel

Jef D. Boeke, Ph.D. The Johns Hopkins University School of Medicine, Baltimore, Maryland

Charles Boone McGill University, Montreal, Quebec, Canada

*K. A. Bostian, Ph.D.** Brown University, Providence, Rhode Island

C. M. Brasier, Ph.D., D.Sc. Forest Research Station, Farnham, Surrey, England

J. A. Bruenn, Ph.D. State University of New York at Buffalo, Buffalo, New York

K. W. Buck, Ph.D., D.Sc. Imperial College of Science and Technology, London, England

Howard Bussey McGill University, Montreal, Quebec, Canada

Daniel Dignard Biotechnology Research Institute, National Research Council of Canada, Montreal, Quebec, Canada

Aleksandra Dmochowska, Ph.D. McGill University, Montreal, Quebec, Canada

J. A. Dodds University of California at Riverside, Riverside, California

J. J. Dowhanick State University of New York at Buffalo, Buffalo, New York

M. El-Sherbeini, Ph.D. Brown University, Providence, Rhode Island

---

**Present affiliation*: Merck Sharp & Dohme Research Laboratories, Rahway, New Jersey

Rosa Esteban, Ph.D. National Institute of Diabetes, Digestive and Kidney Diseases, National Institutes of Health, Bethesda, Maryland

*L. J. Field, Ph.D.** State University of New York at Buffalo, Buffalo, New York

Aliza Finkler, Ph.D. Tel Aviv University, Ramat Aviv, Israel

Tsutomu Fujimura, Ph.D. National Institute of Diabetes, Digestive and Kidney Diseases, National Institutes of Health, Bethesda, Maryland

David J. Garfinkel, Ph.D. Frederick Cancer Research Facility, Frederick, Maryland

Said A. Ghabrial, Ph.D. University of Kentucky, Lexington, Kentucky

Deirdre Greene McGill University, Montreal, Quebec, Canada

Norio Gunge, D.Sc. Kumamoto Institute of Technology, Kumamoto, Japan

Iffat Hussain, Ph.D. University of Medicine and Dentistry of New Jersey—Robert Wood Johnson Medical School, Piscataway, New Jersey

Judith S. Kandel, Ph.D., M.P.H. California State University, Fullerton, California

S.-H. Kho Brown University, Providence, Rhode Island

Walter J. Kleinschmidt, Ph.D. [*ret.*] Lilly Research Laboratories, Indianapolis, Indiana

Yigal Koltin, Ph.D. Tel Aviv University, Ramat Aviv, Israel

M. Lee State University of New York at Buffalo, Buffalo, New York

Michael J. Leibowitz, M.D., Ph.D. University of Medicine and Dentistry of New Jersey—Robert Wood Johnson Medical School, Piscataway, New Jersey

J. L. LeVitre† Brown University, Providence, Rhode Island

Susan W. Liebman, Ph.D. University of Illinois at Chicago, Chicago, Illinois

Susan J. Lolle, Ph.D. McGill University, Montreal, Quebec, Canada

D. M. Mathews, M.S. University of California at Riverside, Riverside, California

---

Present affiliations:
*Cold Spring Harbor Laboratory, Cold Spring Harbor, New York
†Collaborative Genetics, Inc., Waltham, Massachusetts

Russel H. Meints, Ph.D. University of Nebraska, Lincoln, Nebraska

M. E. Nemeroff State University of New York at Buffalo, Buffalo, New York

Donald L. Nuss, Ph.D. Roche Institute of Molecular Biology, Roche Research Center, Nutley, New Jersey

Susan Picologlou, Ph.D. University of Illinois at Chicago, Chicago, Illinois

D. F. Pietras State University of New York at Buffalo, Buffalo, New York

*H. J. Rogers, Ph.D.** Imperial College of Science and Technology, London, England

Anne M. Schuster, Ph.D. University of Nebraska, Lincoln, Nebraska

S. L. Sturley, Ph.D.† Brown University, Providence, Rhode Island

James Tartaglia, Ph.D.‡ Roche Institute of Molecular Biology, Roche Research Center, Nutley, New Jersey

David Y. Thomas, Ph.D.§ Biotechnology Research Institute, National Research Council of Canada, Montreal, Quebec, Canada

R. A. Valverde, Ph.D.‖ University of California at Riverside, Riverside, California

Neal K. Van Alfen, Ph.D. Utah State University, Logan, Utah

James L. Van Etten, Ph.D. University of Nebraska, Lincoln, Nebraska

Thierry Vernet, Ph.D. Biotechnology Research Institute, National Research Council of Canada, Montreal, Quebec, Canada

Reed B. Wickner, M.D. National Institute of Diabetes, Digestive and Kidney Diseases, National Institutes of Health, Bethesda, Maryland

Teresa L. Williams, M.S. University of Medicine and Dentistry of New Jersey—Robert Wood Johnson Medical School, Piscataway, New Jersey

---

Present affiliations:
*Institute of Plant Science Research, Trumpington, Cambridge, England
†University of Wisconsin, Madison, Wisconsin
‡Wadsworth Center for Laboratories and Research, New York State Department of Health, Albany, New York
§McGill University, Montreal, Quebec, Canada
‖Plant Pathology, Ball PanAm Plant Co., Parrish, Florida

Fred Winston Harvard Medical School, Boston, Massachusetts

Hong Zhu McGill University, Montreal, Quebec, Canada

Viruses of Fungi and Simple Eukaryotes

1

Discovery of a Mycovirus: An Indirect Route

Walter J. Kleinschmidt
Lilly Research Laboratories, Indianapolis, Indiana

STATOLON: A FERMENTATION PRODUCT OF *Penicillium stoloniferum* WITH ANTIVIRAL ACTIVITY

The observation in the early 1950s at the Lilly Laboratories that a culture filtrate of *Penicillium stoloniferum* possessed prophylactic activity against several virus infections in mice created some stir of excitement among searchers for antiviral substances (Powell et al., 1952). Early attempts at elucidation of the activity were not rewarding. Our own attempts at isolation of the activity, initiated in the 1960s, were hampered by variable activity of the starting filtrates, loss of activity through aggregation, solubility problems, and long assay time (2 weeks). Eventually we obtained an active preparation by precipitation of the fermentation broth with water-soluble organic solvents and reconstitution of the precipitate in 2% $NaHCO_3$, pH 8.5-9.0. We named this concentrated, active preparation statolon (Kleinschmidt and Probst, 1962).

INDUCTION OF INTERFERON BY STATOLON

The prophylactic nature of the inhibition of viruses by statolon plus the fact that it inhibited increasing number of virus types suggested the possibility that it perhaps acted via the production of another inhibitor, namely interferon, the discovery of which was reported in 1957 (Isaacs and Lindenmann, 1957). On testing statolon as an inducer of interferon in chick embryo cells (Kleinschmidt et al., 1964) we did indeed find a new inhibitor that made its appearance 4-5 hr after the addition of statolon and increased in quantity with time of contact of statolon with the cells (Kleinschmidt and Murphy, 1965). This newly synthesized inhibitor possessed characteristics that were different from statolon and coincided with the properties reported for interferon, including pH and temperature stabilities and susceptibility to proteases. Interferon was found to be the inhibitor induced by statolon in animals as well as in tissue culture. Sera of animals injected with statolon were found to contain an inhibitor again possessing the characteristics of interferon (Kleinschmidt and Murphy, 1967; Kleinschmidt and Murphy, 1965). The time of appearance of circulating interferon was 5-6 hr after administration of statolon, and the time course of decay of interferon activity in the mouse coincided with that seen with viral inducers. The response to statolon was dose-dependent, and production of circulating interferon in mice correlated with protection against lethal viral infection. The phenomenon of resistance, in which a cell culture or animal becomes hyporeactive to further stimulation with an inducer for about a week, was also observed. One injection of statolon was found to afford some protection to mice against a lethal MM virus (a strain of mouse encephalomyocarditis virus) infection for at least two weeks (Kleinschmidt and Murphy, 1967; Kleinschmidt and Streightoff, 1971). Statolon also was found to be capable of inducing circulating interferon in primates, demonstrated in the cynomolgus monkey (Kleinschmidt and Ellis, 1967; Kleinschmidt and Murphy, 1967).

DISCOVERY OF A VIRUS IN STATOLON

The finding that statolon had the capability of inducing interferon permitted more extensive isolation of statolon. With production of circulating interferon by the test animal as the end-point measurement, rather than death or survival of a virus-infected animal, the rapidity of assays was increased considerably. Mice were injected with statolon samples and bled about 16 hr later. The interferon potency of the sera of these mice was then measured on tissue cultures of mouse L cells, challenged with vesicular stomatitis virus (VSV). Greater solubility of statolon, obtained with adjustments of salt and pH, made further purification possible, including attempts at zonal centrifugation on sucrose gradients. These studies were done in collaboration with Dr. Lee Ellis at Lilly (Ellis and Kleinschmidt, 1967; Kleinschmidt and Ellis, 1967). When dissolved statolon samples were placed on sucrose gradients and centrifuged for 16-24 hr, two bands visible by reflected were formed; a light zone (30% sucrose) and a heavy zone (45% sucrose). Fractions were collected and assayed for interferon-inducing capacity as outlined above by injecting mice with the fractions and determining the interferon in their sera.

Both of the two visible bands were found to possess activity. All fractions other than those derived from these two bands were negative. The heavier band received attention first because it contained the major portion of the activity. Electronmicroscopic examination of this band disclosed numerous particles possessing the typical morphology of viruses (Ellis and Kleinschmidt, 1967; Kleinschmidt and Ellis, 1967). Examination of negatively stained preparations under high magnification tion revealed hexagonal particles measuring 30 nm in diameter (Figure 1).

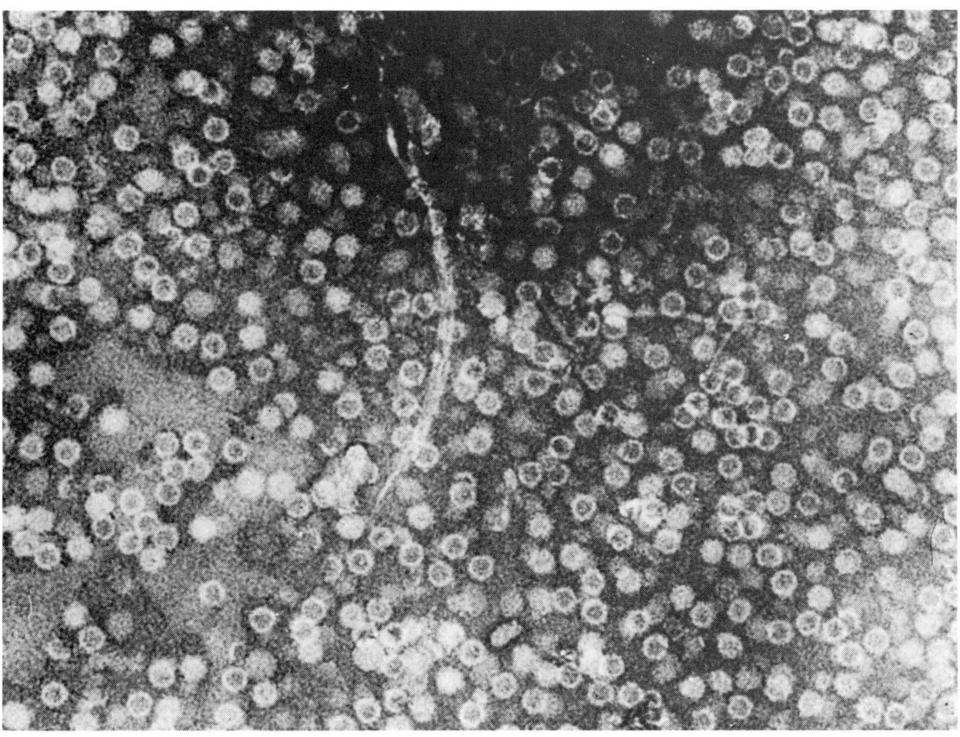

Figure 1 An early electron micrograph of the mycovirus from *Penicillium stoloniferum* purified by isopycnic and rate zonal density gradient centrifugation. Negative stain with phosphotungstic acid.

Our reports (Ellis and Kleinschmidt, 1967; Kleinschmidt and Ellis, 1967) that there were virus particles in statolon and that these particles were responsible for statolon's interferon-inducing activity were met with extreme skepticism. Of course, little if any lysis of the mycelia had been observed, and we had not applied Koch's postulate of infecting a host cell with the virus. Nevertheless, all evidence indicated that virus particles were present in statolon and that these particles were associated with antiviral activity in animals and in tissue culture. Uninoculated media, which admittedly contained some complex ingredients, did not produce particles or interferon-stimulating activity after an incubation for the same period as the normal statolon fermentation (5-7 days). Both the particles and interferon-stimulating activity were detected in *P. stoloniferum*-inoculated media only after an incubation time of 4 days had ensued. No particles were found in inactive fractions obtained by sucrose gradient centrifugation, and there was a direct correlation of interferon-inducing activity with numbers of particles, counted with the aid of the electron microscope. Electron microscopic examination of thin sections of mycelia of *P. stoloniferum* demonstrated particles possessing identical morphology to those obtained by gradient centrifugation. Particles were not detected in the other band of interferon-inducing activity (30% sucrose); at that time it was not known whether the activity of this fraction was mold-derived or virus-derived.

The finding of the mycovirus particles in *P. stoloniferum* was the first time viruses were found associated with a mold belonging to the class Fungi Imperfecti. That these particles were indeed viruses was subsequently demonstrated (Borré et al., 1971) by producing lytic plaques on *Penicillium variabile*. The criterion of infectivity of the penicillium viruses had then clearly been met. Hollings (1962) had earlier isolated and identified a virus that was responsible for the disease of the cultivated mushroom *Agaricus bisporus* (Homobasidiomyces).

PROPERTY OF DOUBLE-STRANDEDNESS OF MYCOVIRUS RNA

By the use of adjuvants composed of double-stranded complexes of polyribonucleotides such as poly rA:rU, Braun and Nakano (1965, 1967) were able to stimulate increased production of antibodies in mice over that with antigen alone. Their results suggested to Field et al. (1967) that such double-stranded polyribonucleotides might serve as adjuvants toward increasing the production of interferon. When they tested their hypothesis with poly rI:rC, these authors made the important discovery that this polyribonucleotide itself could act as a direct stimulator of interferon. An active dsRNA was also extracted by these investigators from the mycelia of *Penicillium funiculosum*

(Lampson et al., 1967), the source of a product called helenine (Shope, 1953a,1953b) which was found capable of inducing interferon (Rytel et al., 1966). We examined the virus particles of *P. stoloniferum* and found the nucleic acid present in these particles to be double-stranded in nature (Kleinschmidt et al., 1968). The light, active fraction obtained by sucrose gradient centrifugation of statolon also was identified as dsRNA, derived from the mycovirus.

DOUBLE-STRANDED RNA, A POTENT INDUCER OF INTERFERON

Present Status of Double-Stranded RNA as Inducer

Double-stranded RNA is the ultimate effective inducer of interferon. Infective single-stranded RNA viruses, such as Sendai or Newcastle disease viruses, are capable of inducing interferon, apparently because of the synthesis of a replicative dsRNA.

A word on the structural essentials of dsRNA necessary for activity may be appropriate here. Of the dsRNAs, the synthetic double-stranded polyribonucleotide poly rI:rC has received the most attention. A helical double-stranded structure is essential for activity, for single-stranded homopolymers are inactive. The sequence of bases in poly rI:rC apparently is of little consequence, for the associated homopolymers poly rI:rC and the complexed, alternating polymer poly (rIrC) were equally efficient as inducers of interferon, and in the same study poly dI:dC was found inactive, indicating the importance of the 2'-OH of the sugar moiety (Colby and Chamberlin, 1969). Double-stranded RNA:DNA hybrids are also not active as inducers. Poly rA:rU posseses little, if any, activity. A specific conformation of the dsRNA, or, more likely, the capability of transitional conformations in assuming the active state, may offer an explanation for the activity of dsRNAs (Yang and Samejima, 1969; Arnott et al., 1968; Mahendrasingham et al., 1986).

Molecular weight of the dsRNA is an important determinant of activity. Lampson et al. (1970), in their studies with poly rI:rC, showed that decreasing the size of active high molecular weight preparations of 7.8×10^3 kDa by sonic radiation to 4.6×10^2 kDa reduced the capacity of the poly rI:rC to protect mice against a pneumonia virus. Others (Tytell et al., 1970; Niblack and McCreary, 1971) showed that decreasing the lengths of the separate homopolymers of poly rI:rC to less than $1\text{-}2 \times 10^2$ kDa resulted in decreased activity of the polynucleotide:oligonucleotide complex. In collaborative studies with Dr. Van Etten and Dr. Vidaver (Kleinschmidt et al., 1974) we examined the three naturally occurring dsRNA segments of bacteriophage ϕ6 of *Pseudomonas phaseolicola* for their interferon-inducing capacity in mice. These three dsRNAs of bacteriophage ϕ6 are of different size, possessing molecular weights 2.3, 2.1, and 5.0 × 10^3 kDa, and their inducing capabilities were found to increase with increasing molecular weight, producing 990, 1780, and 2190 units of interferon, respectively.

Double-stranded RNA, both synthetic and natural, from whatever source, certainly proved effective as an inducer of interferon and held promise of clinical application. When a dsRNA-mycovirus was discovered in *Penicillium chrysogenum*, the penicillin-producing mold (Banks et al., 1969; Lemke and Ness, 1970), we turned our attention to this source of dsRNA. The dsRNA was extracted readily directly from the mycelia, and we proceeded to prepare adequate quantities. We were about ready to undertake studies leading to clinical testing when reports of toxicity and tumorigenicity appeared (Leonard and Eccleston, 1972; Jahiel et al., 1971), dampening hopes somewhat. Then there were reports that dsRNA was not active in primates, particularly humans, because of a dsRNase present in their sera (Nordlund et al., 1970; Lampson et al., 1969). We had earlier demonstrated that statolon was capable of inducing interferon in monkeys, resulting in protection against subcutaneously injected Yaba virus (Kleinschmidt and Ellis, 1967; Kleinschmidt and Murphy, 1967). We thought perhaps that the coat of the intact mycovirus of statolon furnished protection to the dsRNA against the dsRNase, thus permitting interferon induction.

The presence of the degradative dsRNase in the blood of humans did offer the salutary promise that toxicity of the dsRNA might be mitigated somewhat in humans. Humans may have evolved this dsRNase activity as a built-in protection against this noncompatible macromolecule. A potential cumulative toxic effect was eliminated by this protection provided by the nuclease. Little, however, would be served by a lessened hazard of toxicity if along with this decrease toxicity there was a parallel loss in activity. Nevertheless, if the dsRNA could be protected (as it appeared to be) by the intact mycovirus in our Yaba virus experiment in monkeys, perhaps activity could be generated by the inducing dsRNA prior to manifestation of toxicity. A search was then initiated in a number of laboratories to potentiate interferon induction by dsRNA in primates, using a variety of complexing agents including polyamines (Lampson et al., 1969), carboxymethyl cellulose in conjunction with polylysine (Levy et al., 1975), and polylysine (Lampson et al., 1981).

Our own effort was directed toward forming complexes of dsRNA with relatively small, nonpolymeric molecules as complexing agents. An excellent candidate was uncovered in experiments by Dr. Richard J. Douthart at Lilly, when he observed complexing of phosphatides with the aminoglycoside antibiotic tobramycin, the structure of which is given in Figure 2. When he applied tobramycin to dsRNA extracted from *P. chrysogenum* mycelium (PCMdsRNA) he found a stable complex had formed, with a binding constant of $K_{ap} \cong 6.7 \times 10^5\ M^{-1}$ (R. J. Douthart, unpublished observation). Collaborative studies were under-

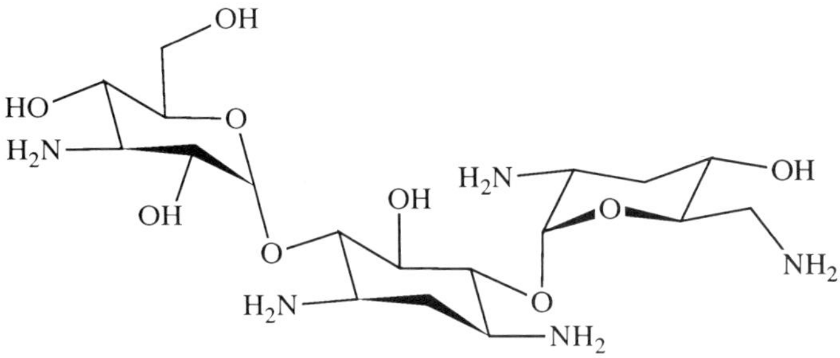

Figure 2 Structure of tobramycin.

taken to determine the interferon induction capacity of the complex and the degree of protection afforded to the dsRNA against human serum dsRNase (Douthart et al., 1982). The tobramycin-dsRNA complex was formed by mixing aliquots of dsRNA and a tobramycin stock solution together in 0.1 M NaCl, with a 10^{-2} M tris buffer (pH 6.8-7.0), the pH of the tobramycin solution having been adjusted to 6.8-7.0 prior to forming the complex. After mixing, the solution was allowed to incubate for 1 hr at ambient temperature to insure completion of complex formation. The formulations were characterized by the ratio of moles of tobramycin as free base to the moles of RNA phosphorus (B/P), and the concentration of the dsRNA in μg. Preliminary experiments in mice had demonstrated that the dsRNA-tobramycin complex was an active inducer of interferon. The tobramycin bound to the dsRNA therefore, did not alter the capacity of the dsRNA to induce interferon.

We determined the protection exerted by the tobramycin as a complexing agent with dsRNA against human serum dsRNase. Human serum at 0.12%, was incubated with the free dsRNA or the dsRNA-tobramycin complex (B/P = 1) for 2 hours at 37°C, the total dsRNA concentration (both free and in the complex) was 20 μg/ml. The reaction was quenched in an ice bath and aliquots were taken for immediate injection into mice for subsequent interferon assays. From the results, given in Table 1, it is clear that the mouse system responded to the degree of degradation experience by the dsRNA and the dsRNA-tobramycin complex during preincubation with human serum in an expected way. Human serum treatment of non-complexed dsRNA decreased the interferon response by 57%. In contrast, there was no significant effect of serum observed on the interferon response induced by the dsRNA-tobramycin complex. Further, the increased level of interferon induction by the complex as compared to dsRNA alone, suggests a protection against the low levels of degradative activity in the mouse.

The small amounts of human serum used in the above experiment are, however, far removed from what would be experienced during

Table 1 Effect of Human Serum on Interferon Production by PCMdsRNA and PCMdsRNA-Tobramycin (B/P = 1)

Sample[a]	Human serum pretreatment[b]	IFN units[c]
PCMdsRNA	-	1050
PCMdsRNA/tobramycin	-	3200
PCMdsRNA	+	447
PCMdsRNA/tobramycin	+	3470
Tobramycin	+	30
Tobramycin	-	<30
Buffer	-	<30

[a]PCMdsRNA (*P. chrysogenum* dsRNA) was administered i.p. at 10 μg/13-15-gm mouse.
[b]Samples to be injected were pretreated with 3.6 μl of human serum in a reaction volume of 3.0 ml for 2 hr at 37°C immediately prior to injection into mice.
[c]Serum levels of IFN were determined 16 hr after i.p. treatment.
Source: Douthart et al. (1982).

clinical use of the complex. In the next experiment we increased the volume of human serum in the test system to 83% (2.5 ml serum in a total volume of 3.0 ml), and the incubation period with the serum was again 2 hours at 37°C. Complexes were prepared to give a final concentration of 20 μg/ml of dsRNA in the reaction tube, and the B/P ratio was varied from 4 to 40. The results of the interferon assays are given in Figure 3. Interferon induction is seen to increase as the B/P increases and begins to plateau at about 350 units at a B/P ≅ 12. The untreated dsRNA control yielded 1380 units of interferon, indicating a 75% loss of interferon induction capacity at the plateau with the complex; but a complete loss of activity was seen in the serum-treated uncomplexed dsRNA control. So, in summary, exposure of the tobramycin-dsRNA complex to human serum results in some degradation but still permits the production of significantly high titers of interferon, while uncomplexed dsRNA is completely vulnerable to the enzyme with no capacity to induce interferon after exposure. The aminoglycoside, tobramycin, would appear to be an excellent choice as a complexing agent for dsRNA. Its structural configuration with its strategic five amino groups and its excellent fit into the shallow groove

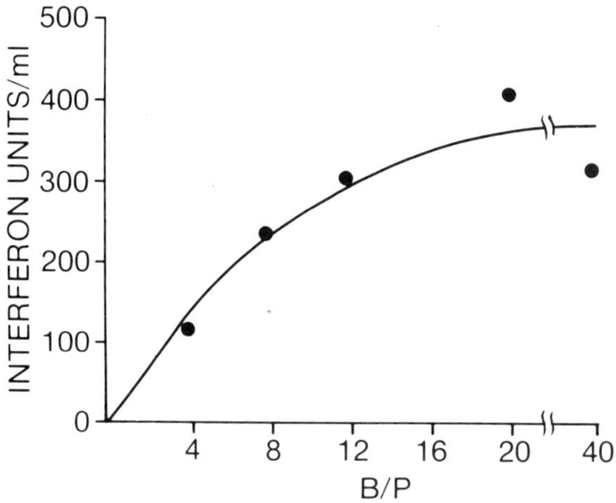

Figure 3 Effect of increasing B/P ratio on the ability of human serum PCMdsRNA-tobramycin complex to induce interferon. (Douthart et al., 1982).

of the dsRNA permits a binding to the dsRNA that is sufficiently strong to produce protection against the dsRNase of human serum but not too tight and/or encompassing to block with interferon-inducing activity.

We wished to compare the efficacy of tobramycin-complexed *P. chrysogenum* mycovirus dsRNA versus that of uncomplexed dsRNA in induction of interferon in a primate animal. The dsRNA-tobramycin complex was prepared as previously at B/P = 7 and injected intravenously into each of four rhesus monkeys at a dosage concentration of 1 mg dsRNA/kg of body weight. Controls of uncomplexed dsRNA free tobramycin, and buffer were also employed at equivalent concentrations as present in the complex solution, using 4, 2, and 2 monkeys, respectively. After 8 hr the monkeys were bled and the sera prepared and assayed for interferon content on Vero cells (African green monkey) by challenging with vesicular stomatitis virus 48 hr after application of sera to cells. The results are given in Table 2 (W. J. Kleinschmidt, E. B. Murphy, and R. M. Schultz, 1982, unpublished observations).

Again we see that the tobramycin-dsRNA complex was protected against the nucleolytic activity of the nuclease in the blood of the monkey and induced significant levels of interferon, while the uncomplexed dsRNA was completely degraded with no production of interferon observed. Tobramycin continued to fulfill its potential as an excellent choice as a blocker of the nuclease action, thereby permitting the production in a primate of adequate quantities of interferon approaching desirable blood concentrations for clinical application. There probably was some degradation of dsRNA as was previously seen in the human

Table 2 Efficacy of PCMdsRNA-Tobramycin Complex to Induce Interferon in the Monkey

Monkey	Sample	Units of interferon/ml serum
1	dsRNA	30
2	dsRNA	<30
3	dsRNA	30
4	dsRNA	<30
5	dsRNA-Tobramycin	320
6	dsRNA-Tobramycin	1000
7	dsRNA-Tobramycin	320
8	dsRNA-Tobramycin	1000
9	Tobramycin	0
10	Tobramycin	0
11	Buffer	0
12	Buffer	0

sera/mouse experiment (Douthart et al., 1982) but the protection was still satisfactory. Complete protection probably is not desirable at any rate, since it could lead to eventual toxicity. It should be noted that no visible indications of toxicity were observed in any of the monkeys during the experiment.

The initial biological activity of interferon upon its discovery was that of an antiviral substance; later it also was found to possess activity against certain neoplasms. The recent reports on the efficacy of interferon against hematologic malignancies, chronic myelogenous leukemia, and hairy cell leukemia (Bonnem and Spiegel, 1984), have stimulated increased interest in the use of interferon and possibly its inducer(s) in the clinic. The availability of an effective inducer, capable of producing endogenous interferon in humans, is highly desirable. An inducer would have the potential advantage over exogenous interferon of inducing a variety of interferons, each with its specific regulatory activities. Endogenously induced interferon also precludes the production of antibody, as has been seen with exogenous interferon (Gutterman et al., 1981). Also, the inducer would have the potential advantage of a longer duration of activity of endogenously induced interferon over that of exogenously injected interferon. Our experiments in mice indicate weekly administration should be sufficient, while exogenous interferon must be injected daily. Exogenous, recombinant DNA-produced interferon is presently occupying the stage, but an inducer is waiting in the wings.

EXTENSIVE OCCURRENCE OF VIRUSES OF FUNGI AND SIMPLE EUKARYOTES

Since the discovery of the presence of virus particles in *P. stoloniferum* the list of viruses in mold species alone had grown to more than 50 by 1974 and more than double that number since (Lemke and Nash, 1974). Other simple eukaryotes, such as the yeasts, have also been found to contain such virus particles, most of which contain a dsRNA as the genome. It is of interest to note that in 1953 O'Dell et al. reported that a yeast RNA was effective in inhibiting MM virus in mice. They had an active RNA in hand, and today there is little doubt as to the nature of their RNA. Availability of better methods for nucleic acid study at that time might have indicated the source of the nucleic acid as viral, the unique double-stranded structure of the protecting agent, even the existence of interferon itself 5 years prior to the initial report of its existence as an endogenous antiviral substance (Isaacs and Lindenmann, 1957).

REFERENCES

Arnott, S., Fuller, W., Hodgson, A., and Prutton, I. (1968). Molecular conformations and structure transitions of RNA complementary helices and their possible biological significance. *Nature 220*:561-564.

Banks, G. T., Buck, K. W., Chain, E. B., Darbyshire, J. E., and Himmelweit, F. (1969). Virus-like particles in pennicillin-producing strains of *Penicillium chrysogenum*. *Nature 222*:89-90.

Bonnem, E. M., and Spiegal, R. J. (1984). Interferon-α: Current status and future promise. *J. Biologic Response Modifiers 3*:580-958.

Borré, E., Morgantini, L. E., Ortali, V., and Tonolo, A. (1971). Production of lytic plaques of viral origin in *Penicillium*. *Nature* (London) *229*:568-569.

Braun, W., and Nakano, M. (1965). Influence of oligodeoxyribonucleotides on early events in antibody formation. *Proc. Soc. Exp. Biol. Med. 119*:701-707.

Braun, W., and Nakano, M. (1967). Antibody formation: Stimulation by polyadenylic and polycytidylic acids. *Science 157*:819-821.

Colby, C., and Chamberlin, M. J. (1969). The specificity of interferon induction in chick embryo cells by helical RNA. *Proc. Nat. Acad. Sci. USA 63*:160-167.

Douthart, R. J., Kleinschmidt, W. J., Murphy, E. B., Beasley, F. W., and Schultz, R. M. (1982). Interferon induction and macrophage activation by a mycovirus double-stranded RNA/tobramycin complex following treatment with human serum. *J. Interferon Res. 2*:493-499.

Ellis, L. F., and Kleinschmidt, W. J. (1967). Virus-like particles of a fraction of statolon, a mold product. *Nature 215*:649-650.

Field, A. K., Tytell, A. A., Lampson, G. P., and Hilleman, M. R. (1967). Inducers of interferon and host resistance. II. Multistranded synthetic polynucleotide complexes. *Proc. Nat. Acad. Sci. USA 58*:1004-1010.

Gutterman, J. U., Horning, S., Merigan, T., Quesada, J., Fein, S., and Dziewanowska, Z. (1981). Recombinant human leukocyte interferon (IFL-rA): A phase I clinical study of pharmacokinetics, single dose tolerance, and biological effects in cancer patients. Abstract. *The Second Annual International Congress of Interferon Research*, San Francisco, California.

Hollings, M. (1962). Viruses associated with a die-back disease of cultivated mushrooms. *Nature* (London) *196*:962-965.

Isaacs, A., and Lindenmann, J. (1957). The interferon. *Proc. Roy. Soc.* (London) *147*:258-267.

Jahiel, R. I., Taylor, D., Rainford, N., Hirschberg, S. E., and Kroman, R. (1971). Inducers of interferon inhibit the mitotic response of liver cells to partial hepatectomy. *Proc. Nat. Acad. Sci. USA 68*:740-742.

Kleinschmidt, W. J., and Ellis, L. F. (1967). Statolon as an inducer of interferon. *Interferon* (G.E.W. Wolstenholme and M. O'Connor, Eds.), Ciba Foundation Symposium, Churchill, London, pp. 39-46.

Kleinschmidt, W. J., and Murphy, E. B. (1965). Investigation on interferon induced by statolon. *Virology 27*:484-489.

Kleinschmidt, W. J., and Murphy, E. B. (1967). Interferon induction with statolon in the intact animal. *Bacteriol. Rev. 31*:132-137.

Kleinschmidt, W. J., and Probst, G. W. (1962). The nature of statolon, an antiviral agent. *Antibiot. Chemother. 12*:289-309.

Kleinschmidt, W. J., and Streightoff, F. (1971). Inhibition of influenza virus and interferon response intranasally with statolon. *Infect. Immun. 4*:738-741.

Kleinschmidt, W. J., Cline, J. C., and Murphy, E. B. (1964). Interferon production induced by statolon. *Proc. Nat. Acad. Sci. USA 52*:741-74.

Kleinschmidt, W. J., Van Etten, J. L., and Vidaver, A. K. (1974). Influence of molecular weights of bacteriophage $\phi 6$ double-stranded ribonucleic acids on interferon induction. *Infect. Immunol. 10*: 284-285.

Lampson, G. P., Tytell, A. A., Field, A. K., Nemes, M. M., and Hilleman, M. R. (1967). Inducers of interferon and host resistance. I. Double-stranded RNA from extracts of *Penicillium funiculosum*. *Proc. Nat. Acad. Sci. USA 58*:782-789.

Lampson, G. P., Tytell, A. A., Field, A. K., Nemes, M. M., and Hilleman, M. R. (1969). Influence of polyamines on induction of interferon and resistance to viruses by synthetic polynucleotides. *Proc. Soc. Exp. Biol. Med. 132*:212-218.

Lampson, G. P., Field, A. K., Tytell, A. A., Nemes, M. M., and Hilleman, M. R. (1970). The relationship of molecular size of rIn:rCn (poly I:C) to induciton of interferon and host resistance. *Proc. Soc. Exp. Biol. Med. 135*:911-916.

Lampson, G. P., Field, A. K., Tytell, A. A., and Hilleman, M. R. (1981). Poly I:C/poly L-lysine: Potent inducer of interferons in primates. *J. Interferon Res. 1*:539-549.

Lemke, P. A., and Nash, C. H. (1974). Fungal viruses. *Bacteriol. Rev. 38*:29-56.

Lemke, P. A., and Ness, T. M. (1970). Isolation and characterization of a double-stranded ribonucleic acid from *Penicillum chrysogenum*. *J. Virol. 6*:813-819.

Leonard, B. J., and Eccleston, E. (1972). Autoimmune disease and malignancies induced by poly IC transmitted through four generations of mice. *Proc. Eur. Soc. Study Drug Toxicity. 13*:274-288.

Levy, H. B., Baer, G., Baron, S., Buckler, E. E., Gibbs, C. J., and Iadarola, M. J., London, W. T., and Rice, J. (1975). A modified polyinosinic-polyribocytidylic acid comples that induces interferon in primates. *J. Inf. Dis. 132*:434-439.

Mahendrasingham, A., Forsythe, V. T., Hussain, R., Greenall, R. J., Pigram, W. J., and Fuller, W. (1986). Time resolved x-ray diffraction studies of the B $\rightleftarrows$ D structural transition in the DNA double helix. *Science 233*:195-197.

Niblack, J. F., and McCreary, M. B. (1971). Relationship of biological activities of poly I:poly C to homopolymer molecular weights. *Nature New Biol. 233*:52-53.

Nordlund, J. J., Wolf, S. M., and Levy, H. B. (1970). Inhibition of biological activity of poly I:poly C by human plasma. *Proc. Soc. Exp. Biol. Med. 133*:439-444.

O'Dell, T. B., Wright, H. N., and Bieter, R. N. (1953). Chemotherapeutic activity of nucleic acids and high-protein diets against the infection caused by the MM virus in mice. *J. Pharmacol. Exp. Therap. 107*:232-240.

Powell, H. M., Culbertson, C. G., McGuire, J. M., Hoehn, M. M., and Baker, L. A. (1952). A filtrate with chemophrophylactic and chemotherpaeutic action against MM and Semliki Forest viruses in mice. *Antibiot. Chemother. 2*:432-437.

Rytel, M. W., Shope, R. E., and Kilbourne, E. C. (1966). An antiviral substance from *Penicillium funiculosum*. *J. Exp. Med. 123*: 577-584.

Shope, R. E. (1953a). An antiviral substance from *Penicillium funiculosum*. I. Effect upon infection in mice with swine influenza virus and Columbia SK encephalomyelitis viruses. *J. Exp. Med. 97*:601-625.

Shope, R. E. (1953b). An antiviral substance from *Penicillium funiculosom*. II. Effect of helenine upon infection in mice with Semliki Forest virus. *J. Exp. Med. 97*:627-638.

Tytel, A. A., Lampson, G. P., Field, A. K., Nemes, M. J., and Hilleman, M. R. (1970). Influence of size of individual homopolynucleotides on the physical and biological properties of complexed rI_n:rC_n (poly I:C). *Proc. Soc. Exp. Biol. Med. 135*:917-921.

Yang, J. T., and Samejima, T. (1969). Optical rotary dispersion and circular dichroism of nucleic acids. *Prog. Nucleic Acid Res. Mol. Biol. 9*:223-300.

2

Yeast Ty Elements as Retroviruses

Jef D. Boeke
The Johns Hopkins University School of Medicine, Baltimore, Maryland

David J. Garfinkel
Frederick Cancer Research Facility, Frederick, Maryland

INTRODUCTION

The Ty elements of the yeast *Saccharomyces cerevisiae* are members of a widely distributed family of eukaryotic, retrovirus-like transposons (retrotransposons). Ty elements are related structurally and functionally to retroviruses (Table 1): their structural similarities are a consequence of the fact that both Ty elements and retroviruses use reverse transcription of an end-to-end transcript as the means of replication or transposition (Boeke et al., 1985; Fink et al., 1986). Remarkably, Ty elements direct the synthesis of virus-like particles (VLPs, Figure 1), which are morphologically reminiscent of certain retrovirus particles, such as the intracisternal A-particle (IAP) of the mouse (Garfinkel et al., 1985; Mellor et al., 1985c). Ty elements may also utilize a mechanism similar to that employed by retroviruses in integrating the new DNA copies into the host's DNA.

The major difference between yeast Ty elements and retroviruses is that Ty elements appear not to leave the cell in undergoing the transposition process; the entire transposition cycle is intracellular. In this respect, yeast Ty element-encoded particles are similar to other virus-like particles described from saccharomyces; there is no evidence

Table 1 Similarities Between Ty Elements and Retroviral Proviruses

DNA structure
LTR sequences, direct orientation, similar size (334 bp)[a]
Small duplication of target DNA (5 bp)
TG. . .CA terminal dinucleotides
Homology to tRNA just inside 5' LTR (initiator met tRNA)
Polypurine stretch just inside 3' LTR (6 purines in 8 bp)
Enhancer sequence present
Circular DNA forms exist
RNA structure
Abundant end-to-end transcript
RNA is terminally repetitious (∿45 nucleotides)
Open reading frames
Overlapping reading frames (*tyb* made as readthrough from *tya*)
Predicted nucleic acid binding site in *tya*
Homology to retroviral reverse transcriptase/integrase/protease in *tyb*
Virus-like particles
Contain reverse transcriptase activity
Contain element-specific RNA and primer
Contain element-specific proteins

[a]Items in parentheses refer to the properties of Ty elements.

for the infectious nature of the element, and extracellular particles are not seen, even in cultures in which the cells are packed with hundreds of *Ty-VLPs* (Garfinkel et al., 1985). Retroviruses, on the other hand, shuttle between cells during each cycle of transposition or infection; extracellular particles are readily demonstrated. Other obvious differences between Ty elements and retroviruses are summarized in Table 2.

Ty elements exist in essentially all strains of saccharomyces; in most laboratory strains there are about 35 copies per haploid genome. In the *S. cerevisiae* strains used in our laboratories, these consist of about 25 copies of Ty1 and 10 copies of Ty2 (J. D. Boeke, unpublished results, 1986). Within each Ty family there is considerable structural heterogeneity, but the overall structure is that shown in Figure 2. The elements are about 6 kilobases in length, and contain at their ends δ or LTR sequences of about 330 base pairs; these are often but not always identical in sequence. Outside the LTRs lies a 5 base pair duplication of the host DNA, into which the Ty element is inserted. Between the LTRs lies a central (ε) sequence, which contains two long open reading frames called *tya* and *tyb*. These apparently correspond to *gag* (core virion proteins) and *pol* (reverse transcriptase/ endonuclease) genes of retroviruses; there is no analog to the retroviral *env* gene, consistent with the noninfectious nature of *Ty-VLPs* (Clare & Farabaugh, 1985; Hauber et al., 1985; Warmington et al., 1985). The major transcript of Ty elements is an end-to-end transcript, terminally repetitious by about 40 nucleotides; this is very similar to the structure of retroviral genomic RNA (Elder et al., 1983).

Within the central segment of the Ty element, near the 5' LTR, lies an enhancerlike element necessary for the efficient transcription of the Ty element and also capable of activating the gene lying 5' to the Ty element (Roeder et al., 1985; Errede et al., 1985). The transcription of both the Ty element and the adjacent gene are regulated by the mating type of the host; in mating-proficient cells, transcription is 5-10-fold higher than in corresponding nonmater (a/α diploid) cells (Elder et al., 1983; Errede et al., 1980; Taguchi et al., 1984). This effect is apparently due to sequences adjacent to or overlapping with the enhancer sequences, which have homology to other yeast genes known to be under a1/α2 regulation.

MECHANISM OF TY TRANSPOSITION

Many clues to the mechanism of Ty element transposition lay in details of its structure. Perhaps most telling among these were the presence of LTR sequences (Cameron et al., 1979), and the structure of the major transcript (Elder et al., 1983). Finally, the DNA sequence of the LTR sequences and their flanking DNA revealed additional important similarities to the structure of retroviral proviruses: The terminal nucleotides (TG. . .CA) form an inverted repeat and are extremely well conserved among retroviruses and many transposable elements (reviewed by Varmus, 1983). Additionally, just inside the 5' LTR lies a 10-nucleotide stretch of perfect homology to the 3' end of yeast initiator methionine tRNA (Simsek and RajBhandary, 1972; Eibel et al., 1981; A. Bystrom and G. R. Fink, unpublished data, 1985). This is significant because all known retroviruses have a tRNA homology at essentially the same position. A host tRNA primes retroviral DNA synthesis at this position, leading to the (-) strong stop DNA molecule, which is the first product of reverse transcription of the retroviral genomic RNA (Figure 3). The (-) strong stop DNA extends from the 5' δ/ε boundary back to the nucleotide corresponding to the 5' end of the Ty transcript. Just inside the Ty 3' LTR lies another sequence that is well-conserved among retroviruses and retrotransposons: A

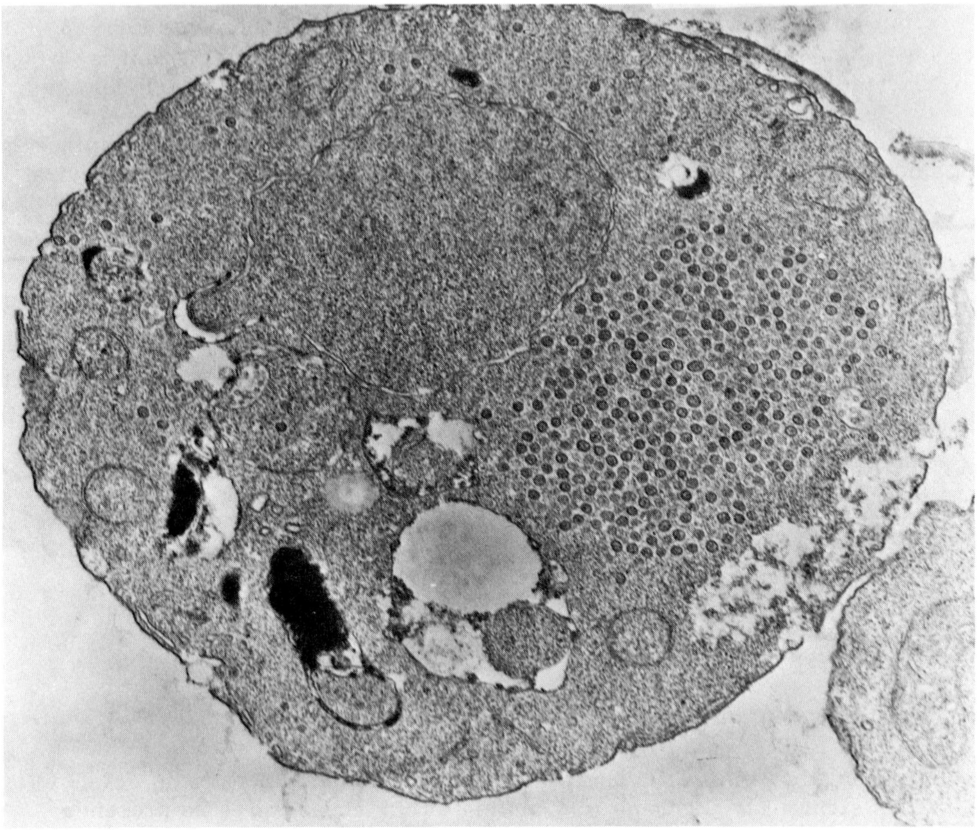

(a)

Figure 1 Virus-like particles (VLPs) are produced by retrotransposons. (a) Ty-VLP particles produced in yeast cells in response to overproduction of Ty RNA from a recombinant pGTyH3 plasmid bearing a wild-type Ty (TyH3). Note the cytoplasmic location of the particles. (b) Morphologically aberrant Ty-VLP particles formed in cells carrying a mutant derivative of pGTyH3 (pJEF1293). The mutation in pJEF1293 is an 8 bp *Not*I linker inserted into the filled-in *Bgl*II site at position 1702 [Figure 6(a)]. The mutation, which does not alter the reading frame of *tyb*, does not affect the reverse transcriptase activity, but lies in the "protease" domain of *tyb*. The mutation destroys the ability of this Ty element to transpose (J. D. Boeke, unpublished data, 1986).

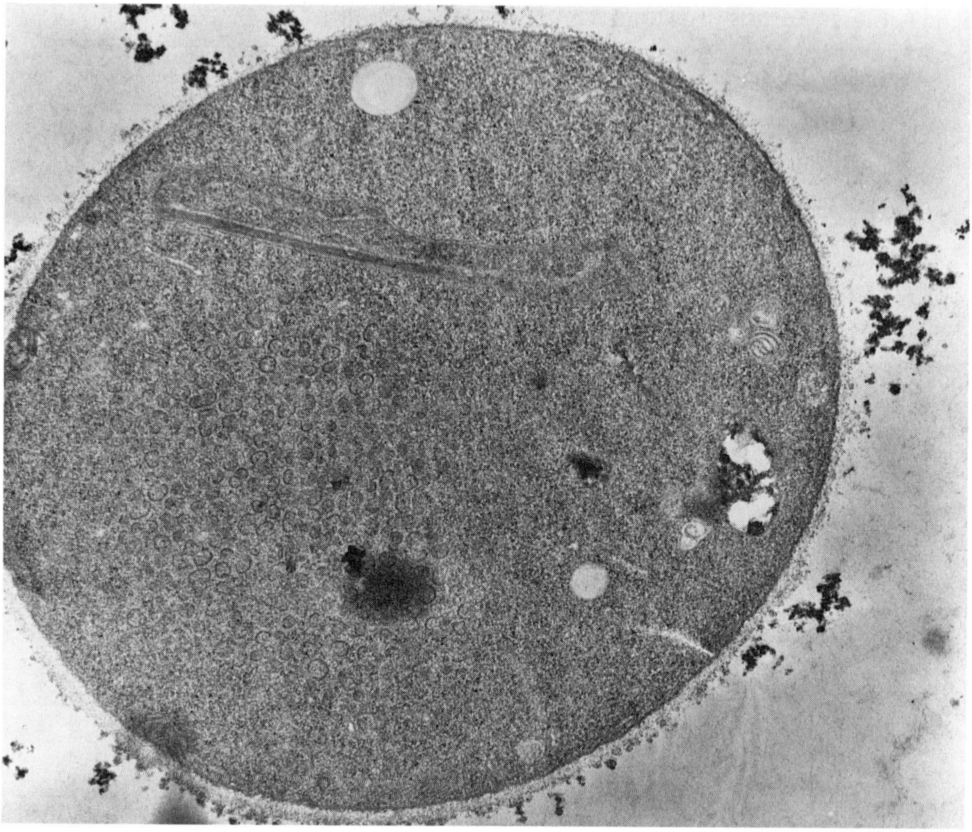

(b)

Figure 1 (Continued)

"purine-rich" segment believed to be important in priming (+) strong stop DNA synthesis (see Figure 3) (Gilboa et al., 1979). The (+) strong stop DNA extends from the 3' δ/ε boundary to the 3' end of the element. Thus, Ty elements and most retrotransposons share many potential cis-acting sequences with retroviruses (Boeke, 1987).

A further parallel between Ty and retroviruses may be drawn from the fact that both can be potent activators of adjacent cellular genes (Neel et al., 1981; Errede et al., 1980). In the case of Ty, this is due to an internal sequence with homology to the SV40 and "core" enhancer sequences of higher eukaryotes (Roeder et al., 1985; Errede et al., 1985). In retroviruses, the enhancer usually resides within the LTR.

Table 2 Differences Between Ty Elements and Retroviruses

Retrovirus	Ty element
Loss of terminal dinucleotides upon integration[a]	No loss of terminal dinucleotides
Enhancer within LTR	Enhancer outside LTR
Env gene required for infectivity	No *env* gene present
Infectious	Noninfectious
Order of functional domains within *pol* (*tyb*) reading frame:	
protease-reverse transcriptase-integrase	protease-integrase-reverse transcriptase[b]

[a]Retroviral RNAs contain two bases (UU) between the CA at the 3' end of the 5' LTR and the sequence homologous to the 3' end of the priming tRNA. These uracil residues are the first bases to serve as template for DNA synthesis and indeed, these bases are recovered in the two-LTR circle molecules which are through to be the immediate precursor for integration (Panganiban and Temin, 1984) (i.e., these bases are not found in the provirus). Two bases just inside the 3' LTR (AA) are similarly lost from the spleen necrosis virus (SNV) circular DNA intermediate during the integration process. No such extra nucleotides are found between the CA at the 3' end of the 5' LTR of Ty and the 3' end of the priming tRNA (i.e., in Ty, the conserved CA dinucleotide itself is the first sequence to serve as template for reverse transcription). Hence, it is unlikely that Ty transposition intermediates (if they closely resemble the SNV integration intermediate) lose bases upon integration. Since the structure of Ty circular DNA is not yet known, such conclusions must be considered tentative.

[b]The assignments of functional domain are based on the rather weak amino acid homologies between *tyb* and retroviral *pol* genes. Mount and Rubin (1985) pointed out that *copia* (a drosophila retrotransposon) has the same order of domains as Ty. Other retrotransposons, however, have the classic retroviral order of functional domains (see Boeke, 1987, for review).

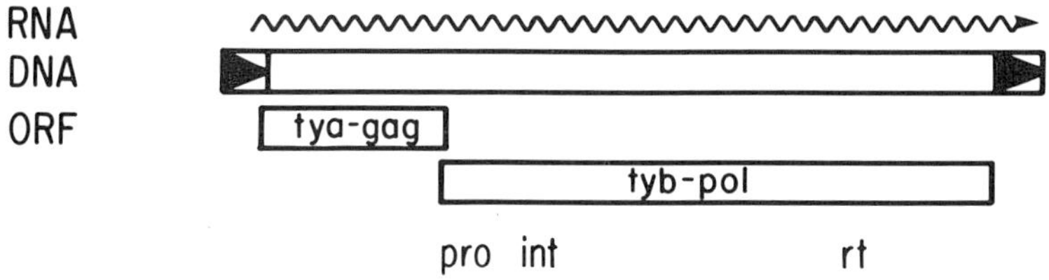

Figure 2 The overall structure of a Ty element. The boxed triangles represent LTR or δ sequences. The central segment, containing the open reading frames, is the open box between them. The wavy line represents the major transcript.

The elucidation of the complete sequence of a Ty element (Clare and Farabaugh, 1985) provided an important additional barrage of evidence suggesting that Ty elements were related to retroviruses: Two long open reading frames (*tya*/*gag* and *tyb*/*pol*) were found. These two reading frames overlapped for about 40 base pairs in a manner reminiscent of the overlap between *gag* and *pol* in Rous sarcoma virus. Moreover, it was shown that *tyb*/*pol* protein was synthesized as a readthrough product of *tya*/*gag* (Clare and Farabaugh, 1985; Mellor et al., 1985b), again, in a way similar to the case of Rous sarcoma virus. All efforts to detect "splicing out" of the overlap region failed. Hence, the mechanism of *gag*/*pol* readthrough is thought to be at the level of frameshifting in both cases.

Circular DNA forms of the Ty element have also been reported, consistent with circular intermediates seen in retrovirus-infected cells, but these have not been extensively characterized (Ballario et al., 1983; Newlon et al., 1982). Similar circular molecules isolated from drosophila, derived from its various retrotransposons, have also been isolated and characterized in much greater depth (Flavell and Ish-Horowicz, 1981,1983). These appear to correspond to the one-LTR and two-LTR circles seen in retrovirus-infected cells (reviewed in Varmus, 1983).

A system allowing the controlled induction of the Ty transposition process has been developed (Boeke et al., 1985; Fink et al., 1986). The system is based on two plasmids, a donor plasmid and a recipient (target) plasmid. The donor plasmid consists of a high-copy-number plasmid backbone carrying an inducible gene fusion in which the transcription of a particular Ty element (TyH3) is driven by the yeast *GAL1* promoter. Importantly, the *GAL1* promoter segment is fused to a point within the 5' LTR of TyH3 such that the resulting transcript is structurally identical to a native Ty transcript (Figure 4). No *GAL1* sequences exist in the transcript (this required the deletion of a substantial portion of the 5' LTR; such a maneuver would incapacitate

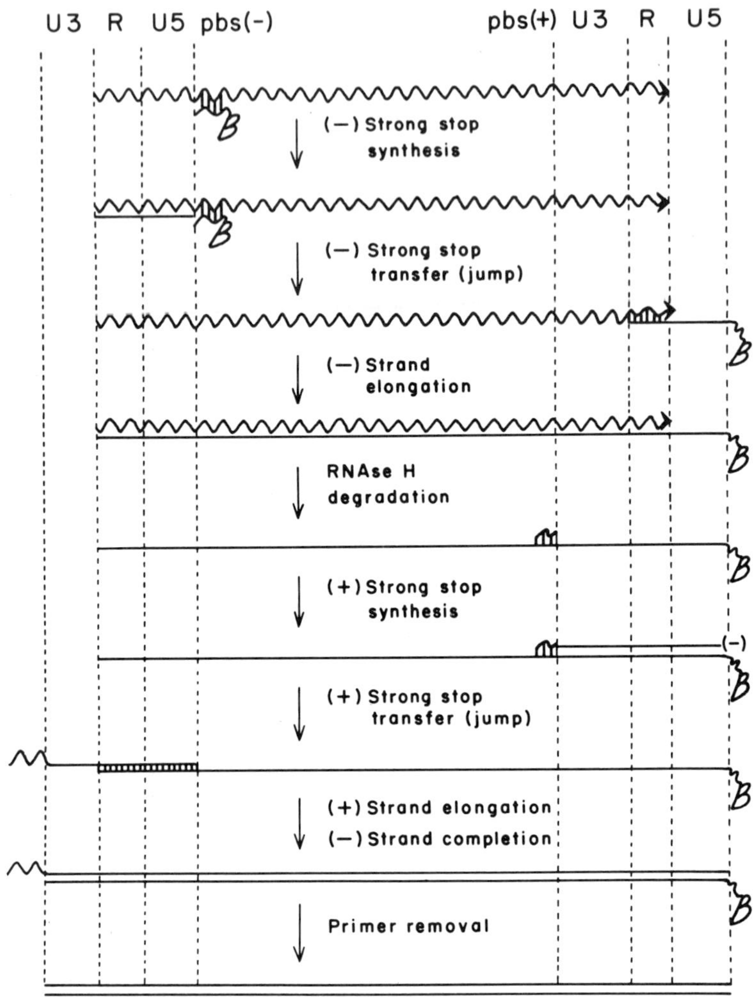

Figure 3 A simplified model for the steps required to generate a full-length DNA copy from the end-to-end transcript by reverse transcription is presented. Straight lines represent DNA, wavy lines RNA. The domains of the LTRs (U3, R, and U5), as well as the (-) and (+) strand primer binding sites [pbs(-)] and [pbs(+)] are separated by dotted lines. The long wavy lines symbolize the end-to-end transcript. Base pairing to the tRNA, to the (+) strand primer, and to the newly transferred strong stop DNAs is indicated by hyphens. For the sake of simplicity, the model assumes that one template RNA is used; however, it is likely that two templates are used and that RNase H degradation begins immediately after (-) strong stop synthesis. See the review by Varmus and Swanstrom (1985) for a review of more detailed models of reverse transcription.

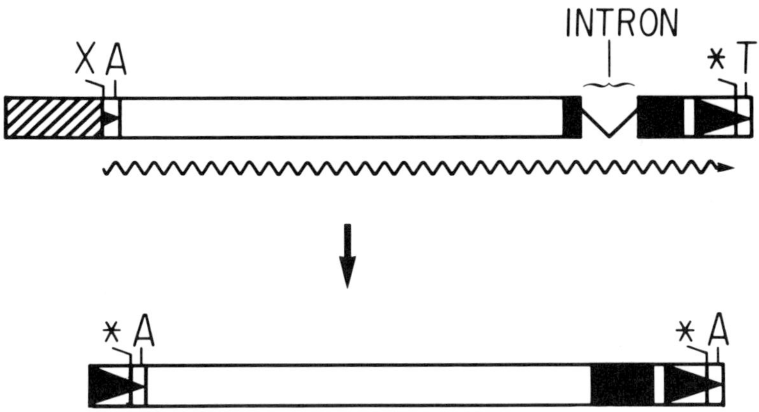

Figure 4 The salient features of a system developed to study the retrotransposition of the yeast Ty elements are summarized. The hatched box represents the *GAL1* inducible promoter. The solid box represents a marker segment introduced into the Ty element; this marker fragment contains an intron (symbolized by a V). A, T, X, and * refer to sequence polymorphisms between the 5' and 3' LTRs. Note that their inheritance is in accord with the predictions of Figure 4. *Source*: Reproduced from Boeke et al., 1985.

most bacterial transposons). These plasmids are called pGTyH3 plasmids. Introduction of such plasmids into appropriate yeast strains and growth of the transformants on inducing (galactose-containing) medium results in a high level of Ty element transposition as measured by several independent assays (Boeke et al., 1985, and unpublished data, 1986). A particularly useful means of assaying Ty transposition involves using the recipient plasmid mentioned above. This centromeric (i.e., single-copy) plasmid bears a promoterless *HIS3* gene in which the promoter sequences are replaced by a segment of phage lambda DNA. Scherer et al. (1982) originally showed that Ty transposition into the lambda DNA segment of such plasmids resulted in a His^+ phenotype. The activation of the previously silent *HIS3* gene is accomplished by the enhancerlike sequence of Ty lying within the first kilobase of the Ty inside the 5' LTR (Roeder et al., 1985; Errede et al., 1985). As a result, nearly all of the Ty elements that activate the *HIS3* gene are oriented such that their transcription proceeds away from the *HIS3* gene. Such is the case for many other yeast genes that have been activated by Ty insertion (Errede et al., 1980; Paquin and Williamson, 1984, 1986).

TyH3 itself was shown to be competent for transposition by marking experiments. Insertion of various marker sequences into the *Bgl*II site near the 3' LTR (but outside *tya/gag* and *tyb/pol*) did not inhibit transposition; moreover, marked TyH3 derivatives could be recovered in the form of His$^+$ revertants in cells containing the recipient plasmid described (Figure 5) (Boeke et al., 1985,1986; Fink et al., 1986). In addition to marked elements, a larger number of *unmarked* elements, probably derived from chromosomal Ty elements, were recovered at a high frequency. Thus, the induction of the pGTyH3 plasmid results in both the transposition of TyH3 itself and the stimulation of transposition of chromosomal Ty elements.

Two key features of the transposed marked Ty elements provided extremely strong support for the idea that reverse transcription of Ty RNA was the mechanism by which they transposed. First, sequence polymorphisms between the 5' and 3' LTR sequences in the pGTyH3 plasmid were inherited in such a way that strongly suggested that they were inherited from the Ty RNA by a reverse transcription process (Figures 3 and 4). The experiment described in Figure 4 shows that the transposed copies derived from the marked Ty element contain in both LTRs right-end markers from the 5' LTR (U5 region) and left-end markers from the 3' LTR (U3/R boundary region), consistent with the jump or transfer mechanism of (-) and (+) strand synthesis shown in Figure 3. Moreover, when an intron-containing marker segment from the yeast *rp51* gene was inserted into the pGTyH3 plasmid, the intron sequences were nearly always spliced out, suggesting directly that RNA was an intermediate in the process (Boeke et al., 1985). Similar experiments with retroviruses had shown that inserted marker genes lost their introns. Interestingly, this process seemed to be much less efficient in the case of retroviruses than in the case of Ty (Sorge and Hughes, 1982). Perhaps this reflects the fact that many retroviral spliced messages are made. Retroviral splicing must be inefficient in order to ensure the synthesis of adequate amounts of progeny full-length (genomic) RNA, but there is no evidence or role for naturally occurring spliced species of Ty RNA.

Thus there is now overwhelming genetic and molecular biological evidence in support of the idea that Ty and related elements (*copia*, etc.) transpose via an RNA intermediate that is reversed transcribed. The basic mechanism of reverse transcription is similar to that utilized by retroviruses (Figure 3).

The *GAL1*/TyH3 system allowed one other important feature of Ty transposition to be tested: the possibility of infectivity. Strains carrying marked and unmarked pGTy plasmids were cocultivated under inducing conditions (i.e., on galactose). Since the strains (both of the same mating type) carried chromosomal auxotrophic markers which enabled them to be distinguished, the two cell types were separated following cocultivation and their chromosomes were examined for the presence or absence of the marker that had been present on the marked Ty. In no case was this marker seen to cross over into the heterologous strain: It was always recovered in the strain in which it resided

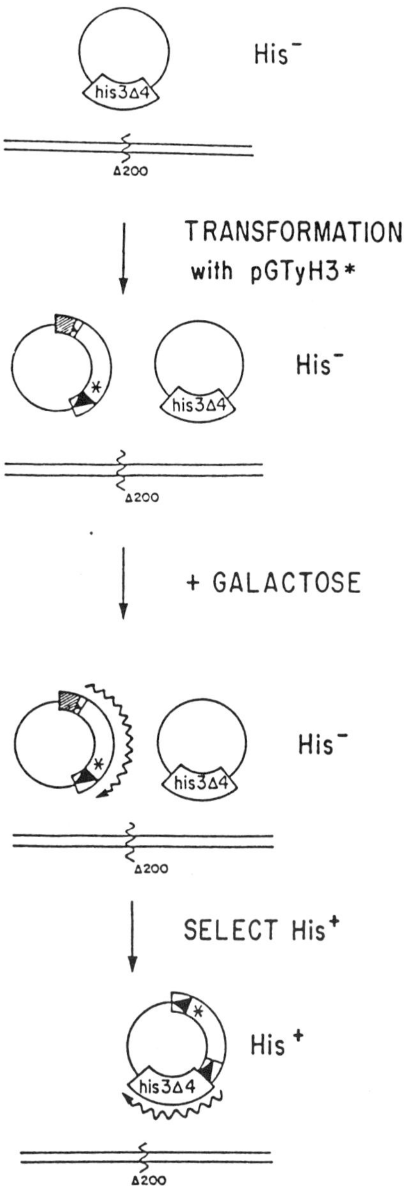

Figure 5 The scheme used to obtain transposition of marked Ty elements is shown. The experiment is performed in a strain carrying a recipient plasmid that becomes His^+ when a Ty element transposes into it. A donor plasmid, carrying a *GAL1*/TyH3 fusion, provides high-frequency transposition. A marker (*) is imbedded within the nonessential region of the Ty in a pGTyH3 plasmid (see text). Its progeny are recovered as His^+ revertants, which are then screened for the presence of the marker. See text and Boeke et al. (1985) for details. *Source*: Reproduced from Fink et al., 1986.

originally (Garfinkel et al., 1985). This shows that, if Ty is infectious, it is only rarely so under normal mitotic growth conditions. The intriguing possibility remains that during schmoo formation (mating) or spore germination (when the cell wall becomes much less significant a barrier) Ty transposition could be infectious. Thus far efforts to "infect" spheroplasts with *Ty-VLPs* (see below) have been unsuccessful (D. J. Garfinkel and J. D. Boeke, unpublished results, 1985,1986).

VIRUSLIKE PARTICLES AND Ty TRANSPOSITION

The relationship between Ty elements and retroviruses was seen as more intimate when viruslike particles (*Ty-VLPs*) were found in cells that were induced for transposition (Figure 1) (Garfinkel et al., 1985; Mellor et al., 1985c). In electron micrographs, dozens to hundreds of uniform particles are present. The *Ty-VLPs* are spherical to ovoid in shape and are about 60 nm in diameter. Upon sucrose gradient ultracentrifugation, the particles have an apparent sedimentation value of 175S (Mellor et al., 1985c) and an estimated bouyant density of 1.21 g/ml (S. Youngren and D. J. Garfinkel, unpublished results, 1986). *Ty-VLPs* are restricted almost entirely to the cytoplasm, but their localization differs between strains. In one strain, the particles are present in large clusters, with up to several hundred particles in a cluster, while in another strain background the *Ty-VLPs* are distributed as single particles throughout the cytoplasm. The accumulation of *Ty-VLPs* is strictly dependent on transposition induction. They are virtually absent in uninduced cells.

Although there is no direct evidence at present, we suspect that *Ty-VLPs* are an essential intermediate in retrotransposition (Garfinkel et al., 1985). Most of the Ty reverse transcriptase activity fractionates as a particle and copurifies with the *Ty-VLPs*. Ty RNA and at least one Ty-encoded protein also cosediment with the *Ty-VLPs*, and this complex specifically synthesizes DNA products that are homologous with the entire Ty element. Recently a tRNA-sized molecule, probably the initiator methionine tRNA, was found to act as the primer for (-) strong stop DNA synthesis, which occurs in preparations of *Ty-VLPs* (A. Bystrom and G. Fink, unpublished results, 1986). It appears, therefore, that most if not all of the factors necessary for Ty reverse transcription are associated with the *Ty-VLPs*.

Particle formation could be important for at least two reasons. First, it could protect the cell from potentially deleterious levels of free reverse transcriptase. Reverse transcription of cellular mRNA and subsequent recombination of the cDNAs with chromosomal genes might lead to unwanted expansion of pseudogenes, gene amplifications, or creation of intronless genes. In yeast, the cDNAs should be highly recombinogenic. Second, it is likely that the particle keeps all of the components required for efficient reverse transcription and possibly for integration at the proper concentrations and levels of activity.

Ty-VLPs are similar morphologically to a variety of retrotransposon and endogenous retroviral particles that have been observed in mammals, avian species, and insects. These particles also resemble and share important functional similarities with the virion core of retroviruses, which is released from the enveloped virus by mild detergent treatment. *Ty-VLPs*, as well as several other retrotransposon particles most likely remain intracellular, lack an envelope protein and are not infectious. Although *Ty-VLPs* are located in the cytoplasm we have never observed *Ty-VLPs* near the cytoplasmic membrane, budding from the cell, or in the extracellular medium. This differs markedly from the behavior of intracytoplasmic A-type particles, which resemble *Ty-VLPs* morphologically, but apparently are precursors to B-type mouse mammary tumor viruses that bud from the plasma membrane. Murine intracisternal A-type particles (IAPs) arise from a class of retrotransposons (Hawley et al., 1982,1984; Kuff et al., 1983; Baltimore, 1985), and are located in the cisternae of the endoplasmic reticulum (Kuff et al., 1968). We do not know if *Ty-VLPs* enter the cisternae of yeast. Particles containing reverse transcriptase activity also accumulate in the cytoplasm of drosophila cells (Heine et al., 1980) or, in the case of the *copia* particles isolated by Shiba and Saigo (1983), in the nucleus. Few *Ty-VLPs* are located in the nucleus as judged by microscopy, however, some form of the *Ty-VLP* may travel to the nucleus where integration must occur.

The available evidence and comparisons presented above clearly distinguish *Ty-VLPs* from another virus-like particle of *S. cerevisiae*, the killer particles discussed elsewhere in this volume. It is interesting to note that both killer particles and *Ty-VLPs* are apparently noninfectious to intact yeast cells under most conditions tested.

Ty PROTEIN PRODUCTS AND THEIR SYNTHESIS

Although we know very little about the biochemistry of Ty proteins, their mode of expression and genetic organization is strikingly similar to the *gag* and *pol* genes of retroviruses. *Tya* almost certainly encodes polypeptides that contribute to the structure of the *Ty-VLP*, and *tyb* has significant amino acid sequence homology with the *pol* gene. Mutations in *tyb* abolish the reverse transcriptase activity associated with the particle (Garfinkel et al., 1985; J. D. Boeke and D. J. Garfinkel, unpublished results, 1986). The detection of reverse transcriptase activity and Ty proteins is strictly dependent on the overexpression of a single Ty element.

Mode of Expression

Many retrotransposons have a discontinuity in the reading frames separating *gag* and *pol*. *Tyb* is in the +1 frame compared to *tya*, whereas the *pol* gene of *RSV* and the retrotransposons *17.6*, *gypsy*, and *297* from drosophila (reviewed by Boeke, 1987) and the proposed *pol* homologue from the putative retrotransposon *DIRS1* from dictyostelium (Cappello et al., 1985) are in the -1 frame with respect to the first open reading frame. In *MuLV*, a single UAG stop codon separates *gag* from *pol*. The *pol* gene also lacks an initiating methionine. For years it was thought but never proven that an RNA splice corrected the phase to produce the *gag-pol* polyprotein. Recently, it has been shown for *MuLV*, *RSV*, and Ty that the discontinuity in the reading frames is reconciled at the translational level. In *MuLV*, readthrough from *gag* and *pol* occurs at a level of about 10%. Glutamine is inserted at the *gag* stop codon (Yoshinaka et al., 1985) in the readthrough product. Readthrough to *pol* also occurs about 10% of the time in Ty (Clare and Farabaugh, 1985; Mellor et al., 1985b). Only with *RSV* has translational frameshifting been demonstrated directly. Jacks and Varmus (1985) showed that in vitro synthesized *SP6/RSV* RNA could direct the synthesis of small but significant amounts of *gag/pol* readthrough products in a eukaryotic (reticulocyte) translation system but that the same synthetic RNA specified only *gag* product in an *Escherichia coli* system. The exact sequence of the *gag/pol* junction in the fusion proteins has not yet been determined, and such sequence data would implicate a specific tRNA as well as the precise site of frameshifting. The site may be the termination codon of the *gag* gene; the balance between action of termination factors and the tRNA might determine the extent of frameshifting. At the present time, however, there is nothing to preclude frameshifting by a frameshift-suppressing tRNA at a site within the *gag/pol* overlap region. The universal nature of *gag/pol* expression (and hence *pol* expression) suggests that the ratio of these gene products may be crucial to re-verse transcription, core particle assembly, and integration.

The *gag-tya* Genes

The *gag* gene encodes a polyprotein of 65-75 kDa in most retroviruses. It is cleaved by a retrovirally-encoded protease to form four or five components that make up most of the core structure of the virus. Some *gag* proteins are post-translationally modified by glycosylation, phosphorylation, or acylation by myristic acid. Recently, the role of acylation in assembly of mammalian type C retroviruses has become clearer. The *gag* gene product of *MuLV*, *Pr65gag*, is modified by covalent attachment of myristic acid to the amino-terminal glycine (Schultz and Oroszlan, 1983). To determine the function of myristylated *Pr65gag*, *MuLV* was mutagenized by site-directed mutagenesis so as to change the amino terminus from glycine to alanine, or to delete the glycine codon. Upon transfection into mammalian cells, these mutants direct the synthesis of *gag* proteins, but these proteins are not myristylated. The mutants do not form virus particles or any recognizable virus-specific structures and the mutant *gag* proteins appear to remain in the cytosol. *MuLV* assembles at the plasma membrane and the *gag* gene products are usually found in particulate fractions of the cell (Rein et al., 1986). It appears that myristate is required for the association of the *gag* protein with the plasma membrane, and that this association is essential for virus assembly. However, the diversity found among retroviruses is striking. Avian type C viruses lack myristic acid but have assembly processes that are morphologically similar to *MuLV*. Some retroviruses, such as mammary tumor virus (type B) or Mason-Pfizer monkey virus (type D), synthesize myristylated *gag* proteins but assemble virion cores within the cell. *Ty-VLPs* assembly also occurs within the cell rather than at the plasma membrane. The idea that covalent protein modifications participate in retrovirus core assembly may also apply to *Ty-VLPs*. A subset of *gag* proteins also interacts with virion RNA. These are usually small (less than 20 kDa) basic proteins that form a ribonucleoprotein complex with virion RNA. They are most likely important for packaging of retroviral RNA into the core. However, the complete function of *gag* proteins with regard to the structure of the virion or to the roles of these proteins in packaging, uncoating, reverse transcription, and RNA processing remains to be understood.

The *gag*-like open reading frame, *tya*, begins within the 5' LTR and extends for at least 440 codons (depending on the particular Ty) and terminates normally. This organization differs slightly from the *gag* coding sequence that begins within the body of the provirus and is preceded by a putative untranslated leader. Both *tya* and *gag* are translated from an unspliced RNA that is apparently indistinguishable from genomic RNA.

It has been reported that *tya* is phosphorylated in vivo and binds to nucleic acid (Mellor et al., 1985a). *Tya* also has homology to the consensus sequence for prokaryotic DNA binding proteins (Clare and Farabaugh, 1985; Hauber et al., 1985). The full *tya* protein has two regions of unusual sequence. A region of 125 amino acids near the amino terminus contains 27 proline residues and a carboxy terminal region (just downstream from the DNA binding protein homology) that is highly basic (Clare and Farabaugh, 1985). *Tya* proteins detected from the two elements that are overexpressed behave anomalously upon electrophoresis. Molecular weight estimates from sodium dodecyl sulfate–polyacrylamide gel electrophoresis are consistently several thousand daltons higher than the sizes predicted by their DNA sequence (Figure 6) (Mellor et al., 1985a). This may be due to protein modification or it may be intrinsic to the primary sequence.

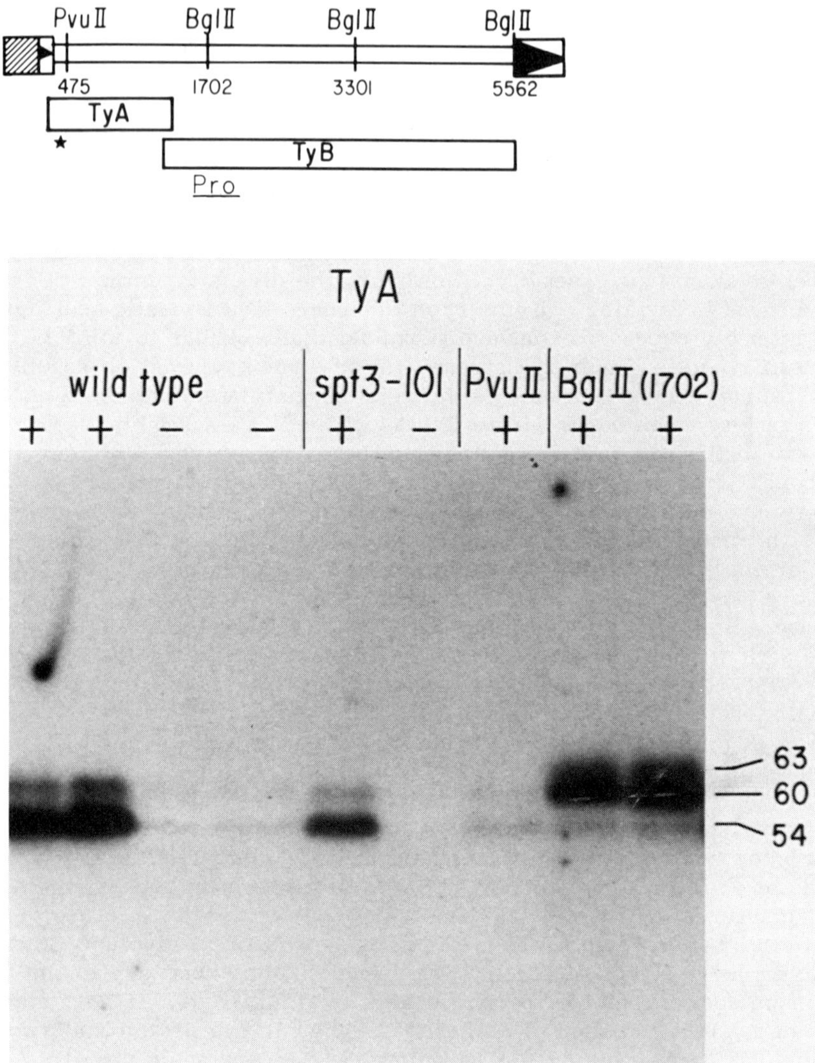

Figure 6(a) The *tya* region of TyH3 encodes a 60 kDa protein that is posttranslationally processed. TyH3 drive by the *GAL1* promoter, denoted by the hatched box, is symbolized by the diagram at the top. Landmark restriction sites and their positions (based on the Ty912 sequence reported by Clare and Farabaugh, 1985) are marked. Arrowheads represent LTR (or δ) sequences; the arrows point in the direc-

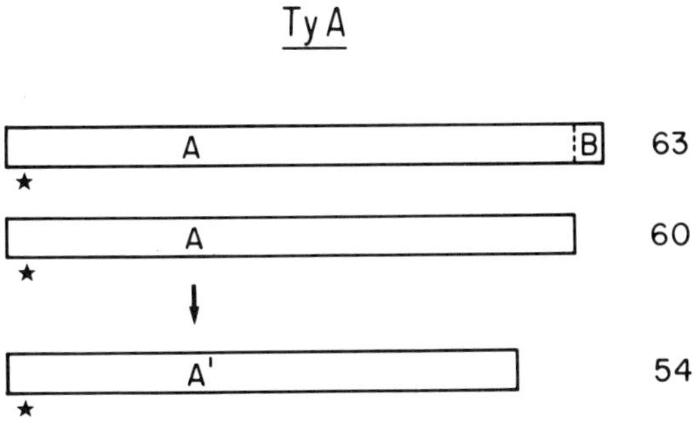

Figure 6(b) Proteolytic cleavage of the *tya* protein by Ty protease is shown. A frameshift mutation was made by filling in the *Bgl*II site (1702) which is immediately adjacent to a core homology region with the retroviral protease (Mount and Rubin, 1985). We detect two Ty proteins (*tya*-54 and *tya*-60) in cells containing the wild-type TyH3. The amount of these proteins increases substantially on galactose (+, induced) compared to glucose (-, noninduced). An additional 63 kDa protein is detected when the protease mutant [*Bgl*II (1702)] is overexpressed. We believe that the 63 kDa protein is a readthrough product from *tya* into *tyb* that terminated prematurely because of the mutation in the protease domain. *Tya*-60 is probably the primary translation product of *tya*, and it is apparently cleaved near its carboxy terminus to form *tya*-54. Note that the relative amounts of the antigen shift in the protease mutant relative to the wild-type TyH3 [Fig. 6(a)]. The wild type has a small amount of the putative primary translation product *tya*-60, while the protease mutant accumulates a large amount of *tya*-60.

---

tion of Ty transcription. The Ty genes *tya* and *tyb* are boxed. The star marks the position of *tya* peptide 136 (aa34asp-val-ser-ala-ser-ala-ser-lys-thr-glu-glu-cysaa43), which was used to raise anti-*tya* antibodies. At the bottom is a Western blot of total yeast protein extracted from strains JB282 (wild-type pGTyH3), JB458 (wild-type pGTyH3 in *spt3-101* strain background), JB288 (pGTyH3 with an insertion in the *Pvu*II site of *tya*), and JB413 (pGTyH3 with frameshift mutation at the *Bgl*II site at position 1702). Plus signs designate protein samples isolated from transposition induced (galactose-grown) cells; minus signs designate samples from uninduced (glucose-grown) cells.

It is likely that *tya* encodes a polyprotein which is then cleaved to form several smaller components. Preliminary results, using an anti-peptide antibody directed toward amino acids 34-43 of *tya*, suggest that it is cleaved approximately 6 kDa from its carboxy terminus to form a protein (*tya*-54), that migrates at a 54 kDa position (Figure 6). Further processing events are likely to occur, as is suggested by our inability to detect *tya*-60 or *tya*-54 in preparations of purified *Ty-VLPs* (D. J. Garfinkel and J. D. Boeke, unpublished results, 1985).

pol-tyb

The retroviral *pol* gene is divided into three functional domains; a protease (*pro*) that is required for cleaving *gag*, the *gag-pol* polyprotein, and the envelope protein into smaller components, the reverse transcriptase/RNase H protein (RT) that is required for DNA synthesis to form the provirus, and an endonuclease or integrase that is thought to be required for proviral integration. However, the gene encoding protease differs in different retroviruses (reviewed by Varmus, 1985). Protease activity has been associated with the carboxy-terminal product of the *ALV gag* polyprotein but not in the *MuLV gag* proteins. The *MuLV* protease appears to span the *gag-pol* junction (and therefore the *gag* stop codon), with four amino acids of protease overlapping the carboxy terminus of *gag* in the same frame (Yoshinaka et al., 1985).

The genetic organization of retrovirus *pol* genes, as deduced by computer searches and intensive biochemical analysis, is NH_2-*pro-RT-int*-COOH. Similar computer-assisted sequence comparisons applied to Ty reveal that the organization of *tyb* is NH_2-*pro-int-RT*-COOH (note that the *tyb int* domain is located amino terminal to the RT domain) (Clare and Farabaugh, 1985; Mount and Rubin, 1985).

The initial *tyb* translation product should appear as a *tya*/*tyb* fusion protein of approximately 190 kDa (Mellor et al., 1985b; Clare and Farabaugh, 1985). Proteins this large have been seen in protein extracts from pulse-labeled cells that are over-expressing a Ty element. Proteins of 50 and 70 kDa dissociated from *Ty-VLPs* are visible on Coomassie-stained gels, but it is unclear which Ty gene encodes these products (Mellor et al., 1985c).

Anti-Ty antibodies directed against a peptide near the RT homology domain cross-react strongly with a 90 kDa *tyb* protein and weakly with *tyb* proteins of 70, 120, 150, 165, and 200 kDa. The *tyb*-90 protein copurifies with *Ty-VLPs* and is not present in mutants that lack reverse transcriptase activity (Garfinkel et al., 1985; J. D. Boeke and D. J. Garfinkel, unpublished results, 1985). Although it is distinctly possible that the *tyb*-90 protein is Ty reverse transcriptase, this enzyme is not sufficiently pure to determine whether it is identical to *tyb*-90. Mutations immediately adjacent to the protease domain in the amino terminal portion of *tyb* affect the sizes of *tya* proteins that can be detected immunologically (Figure 6). Mutations within the protease domain can also affect particle morphology [Figure 1(b)] and reverse transcriptase activity (J. D. Boeke and D. J. Garfinkel, unpublished results, 1985).

Retroviruses or Retrotransposons—Which Came First?

It is clear that retrotransposons and retroviruses are closely related structurally and functionally. Consequently, they must have a common ancestor. The major difference between retrotransposons and retroviruses is that the transposons, sensu strictu, are not infectious but instead seem to undergo an intracellular transposition process. Although it is probably impossible to resolve the question of which arose from which in a convincing way, it may be instructive to review the evidence for both sides of what has by now become a classic argument in light of recent data.

The argument that a cellular reverse transcriptase gene was a progenitor to retroviruses was put forward by Temin (1971,1974,1980). It can be argued that such a gene, if flanked by repeated sequences that could serve as LTR sequences, could rapidly evolve into a Ty-like transposon (but lacking *gag*). Many copies of such a "proto-retrotransposon" could be generated by a reverse transcription replication mechanism similar to that used by Ty. This would undoubtedly lead to a cell with very high levels of free reverse transcriptase. Such high levels of reverse transcriptase might result in unacceptably high levels of reverse information flow in the cell. Selective pressure to reduce the amount of free reverse transcriptase in the cell could lead to the evolution of encapsidation (*gag*) functions that would sequester the enzyme from cellular mRNAs, but leave the transposon's RNA packageable. Thus, it can be argued, particles evolved in response to a stress on the cellular genome. A consequence of particle formation would be that it might allow the transposition process to become more efficient as well as less deleterious through further evolution. These primitive particle-encoding elements could later have been subjected to later selection for the ability to infect sensitive cells and for transmission of genetic information to other cells. The infectious transfer of genetic information is of course a major route to the evolution of new phenotypes, and may explain at least partially the punctuated equilibrium theory of evolution (Erwin and Valentine, 1984).

A weakness in the argument that retroviruses arose from "proto-retrotransposons" is that it is not yet known whether Ty encapsidation is required for transposition, although this seems likely since reverse transcription is required for transposition and reverse transcription takes place in isolated *Ty-VLPs* (Boeke et al., 1985; Garfinkel et al., 1985; Mellor et al., 1985c) as well as *copia* VLPs. A further point that is very difficult to verify is the existence of an ancestral "cellular reverse transcriptase gene." All reverse transcriptases found thus far in integrated form (except caulimovirus and hepadnavirus reverse transcriptases) are flanked by LTR sequences of some type (a possible exception is the L1Md element of mouse, which has a a series of tandem repeats at the 5' end and a polyA segment at the 3' end [Loeb et al., 1986]). Of course, it would be unlikely for an original "cellular reverse transcriptase gene" to persist if, as argued above, it proved harmful to its host.

The counterargument, put forward by Weinberg (1980), is that retroviruses are generally parasitic and can become established as endogenous proviruses, possibly defective ones. Consequently, it would be easy to imagine a retrotransposon such as Ty, which can be viewed as "trapped" inside the virtually impenetrable yeast cell wall, as nothing more than a degenerate endogenous retrovirus that has lost its capability to infect other cells. Indeed, Ty elements have no *env* gene or anything corresponding to it. By this argument, the ability of Ty and *copia* to form VLPs might be considered vestigial and of little functional significance. Of the sequenced retrotransposons, there are four (Ty, *copia* from drosophila, L1Md of the mouse, and IAP from Syrian hamster) that lack *env* genes (Mount and Rubin, 1985; Emori et al., 1985; Loeb et al., 1986; Ono et al., 1985). In the case of *copia* there is very little coding capacity between the end of the ORF that encodes reverse transcriptase and the 3' LTR, an arrangement very similar to that seen in Ty. In IAP, however, there is a ∿1500-nucleotide-long region that contains no open reading frames between the end of *pol* and the 3' LTR. The authors attribute this region to a vestigial *env* gene that has accumulated many point mutations through lack of use. Thus IAP's structure suggests that it may have arisen from an endogenous murine retrovirus. If Ty and *copia* are derived from elements that originally were infectious and had *env* genes, they must have diverged from that ancestor much sooner than IAP did from its infectious ancestor. L1Md does not have LTRs and so it cannot be compared directly to the other elements. It does however have a *gag/pol*-like arrangement of two open reading frames (Loeb et al., 1986).

Ty and *copia* differ from IAP in another significant way: Their *pol* gene is organized differently. Thus, the possibility exists that they are not derived from an infectious retrovirus (as IAP appears to be), but form an evolutionarily independent group, perhaps the progenitors of retroviruses. Indeed, recent dendrogram studies performed on the basis of amino acid sequence homology by Toh and his colleagues (H. Toh, personal communication, 1986) suggest that IAP is closely related to murine and avian retroviruses, whereas Ty and *copia* fall together in a very distantly related group. There are likely to be additional members of this group, because there are several other retrotransposons which, by their overall length, appear not to have room to encode an *env* gene, although this remains to be confirmed by sequencing (Boeke, 1987). *Copia*, unlike Ty, is not "trapped" inside an impenetrable cell wall; in fact, it coexists with many *env*-

containing drosophila retrotransposons. These facts do not prove anything about the origin of *copia* and Ty elements, but certainly leave open the possibility that they are progenitors of, and not merely the degenerate dregs of, infectious retroviruses.

ACKNOWLEDGMENTS

We wish to thank Gerald R. Fink, in whose laboratory some of the work was performed. JB is supported by a Mellon Foundation award and is a Searle Scholar; his research is supported in part by a grant from the National Institutes of Health. DG's research was sponsored in part by the National Cancer Institute, DHHS, under contract No. N01-CO-23909 with Bionetics Research, Inc. The contents of this publication do not necessarily reflect the views or policies of the Department of Health and Human Services, nor does mention of trade names, commercial products, or organizations imply endorsement by the United States government.

REFERENCES

Ballario, P., Filetici, P., Junakovic, N., and Pedone, F. (1983). Ty1 extrachromosomal circular copies in *Saccharomyces cerevisiae*. *FEBS Lett.* *155*:225-229.

Baltimore, D. (1985). Retroviruses and retrotransposons: The role of reverse transcription in shaping the eukaryotic genome. *Cell* *40*:481-482.

Boeke, J. D. (1987). Retrotransposons. In *RNA Genetics*, vol. II, *Retroviruses, Viroids, and RNA Recombination* (E. Domingo, J. Holland, and P. Ahlquist, Eds.), CRC Press, in press.

Boeke, J. D., Garfinkel, D. J., Styles, C. A., and Fink, G. R. (1985). Ty elements transpose through an RNA intermediate, *Cell* *40*:491-500.

Boeke, J. D., Styles, C. A., and Fink, G. R. (1986). The yeast *SPT3* gene is required for transposition and transpositional recombination of chromosomal Ty elements. *Mol. Cell. Biol.* *6*:3575-3581.

Cameron, J. R., Loh, E. Y., and Davis, R. W. (1979). Evidence for transposition of dispersed repetitive DNA families in yeast. *Cell* *16*:739-751.

Cappello, J., Handelsman, K., and Lodish, H. F. (1985). Sequence of dictyostelium DIRS-1: An apparent retrotransposon with inverted terminal repeats and an internal circle junction sequence. *Cell 43*:105-115.

Clare, J., and Farabaugh, P. (1985). Nucleotide sequence of a yeast Ty element: Evidence for an unusual mechanism of gene expression. *Proc. Natl. Acad. Sci. USA 82*:2829-2833.

Eibel, H., Gafner, J., Stotz, A., and Phillipsen, P. (1981). Characterization of the yeast mobile genetic element Ty1. *Cold Spr. Harb. Symp. Quant. Biol. 45*:609-617.

Elder, R. T., Loh, E. Y., and Davis, R. W. (1983). RNA from the yeast transposable element Ty1 has both ends in the direct repeats, a structure similar to retrovirus RNA. *Proc. Natl. Acad. Sci., USA 80*:2432-2436.

Emori, Y., Shiba, T., Kanaya, S., Inouye, S., Yuki, S., and Saigo, K. (1985). The nucleotide sequence of *copia* and *copia*-related RNA in *Drosophila* virus-like particles. *Nature 315*:773-776.

Errede, B., Cardillo, T. S., Sherman, F., Dubois, E., Deschamps, J., and Wiame, J. M. (1980). Mating signals control expression of mutations resulting from insertion of transposable repetitive element adjacent to diverse yeast genes. *Cell 22*:427-436.

Errede, B., Company, M., Ferchak, J. D., Hutchison III, C. A., and Yarnell, W. S. (1985). Activation regions in a yeast transposon have homology to mating type control sequences and to mammalian enhancers. *Proc. Natl. Acad. Sci., USA 82*:5423-5427.

Erwin, D. H., and Valentine, J. W. (1984). "Hopeful monsters," transposons, and metazoan radiation. *Proc. Natl. Acad. Sci., USA 81*:5482-5483.

Fink, G. R., Boeke, J. D., and Garfinkel, D. G. (1986). The mechanism and consequences of retrotransposition. *Trends Genet. 2*:118-123.

Flavell, A. J., and Ish-Horowicz, D. (1981). Extrachromosomal circular copies of the eukaryotic transposable element *copia* in cultured *Drosophila* cells. *Nature 292*:591-595.

Flavell, A. J., and Ish-Horowicz, D. (1983). The origin of extrachromosomal circular *copia* elements. *Cell 34*:415-419.

Garfinkel, D. J., Boeke, J. D., and Fink, G. R. (1985). Ty element transposition: Reverse transcriptase and virus-like particles. *Cell 42*:507-517.

Gilboa, E., Mitra, S. W., Goff, S., and Baltimore, D. (1979). A detailed model of reverse transcription and a test of crucial aspects. *Cell 18*:93-100.

Hauber, J., Nelbock-Hochstetter, P., and Feldman, H. (1985). Nucleotide sequence and characteristics of a Ty element from yeast. *Nucl. Acids Res. 13*:2745-2758.

Hawley, R. G., Shulman, M. J., Murialdo, H., Givson, D. M., and Hozumi, N. (1982). Mutant immunoglobulin genes have repetitive DNA elements inserted into their intervening sequences. *Proc. Natl. Acad. Sci., USA 79*:7425-7429.

Hawley, R. G., Shulman, M. J., and Hozumi, N. (1984). Transposition of two different intracisternal A particle elements into an immunoglobulin kappa-chain gene. *Mol. Cell. Biol. 4*:2565-2572.

Heine, C. W., Kelly, D. C., and Avery, R. J. (1980). The detection of intracellular retrovirus-like entities in *Drosophila melanogaster* cell cultures. *J. Gen. Virol. 49*:385-395.

Jacks, T., and Varmus, H. E. (1985). Expression of the Rous sarcoma virus *pol* gene by ribosomal frameshifting. *Science 230*:1237-1242.

Kuff, E. L., Wivel, N. A., and Lueders, K. K. (1968). The extraction of intracisternal A-particles from a mouse plasma-cell tumor. *Can. Res. 28*:2137-2148.

Kuff, E. L., Feenstra, A., Lueders, K., Smith, L., Hawley, R., Hozumi, N., and Shulman, M. (1983). Intracisternal A-particle genes as movable elements in the mouse genome. *Proc. Natl. Acad. Sci., USA 80*:1992-1996.

Loeb, D. D., Padgett, R. W., Hardies, S. C., Sheehee, W. R., Comer, M. B., Edgell, M. H., and Hutchison, C. A. (1986). The sequence of a large L1Md element reveals a tandemly repeated 5' end and several features found in retrotransposons. *Mol. Cell. Biol. 6*:168-182.

Mellor, J., Fulton, A. M., Dobson, M., Roberts, N. A., Wilson, W., Kingsman, A., and Kingsman, S. M. (1985a). The Ty transposon of *Saccharomyces cerevisiae* determines the synthesis of at least three proteins. *Nucl. Acids Res. 13*:6249-6263.

Mellor, J., Fulton, S. M., Dobson, J. J., Wilson, W., and Kingsman, A. J. (1985b). A retrovirus-like strategy for the expression of a fusion protein encoded by yeast transposon Ty1. *Nature 313*: 243-246.

Mellor, J., Malim, M. H., Gull, K., Tuite, M. F., McCready, S., Dibbayawan, T., Kingsman, S. M., and Kingsman, A. J. (1985c). Reverse transcriptase activity and Ty RNA are associated with virus-like particles in yeast. *Nature 318*:583-586.

Mount, S. M., and Rubin, G. M. (1985). Complete nucleotide sequence of the *Drosophila* transposable element *copia*: Homology between *copia* and retroviral proteins. *Mol. Cell. Biol.* 5:1630-1638.

Neel, B. G., Hayward, W. S., Robinson, H. L., Fang, J., and Astrin, S. M. (1981). Avian leukosis virus-induced tumors have common proviral integration sites and synthesize discreet new RNAs: Oncogenesis by promoter insertion. *Cell 23*:323-334.

Newlon, C. S., Devenish, R. J., and Lipchitz, L. R. (1982). Mapping autonomously replicating segments on a circular derivative of chromosome III, in *Proceedings of the Berkeley Workshop on Recent Advances in Yeast Molecular Biology*: *Recombinant DNA*, Berkeley, California.

Ono, M., Toh, H., Miyata, T., and Awaya, T. (1985). Nucleotide sequence of the Syrian hamster intracisternal A-particle gene: Close evolutionary relationship of type A particle gene to types B and D oncovirus genes. *J. Virol. 55*:387-394.

Panganiban, A. T., and Temin, H. M. (1984). Circles with two tandem LTRs are precursors to integrated retrovirus DNA. *Cell 36*: 673-679.

Paquin, C. E., and Williamson, V. M. (1984). Temperature effects on the rate of Ty transposition. *Science 226*:53-55.

Paquin, C. E., and Williamson, V. M. (1986). Ty insertions at two loci account for most of the spontaneous antimycin A resistance mutations during growth at 15°C of *Saccharomyces cerevisiae* strains lacking ADH1. *Mol. Cell. Biol. 6*:70-79.

Rein, A., McClure, M. R., Rice, N. R., Lufting, R. B., and Schultz, A. M. (1986). The myristylation site in Pr 65*gag* is essential for virus particle formation by Moloney MuLV. *Proc. Natl. Acad. Sci., USA*, in press.

Roeder, G. S., Farabaugh, P. J., Chaleff, D. T., and Fink, G. R. (1980). The origins of gene instability in yeast. *Science 209*: 1375-1380.

Roeder, G. S., Rose, A. B., and Pearlman, R. E. (1985). Transposable element sequences involved in the enhancement of yeast gene expression. *Proc. Natl. Acad. Sci., USA 82*:5428-5432.

Scherer, S., Mann, C., and Davis, R. W. (1982). Reversion of a promoter deletion in yeast. *Nature 298*:815-819.

Schultz, A. M., and Oroszlan, S. (1983). *In vivo* modification of retroviral *gag* gene encoded proteins by myristic acid. *J. Virol 46*:355-361.

Shiba, T., and Saigo, K. (1983). Retrovirus-like particles containing RNA homologous to the transposable element *copia* in *Drosophila melanogaster*. *Nature 302*:119-124.

Simsek, M., and RajBhandary, U. L. (1972). The primary structure of initiator methionine tRNA. *Biochim. Biophys. Res. Commun. 49*:508-515.

Sorge, J., and Hughes, S. (1982). Splicing of intervening sequences introduced into an infectious retroviral vector. *J. Mol. Appl. Genet. 1*:547-559.

Taguchi, A. K. W., Ciriacy, M., and Young, E. T. (1984). Carbon source dependence of transposable element-associated gene activation in *Saccharomyces cerevisiae*. *Mol. Cell. Biol. 4*:61-68.

Temin, H. M. (1971). The protovirus hypothesis. *J. Nat. Cancer Inst. 46*:III-VII.

Temin, H. M. (1974). On the origin of RNA tumor viruses. *Ann. Rev. Genet. 8*:155-177.

Temin, H. M. (1980). Origin of retroviruses from cellular moveable genetic elements. *Cell 21*:599-600.

Varmus, H. E. (1983). Retroviruses, in *Mobile Genetic Elements* (J. A. Shapiro, Ed.), Academic, New York, Chap. 10.

Varmus, H. E. (1985). Reverse transcriptase rides again. *Nature 314*:583-584.

Varmus, H., and Swanstrom, R. (1984). Replication of retroviruses, in *RNA Tumor Viruses*, 2nd ed. (R. Weiss, N. Teich, H. Varmus, and J. Coffin, Eds.), Cold Spring Harbor Laboratory, Cold Spring Harbor, New York, Chap. 5.

Warmington, J. R., Waring, R. B., Newlon, C. S., Indge, K. J., and Oliver, S. G. (1985). Nucleotide sequence characterization of Ty 1-17, a class II transposon from yeast. *Nucl. Acids Res. 13*:6679-6693.

Weinberg, R. (1980). Origins and roles of endogenous retroviruses. *Cell 22*:643-644.

Yoshinaka, Y., Katoh, I., Copeland, T. D., and Oroszlan, S. (1985). Murine leukemia virus protease is encoded by the *gag-pol* gene and is synthesized through suppression of an amber termination codon. *Proc. Natl. Acad. Sci., USA 82*:1618-1622.

3

Transcriptional Regulation of Ty Elements in *Saccharomyces cerevisiae*

Fred Winston
Harvard Medical School, Boston, Massachusetts

INTRODUCTION

Transposable genetic elements have been identified in almost every organism that has been studied at a molecular genetic level. First identified in maize, from genetic and cytological studies by McClintock (reviewed by McClintock, 1965), transposable elements have been studied in greatest detail in prokaryotes (for a review see Kleckner, 1981). Transposable elements define a large and diverse group of elements whose one common characteristic is the ability to transpose to different positions in the host genome by nonhomologous recombination. The mechanism of transposition and the function of transposable elements has already been shown to vary greatly.

The study of transposable genetic elements (Ty elements) in the yeast *Saccharomyces cerevisiae* has been notable for two main reasons. First, these elements are members of a family of transposable elements called retrotransposons (Boeke et al., 1985). Similar to mammalian retroviral proviruses, Ty elements transpose via an RNA intermediate and are likely to encode a reverse transcriptase (Boeke et al., 1985; Garfinkel et al., 1985; Mellor et al., 1985). Second, Ty elements have profound effects on adjacent gene expression when inserted in the 5' region of a gene. These effects are in part due to transcription signals on the Ty elements, both in the long terminal repeats (δ sequences) (Elder et al., 1983) and in the internal ε region of the element (Roeder et al., 1985; Errede et al., 1985). The effects of a Ty insertion on adjacent gene expression is also mediated by unlinked host genes. Given that transposable elements are found throughout the biological kingdom and that retrotransposons are found in many types of eukaryotic organisms, *S. cerevisiae* is a model system for the understanding of these elements, both for their mechanism of transposition and for their mechanisms of gene regulation.

PHENOTYPES OF INSERTION MUTATIONS

Insertion of transposable genetic elements into the 5' regulatory regions of genes can have profound *cis*-acting effects on gene expression, ranging from complete inhibition of expression to constitutive expression. The particular effect depends upon the particular transposable element, the position of the insertion with respect to the adjacent gene, and the genetic background of the organism. The effects, where studied, have been shown to be exerted at the transcriptional level by such alterations as inhibition of transcription initiation, abérrant transcription initiation, or premature transcription termination. In most cases it is suspected (and in some cases it has been demonstrated) that the mutant phenotype is caused by the activity of transcription signals on the transposable element.

The phenotype of insertion mutations can depend upon *trans*-acting genes unlinked to the insertion mutation. Mutations in these genes result in alteration of the phenotype of the insertion mutation, often causing suppression, which results in at least partial restoration of a wild type phenotype. Such suppressor genes are candidates to be important in control of gene expression by transposable elements and for control of transcription of transposable elements.

Only in prokaryotes is the function of any *trans*-acting suppressor of insertion mutations understood at the molecular level. In *Escherichia coli* and salmonella, insertion mutations in operons can cause transcriptional polarity due to the presence of transcription termination signals on the element. Mutations in the transcription termination factor *rho* suppress the transcriptional polarity caused by these insertions (Das et al., 1976; Ciampi et al., 1982). Likewise, the phage λN gene product can also suppress the polarity caused by insertion mutations, also apparently by suppression of transcription termination (Brachet et al., 1970).

In yeast, drosophila and mice, *trans*-acting suppressors of insertion mutations have also been identified (Winston et al., 1984a; Green, 1952, 1955,1959; Modolell et al., 1983; Searles and Voelker, 1986; Levis et al., 1984; Sweet, 1983). In maize, unlinked genetic loci cause a variety of effects on insertion mutations, including enhancement of the mutant phenotype and destabilization of the insertion element (for reviews see McClintock, 1965; Fincham and Sastry, 1974; Peterson, 1977; Federoff, 1983). For suppressors of insertion mutations in eukaryotes, little is known about the mechanisms of suppression or the function of the wild type gene products altered by these mutations; however, there is good evidence implicating them in transcriptional regulation. For example, Zachar et al. (1985) demonstrated a specific alteration of transcription of an insertion mutation at the *white* locus in drosophila, *white-apricot* (w^a), in strains that contain the *suppressor of white-apricot* mutation Su(w^a). Analysis of other classes of suppressors in drosophila, yeast, and mice strongly suggests that suppression can occur by several different mechanisms.

INSERTION MUTATIONS IN YEAST

In the yeast, *Saccharomyces cerevisiae*, insertion mutations caused by Ty elements have been isolated and studied. Ty elements are a dispersed set of repetitive elements, approximately 6000 base pairs in length. They are members of the retrotransposon family of eukaryotic transposable elements which also includes retroviruses of mammals and *copia*-like elements of drosophila. Ty elements are flanked by direct terminal repeats (δ sequences) (Cameron et al., 1979), transpose via an RNA intermediate (Boeke et al., 1985), contain two open reading frames (Clare and Farabaugh, 1985), one of which appears to encode a reverse transcriptase, and are transcribed from δ to δ (Elder et al., 1983). For common laboratory strains of *Saccharomyces cerevisiae* there are approximately 35 Ty elements per haploid genome (Cameron et al., 1979).

Ty transcription occurs at a high level in a haploid strain, comprising approximately 10% of the polyA$^+$ RNA (Elder et al., 1980). Ty transcripts initiate in the 5' δ sequence and terminate in the 3' δ sequence (Figure 1). The transcripts are terminally redundant for 45 bases, similar to those of mammalian retroviral proviruses (Elder et al., 1983). Since Ty transcription initiates and terminates in δ sequences, these sequences must contain sites important for both of these events. However, the site for termination is utilized only in the 3' δ sequence and must be virtually ignored in the 5' δ sequence. Likewise, the initiation site is utilized in the 5' δ sequence and not in the 3' δ sequence. This latter observation can be explained if the δ sequence is not sufficient for transcription initiation and the Ty sequences preceding it are not adequate to allow the initiation event to occur. Alternatively, transcription entering the 3' δ sequence may

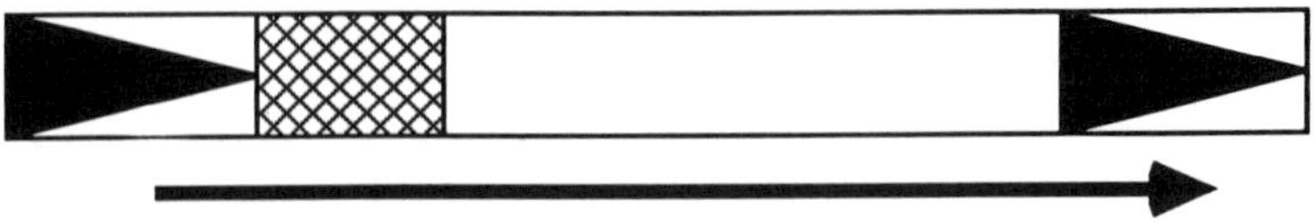

Figure 1 Transcription of Ty elements. Ty transcription initiates in the 5' δ sequence and terminates in the 3' δ sequence. The transcript is terminally redundant by 45 bases (Elder et al., 1983). The arrows at the end represent the δ sequences. The crosshatched region represents the approximate position of the Ty enhancer-like region, also believed to affect the level of both Ty transcription and adjacent gene expression (Roeder et al., 1985; Errede et al., 1985). The arrow below the Ty element represents the transcript.

somehow block transcription initiation as suggested by Cullen et al. (1984), for avian leukosis virus.

Insertion of a Ty element into the 5' noncoding region of a gene can have a variety of effects on expression of that gene, ranging from complete inhibition of expression to constitutive expression (for examples see Roeder et al., 1980; Errede et al., 1980; Williamson et al., 1981). Solo δ derivatives of Ty insertions, believed to arise by homologus recombination between the δ sequences flanking a Ty, also cause mutant phenotypes (Figure 2). The phenotype of a solo δ derivative of a Ty insertion is sometimes distinct from that of the complete Ty insertion (Chaleff and Fink, 1980; Ciriacy and Williamson, 1981). The phenotype of a Ty or solo δ insertion mutation is likely to be caused by the effects of transcriptional signals in the Ty element or δ sequence. For example, a Ty insertion mutation that causes constitutive expression in a haploid strain is regulated by the yeast mating type locus (*MAT*) (Errede et al., 1980). In haploid a or α strains, or diploid a/a or α/α strains, the constitutive phenotype is observed. However, in a/α diploid strains, the constitutive expression is greatly reduced. This regulation mimics control of Ty transcription itself, which is also under the same type of mating type control (Elder et al., 1980).

Ty and solo δ insertion mutations that inhibit adjacent gene expression can also do so by the influence of transcription signals on the insertion element. For example, in some cases of solo δ insertion mutations, transcription initiates in the solo δ instead of the wild type position for the adjacent gene and produces a nonfunctional transcript (Silverman and Fink, 1984; Winston et al., 1984b). In other cases, insertion of a complete Ty inhibits adjacent gene transcription (Winston et al., 1984b). For some classes of Ty insertion mutations, the inhibition is at least partially caused by a region in the internal portion of the Ty element (Roeder et al., 1985).

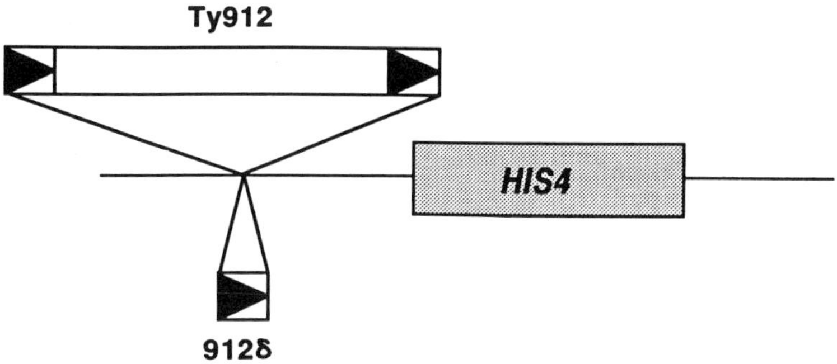

Figure 2 A complete Ty insertion and its solo δ derivative. The mutation *his4-912* contains Ty912 inserted at a position -98 from the start of wild type *HIS4* transcription. This mutation has a nonconditional His⁻ phenotype. The mutation *his4-912δ* contains the solo δ derivative of Ty912, inserted at the identical position. This mutation confers a cold-sensitive His⁻ phenotype.

SUPPRESSORS OF INSERTION MUTATIONS IN YEAST

Selection for suppressors of some Ty and solo δ insertion mutations that inhibit adjacent gene expression has yielded mutations in several genes, called *SPT* genes (SPT = suppressor of Ty; Winston et al., 1984a). The *spt* mutations alter transcription of the adjacent gene and, in many cases, of or from the insertion element as well. *SPT* genes are candidates to encode transcription factors that regulate transcription of Ty and δ sequences and may also be important for general cellular transcription. One class of *SPT* genes is required for normal transcription of Ty elements (Winston et al., 1984a; Winston et al., 1986).

To select for *spt* mutations, strains containing Ty and solo δ insertion mutations at *HIS4* and *LYS2* were used that cause His⁻ and Lys⁻ phenotypes, respectively. The insertion mutation used for the majority of the mutant isolations was *his4-912δ*, a solo δ insertion mutation that causes a cold sensitive His⁻ phenotype. This insertion is located 98 base pairs upstream of the *HIS4* transcription start site, between the *HIS4* upstream activator sequence (UAS) and the *HIS4* TATA region (Donahue et al., 1983; Nagawa and Fink, 1986). In a strain that contains *his4-912δ* that is grown at low temperature, *HIS4* transcription initiates in the δ sequence rather than at the wild type *HIS4* position. Since the strain has a His⁻ phenotype, the *his4-912δ* mRNA must be nonfunctional. Examination of the *his4-912δ* DNA sequence reveals that the mRNA contains an AUG and several termination codons in frame upstream of the normal *HIS4* AUG. Therefore, the *his4-912δ* mRNA is probably unable to initiate translation at the *HIS4* AUG. Selection for His^{+}, then, selects for suppressor mutations that allow production of the wild type *HIS4* mRNA.

Over 200 independent, spontaneous *spt* mutants were isolated, primarily as revertants of *his4-912δ*. These mutants were characterized by dominance and complementation analysis. Representative alleles were also tested for linkage, additional mutant phenotypes, and for suppression patterns with respect to different Ty and solo δ insertion mutations.

By complementation and dominance tests, the mutations fell into seven complementation groups (*spt1-spt7*). Some members of the *spt2* group were partially or fully dominant and all others were recessive. The fully dominant *spt2* mutants were classified by crosses with known *spt2* strains. Selection for *his4-912δ* revertants yields mostly *spt2* mutants (Winston et al., 1984a).

To examine linkage relationships and to rule out the possibility of intragenic complementation, pairwise crosses were performed between members of each *spt* group. From these crosses, it was shown that the seven complementation groups represent seven unlinked genes. Furthermore, the crosses revealed that double mutants between *spt4*, *spt5* and *spt6* are inviable. This suggests that these *SPT* gene products may interact.

Since most of the *spt* mutations were isolated as suppressors of *his4-912δ*, they were tested for their ability to suppress other insertion mutations at *HIS4* and *LYS2*. The results (Table 1) show that mutations in different *SPT* genes have different suppression patterns. This suggests that the seven *SPT* genes identified by this selection are not involved in the same process and that perhaps different *SPT* gene products recognize distinct Ty or δ transcription signals or act in different ways at these sites. That the different suppression patterns are not merely the result of allele specificity is shown by the result that null alleles of *SPT2*, *SPT3* and *SPT4* have distinct suppression patterns from each other (Winston et al., 1984a; J. Fassler and F. Winston, unpublished).

Other evidence that different *SPT* gene products participate in distinct functions came from examination of other phenotypes of *spt* mutants. All of the *spt* mutants isolated were screened for temperature sensitivity, sensitivity to ultraviolet light, and sensitivity to methylmethane sulfonate (MMS). From these tests, it was discovered that some alleles of *SPT6* are temperature-sensitive for growth and some alleles of *SPT4* are hypersensitive to MMS. These results suggest that *SPT6* is an essential gene and that *SPT4* may be involved in some aspect of DNA repair.

In addition, the mating and sporulation proficiencies of representative mutants were tested. Strains containing *spt3* mutants showed

Table 1 Suppression of Insertion Mutations by *spt* Mutations

spt allele	*his4-912δ*	*his4-917δ*	*his4-917*	*lys2-173R2*
SPT$^+$	cs[a]	-	-	+
spt1-1	+	-	+/-	+
spt2-150	+	-	+/-	+
spt3-202	+	+	+	-
spt4-3	+	-/+	-	+
spt5-194	+	-	-	+
spt6-140	+	-	-/+	+
spt7-159	+	-	-	+

[a]Designations are as follows: cs, cold sensitive; -, His$^-$ or Lys$^-$ phenotype; +, His$^+$ or Lys$^+$ phenotype; +/- and -/+, intermediate phenotypes.

a greatly reduced frequency of mating; furthermore, *spt3* homozygous diploids are unable to sporulate. The *spt3* mating defect is distinct from those of yeast sterile (*ste*) mutants in that the mutant phenotype is only strongly manifested when both parents contain the mutation. For *ste* mutants, the mating defect is observed when only one parent contains the *ste* mutation (see Sprague et al., 1983 for a review). Diploid strains homozygous for *spt3* null mutations fail to initiate sporulation and may be defective in an early step in this process.

Mutations in *SPT* genes are also distinct from each other in the transcriptional alterations associated with suppression. An initial study on the transcriptional effects of *spt* mutations on *his4-912δ* transcription by Silverman and Fink (1984) showed that *spt1*, *spt2*, and *spt3* all suppressed this allele at the transcriptional level, but in different ways. At low temperature in wild type (*SPT*) strains, transcription initiates in the δ sequence and not from the wild type *HIS4* position. In *spt1* and *spt2* mutants, both the δ-initiated transcript and a wild type length *HIS4* transcript are made. In *spt3* mutants, however, the δ-initiated transcript is gone and only a wild type *HIS4* transcript is made.

This initial isolation of *spt* mutations, then, identified seven distinct genes, mutant alleles of which can suppress Ty and δ insertion mutations. The different suppression patterns conferred by these mutations, taken together with the additional mutant phenotypes, strongly suggests that the different *SPT* genes identified act by distinct mechanisms. Furthermore, since the additional *spt* mutant phenotypes are not apparently related to Ty elements, the roles of *SPT* gene products are probably not limited to affecting Ty-mediated gene expression, but rather, may be more general.

SPT3 AND TRANSCRIPTION INITIATION IN δ SEQUENCES

SPT3 was chosen as the focus of further study for three reasons. First, unlike most of the classes of *spt* mutations found, *spt3* mutations suppress both complete Ty and solo δ insertion mutations; second, the mating and sporulation defects of *spt3* mutants suggested that the gene is required for other cellular functions; and third, and most importantly, *spt3* mutations alter Ty transcription both quantitatively and qualitatively. The analysis of transcriptional alterations of Ty elements and of genes adjacent to solo δ insertions in *spt3* mutants (reviewed below) has led to the model that *SPT3* is a positive activator required for transcription initiation in δ sequences.

After analysis of the *spt3* mutations isolated in vivo, two related questions remained to be answered. First, is the suppression observed due to loss of function or due to alteration of function of the *SPT3* gene product? Second, what is the phenotype of an *spt3* null mutant? To analyze *spt3* mutants in greater detail, the *SPT3* gene was cloned and two mutations believed to be null were constructed in vitro and were then recombined back into the yeast genome, replacing the wild type *SPT3* allele. The mutations constructed were *spt3-101*, a frameshift mutation in the middle of the *SPT3* gene (Winston et al., 1984b) and *spt3-202*, a deletion mutation that removes approximately half of the *SPT3* gene from the middle of the coding region (Winston and Minehart, 1986).

The results of the null mutant analysis were that (1) *spt3* null mutants are viable, although quite sick (the generation time is 1.3 times greater than wild type and the cells are very clumpy) and (2) they show essentially the same suppression pattern as the spontaneous alleles previously analyzed. Therefore, suppression is due to loss of function and *SPT3*, although not essential for growth, is important for normal growth. Construction of strains containing *spt3* null alleles allowed examination of the effects of *spt3* mutations in the absence of any functional gene product.

Ty transcription was examined in both wild type and *spt3* null mutants by Northern hybridization analysis and nuclease S1 protection analysis. These experiments demonstrated that normal δ-δ Ty transcription is virtually abolished in *spt3* mutants. Instead, at about one-fifth the level of the normal Ty transcript seen in wild type cells, a shorter Ty transcript that initiates approximately 800 bases farther

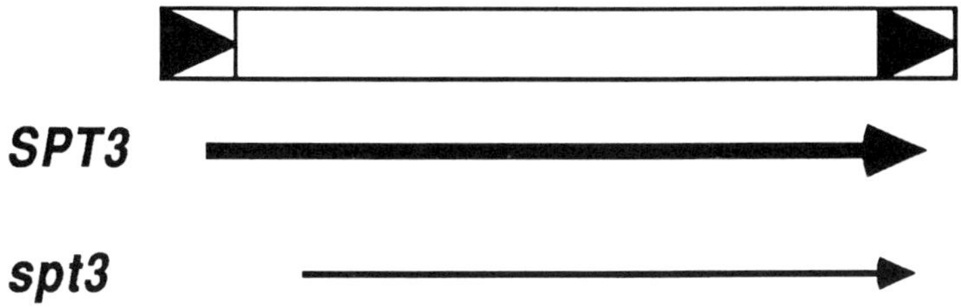

Figure 3 *SPT3* is required for normal Ty transcription. In wild type cells, Ty transcription initiates in the 5' δ sequence. In *spt3* mutants, Ty transcription is reduced approximately fivefold and it initiates 800 bases farther downstream, in the internal region of the Ty.

downstream in the internal portion of the Ty is observed (Figure 3; Winston et al., 1984b). Since these experiments studied the total population of Ty transcripts, the results indicate that an *spt3* mutation affects virtually every Ty in the cell (or at least every Ty that is transcribed) such that transcription no longer initiates in the 5' δ sequence. Recently, Boeke et al. (1986) demonstrated that in *spt3* mutants, Ty transposition no longer occurs, presumably because of the dramatic reduction in the level of Ty transcripts.

Transcription of genes adjacent to Ty and δ insertion mutations was also examined to attempt to elucidate the transcriptional effect of *spt3* mutations on adjacent gene expression. This was done for four insertion mutations, three at *HIS4* and one at *LYS2*: *his4-912δ*, *his4-917δ*, *his4-917*, and *lys2-173R2*. All of those examined exhibit transcriptional alterations in *spt3* mutants; for three of the four insertion mutations, the alterations are most easily explained by the model that *SPT3* is required for transcription initiation in δ sequences and in its absence transcription initiates farther downstream. For these three insertion mutations, *his4-912δ*, *his4-917δ*, and *lys2-173R2*, transcription initiates in a δ sequence in the wild type *SPT3*$^+$ background; however, in the *spt3* mutant, transcription no longer begins in the δ, but rather begins farther downstream.

The transcriptional patterns of *his4-912δ* and *his4-917δ* are virtually identical even though these two solo δ insertion mutations differ with respect to both position and orientation with respect to *HIS4* (Figure 4). The *his4-912δ* insertion is located between the *HIS4* UAS and the *HIS4* TATA region. This δ is oriented transcriptionally toward *HIS4*; therefore, it is not surprising that transcription from 912δ proceeds across *HIS4*. Transcription initiates in 912δ at the same position in the δ sequence as does Ty transcription. In some manner, perhaps by competition, this δ-initiated transcript inhibits transcription from the normal *HIS4* initiation site. Suppression by *spt3* mutations inhibits

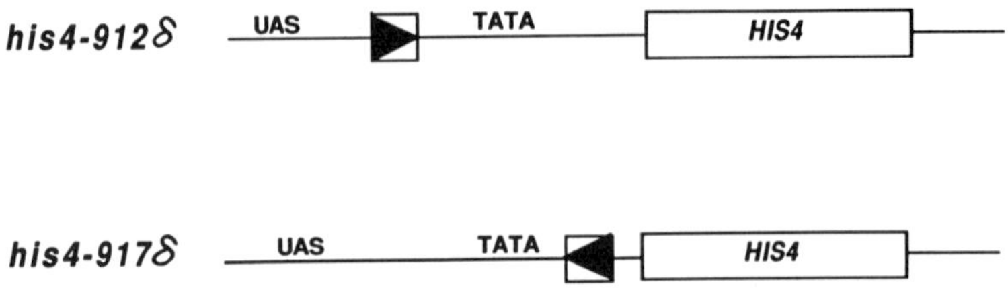

Figure 4 Comparison of *his4-912δ* and *his4-917δ*. These two solo δ insertion mutations differ with respect to position, orientation, and phenotype. The *his4-912δ* mutation causes a cold-sensitive His⁻ phenotype and *his4-917δ* causes an unconditional His⁻ phenotype.

transcription from the δ, thereby allowing transcription to initiate from the wild type *HIS4* position (Figure 5).

The behavior of *his4-917δ* with respect to *spt3* is not as easy to understand. This solo δ insertion is downstream of both the *HIS4* UAS and TATA regions and is just seven base pairs upstream of the start of *HIS4* transcription. Furthermore, 917δ is oriented away from *HIS4*. Surprisingly, in *his4-917δ* mutants, a nonfunctional *HIS4* transcript is still made, initiating from the δ sequence. Transcripts in both directions have been demonstrated to initiate from the LTR of the retrotransposon TED (Friesen et al., 1986). The production of this nonfunctional *HIS4* transcript may be dependent upon the *HIS4* UAS sequence, with the δ only providing an initiation site. In the *spt3* mutant, as for *his4-912δ*, a wild type length *HIS4* transcript is produced (Figure 5). This result demonstrates that *SPT3* is required for transcription initiation from a δ regardless of the direction of that transcription.

For *lys2-173R2*, there is a similar transcriptional alteration as occurs for *his4-912δ* even though the phenotype is the opposite from other suppressible insertion mutations; that is, Lys^+ in an $SPT3^+$ background and Lys^- in an *spt3* mutant background (Table 1). This mutation was isolated as a Lys^+ revertant of the Lys^- Ty insertion *lys2-173* (Simchen et al., 1984). The reversion event arose from a complicated rearrangement of the Ty element, including an inversion, such that the Ty is oriented in the same transcriptional orientation as *LYS2* in *lys2-173R2*. The phenotype of *lys2-173R2* can be explained by transcription initiating in the 3' δ sequence of *lys2-173R2* in the $SPT3^+$ background, producing a functional transcript for *LYS2* expression; a transcript is observed by Northern hybridization analysis whose size is consistent with this interpretation. Also as determined by Northern analysis, a shorter transcript is made in the *spt3* mutant; this transcript must by nonfunctional, either because it is too short to encode a functional *LYS2* product, or because there is some problem with its

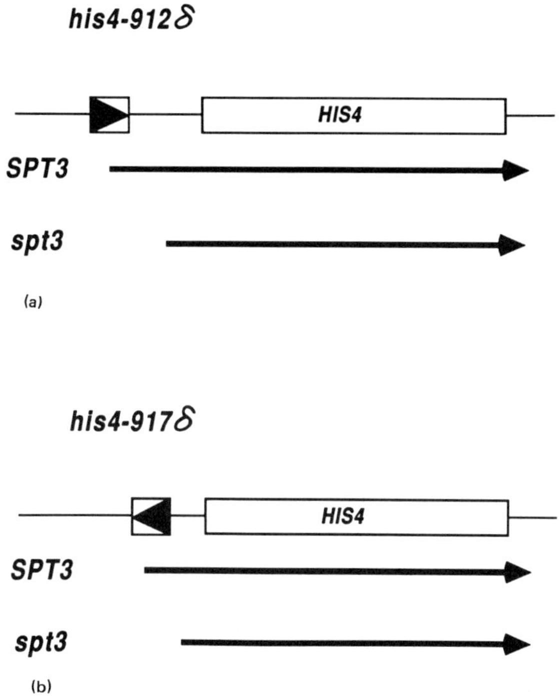

Figure 5 Suppression of *his4-912δ* and *his-917δ* by *spt3* mutations. For both insertion mutations, in wild type cells, transcription initiates in the δ sequence, producing a nonfunctional *HIS4* transcript. In *spt3* mutants, transcription cannot initiate in the δ and instead initiates farther downstream, at or near the wild type *HIS4* start site, resulting in a His$^+$ phenotype.

structure, for example, the first AUG is out of frame. Thus, opposite phenotypic effects of *spt3* mutations on different insertion mutations can be explained by the same transcriptional defect. The phenotype is determined by the position of the insertion.

Suppression of a complete Ty insertion, such as *his4-917*, cannot be explained by the identical mechanism. Northern hybridization results demonstrate that, in the *SPT3*$^+$ background, there is no *HIS4* transcription for strains containing this *HIS4* allele. However, in an *spt3* mutant, there is a wild type length *HIS4* transcript (Figure 6). Therefore, we cannot invoke an explanation where the *spt3* mutation blocks production of a nonfunctional *HIS4* transcript that initiates in a δ sequence. Rather, the effect must be mediated by a less direct mechanism, probably via an effect on the adjacent Ty element. Perhaps

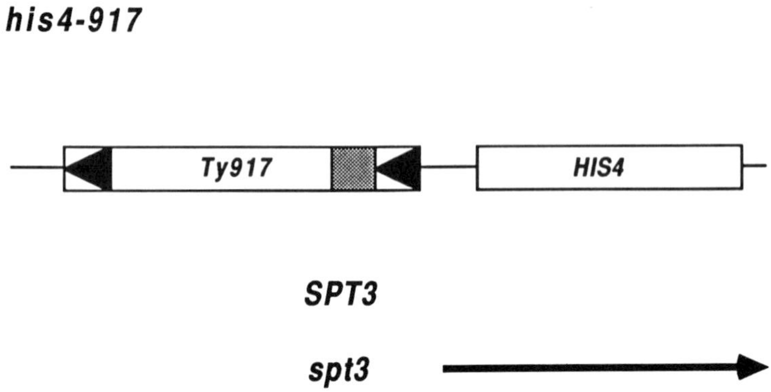

Figure 6 Suppression of *his4-917* by *spt3* mutations. In wild type cells, *his4-917* inhibits transcription of *HIS4*. In *spt3* mutants, a wild type length *HIS4* transcript is made, resulting in a His^+ phenotype.

alteration of transcription of the Ty "opens up" the region for transcription of *HIS4*.

Alternatively (or additionally) the enhancer-like sequences in the Ty element (Roeder et al., 1985; Errede et al., 1985) may play a role in suppression of *his4-917*. Roeder and Fink (1982) isolated gene convertants in which sequences in Ty917 were changed such that the Ty now allows a His^+ phenotype. Therefore, different Ty elements in the same position and orientation can cause opposite phenotypes. Roeder et al. (1985) demonstrated that this is caused by differences in an enhancer-like element in the internal region of the Ty element near the 5' end. Likewise, *spt3* mutations cause a Ty that turns off adjacent gene expression to allow adjacent gene expression. This phenotype suggests that *spt3* mutations may alter the ability of the enhancerlike element in Ty elements to function.

PHYSICAL STUDIES OF *SPT3*

To understand the nature of the *SPT3* function, the *SPT3* gene was characterized by DNA sequence analysis. To define the boundaries of the gene, *Bam*HI linker insertion mutations were made throughout a restriction fragment that contained the *SPT3* gene. The linker insertion mutations were then screened in yeast for those that disrupted *SPT3* function in order to define the region that is essential for *SPT3* function. This region was sequenced and an open reading frame of 1011 base pairs was found. This correlated well with the size of the *SPT3* mRNA, measured to be approximately 1300 bases long. A homology search failed to reveal any significant homology between the predicted amino acid sequence of the *SPT3* gene product and the amino acid sequences in the NBRF data base (Winston and Minehart, 1986).

To positively identify this open reading frame as the *SPT3* gene, an *spt3* ochre mutation was identified and sequenced. The *spt3* ochre mutation was identified after mutagenesis of plasmid pCC1, which contains a wild type *SPT3* gene and the yeast *URA3* gene. For mutagenesis, the plasmid was grown through an *E. coli mutD* strain. Plasmid was then prepared and transformed into an *spt3-101* host, which also contained the insertion mutations *his4-912δ* and *lys2-128δ*. Transformants that received a wild type *SPT3* gene will be His^- Lys^-. However, transformants that received a plasmid with an *spt3* mutation were His^+ Lys^+. To identify *spt3* ochre mutations, yeast strains containing the *spt3* mutant plasmids were transformed with plasmid pPM702 (Shaw and Olson, 1984) which contains the ochre suppressor *SUP4-o*. One mutant, *spt3-269*, was shown to be suppressed by *SUP4-o* (C. D. Clark-Adams and F. Winston, unpublished results). The position of the ochre mutation was localized to approximately 300 base pairs by recombination experiments between a strain containing *spt3-269* in the genome and plasmids deleted for various restriction fragments within *SPT3*. This region was sequenced and *spt3-269* was identified as a G→A transition, resulting in a UAA codon in the 1011 base pair open reading frame (Winston and Minehart, 1986).

To identify the *SPT3* gene product, expected to be a protein of 337 amino acids with a predicted molecular weight of 38,748, we constructed a gene fusion between *SPT3* and the *E. coli lacZ* gene, identified the hybrid protein from extracts of *E. coli*, and used the hybrid protein to immunize rabbits (Winston and Minehart, 1986). Antisera immunoprecipitated a protein from yeast extracts of approximately 40,000 M.S. This protein is believed to be *SPT3* because it is present in extracts of strains that contain SPT3 on a high copy number plasmid, but not in strains that contain a high copy number plasmid with an *SPT3* deletion. *SPT3* protein is barely detectable in extracts of wild type strains, suggesting that it is normally present at very low level. The fact that a high copy number *SPT3* plasmid increases the level of the gene product suggests that *SPT3* does not regulate its own synthesis.

IDENTIFICATION OF TWO ADDITIONAL GENES REQUIRED FOR TRANSCRIPTION FROM δ SEQUENCES

Among the seven classes of *spt* mutants identified, *spt3* mutants were unique in terms of their suppression pattern of Ty and δ insertion mutations and in terms of their effect on Ty transcription. Representative alleles of the other *SPT* genes did not have any qualitative or quantitative effect on Ty transcription. Therefore, we reasoned that new *spt* mutations that had the same suppression pattern as *spt3* would be strong candidates to affect Ty transcription in the same way as *spt3* mutations. This could be either because these gene products regulate *SPT3*, are regulated by *SPT3*, or interact directly with *SPT3* to allow initiation in δ sequences. The result that an increase in the level of the *SPT3* gene product did not increase the level of Ty transcription (Winston and Minehart, 1986) suggested that *SPT3* is not the only limiting factor in the cell for Ty transcription.

To isolate such mutants we took advantage of the unique *spt3* suppression pattern (Table 1). We selected for suppressors of the insertion *his4-917*δ and screened among those for suppressors which caused a Lys^- phenotype in the presence of *lys2-173R2*. This selection and screen identifies mainly *spt3* mutants; however, we also found mutations in two other genes that confer the same phenotype and cause the same transcriptional defect in Ty transcription.

By complementation analysis and linkage analysis, we found that one set of mutations was in *SPT7*, a gene already identified in the first round of isolation of *spt* mutants. In addition, we identified a new gene, *SPT8*. By examination of different alleles in each gene, we found that mutations in *SPT7* and *SPT8* cause the identical change in Ty transcription as do *spt3* mutations: δ-δ Ty transcription is greatly reduced and transcripts approximately 800 base pairs shorter at the 5' end are present (Winston et al., 1986). The original isolates of *spt7* mutations were too leaky to show this suppression pattern or transcriptional defect.

These results demonstrate that three genes are required for *trans*-activation of transcription from δ sequences. The degree of the requirement was measured by assay of a Ty-*lacZ* fusion, integrated in single copy in the yeast genome. This fusion contains only a short amount of the Ty element, such that the shorter transcript made in the *spt3* host would not encode a functional β-galactosidase. The assays, then, reflect the level of transcription initiation from the δ sequence. These results (Table 2) showed that there is a 15-30-fold reduction in β-galactosidase levels in *spt3, spt7*, or *spt8* mutants, as compared to wild type strains (Winston et al., 1987). Assay of an *spt3, spt7, spt8* triple mutant gave a comparable reduction in β-galactosidase levels, suggesting that these three genes participate in a common function.

Given these results, it is not yet clear if there is a regulatory interaction between these three gene products or if they directly interact to promote transcription from δ sequences. We can rule out that *spt7* or *spt8* is a positive activator of *SPT3* at the transcriptional level, since the level of *SPT3* mRNA is not altered in either of these mutants (Winston et al., 1987). Continued analysis of *SPT7* and *SPT8*, including cloning of the genes, construction of null alleles, identification of

Table 2 β-galactosidase Assays of Ty-*lacZ* Fusions

spt allele	β-galactosidase units
SPT^+	860
spt3-202	34
spt7-217	66
spt8-113	83

Source: Winston et al. (1987).

their mRNAs and the phenotype of *spt3, spt7, spt8* triple null mutants will aid in elucidation of the nature of these functions. Additionally, isolation and analysis of pseudorevertants of these mutations will help to identify other gene products with which these may interact to specify transcription initiation in δ sequences.

POSSIBLE FUNCTIONS FOR *SPT3, SPT7,* AND *SPT8*

The function of *SPT3, SPT7*, and *SPT8* is to specify transcription initiation in δ sequences. From studies of eukaryotic transcription initiation, one *cis*-acting region important for specificity of initiation is the TATA region. For SV40, both in vivo and in vitro studies (Benoist and Chambon, 1981; Mathis and Chambon, 1981) demonstrated a requirement for the TATA region for accurate initiation of transcription. In yeast, the results of several studies (Nagawa and Fink, 1985; Hahn et al., 1985; Chen and Struhl, 1985; McNeil and Smith, 1985) suggest that both the TATA region and specific sequences at the initiation site determine the pattern of transcription initiation. These studies demonstrated some flexibility in the distance between the TATA and the initiation site, ranging between 40 and 100 base pairs. The δ TATA region, 72 base pairs from the site of initiation, falls within this range.

SPT3, SPT7, and *SPT8*, then, may act, directly or indirectly at either the TATA region or the initiation site. If these two *cis*-acting sequences are both required for initiation, they may interact in some way. One piece of evidence that these *SPT* functions act near TATA is a mutation in *his4-912*δ that mimics the transcriptional phenotype of *spt3*; that is, transcription no longer initiates in the δ sequence, but rather initiates at the wild type *HIS4* sequence (Durbin, 1985). The mutation causes an AT→GC base change of the base pair immediately preceding the sequence TATAAAA region in the δ. Further experiments may elucidate the relationship of this region to the activity of *SPT3*, *SPT7*, and *SPT8*.

The existence of additional mutant phenotypes observed in *spt3*, *spt7*, and *spt8* mutants, such as sporulation and mating defects, suggests that these gene products are important for expression of other genes in addition to those of Ty elements. Perhaps δ sequences are one member of a class of genes whose initiation is controlled by these *SPT* gene products. Further analysis of defects in gene expression in *spt3*, *spt7*, and *spt8* mutants will address this possibility.

RELATIONSHIP OF *SPT* MUTATIONS TO OTHER SYSTEMS

Ty elements of yeast are closely related to other eukaryotic retrotransposons. Insertion by these elements can cause mutant phenotypes, including amino acid auxotrophy in yeast (Chaleff and Fink, 1980), eye color alterations in Drosophila (Bingham and Judd, 1981), and coat color changes (Copeland et al., 1983) and oncogenic transformation (Payne et al., 1982; Corcoran et al., 1984) in mice. In all three of these organisms, unlinked suppressor mutations have been identified. Work on two such suppressors in drosophila, *suppressor of white apricot* and *suppressor of hairy-wing*, has shown that both alter transcription (Levis et al., 1984; Parkhurst and Corces, 1985;1986a,b; Zachar et al., 1985). Parkhurst and Corces (1986a) reported that, like *SPT3*, Su(Hw) is required for a normal level of transcription of the transposable element gypsy in drosophila. In contrast, Su(w^a) may be required for transcription termination in copia elements (Zachar et al., 1985) so suppression by this mutation may be analogous in some respects to suppression by *E. coli rho* mutations.

In addition, *SPT3*, *SPT7*, and *SPT8* are functionally analogous to some retroviral functions. Given the similarity between Ty elements and retroviruses, transcription from the δ sequences of Ty elements is equivalent to that from LTRs of retroviral proviruses. Therefore, *SPT3* may be similar to *trans*-acting factors encoded by HTLV viruses (Sodroski et al., 1984) and Rous sarcoma virus (Broome and Gilbert, 1985) that are necessary for transcription from the viral LTRs. Others have suggested that the HTLV I *trans*-acting protein is the modulator of transformation, perhaps by activation of specific cellular genes (Aldovini et al., 1986). We have identified three genes involved in specifying transcription from δ sequences. In the case of these retroviruses, cellularly encoded gene products are also likely to be involved. Analysis of *SPT* genes may help lead us to identification and understanding of these mammalian products.

ACKNOWLEDGMENTS

This work was supported by NIH grant GM32967, NSF grant DCB-84-51649, and grants from The Stroh Brewery Company and Monsanto Company.

REFERENCES

Aldovini, A., De Rossi, A., Feinberg, M. B., Wong-Staal, F., and Franchini, G. (1986). Molecular analysis of a deletion mutant provirus of type I human T-cell lymphotropic virus: Evidence for a doubly spliced *x-lor* mRNA. *Proc. Nat. Acad. Sci. USA 83*:38-42.

Benoist, C., and Chambon, P. (1981). *In vivo* sequence requirements of the SV40 early promoter region. *Nature 290*:304-310.

Bingham, P. M. and Judd, B. (1981). A copy of the *copia* transposable element is very tightly linked to the W^a allele at the *white* locus of *D. melanogaster*. *Cell 25*:705-711.

Boeke, J. D., Garfinkel, D. J., Styles, C. A., and Fink, G. R. (1985). Ty elements transpose through an RNA intermediate. *Cell 40*:491-500.

Boeke, J. D., Styles, C. A., and Fink, G. R. (1986). *Saccharomyces cerevisiae SPT3* gene is required for transposition and transpositional recombination of chromosomal Ty elements. *Mol. Cell. Biol. 6*:3575-3581.

Brachet, P., Eisen, H., and Rambach, A. (1970). Mutations of coliphage λ affecting the expression of replicative functions O and P. *Mol. Gen. Genet. 108*:266-276.

Broome, S. and Gilbert, W. (1985). Rous sarcoma virus encodes a transcriptional activator. *Cell 40*:537-546.

Cameron, J. R., Loh, E. Y., and Davis, R. W. (1979). Evidence for transposition of dispersed repetitive DNA families in yeast. *Cell 16*:739-751.

Chaleff, D. T. and Fink, G. R. (1980). Genetic events associated with an insertion mutation in yeast. *Cell 21*:227-237.

Chen, W. and Struhl, K. (1985). Yeast mRNA initiation sites are determined primarily by specific sequences, not by the distance from the TATA element. *EMBO J. 4*:3273-3280.

Ciampi, M. S., Schmid, M. B., and Roth, J. R. (1982). Transposon Tn10 provides a promoter for transcription of adjacent sequences. *Proc. Nat. Acad. Sci. USA 79*:5016-5020.

Ciriacy, M. and Williamson, V. M. (1981). Analysis of mutations affecting Ty-mediated gene expression in *Saccharomyces cerevisiae*. *Mol. Gen. Genet.* *182*:159-163.

Clare, J., and Farabaugh, P. (1985). Nucleotide sequence for a yeast Ty element: Evidence for an unusual mechanism of gene expression. *Proc. Nat. Acad. Sci. USA 82*:2899-2833.

Copeland, N. G., Hutchison, K. W., and Jenkins, N. A. (1983). Excision of the DBA ecotropic provirus in dilute coat-color revertants of mice occurs by homologous recombination involving the viral LTRs. *Cell 33*:379-387.

Corcoran, L. M., Adams, J. M., Dunn, A. R., and Cory, S. (1984). Murine T lymphomas in which the cellular *myc* oncogene has been activated by retroviral insertion. *Cell 37*:113-122.

Cullen, B. R., Lomedico, P. T., and Ju, G. (1984). Transcriptional interference in avian retroviruses—implications for the promoter inserting model of leukaemogenesis. *Nature 307*:241-245.

Das, A., Court, D., and Adhya, S. (1976). Isolation and characterization of conditional lethal mutants of *Escherichia coli* defective in transcription termination factor rho. *Proc. Nat. Acad. Sci. USA 73*:1959-1963.

Donahue, T. F., Daves, R. S., Lucchini, G., and Fink, G. R. (1983). A short nucleotide sequence required for regulation of *HIS4* by the general control system of yeast. *Cell 32*:89-98.

Durbin, K. J. (1985). Transcription of Ty elements: Control by the *SPT3* gene. Ph.D. thesis, Cornell University, Ithaca, New York.

Elder, R. T., St. John, T. P., Stinchcomb, D. T., and Davis, R. W. (1980). Studies on the transposable element Ty1 of yeast. I. RNA homologous to Ty1. *Cold Spring Harbor Symp. Quant. Biol. 45*:581-584.

Elder, R. T., Loh, E. Y., and Davis, R. W. (1983). RNA from the yeast transposable element Ty1 has both ends in the direct repeats, a structure similar to retrovirus RNA. *Proc. Nat. Acad. Sci. USA 80*:2432-2436.

Errede, B., Cardillo, T. S., Sherman, F., Dubois, E., Deschamps, J., and Wiame, J.-M. (1980). Mating signals control expression of mutations resulting from insertion of a transposable repetitive element adjacent to diverse yeast genes. *Cell 22*:427-436.

Errede, B., Company, M., Ferchak, J. D., Hutchison, C. A., III, and Yarnell, W. S. (1985). Activation regions in a yeast transposon have homology to mating type control sequences and to mammalian enhancers. *Proc. Nat. Acad. Sci. USA 82*:5423-5427.

Federoff, N. (1983). Controlling elements in maize. In *Mobile Genetic Elements* (J. A. Shapiro, Ed.), Academic, New York, pp. 1-63.

Fincham, J. R. S., and Sastry, G. R. K. (1974). Controlling elements in maize. *Ann. Rev. Genet. 8*:15-50.

Friesen, P. D., Rice, W. C., Miller, D. W., and Miller, L. K. (1986). Bidirectional transcription from a solo long terminal repeat of the retrotransposon TED: Symmetrical RNA start sites. *Mol. Cell. Biol. 6*:1599-1607.

Garfinkel, D. J., Boeke, J. D., and Fink, G. R. (1985). Ty element transposition: Reverse transcriptase and virus-like particles. *Cell 42*:507-517.

Green, M. M. (1952). Mutant isoalleles at the *vermilion* locus in *Drosophila melanogaster*. *Proc. Nat. Acad. Sci. USA 38*:300-305.

Green, M. M. (1955). Phenotypic variation and pseudoallelism at the *forked* locus in *Drosophila melanogaster*. *Proc. Nat. Acad. Sci. USA 41*:375-379.

Green, M. M. (1959). Spatial and functional properties of pseudoalleles at the *white* locus in *Drosophila melanogaster*. *Heredity 13*:303-315.

Hahn, S., Hoar, E. T., and Guarente, L. (1985). Each of three "TATA elements" specifies a subset of the transcription initiation sites at the *CYC-1* promoter of *Saccharomyces cerevisiae*. *Proc. Nat. Acad. Sci. USA 82*:8562-8566.

Kleckner, N. (1981). Transposable elements in prokaryotes. *Ann. Rev. Genet. 15*:341-404.

Levis, R., O'Hare, K., and Rubin, G. M. (1984). Effects of transposable element insertions on RNA encoded by the *white* gene of *Drosophila*. *Cell 38*:471-481.

Mathis, D. J. and Chambon, P. (1981). The SV40 early region TATA box is required for accurate *in vitro* initiation of transcription. *Nature 290*:310-315.

McClintock, B. (1965). The control of gene action in maize. *Brookhaven Symp. Biol. 18*:162-184.

McNeil, J. B. and Smith, M. (1985). Transcription initiation of the *Saccharomyces cerevisiae* iso-1-cytochrome c gene. Multiple, independent T-A-T-A sequences. *J. Mol. Biol. 187*:363-378.

Mellor, J., Malim, M. H., Gull, K., Tuite, M. F., McCready, S., Dibbayawan, T., Kingsman, S. M., and Kingsman, A. J. (1985).

Reverse transcriptase activity and Ty RNA are associated with virus-like particles in yeast. *Nature 318*:583-586.

Modolell, J., Bender, W., and Meselson, M. (1983). *D. melanogaster* mutations suppressible by the *suppressor of Hairy-wing* are insertions of a 7.3 kb mobile element. *Proc. Nat. Acad. Sci. USA 80*:1678-1682.

Nagawa, F., and Fink, G. R. (1985). The relationship between the "TATA" sequence and transcription initiation sites at the *HIS4* gene of *Saccharomyces cerevisiae*. *Proc. Nat. Acad. Sci. USA 82*:8557-8561.

Parkhurst, S. M., and Corces, V. G. (1985). *forked*, gypsys and suppressors in Drosophila. *Cell 41*:429-437.

Parkhurst, S. M., and Corces, V. G. (1986a). Interactions among the gypsy transposable element and the yellow and the suppressor of Hairy-wing loci in *Drosophila melanogaster*. *Mol. Cell. Biol. 6*:47-53.

Parkhurst, S. M., and Corces, V. G. (1986b). Mutations at the suppressor of forked locus increase the accumulation of gypsy-encoded transcripts in *Drosophila melanogaster*. *Mol. Cell. Biol. 6*:2271-2274.

Payne, G. S., Bishop, M. G., and Varmus, H. E. (1982). Multiple rearrangement of viral DNA and an activated host oncogene in bursal lymphomas. *Nature 295*:209-214.

Peterson, P. A. (1977). The position hypothesis for controlling elements in maize, in *DNA Insertion Elements, Plasmids, and Episomes* (A. I. Bukhari, J. A. Shapiro, and S. L. Adhya, Eds.), Cold Spring Harbor Laboratory, Cold Spring Harbor, New York, pp. 429-435.

Roeder, G. A., and Fink, G. R. (1982). Movement of yeast transposable elements by gene conversion. *Proc. Nat. Acad. Sci. USA 79*:5621-5625.

Roeder, G. S., Farabaugh, P. J., Chaleff, D. T., and Fink, G. R. (1980). The origins of gene instability in yeast. *Science 209*: 1375-1380.

Roeder, G. S., Rose, A. B., and Pearlman, R. E. (1985). Transposable element sequences involved in the enhancement of yeast gene expression. *Proc. Nat. Acad. Sci. USA 82*:5428-5432.

Searles, L. L., and Voelker, R. A. (1986). Molecular characterization of the *Drosophila* vermilion locus and its suppressible alleles. *Proc. Nat. Acad. Sci. USA 83*:404-408.

Shaw, K. J., and Olson, M. V. (1984). Effects of altered 5'-flanking sequences on the in vivo expression of a *Saccharomyces cerevisiae* $tRNA^{tyr}$ gene. *Mol. Cell. Biol.* *4*:657-665.

Silverman, S. J., and Fink, G. R. (1984). Effects of Ty insertions on *HIS4* transcription in *Saccharomyces cerevisiae.* *Mol. Cell. Biol.* *4*:1246-1251.

Simchen, G., Winston, F., Styles, C. A., and Fink, G. R. (1984). Ty-mediated expression of the *LYS2* and *HIS4* genes of *Saccharomyces cerevisiae* is controlled by the same *SPT* genes. *Proc. Nat. Acad. Sci. USA* *81*:2431-2434.

Sodroski, J. G., Rose, C. A., and Haseltine, W. A. (1984). *Trans*-acting transcriptional activation of the long terminal repeat of human T lymphotrophic viruses in infected cells. *Science* *225*:381-385.

Sprauge, G. F., Jr., Blair, L. C., and Thorner, J. (1983). Cell interactions and regulation of cell type in the yeast *Saccharomyces cerevisiae.* *Ann. Rev. Microbiol.* *37*:623-660.

Sweet, H. O. (1983). *Dilute suppressor*, a new suppressor gene in the house mouse. *J. Heredity* *74*:304-306.

Williamson, V. M., Young, E. T., and Ciriacy, M. (1981). Transposable elements associated with constitutive expression of yeast alcohol dehydrogenase II. *Cell* *23*:605-614.

Winston, F., and Minehart, P. L. (1986). Analysis of the yeast *SPT3* gene and identification of its product, a positive regulator of Ty transcription. *Nucl. Acids Res.* *14*:6885-6900.

Winston, F., Chaleff, D. T., Valent, B., and Fink, G. R. (1984a). Mutations affecting Ty-mediated expression of the *HIS4* gene of *Saccharomyces cerevisiae.* *Genetics* *107*:179-197.

Winston, F., Durbin, K. J., and Fink, G. R. (1984b). The *SPT3* gene is required for normal transcription of Ty elements in *S. cerevisiae.* *Cell* *39*:675-682.

Winston, F., Dollard, C., Malone, E., Clare, J., Kapakos, J. G., Farabaugh, P., and Minehart, P. L. (1987). Three genes are required for *trans*-activation of Ty transcription in yeast. *Genetics* *115*:649-656.

Zachar, Z., Davison, D., Garza, D., and Bingham, P. M. (1985). A detailed developmental and structural study of the transcriptional effects of insertion of the *copia* transposon into the white locus of *Drosophila melanogaster.* *Genetics* *111*:495-515.

4

Recombination Associated with Yeast Retrotransposons

Susan W. Liebman and Susan Picologlou
University of Illinois at Chicago, Chicago, Illinois

INTRODUCTION

Although the DNA of most organisms is remarkably stable, it has now become clear that some genes can move about in the genome. These movable genes or transposable elements are found in a variety of different organisms including bacteria, yeast, drosophila, and maize, and have been the subject of recent reviews (Shapiro, 1983). The function of these elements is unknown, but they are frequently associated with increased mutability and the formation of deletions and genetic rearrangements, and they can affect the regulation of neighboring genes. Some of these movable elements are strikingly similar to retroviruses in their structure and mode of transposition, and have been called retrotransposons (Baltimore, 1985; Boeke et al., 1985).

The Ty elements of the yeast, *Saccharomyces cerevisiae*, are well studied retrotransposons that are similar to the endogenous proviruses found in vertebrate chromosomes (see Chapter 3 of this book for details, and Roeder and Fink, 1983; and Williamson, 1983 for reviews). The 6 kilobase pair yeast Ty elements consist of a large central region called epsilon (ε), flanked by direct repeats of about 330 base pairs called delta (δ). Deltas are analogous to the long terminal repeats (LTR) found at the ends of retroviruses.

There are striking sequence similarities between Ty elements and retroviruses (Gafner and Philippsen, 1980; Varmus, 1983; Clare and Farabaugh, 1985; Hauber et al., 1985; Mellor et al., 1985; Warmington et al., 1985). Both share a common mechanism of transposition involving an RNA intermediate and reverse transcriptase encoded by the element (Boeke et al., 1985; Garfinkel et al., 1985). Low levels of Ty DNA circles detected in yeast (Ballario et al., 1983) could be analogous to retroviral DNA circles that result from reverse transcription of the retroviral RNA. These retroviral circles, which contain 2 nearly tandem directly oriented LTRs at the circle junction (Panganiban and Temin, 1984a), are available to integrate into the genome at nonhomologous sites. Such integration or transposition causes a short duplication of the target sequence. When a Ty element transposes it likewise causes a 5 base pair target duplication (Farabaugh and Fink, 1980; Gafner and Philippsen, 1980). An endonucleolytic activity associated with the reverse transcriptase is believed to cut the circles at the LTR junction sequence as part of the retroviral integration process (Duyk et al., 1983; Panganiban and Temin, 1984b). Finally, Ty transposition into 5' noncoding regions alters the regulation of the nearby genes (Bach, 1984; Dubois et al., 1982; Errede et al., 1980; Roeder et al., 1980; Scherer et al., 1982; Williamson et al., 1981) in analogy to proviral insertion mutagenesis.

Ty elements have been divided into two main classes, Ty1 and Ty2. Restriction maps, along with Southern and heteroduplex analyses, showed these classes differ by substitution mutations covering about two-thirds of the epsilon region (Kingsman et al., 1981; Williamson et al., 1983). Sequence analyses of two Ty1 elements (Clare and Farabough, 1985; Hauber et al., 1985) and one Ty2 element (Warmington et al., 1985) have confirmed these differences. There are about 30 dispersed copies of Ty1 and about 6-12 copies of Ty2 in laboratory yeast strains. Restriction site polymorphisms are common even among Ty elements in a given class (see Williamson, 1983; Roeder and Fink, 1983).

LTR-like δ elements are present in at least 100 copies per haploid genome (Cameron et al., 1979), many of which are not associated with Ty elements (solo δ). Solo δ sequences are thought to arise from a recombination event between the directly repeated deltas surrounding a Ty element. The sequences of several δ elements have been compared (Williamson, 1983; Roeder and Fink, 1983; Bowen et al., 1984) and considerable sequence heterogeneity exists. In general, the two δs flanking a single Ty are identical or nearly so, while solo δ sequences are much more divergent.

Ty elements are found in different locations in different strains (see below, on the distribution of Ty elements). Their movement occurs most frequently (10^{-4}-10^{-5}) by homologous recombination, and more rarely (10^{-7}-10^{-8}) by the nonhomologous transposition events discussed above. Homologous recombination, which may include intra- or interchromosomal reciprocal recombination, unequal sister-strand recombination, or gene conversion, has been shown to cause chromosomal rearrangements such as deletions, duplications, inversions, and reciprocal translocations. Rearrangements of this type will be examined in detail in this chapter. Possible mechanisms involved in Ty/δ recombination will be considered. These may be the same as those governing recombination between any sequences in nonhomologous positions. Alternatively, there may be specific controls adapted for recombination between dispersed repeated sequences. Finally, there may be some δ- and Ty-specific effects. For example, the reverse transcribed free Ty DNA circles or a δ-specific endonuclease may be involved in delta recombination events. Since sequences analogous to Tys and solo δ elements exist in other eukaryotes, these types of rearrangements may have general significance (Mager and Henthorn, 1984; Rotman et al., 1984; Schmidt et al., 1984).

GENOMIC LOCATION OF Ty AND δ ELEMENTS

Distribution of Ty Elements

Ty elements were first described as a moderately repeated DNA sequence found adjacent to the *SUP4* locus in one strain but not in others (Cameron et al., 1979). In addition, Southern analysis of various laboratory yeast strains using a Ty-specific probe indicated that the location of the elements differed from strain to strain.

Since then, others have used a wide variety of techniques to analyze the distribution of Ty elements throughout the yeast genome. Oyen and Gabrielsen (1983) examined the distribution of sequences homologous to Ty within nuclear DNA fractionated according to A+T content. Ty homologous sequences were found throughout the main DNA fraction with a bias toward the A+T rich portion. The frequency of Ty homologous sequences in the satellite DNA, which contains all the rDNA and other G+C rich nuclear DNA, was low but detectable.

Klein and Petes (1984) mapped Ty elements to the yeast chromosomes using the technique of integrative transformation. A vector containing the selectable marker *LEU2* and a portion of a Ty element was used to obtain transformants in which plasmid integration had occurred at the various resident Ty elements present in their host strain. The positions of the Ty elements in these transformants were then mapped by standard techniques using the integrated *LEU2* marker as a tag. Eleven were found to be centromere-linked and these were mapped to six sites on five different chromosomes. The remaining transformants were found to represent 12 different sites. These genetic data showed Ty elements to be dispersed throughout the genome.

Recently, Fink et al., (1986) confirmed these genetic results using orthogonal field alternating gel electrophoresis (OFAGE), which separates intact DNA isolated from whole yeast chromosomes on an agarose gel. Southern blots of an OFAGE gel hybridized to Ty1-specific and Ty2-specific probes showed that both Ty elements are randomly distributed on the chromosomes of *S. cerevisiae*, and that different labortory strains exhibit differences in band intensities. Interestingly, *Saccharomyces norbensis*, which is closely related to *S. cerevisiae* and produces viable spores when crossed to *S. cerevisiae* (Fink et al., 1981), appeared to have no sequences which hybridized to the Ty1-specific probe, but many which hybridized to the Ty2-specific probe.

In similar work Eibel et al., (1981) examined 21 natural isolated of *S. cerevisiae* from all parts of the world for the presence of Ty elements and found these isolates to harbor between 4 and 20 copies. Thus, these natural isolates contain fewer Ty elements than do the laboratory strains that have been examined. Other yeast genera were also examined for the presence of Ty elements (Philippsen et al., 1983). Kluveromyces, debaromyces, candida, and schizosaccharomyces did not show any detectable homology to either Ty of δ sequences, so that Tv elements seem to be confined to saccharomyces.

Target Site Selection of Ty Elements

Although Ty elements are dispersed throughout the genome, there may be specific sequences that are preferred sites of transposition. The original Ty element isolated by Cameron et al., (1979) was found adjacent to *SUP4* ($tRNA^{tyr}$). Several other groups have also found Ty or solo δ elements adjacent to various tRNA genes (Eigel and Feldmann, 1982; Tschumper and Carbon, 1982; Cummins et al., 1982; Andreadis et al., 1982; Gafner et al., 1983; del Ray et al., 1983; Chisholm et al., 1984; Genbauffe et al., 1984; Breilmann et al., 1985). The Ty elements that we have been investigating (see the section on the *CYC1-sup4* region) are also found adjacent to tRNA genes (see Figure 1). In summary, a total of nine Ty elements have inserted within 89 to about 500 base pairs 5' to eight different tRNAs, and a total of ten solo δ are found within 52 to 485 base pairs 5' to ten different tRNAs. Gafner et al. (1983) estimated an average tRNA-delta distance of approximately 30 kilobases if the 100 delta sequences and the 300 tRNA genes are randomly distributed throughout a haploid genome of 10^7 base pairs. If solo delta elements are remnants of former transposition events, then the regions 5' to tRNA genes certainly appear to be "hot spots" for Ty transpositions.

Another hot spot for Ty transpositions seems to be δ elements themselves, as suggested by Gafner and Philippsen (1980) who reported the insertion of a complete Ty element into a solo δ in the opposite orientation. More recently, Breilmann et al. (1985) described a

Ty element that was inserted into two nested δ elements. The Ty and both δ sequences appeared to be the results of transposition events. It seems the initial Ty transposition was followed by Ty excision leaving a solo δ. A second Ty transposition into the solo δ occurred in the opposite orientation, again followed by Ty excision. A third Ty transposition occurred into the nested δ elements (in the same orientation as the internal δ), resulting in the observed structure. Warmington et al. (1986a) also reported finding nested δ elements. Both δ elements had 5 base pair duplications diagnostic of transposition events. These nested δ elements also presumably arose via sequential transposition and excision events. This apparent Ty transposition into a δ and the one reported by Gafner and Philippsen (1980) both inserted within a 15-base pair region near the transcription start site of the solo δ. These findings support the idea that δ elements are preferred sites for Ty transposition.

Numerous investigators have reported Ty insertions in the 5' noncoding region of certain structural genes, suggesting that these regions may also be preferred sites for Ty transposition. However, many of these insertions occurred during selection for alterations in regulation of the structural genes in question and therefore transposition into the structural gene would not have been detected. Two groups examined Ty transposition by selecting for loss of function of *LYS2* (at 30°C) thus avoiding this bias (Eibel and Philippsen, 1984; Simchen et al., 1984). A total of ten Ty insertions were found in the *LYS2* gene: eight mapped within a 0.6 kilobase segment in the 5' noncoding region, only two mapped within the 4.7 kilobase structural gene, and one of these was within 220 base pairs of the translation start site. These results suggest a preference for the 5' noncoding region at this locus.

However, analysis of Ty transpositions into the *CAN1* gene (arginine permease) indicates a more random distribution of Ty insertions at this locus (Wilke et al., 1987). CAN^R (canavanine-resistant) mutants were isolated at both 20°C and 30°C. Southern blot analyses indicated that 0/34 CAN^R mutants isolated at 30°C were due to Ty transpositions into the *CAN1* gene, while 13/54 CAN^R mutants isolated at 20°C were due to Ty transpositions into *CAN1*. This corroborates the observation that Ty transposition is increased 100-fold at low temperatures (Paquin and Williamson, 1984), probably due to the increased activity at low temperatures of the reverse transcriptase isolated from yeast (Garfinkel et al., 1985). The 13 independent Ty transpositions into the *CAN1* gene that we (Wilke et al., 1987) found were distributed throughout the locus: six were in the 5' noncoding region within a few hundred base pairs of the gene; and seven were found throughout the coding region, one of these near the 5' end, three throughout the middle, and three near the 3' end of the gene. Clearly Ty transpositions into the *CAN1* locus are not limited to the 5'

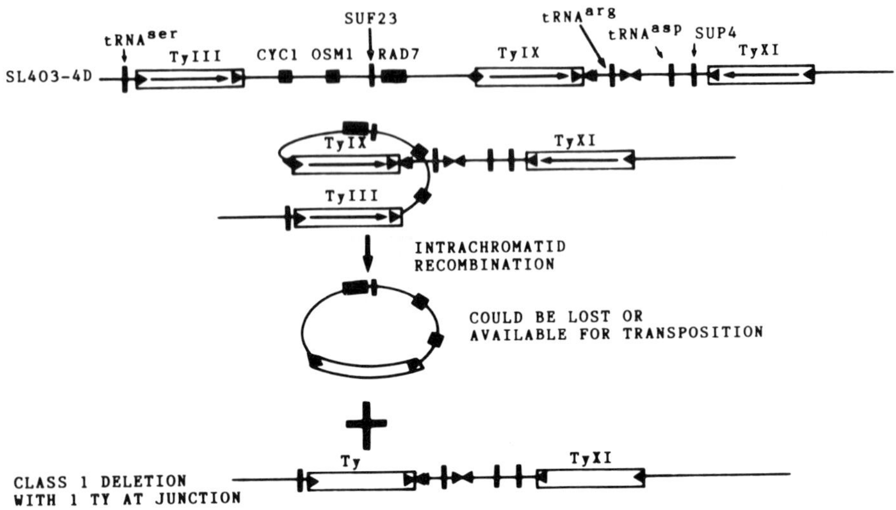

(a)

Figure 1 Hypothetical mechanisms for generation of different classes of chromosome X *cycl*-deletions from haploid *DEL1* strain SL403-4D. (a) Intrachromatid recombination resulting in a class 1 deletion. (b) Gene conversion resulting in a class 4 deletion. (c) Gene conversion resulting in a class 3 deletion. Triangles represent δ or pieces of δ. Filled rectangles indicate genes. The different sizes of symbols were used for illustrative purposes only, and do not reflect actual size.

noncoding region. Perhaps the promoter regions of these two genes (*LYS2* and *CAN1*) have different structures, and this accounts for the difference in the pattern of Ty insertions, or perhaps the temperature at which the selections were done accounts for the difference.

Aberrant Ty Elements

Errede et al. (1986) recently described an aberrant Ty element in which the 5' δ is truncated and consists entirely of an unusual inverted repeat structure. This δ is only 136 base pairs long and the inverted repeat is centered near the site that corresponds to the 5' end of the Ty RNA. The 3' δ is normal. Insertion of this Ty element into the genome apparently occurred normally because it is flanked by 5 base pair direct repeats as expected for a Ty transposition event. This aberrant Ty may have arisen through an error in reverse transcription.

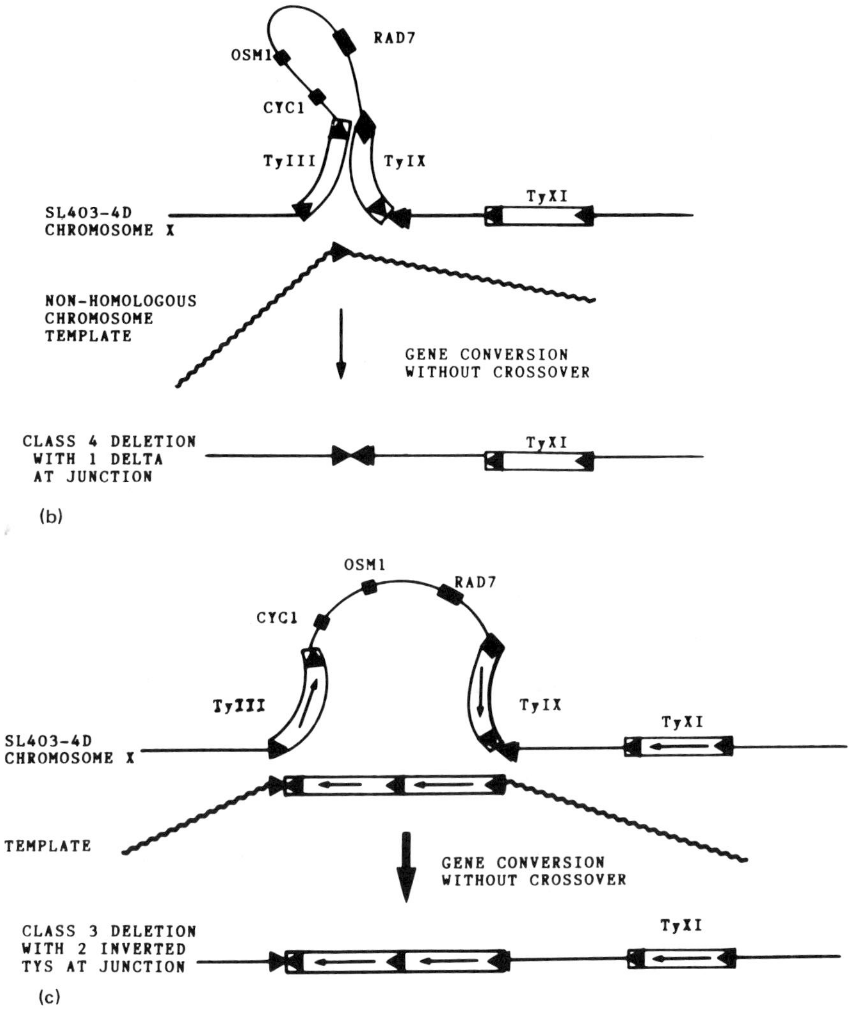

Figure 1 (Continued)

We have also found an unusual Ty element in the *CYC1-sup4* region of a *DEL1* strain (see the section on *CYC1-sup4* region). In this case, the 5' δ of TyIX in strain SL403-4D (see Figure 1) is normal and is flanked by the 5 base pair repeat ATTTC. The 3' end of the Ty abuts another δ in the opposite orientation. The inverted δ is then adjacent to another 15 base pairs of a δ. This 15 base pair sequence is followed by the 5 base pair repeat ATTTC (C. Wilke and S. W. Liebman, unpublished). Because this entire structure is flanked by the same 5 base pair repeats, it appears that it was inserted into the genome in a single transposition event.

Other Repeated Sequences Similar to δ Elements

Two other repeated sequences that resemble δ elements have been isolated from yeast: sigma (σ) and tau (τ). The σ elements have always been found within 16 to 18 base pairs 5' to tRNA genes (del Ray et al., 1982; Sandmeyer and Olson, 1982; del Ray et al., 1983; Brodeur et al., 1983). They are approximately 340 base pairs, have 8 base pair inverted repeats at their ends, and are flanked by 5 base pair direct repeats of a genomic sequence. The σ sequences that have been examined are highly conserved. There are 20-25 copies of σ per haploid genome, and the hybridization patterns are different even between closely related strains. σ+ and σ- alleles of the same tRNA gene have been isolated. These traits are characteristic of transposable elements; however, so far there is no direct evidence for sigma transposition.

The two τ elements that have been described were also found 5' to tRNA genes, although not as closely associated as σ elements (Chisholm et al., 1984; Genbauffe et al., 1984). The τ elements are 371 base pairs, have 5 base pair inverted repeats at their ends, and are also flanked by 5 base pair direct repeats of genomic sequence. The sequence of the two τ elements is highly conserved. There appear to be 20-30 copies of τ per haploid genome, and τ+ and τ- alleles of the same tRNA gene have been isolated. Thus, τ also has the characteristics of a transposable element, although transposition has not been detected. As yet, neither σ nor τ have been found in association with a larger element like Ty, and neither is large enough to encode proteins necessary for transposition. So the mechanism of transposition for σ and τ is still an open question.

σ, τ, and δ do not cross-hybridize, and computer-assisted comparisons (dot-matrix homology search) of σ, τ, and δ sequences fail to show homology between them. However, detailed sequence analysis revealed 6 to 11 regions of short homologies (5 to 10 base pairs each) that occur in the same order and are similarly spaced across the elements, implying that δ, σ, and τ may be derived from a common ancestor (Genbauffe et al., 1984), or may have evolved convergently. Interestingly, one isolate of τ was found inserted into a δ element which was inserted into a truncated σ, resulting in a nested set of elements (Chisholm et al., 1984).

RECOMBINATION INVOLVING TyS AND δ ELEMENTS

Overview

In this section, we review documented genetic rearrangements that could have arisen solely by recombination involving Tys and δ elements. Such rearrangements have now been detected in a variety of experimental systems including those involving *CYC1, SUP4, HIS4,* and *ADR3.* Unless specifically noted, the rearrangements discussed below have all arisen during vegetative (mitotic) growth.

The *CYC1-SUP4* Region

The *DEL1* Mutator

Some years ago, using classical genetics, we found that certain strains (*DEL1*) of the yeast, *Saccharomyces cerevisiae*, gave rise to multisite mutations at an unusually high rate of 10^{-5} to 10^{-6} per cell division (Liebman et al., 1979). These mutations included three closely linked genes of chromosome X: *CYC1*, which determines the structure of iso-1-cytochrome *c*; *OSM1*, which controls osmotic sensitivity; and *RAD7*, which controls sensitivity to ultraviolet irradiation. A fourth linked gene, *sup4*, was not included in these multisite mutations. Normal strains (*del1+*) did not produce such multisite mutations even after mutagenic treatment and produced all types of *cyc1* mutants at a low rate of approximately 10^{-8} to 10^{-9} per cell division. Classical genetic analysis indicated that *DEL1* segregated as a single Mendelian gene that was closely linked to *CYC1* and appeared to be both *cis-* and *trans*-dominant.

Ty-Ty Recombination

The molecular basis of the high frequency of multisite mutations was determined using Southern blot analysis (Liebman et al., 1981). A haploid *DEL1* strain (SL403-4D) was shown to differ from *del1+* strains in the *CYC1-sup4* region by the presence of three nontandem copies of the then newly discovered yeast transposable element Ty1 (TyIII, TyIX, and TyXI; see Figure 1). Each of ten independently isolated multisite mutations from this *DEL1* strain involved the loss of 13 kilobases of DNA between two directly oriented Ty elements, TyIII and TyIX. The deleted DNA included the genes *CYC1, OSM1,* and *RAD7.* The third Ty element (TyXI), which was located 6 kilobases distal and was inversely oriented relative to the other two, was not involved. None of these 10 deletions was associated with a translocation or inversion. Despite these similarities, the ten deletions examined were not identical. They fell into five distinct classes differing in the number, orientation, or type of Ty(s) remaining at the deletion junction point. These results suggested that the Ty elements participate directly in the formation of the deletions. Thus, the *DEL1* mutator was not a gene, as previously thought, but instead was a particular configuration of Ty elements in the *CYC1* region.

One possible origin for the class 1 deletions, where a single Ty remains at the deletion junction point, is an intrastrand mitotic recombination event between TyIII and TyIX [Figure 1(a)]. An alternative mechanism for the generation of the class 1 deletions is unequal sister-strand recombination between TyIII and TyIX. The origin of the class 2 and class 4 deletions, where 2 or no Tys remain at the junction, can be seen as special cases of class 1 in which the inner or outer δ elements of TyIII and TyIX, respectively, would be involved in the recombination. The class 5 deletion, where an altered Ty remains, could result from mitotic gene conversion in which TyIII, the 13 kilobases of DNA between the Tys, and TyIX, are replaced with an unlinked heterologous template Ty. All that is necessary for this conversion event between nonhomologous chromosomes is that the ends of the template be homologous to the ends of the segment being replaced. Similar converison events could also explain the origin of the class 1, 2, and 4 deletions [see Figure 1(b)].

The class 3 deletion contained two directly repeated abutting Tys at the junction point which were in the opposite orientation compared to the parent Tys. This unusual deletion could result from a class 2 event followed by an inversion of the resulting tandem Ty elements, involving an inversely oriented δ known to be adjacent to TyIX (C. Wilke and S. Liebman, unpublished results; see Figure 1). Alternatively, the class 3 deletion could arise via a single gene conversion event in which 2 tandem Ty elements in the opposite orientation from some other site in the genome would serve as a template to replace TyIII, the 13 kilobases, and TyIX. For this to occur, there would have to be an extra δ located at the periphery of the template region to provide correctly oriented homologous recombination sites for the gene conversion [Figure 1(c)].

Ty-δ Recombination

The frequency and types of rearrangements described above, resulting from Ty-Ty recombination, are remarkably similar to those resulting from Ty-δ recombination. Downs et al. (1985) obtained Ty-delta recombinants in strain A364A, containing only one Ty element in the *CYC1* region (Figure 2). This Ty element corresponds in position to TyIII in SL403-4D. The δ elements associated with this Ty do not contain *Xho*I sites. Distal to *CYC1*, A364A contains several solo δ

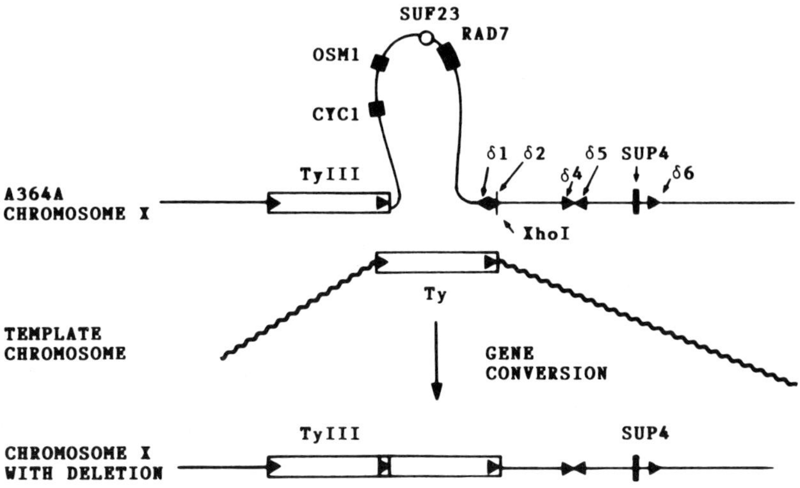

Figure 2 Hypothetical model of the origin of the deletion from A364A with 2 Tys at the junction. The chromosome X TyIII-δ2 region is replaced with a template Ty by gene conversion.

elements, one of which, δ2, corresponds to the position and orientation of TyIX in SL403-4D and contains an *Xho*I site.

Most of the deletions obtained from A364A resulted from recombination involving TyIII and the directly oriented δ2. These deletions differed by the number of Ty1 elements (0, 1, or 2) present at the junction point, and, in two cases by restriction site heterogeneities associated with these elements. As described above, the altered Ty elements at the junctions could result from gene conversion. Gene conversion could also explain the one case in which there were two tandem Ty elements at the junction point. In this case, a template Ty may have aligned with the right δ of TyIII and the solo δ2 (Figure 2). As found above for the Ty-Ty recombinants, none of the Ty-delta recombinants involved translocations.

An unexpected finding was that a single unaltered TyIII element remained at the deletion junction point much more frequently (14 cases) than no intact Ty1 remained at that site (1 case). Apparently the TyIII-associated delta element nearer to the solo delta was involved more often in recombination than the more distal TyIII-associated delta. The preferential involvement of one delta of a Ty in initiating Ty-Ty recombination events was also observed by Roeder et al. (1984). Possibly, these δ elements contain sequences that make them more recombinogenic. Another possibility, is that nearby δ elements recombine more frequently than more distal δ elements, since in both cases the recombinogenic δ is closest to the δ with which it is recombining. However, distance did not affect the frequency of δ-δ recombination observed in the *DEL1* strain SL403-4D (described above). Here, recombination between the directly oriented TyIII and TyIX involved the two external δ elements that are about 25 kilobases apart as frequently as the two internal deltas that are only about 13 kilobases apart (Liebman et al., 1981; P. Sutton, K. Downs, P. Kovarik, S. Liebman, unpublished data).

The fortuitous *Xho*I restriction site difference in A364A between the TyIII-associated δ elements and the solo δ2 allowed us to investigate the position of crossing-over within the δ elements. Recombination anywhere in the 240 base pairs to the left of the *Xho*I in δ2 would lead to retention of the *Xho*I, while recombination in the remaining 90 base pairs in the δ would result in its loss. The results were consistent with a nonspecific reciprocal recombination model, since the δ elements at the deletion junction points generally, but not always, contained an *Xho*I site, indicating multiple sites at which homologous δ-δ recombination can occur. This shows that site-specific recombination within the δ elements, if occurring, is not the only mechanism of δ-δ recombination.

Other *cyc1-* mutations isolated from A364A bore deletions that began at the end of TyIII and extended into regions not bearing Ty1 or known δ sequences. These will be described in the section on recombination involving Tys, deltas, and nonhomologous sequences.

δ-δ Recombination

The tyrosine tRNA gene, *SUP4*, is located about 6 kilobases distal to the 13 kilobases of frequently deleted DNA described above. Working with a strain that contains only δ elements and no Tys in this region, Rothstein (1979) showed that the *SUP4* gene was frequently deleted. The sizes of the deletions suggested that they resulted from intragenic recombination between the δ elements known to be in the region. Sequencing results have verified that the deletions result from δ-δ recombination that takes place randomly within the delta sequence (Rothstein et al., 1987).

Transposition

Suggestive evidence for Ty-mediated transposition comes from an ultraviolet-induced transposition of the chromosome X gene cluster *CYC1-OSM1-RAD7*, isolated in a diploid strain containing Tys flanking the cluster (Stiles et al., 1981). The segment was transposed to chromosome VII and was indeed shown to be bounded by a Ty on one end. The position of the other end of the transposition was not determined. It was proposed that the transposition did not cause the loss of the gene cluster at its normal position since segregants with higher than normal levels of iso-1-cytochrome *c* were recovered from the transposition strain, while segregants deleted for the transposed region (assumed to be viable) were not.

On possible mechanism to explain this type of transposition is the gene conversion of a Ty on another chromosome using the unique DNA and the surrounding Tys as a template. In this case, the Ty DNA would provide the required anchors of homology and the transposed DNA would also remain at its original position. It is interesting to note that Kline and Petes (1984) found a Ty element on chromosome VII at the map position corresponding to the transposed *CYC1-OSM1-RAD7* DNA.

The striking similarity between the circularly permuted gene clusters, *CYC1-OSM1-RAD7* and *CYC7-RAD23-ANP1*, located on different chromosomes, suggests that these clusters may be related by an ancient transposition event involving a circular intermediate (McKnight et al., 1981). The circular intermediate could have been formed by intrachromosomal recombination between directly oriented Ty elements surrounding the original gene cluster. Reintegration of the circle by recombination at another position would give rise to the transposition. Such an event would cause the loss of the unique DNA at the normal chromosomal position concurrent with its transposition to a new position. If the crossover leading to the integration did not occur at the Ty, the transposed gene cluster would be circularly permuted. Finally, according to this hypothesis, an intact copy of chromosome X would have replaced the deleted copy in matings soon after the original transposition event.

HIS4

The Ty917(Ty2) and Ty912(Ty1) insertions into the 5' noncoding region of *HIS4* inactivate *HIS4* and have been involved in a number of different types of genetic rearrangements (Roeder and Fink, 1980; Roeder et al., 1980; Farabaugh and Fink, 1980; Chaleff and Fink, 1980; Roeder and Fink, 1982).

Reversion back to HIS+ occurs most often by excision of the Ty element, leaving behind a solo δ. This could occur via recombination between the directly repeated δ elements, leading to the excision of a Ty circle. Alternatively, the Ty could be gene converted into a δ. No true revertants arising from recombination between the 5 base pair target duplications have ever been isolated.

Other HIS+ revertants did not involve loss of the Ty. Instead, Ty917 or Ty912 were replaced by Tys that permit HIS+ expression. In some cases, this occurred via gene conversion of the *HIS4* proximal Ty into the altered Ty without other chromosomal aberrations (Roeder and Fink, 1982). In other cases, crossovers occurred between the *HIS4* Ty and the other Ty. For example, an inversion arose from recombination between Ty912 and a Ty element in the opposite orientation on the same chromosome; and translocations arose from recombination of the *HIS4* Ty with Tys on other chromosomes (Chaleff and Fink, 1980; Roeder and Fink, 1980).

Movement of the Ty912 element was detected by genetically marking the element with the *URA3+* gene (Roeder et al., 1984). Ty912(*URA3*) was found to move by homologous recombination with other Ty elements and not by transposition. It recombined most frequently with nearby Tys, less frequently with more distal Tys, and least frequently with Tys on other chromosomes.

Unequal crossing over between Ty917 and a directly oriented Ty about 40 kilobases apart has been detected in diploids in about 1% of meioses (Roeder, 1983). This event gives rise to one sister spore with a 40-kilobase deletion and another with a 40-kilobase duplication. Deletions of this type also arise in mitotically dividing haploids, but with a frequency of less than 10^{-7}.

ADH2

Ty insertions in front of *ADH2* lead to constitutive synthesis of alcohol dehydrogenase II (ADHII) (Williamson et al., 1981). These constitutive mutants never revert back to wild type, regulated, ADHII activity, but they can mutate to complete loss of ADHII. Most frequently this occurs by excision of the Ty element leaving behind a solo δ, as described above for *HIS4*. In other cases, recombination with other Ty elements leads to both the ADHII- phenotype and translocations (Ciriacy and Williamson, 1981). One of these translocations has now been sequenced (Breilmann et al., 1985). In this case, reciprocal recombination between the Ty proximal to *ADH2* on chromosome XIII, and a Ty on chromosome VII, was accompanied by gene conversion in the region of the crossover. The gene-converted segment included the δ sequence closest to *ADH2*. The anchor of homology to the right of the converted segment was limited to a short stretch of 66 base pairs of δ sequence, and mismatches present in this region were not repaired.

In addition to these Ty-promoted rearrangements, Ciriacy and Breilmann (1982) showed that a solo δ element 5' to *ADH2* would also promote DNA rearrangements. This solo δ causes an ADHII- phenotype that can be reverted to ADHII+ if the δ is replaced by a Ty. Most frequently, this occurred by the reverse of a Ty excision event, probably by gene conversion. In other cases an inversion, possibly resulting from Ty-δ recombination, caused a Ty to replace the δ in front of the *ADH2* gene. Finally, in other revertants, *ADH2* and 3' flanking sequence appeared to be duplicated and transposed to a remote Ty.

Other Cases of Ty-Ty and Ty-δ Recombination

Recombination between directly oriented Tys (on opposite chromosome arms) resolved a dicentric chromosome III into a 60-kilobase ring chromosome, with all the sequences between the two Tys, and a linear chromosome with the remaining sequences (Surosky and Tye, 1985). Recombination between directly oriented Tys elsewhere on a monocentric chromosome III lead to the deletion of the unique intervening DNA (Warmington et al., 1986b). In addition, Ty-containing plasmids can integrate into the genome via homologous recombination with a chromosomal Ty element. The plasmid sequences may then be lost by recombination between the duplicated Tys (Scherer and Davis, 1980a,b).

RECOMBINATION INVOLVING TyS, δs AND NONHOMOLOGOUS SEQUENCES

Some rearrangements associated with Tys and δ elements can not be explained simply by recombination between homologous dispersed copies of the repeated Ty and δ elements. Several revertants of the *his4-912* mutant resulted from recombination between an internal region of Ty912 and a nonhomologous sequence within the *HIS4* coding region. These revertants regained one or two of the three enzymatic activities associated with the complete *HIS4* gene. This type of deletion occurred much more frequently in the *his4-912* mutant than in *HIS4+* strains, and one endpoint always lay within Ty912, suggesting that the Ty element stimulated the nonhomologous recombination event (Chaleff and Fink, 1980; Roeder and Fink, 1980,1983).

Two of the chromosome X deletions we isolated from strain A364A extend from the 3' end of the single Ty1 element in the *CYC1* region, into sequences that do not contain any δ homology (Downs et al., 1985; Sutton et al., 1986). These deletions eliminate approximately the same 7 kilobases of chromosome X DNA, including *CYC1* and *OSM1*. Since the Ty1s at the deletion junctions picked up new restriction sites, gene conversion with other Tys or δ elements was probably involved. One deletion ends 170 base pairs 5' to a $tRNA^{gly}$ gene, *SUF23*. This endpoint is in the middle of a $\sim$300 base pair element that shares spatially conserved sequence motifs with the repetitive σ element. This has been termed SLSI (σ-like sequence) by L. Melnick and F. Sherman (personal communication). Although it remains to be determined whether SLSI is repeated in A364A, this deletion may have arisen by gene conversion with an unlinked δ-SLSI or δ-SLSI-$tRNA^{gly}$ serving as a template. Indeed, genetic evidence suggests that there are 11 dispersed repeats of the *SUF23* $tRNA^{gly}$ gene (Gaber and Culbertson, 1982), and tRNAs are often associated with δ and other repeated sequences (see the section on the genomic location of Ty and δ elements). Alternatively, the deletion may have arisen by nonhomologous recombination between the 3' end of the Ty1 and SLSI, or by a two-step event involving transposition of a Ty1 into SLSI followed by homologous recombination between TyIII and the SLSI Ty.

The other deletion ends only 1 base pair 5' to the *SUF23* $tRNA^{gly}$, and it is associated with a reciprocal translocation at the deletion breakpoint. OFAGE analysis has confirmed that chromosomes X and XVI are involved in the translocation, since the sizes of these chromosomes are altered in the translocation strain (Sutton et al., 1986). It is interesting to note that another copy of $tRNA^{gly}$, *SUF21*, is known to map on chromosome XVI (Gaber and Culbertson, 1982). These data are consistent with a model involving gene conversion of the chromosome X gene cluster TyIII-*CYC1-OSM1-SUF23* with a template consisting of δ-*SUF23* from chromosome XVI, associated with a reciprocal exchange within the $tRNA^{gly}$ genes (Figure 3).

GENES AFFECTING Ty AND δ RECOMBINATION

To date, few genes have been identified that affect the frequency or types of Ty and δ recombination. *EDR1* (enhanced δ recombination) is one such gene (J. W. Wallis, G. Chrebet, A. Beniaminovitz, and R. Rothstein, personal communication). A mutation at this locus (*edr1-1*) increases the frequency of δ-mediated *sup4* deletions tenfold (see section on the *CYC1-sup4* region). The wild type *EDR1* gene has been cloned, and disrupted allele has been reintroduced into the genome. This *edr1*-disruption increases the frequency of delta-mediated *sup4* deletions by 100-fold. This mutation seems to have no effect on non-δ recombination, and thus defines a locus whose function in wild type cells appears to specifically inhibit δ-mediated recombination.

We have also found other genetic factors that affect the frequency of Ty- and δ-mediated recombination. Some *DEL1* strains were found to have an extremely high rate of recombination between the inversely oriented Tys IX and XI (see Figure 1). This high rate of recombination results in the inversion/reinversion of the 6 kilobase *sup4* region between the Tys. In these unstable strains, inversion/reinversion occurs at a rate of about 1 in 50 to 100 unselected mitotically growing cells! Additional rearrangements, all of which can be explained by recombination between known Tys and δ elements in the *CYC1-sup4* region also occur in these strains at a high rate. The rates of rearrangements in other *DEL1* strains having the same Ty and δ configuration in the region are similar to previously reported values ($1/10^4$ to $1/10^7$). The factors responsible for this high rate of Ty and δ

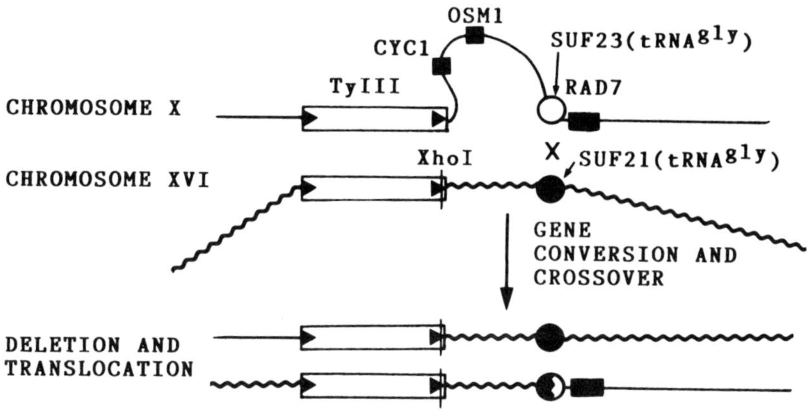

Figure 3 Hypothetical single-step mechanism to explain the origin of the chromosome X deletion and reciprocal translocation between chromosomes X and XVI that arose in strain A364A. A chromosome XVI δ and $tRNA^{gly}$ (*SUF21*) serve as template for a gene conversion of chromosome X's TyIII-$tRNA^{gly}$ (*SUF23*) region. Associated with this conversion event is a reciprocal exchange between the $tRNA^{gly}$ genes.

recombination are as yet uncharacterized, but they are unlinked to the *CYC1-sup4* region, and do not segregate as a single Mendelian gene (Picologlou et al., in press). It is not known whether the high rate of recombination in these strains is specific for Ty or δ elements, or whether a high rate of recombination would also occur between other repeated sequences.

Winston et al. (1984) reported that δ-δ recombination, specifically Ty excision, was reduced 25-fold in strains containing the pleiotropic recombination and repair mutation *rad52*. In addition, the frequency of δ-mediated deletions of the *sup4* locus is decreased more than three orders of magnitude in *rad52-* strains (Rothstein et al., 1987). These results imply that the general recombination system does function in δ-δ recombination.

We examined the effect of genes from the three DNA repair groups of yeast on the frequency of Ty-mediated *CYC1-OSM1-RAD7* deletions in *DEL1* strains (Liebman and Downs, 1980; Downs et al., 1983). None of these genes (including *rad52*) significantly affected the deletion frequency. We did not analyze the molecular basis of the deletions isolated from *rad-DEL1* strains. Other mechanisms beside δ-δ recombination may be responsible for these deletions.

Other genes originally thought to have an effect on δ-δ recombination were the *SPT* genes (suppressor of Ty), specifically *spt2* and *spt3* (Roeder et al., 1980). However, Winston et al. (1984) later demonstrated that these two mutations had no significant effect on δ-δ recombination (Ty excision).

CONCERTED EVENTS

In their analysis of revertants of a Ty insertion mutation in the 5' noncoding region of the *HIS4* gene, Chaleff and Fink (1980) found that many of the revertants had additional mutational alterations unrelated to the *HIS4* reversion event. These included appearance of a *leu2-* phenotype, reversion of both *his4* copies in a diploid strain, and aneuploidy. Since the frequency of these multiple events seemed too high for them to have been independent, it was proposed that a cell undergoing one Ty-mediated recombination event may have a higher probability of undergoing other Ty-mediated events (Roeder and Fink, 1983). They called these "concerted events."

We have also found evidence for Ty- or δ-mediated multiple events in certain strains. Those *DEL1* strains that exhibit a high rate (1 in 50 to 100 cells) of inversion/reinversion of the 6 kilobase *sup4* region between TyIX and TyXI also have a high frequency of other unselected rearrangements in the *CYC1-sup4* region involving recombination between known Ty and δ elements (see the section on genes affecting Ty and δ recombinations; Picologlou et al., in press). In these unstable strains, we have often observed deletions, duplications, and inversions in a single subclone. However, since the rate of single rearrangements is so high in these strains, multiple rearrangements probably arose independently by chance, rather than by a concerted mechanism.

Other Ty- and δ-mediated events which may appear to be concerted could, instead, be due to complex single events. Several of the *sup4* deletions resulting from δ-δ recombination (see the section on the *CYC1-sup4* region) are accompanied by adjacent inversions between inversely oriented δ elements in the region (Rothstein et al., 1987). In addition, we have found adjacent inversions concomitant with deletions isolated from some, but not other, *DEL1* strains (Picologlou et al., in press). Certain *DEL1* strains have the 6 kilobases of unique DNA between TyIX and TyXI, which includes the *sup4* gene in an inverted configuration. *CYC1-OSM1-RAD7* deletions between Tys III and IX from these inverted *DEL1* strains fall into classes similar to deletions isolated from normal *DEL1* strains, except that the deletions from some inverted *DEL1* strains are frequently (30%) associated with adjacent reinversions of the 6 kilobase *sup4* region between Tys IX and XI. In these strains, the deletion frequency is about 1 in 10^6, and, in the absence of deletions, inversions were never observed.

Both the δ-mediated deletions with adjacent inversions observed by Rothstein and the Ty-mediated deletions with adjacent inversions we have seen could be viewed as concerted events, because the double event occurs much more frequently than the product of the deletion and inversion frequencies. However, models involving gene conversion (Rothstein et al., 1987) suggest another alternative: Concurrent deletions and inversions could arise from a single event. Using the Ty-mediated deletions isolated from inverted *DEL1* strains as an example, Figure 4 depicts these models. The first model [Figure 4(a)] involves intrachromatid gene conversion of TyIII, TyXI, and the intervening 13 kilobases containing *CYC1-OSM1-RAD7* by TyXI. If the conversion event were associated with a crossover, a reinversion of the DNA between TyIX and TyXI would result. Since gene conversion is often associated with crossovers, this could be considered a single event. In the second model [Figure 4(b)], sister chromatids in the region of interest align in opposite orientation. This allows inversely oriented δ segments surrounding the *sup4* segment to pair with a TyIII δ and an inversely oriented TyXI δ. The *sup4* segment (with or without the adjacent TyXI) could then serve as a template to replace the paired region on the sister chromatid between TyIII and TyXI, resulting in a a gene conversion between sister chromatids. If either of these models were correct, the occurrence of deletions with associated inversions may may, in fact, be due to a single event rather than two concerted events.

Likewise, another unusual rearrangement isolated from strain A364A (see the section on the *CYC1-sup4* region) (Downs et al., 1985) could be viewed as either a concerted or a single event. A364A has one Ty element near the *CYC1* gene (TyIII) and several solo δ elements in the *CYC1-sup4* interval (see Figure 2), and gives rise to deletions at a rate of about 1 in 10^7 cells. The rearrangement in question, a deletion with an overlapping inversion, may have arisen by two events, an inversion between the inversely oriented δ elements 2 and 5, followed by a deletion beginning at TyIII and extending into the inverted region. Alternatively, the rearrangement could be explained by a single event analogous to that described in Figure 4(b).

Another event that at first appeared to be concerted is really due to sequential events. Classical genetic analysis of *DEL1* strains indicated that heterozygous *DEL1* diploids could give rise to *CYC1-OSM1-RAD7* deletions in the *trans*-configuration (relative to *DEL1*) at frequencies similar to those observed in haploid *DEL1* strains (Liebman et al., 1979; Picologlou and Liebman, 1985). Molecular analysis of these "*trans*" deletions revealed that they probably arose in two steps, a mitotic crossover or gene conversion making the *DEL1* chromosomal configuration of Tys homozygous, followed by a deletion in one of the chromosomes. A subtle selection for diploids containing the deletion in *trans* was responsible for the observed high frequency of the double event (Picologlou and Liebman, 1985).

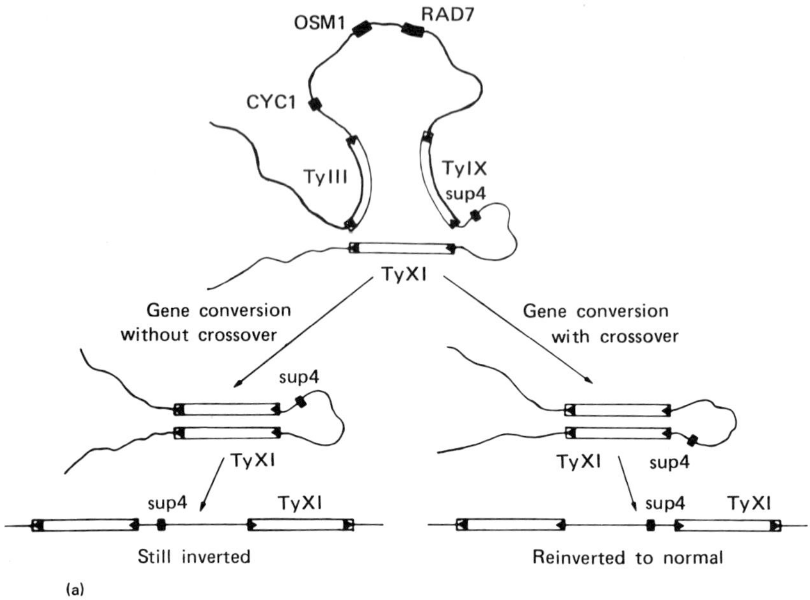

Figure 4 Single-step models for the formation of a deletion with an associated adjacent inversion. The models are described in detail in the text and involve (a) intrachromatid gene conversion and an associated crossover, and (b) gene conversion between sister chromatids. The drawings are not to scale and the direction of the triangles representing the deltas is the opposite of the convention used in Figures 1 through 3.

These examples suggest caution in viewing apparent multiple events as concerted. Other factors, such as an overall increase in recombination rate, complex single events, or subtle selection pressures, may be responsible for some of the observed multiple events.

DISCUSSION

In this chapter we have described a number of Ty- and δ-mediated recombination events in several regions of the genome. These data suggest that δ elements may be hotspots for recombination. Indeed, δ sequences are more recombinogenic than the other repeated sequences associated with Tys, the ε regions. However, it is possible that δ elements only appear to be recombinogenic because there are so many more δ elements than ε regions present in the genome to serve as templates in recombination events.

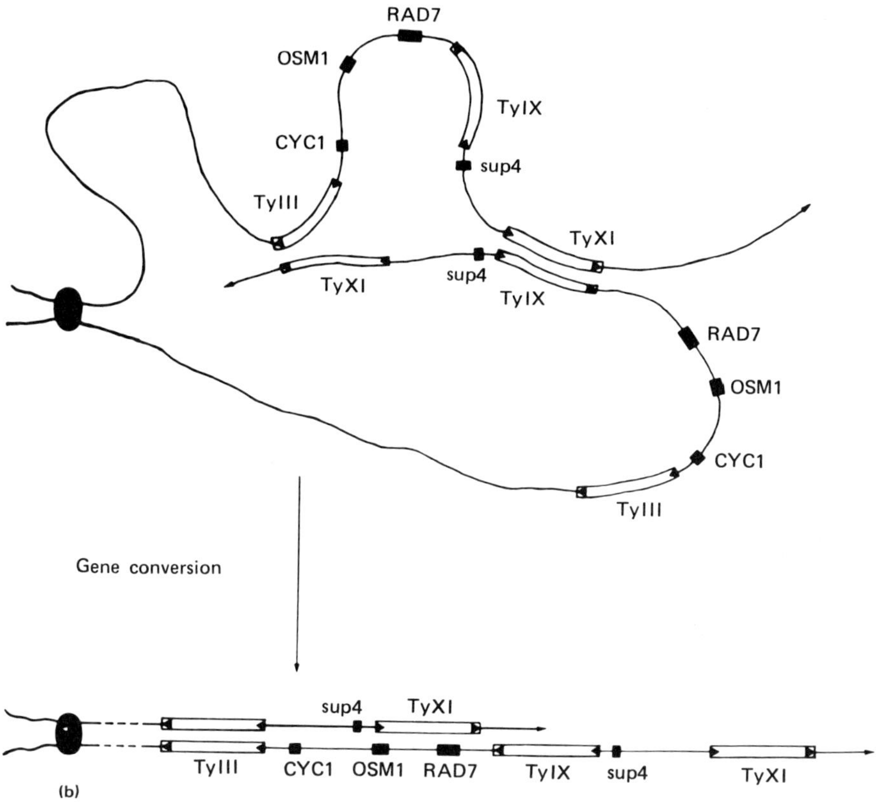

Figure 4 (Continued)

Trans-acting factors have been found that control the rate of δ-δ recombination. In some strains rearrangements resulting from δ-δ recombination occur in about 1 in 50 to 100 mitotically growing cells, while in others the rate is at least 2 orders of magnitude lower. These factors may act specifically on δ sequences or they may affect recombination in general. Indeed, δ-δ recombination is not limited to a site-specific mechanism, and may instead occur via the normal homologous recombination pathways. *RAD52*, which is required for general recombination, is likewise needed for δ-δ recombination. The finding of a gene, *edr1*, specific for δ-δ recombination, is the strongest evidence suggesting a difference between normal homologous recombination and δ-δ recombination. Whatever the mechanism, recombination between Ty1 and δ elements does frequently generate genomic rearrangements, and these are likely to play an important role in evolution.

ACKNOWLEDGMENTS

We wish to thank A. Beniaminovitz, G. Chrebet, C. Helms, N. Rosenberg, J. W. Wallis, R. Rothstein, L. Melnick, F. Sherman, C. Newlon, J. R. Warmington, S. G. Oliver, G. Brennan, M. E. Dicig, K. Downs, S. Heidler, P. Kovarik, P. Sutton, and C. Wilke for allowing us to quote unpublished results, and for helpful comments and discussions. This work was supported by NSF grant PCM 82-06806 and by Research Career Development Award CA00780 to S. W. L. from the National Cancer Institute.

REFERENCES

Andreadis, A., Hsu, Y.-P., Kohlhow, G. B., and Schimmel, P. (1982). Nucleotide sequence of yeast *LEU2* shows 5'-noncoding region has sequences cognate to leucine. *Cell 31*:319-325.

Bach, M.-L. (1984). Ty1-promoted expression of aspartate transcarbamylase in the yeast *Saccharomyces cerevisiae*. *Mol. Gen. Genet. 194*:395-401.

Ballario, P., Filetici, P., Junakovic, N., and Pedone, F. (1983). Ty1 extrachromosomal circular copies in *Saccharomyces cerevisiae*. *FEBS Lett. 155*:225-229.

Baltimore, D. (1985). Retroviruses and retrotransposons: The role of reverse transcription in shaping the eukaryotic genome. *Cell 40*:481-482.

Boeke, J. D., Garfinkel, D. J., Styles, C. A., and Fink, G. R. (1985). Ty elements transpose through an RNA intermediate. *Cell 40*:491-500.

Bowen, B. A., Fulton, A. M., Tuite, M. F., Kingsman, S. M., and Kingsman, A. J. (1984). Expression of Ty-lacZ fusions in *Saccharomyces cerevisiae*. *Nucleic Acids Res. 12*:1627-1640.

Breilmann, D., Gafner, J., and Ciriacy, M. (1985). Gene conversion and reciprocal exchange in a Ty-mediated translocation in yeast. *Curr. Genet. 9*:553-560.

Brodeur, G. M., Sandmeyer, S. B., and Olson, M. V. (1983). Consistent association between *sigma* elements and tRNA genes in yeast. *Proc. Nat. Acad. Sci. USA 80*:3292-3296.

Cameron, J. R., Loh, E. Y., and Davis, R. W. (1979). Evidence for transposition of dispersed repetitive DNA families in yeast. *Cell 16*:739-751.

Chaleff, D. T., and Fink, G. R. (1980). Genetic events associated with an insertion mutation in yeast. *Cell 21*:227-237.

Chisholm, G. E., Genbauffe, F. S., and Cooper, T. G. (1984). *tau*, a repeated DNA sequence in yeast. *Proc. Nat. Acad. Sci. USA 81*:2965-2969.

Ciriacy, M., and Breilmann, D. (1982). Delta sequences mediate DNA rearrangements in *Saccharomyces cerevisiae. Curr. Genet. 6*:55-61.

Ciriacy, M., and Williamson, V. M. (1981). Analysis of mutations affecting Ty-mediated gene expression in *Saccharomyces cerevisiae. Mol. Gen. Genet. 182*:159-163.

Clare, J., and Farabaugh, P. (1985). Nucleotide sequence of a yeast Ty element: Evidence for an unusual mechanism of gene expression. *Proc. Nat. Acad. Sci. USA 82*:2829-2833.

Cummins, C. M., Donahue, T. F., and Culbertson, M. R. (1982). Nucleotide sequence of the *SUF2* frameshift suppressor gene of *Saccharomyces cerevisiae. Proc. Nat. Acad. Sci. USA 79*:3565-3569.

del Rey, F., Donahue, T. F., and Fink, G. R. (1982). Sigma, a repetitive element found adjacent to tRNA genes of yeast. *Proc. Nat. Acad. Sci. USA 79*:4138-4142.

del Rey, F., Donahue, T. F., and Fink, G. R. (1983). The histidine tRNA genes of yeast. *J. Biol. Chem. 258*:8175-8182.

Downs, K. M., Szwast, A. E., and Liebman, S. W. (1983). Analysis of the effect of radiation repair mutations on the *DEL1* mutator region of *Saccharomyces cerevisiae. Curr. Genet. 7*:57-61.

Downs, K. M., Brennan, G., and Liebman, S. W. (1985). Deletions extending from a single Ty element in *Saccharomyces cerevisiae. Mol. Cell. Biol. 5*:3451-3457.

Dubois, E., Jacobs, E., and Jauniaux, J.-C. (1982). Expression of the ROAM mutations in *Saccharomyces cerevisiae*: Involvement of *trans*acting regulatory elements and relation with the Ty1 transcription. *EMBO J. 1*:1133-1139.

Duyk, G., Leis, J., Longiaru, M., and Skalka, A. M. (1983). Selective cleavage in the avian retroviral long terminal repeat sequence by the endonuclease associated with the $\alpha\beta$ form of avian reverse transcriptase. *Proc. Nat. Acad. Sci. USA 80*:6745-6749.

Eibel, H., and Philippsen, P. (1984). Preferential integration of yeast transposable element Ty into a promoter region. *Nature 307*:386-388.

Eibel, H., Gafner, J., Stotz, A., and Phillippsen, P. (1981). Characterization of the yeast mobile element Ty1. *Cold Spring Harbor Symp. Quant. Biol. 45*:609-617.

Eigel, A., and Feldmann, H. (1982). Ty1 and delta elements occur adjacent to several tRNA genes in yeast. *EMBO J. 1*:1245-1250.

Errede, B., Cardillo, T. S., and Sherman, F. (1980). Mating signals control expression of mutations resulting from insertion of a transposable repetitive element adjacent to diverse yeast genes. *Cell 22*:427-436.

Errede, B., Company, M., and Swanstrom, R. (1986). An anomalous Ty1 structure attributed to an error in reverse transcription. *Mol. Cell. Biol. 6*:1334-1338.

Farabaugh, P. J., and Fink, G. R. (1980). Insertion of the eukaryotic transposable element Ty1 creates a 5-base pair duplication. *Nature 286*:352-356.

Fink, G. R., Farabaugh, P. J., Roeder, G. S., and Chaleff, D. (1981). Transposable elements (Ty) in yeast. *Cold Spring Harbor Symp. Quant. Biol. 45*:575-580.

Fink, G. R., Boeke, J. D., and Garfinkel, D. J. (1986). The mechanism and consequences of retrotransposition. *Trends Genet. 2*: 118-123.

Gaber, R. F., and Culbertson, M. R. (1982). Frameshift suppression in *Saccharomyces cerevisiae*. IV. New suppressors among spontaneous co-revertants of the group II *HIS4-206* and *LEU2-3* frameshift mutations. *Genetics 101*:345-367.

Gafner, J., and Philippsen, P. (1980). The yeast transposon Ty1 generates duplications of target DNA on insertion. *Nature 286*: 414-418.

Gafner, J., De Robertis, E. M., and Philippsen, P. (1983). Delta sequences in the 5' non-coding region of yeast tRNA genes. *EMBO J. 2*:583-591.

Garfinkel, D. J., Boeke, J. D., and Fink, G. R. (1985). Ty element transposition: Reverse transcriptase and virus-like particles. *Cell 42*:507-517.

Genbauffe, F. S., Chisholm, G. E., and Cooper, T. G. (1984). Tau, sigma, and delta. A family of repeated elements in yeast. *J. Biol. Chem. 259*:10518-10525.

Hauber, J., Nelböck-Hochstetter, P., and Feldmann, H. (1985). Nucleotide sequence and characteristics of a Ty element from yeast. *Nucleic Acids Res. 13*:2745-2757.

Kingsman, A. J., Gimlich, R. L., Clarke, L., Chinault, A. C., and Carbon, J. (1981). Sequence variation in dispersed repetitive sequences in *Saccharomyces cerevisiae*. *J. Mol. Biol.* *145*:619-632.

Klein, H. L., and Petes, T. D. (1984). Genetic mapping of Ty elements in *Saccharomyces cerevisiae*. *Mol. Cell. Biol.* *4*:329-339.

Liebman, S. W., and Downs, K. M. (1980). The *RAD52* gene is not required for the function of the *DEL1* mutator gene in *Saccharomyces cerevisiae*. *Mol. Gen. Genet.* *179*:703-705.

Liebman, S. W., Singh, A., and Sherman, F. (1979). A mutator affecting the region of the iso-1-cytochrome *c* gene in yeast. *Genetics* *92*:783-802.

Liebman, S., Shalit, P., and Picologlou, S. (1981). Ty elements are involved in the formation of deletions in *DEL1* strains of *Saccharomyces cerevisiae* *Cell* *26*:401-409.

Mager, D. L., and Henthorn, P. S. (1984). Identification of a retrovirus-like repetitive element in human DNA. *Proc. Nat. Acad. Sci. USA* *81*:7510-7514.

McKnight, G. L., Cardillo, T. S., and Sherman, F. (1981). An extensive deletion causing overproduction of yeast iso-2-cytochrome *c*. *Cell* *25*:409-419.

Mellor, J. Fulton, S. M., Dobson, M. J., Wilson, W., Kingsman, S. M., and Kingsman, A. J. (1985). A retrovirus-like strategy for expression of a fusion protein encoded by yeast transposon Ty1. *Nature* *313*:243-246.

Oyen, T. B., and Gabrielsen, O. S. (1983). Non-random distribution of the Ty1 elements within nuclear DNA of *Saccharomyces cerevisiae*. *FEBS Lett.* *161*:201-206.

Panganiban, A. T., and Temin, H. M. (1984a). Circles with two tandem LTRs are precursors to integrated retrovirus DNA. *Cell* *36*:673-679.

Panganiban, A. T., and Temin, H. M. (1984b). The retrovirus *pol* gene encodes a product required for DNA integration: Identification of a retrovirus *int* locus. *Proc. Nat. Acad. Sci. USA* *81*:7885-7889.

Paquin, C. E., and Williamson, V. M. (1984). Temperature effects on the rate of Ty transposition. *Science* *226*:53-55.

Philippsen, P., Eibel, H., Gafner, J., and Stotz, A. (1983). Ty elements and the stability of the *Saccharomyces cerevisiae* genome. Gene Expression in Yeast. Proceedings of the Alko Yeast Symposium, Helsinki (M. Korhola and E. Väisänen, Eds.), *Found. Biotech. Indust. Fermentation Res.* *1*(1983):189-200.

Picologlou, S., and Liebman, S. W. (1985). The *DEL1* mutator gene in *Saccharomyces cerevisiae* does not act in *trans*. *Curr. Genet.* *9*:259-262.

Picologlou, S., Dicig, M. E., Kovarik, P., and Liebman, S. W. The same configuration of Ty elements promotes different types and frequencies of rearrangements in different yeast strains. *Mol. Gen. Genet.* (in press).

Roeder, G. S. (1983). Unequal crossing-over between yeast transposable elements. *Mol. Gen. Genet.* *190*:117-121.

Roeder, G. S., and Fink, G. R. (1980). DNA rearrangements associated with a transposable element in yeast. *Cell* *21*:239-249.

Roeder, G. S., and Fink, G. R. (1982). Movement of yeast transposable elements by gene conversion. *Proc. Nat. Acad. Sci. USA* *79*:5621-5625.

Roeder, G. S., and Fink, G. R. (1983). Transposable elements in yeast. In *Mobile Genetic Elements*. Academic, New York, pp. 299-328.

Roeder, G. S., Farabaugh, P. J., Chaleff, D. T., and Fink, G. R. (1980). The origins of gene instability in yeast. *Science* *209*:1375-1380.

Roeder, G. S., Smith, M., and Lambie, E. J. (1984). Intrachromosomal movement of genetically marked *Saccharomyces cerevisiae* transposons by gene conversion. *Mol. Cell. Biol.* *4*:703-711.

Rothstein, R. (1979). Deletions of a tyrosine tRNA gene in *S. cerevisiae*. *Cell* *17*:185-190.

Rothstein, R., Helms, C., and Rosenberg, N. (1987). Concerted deletions and inversions are caused by mitotic recombination between delta sequences in *Saccharomyces cerevisiae*. *Mol. Cell. Biol.* *7*:1198-1207.

Rotman, G., Itin, A., and Keshet, E. (1984). "Solo" large terminal repeats (LTR) of an endogenous retrovirus-like gene family (VL30) in the mouse genome. *Nucleic Acids Res.* *12*:2273-2282.

Sandmeyer, S. B., and Olson, M. V. (1982). Insertion of a repetitive element at the same position in the 5'-flanking regions of two dissimilar yeast tRNA genes. *Proc. Nat. Acad. Sci. USA* *79*:7674-7678.

Scherer, S., and Davis, R. W. (1980a). Recombination of dispersed repeated DNA sequences in yeast. *Science* *209*:1380-1384.

Scherer, S., and Davis, R. W. (1980b). Studies on the transposable element Ty1 of yeast. II. Recombination and expression of Ty1 and adjacent sequences. *Cold Spring Harbor Symp. Quant. Biol. 45*:584-591.

Scherer, S., Mann, C., and Davis, R. W. (1982). Reversion of a promoter deletion in yeast. *Nature 298*:815-819.

Schmidt, M., Gloggler, K., Wirth, T., and Horak, I. (1984). Evidence that a major class of mouse endogenous long terminal repeats (LTRs) resulted from recombination between exogenous retroviral LTRs and similar LTR-like elements (LTR-IS). *Proc. Nat. Acad. Sci. USA 81*:6696-6700.

Shapiro, J. (Ed.) (1983). *Mobile Genetic Elements.* Academic, New York.

Simchen, G., Winston, F., Styles, C. A., and Fink, G. R. (1984). Ty-mediated gene expression of the *LYS2* and *HIS4* genes of *Saccharomyces cerevisiae* is controlled by the same *SPT* genes. *Proc. Nat. Acad. Sci. USA 81*:2431-2434.

Stiles, J. I., Friedman, L. R., Helms, C., Consaul, S., and Sherman, F. (1981). Transposition of the gene cluster *CYC1-OSM1-RAD7* in yeast. *J. Mol. Biol. 148*:331-346.

Surosky, R. T., and Tye, B.-K. (1985). Resolution of dicentric chromosomes by Ty-mediated recombination in yeast. *Genetics 110*:397-419.

Sutton, P. R., Brennan, G., and Liebman, S. W. (1986). Sequence characterization of deletions extending from a single Ty element in *S. cerevisiae*. *Yeast 2*:S376.

Tschumper, G., and Carbon, J. (1982). Delta sequences and double symmetry in a yeast chromosomal replicator region. *J. Mol. Biol. 156*:293-307.

Varmus, H. E. (1983). Retroviruses. In *Mobile Genetic Elements* (J. A. Shapiro, Ed.), Academic, New York, pp. 411-503.

Warmington, J. R., Waring, R. B., Newlon, C. S., Indge, K. J., and Oliver, S. G. (1985). Nucleotide sequence characterization of Ty1-17, a class II transposon from yeast. *Nucleic Acids Res. 13*(18):6679-6693.

Warmington, J. R., Anwar, R., Newlon, C. S., Waring, R. B., Davies, R. W., Indge, K. J., and Oliver, S. G. (1986a). A "hot-spot" for Ty transposition on the left arm of yeast chromosome III. *Nucleic Acids Res. 14*(8):3475-3485.

Warmington, J. R., Waring, R. B., Newlon, C. S., and Oliver, S. G. (1986b). Distribution and sequence organization of transposable elements on yeast chromosome III. *Genetics 113*:S13.

Wilke, C. M., Heidler, S. H., and Liebman, S. W. (1987). Analysis of Ty insertions at the *CAN1* locus of *Saccharomyces cerevisiae*. *Genetics 116*:S25.

Williamson, V. M. (1983). Transposable elements in yeast. *Int. Rev. Cytol. 83*:1-25.

Williamson, V. M., Young, E. T., and Ciriacy, M. (1981). Transposable elements associated with constitutive expression of yeast alcohol dehydrogenase II. *Cell 23*:605-614.

Williamson, V. M., Cox, D., Young, E. T., Russell, D. W., and Smith, M. (1983). Characterization of transposable element-associated mutations that alter yeast alcohol dehydrogenase II expression. *Mol. Cell. Biol. 3*:20-31.

Winston, F., Chaleff, D. T., Valent, B., and Fink, G. R. (1984). Mutations affecting Ty-mediated expression of the *HIS4* gene of *Saccharomyces cerevisiae*. *Genetics 107*:179-197.

5

Double-Stranded RNA Replication in *Saccharomyces cerevisiae*: Genetic Control and In Vitro Replication

Tsutomu Fujimura, Rosa Esteban, and Reed B. Wickner
National Institute of Diabetes, Digestive and Kidney Diseases, National Institutes of Health, Bethesda, Maryland

INTRODUCTION

There are five families of dsRNA found in *Saccharomyces cerevisiae*, called L-A, L-BC, M, T, and W (Bevan et al., 1973; Vodkin et al., 1974; Sommer and Wickner, 1982a; Wesolowski and Wickner, 1984; reviewed in Wickner, 1986; Tipper and Bostian, 1984) (Table 1). Of these, L-A, L-BC, and M are present in intracellular noninfectious VLPs, while the intracellular location of T and W is not yet known (Herring and Bevan, 1974; Sommer and Wickner, 1982a). The components of the *S. cerevisiae* killer systems are compiled in Table 2.

All of these dsRNAs are transmitted by cytoplasmic mixing like that occurring during mating or heterokaryon formation. This is the mode of transmission of all the so-called "mycoviruses," and it has long been puzzling why no natural extracellular route of transmission has been found for members of this group. In vertebrates, cellular fusion between cells of different individuals is a rare event, generally only occurring during zygote formation. Within an individual, heterokaryon formation occurs in the formation of muscle cells and a few other cell types. Thus, a virus whose transmission could only occur by mating would have to wait years between opportunities to spread.

Table 1 dsRNA Families of *Saccharomyces cerevisiae*

dsRNA family	Size (kbases)	Copy No.	In VLPs	Genes	Refs.
L-A	4.5	1000	+	[HOK], [NEX], [EXL] in various combinations. Major VLP protein of L-A and M VLPs	a-d
L-BC	4.5	100	+	Major VLP protein of L-BC VLPs	a, d
M_1, M_2, . .	1.8, 1.5, . .	180	+	Precursor protein of secreted toxin and immunity	e, f
T	2.7	20	?	?	g
W	2.25	20	?	?	g

Sources: (a) Sommer and Wickner, 1982a,1982b; (b) Wickner, 1980; (c) Hopper et al., 1977; (d) El-Sherbeini et al., 1984; (e) Bostian et al., 1980; (f) Boone et al., 1986; (g) Wesolowski and Wickner, 1984.

Table 2 Components of the *Saccharomyces cerevisiae* Killer Systems

dsRNA	
M_1	A 1.8 kilobase dsRNA that encodes the K_1 preprotoxin-immunity protein. Strains carrying M_1 have the genotype [KIL-k_1]. Secretion of an active K_1 toxin is the K_1^+ phenotype. Resistance to the K_1 toxin is the R_1^+ phenotype.
M_2	A 1.5 kbase dsRNA that codes for a second toxin which kills M-o- and M_1-containing cells. M_2, [KIL-k_2], K_2^+, and R_2^+ are analogous to M_1, [KIL-k_1], K_1^+, and R_1^+.
L-A	A 4.5 kbase dsRNA that is noninfectious and stably maintained like a plasmid, but encapsidated like a virus. It encodes its major capsid protein (81 kDa). The genes [EXL], [NEX], and [HOK] (see below) are present in various combinations on various forms of L-A, denoted L-A-E ([EXL] alone), L-A-HN ([HOK] and [NEX]; found in all K_1 killers), L-A-HE ([HOK] and [EXL]), or L-A-H ([HOK] only). L-A depends on *MAK3, MAK10,* and *PET18* for replication.
L-B and L-C	Are 4.5 kbase dsRNAs unrelated to L-A and present in VLPs with different major proteins. L-B and L-A or L-C and L-A are compatible. L-B and L-C show some sequence homology. L-(BC) means an L dsRNA related to L-B and L-C type.
T and W	Are 2.76 kbase and 2.25 kbase minor dsRNAs. They do not cross-hybridize with each other, other dsRNAs, or cell DNA. They are cytoplasmically inherited. The copy number of T and W is induced 10-fold by growth at 37°C.
Cytoplasmic genes	
[HOK]	*h*elper *o*f *k*iller. This non-Mendelian gene supplies the helper function needed by M_1 for replication in a wild type strain. It is located on certain forms of L-A dsRNA, namely, L-A-HN, L-A-HE, or L-A-H. To test for [HOK], the combination of L-A-E + M_1 is introduced from a *ski2* strain into the *SKI*$^+$ M-o strain to be tested. L-A-E cannot support M_1 replication in a *SKI*$^+$ host, but if [HOK] is present, M_1 will be maintained.

Table 2 (Continued)

[EXL]	*exc*luder of M_2 dsRNA. This non-Mendelian trait prevents the replication of M_2 if [NEX] is absent, but not if [NEX] is present. [EXL] is located on certain forms of L-A, namely, L-A-E and L-A-HE. L-A-E acts by lowering the copy number of the L-A-H in the M_2 strain. It can exclude M_1 from a strain carrying M_1 + L-A-H.
[NEX]	M_2 *nonex*cludable by [EXL], but does not prevent exclusion of M_2 by strains carrying M_1. [NEX] is located on L-A-HN, the form of L-A found in wild type K_1 killer strains.
[B]	A cytoplasmic gene that allows M_1 to dispense with certain of the *MAK* gene products and elevates the copy number of M_1, making the strain a superkiller.
[D]	A cytoplasmic gene that, when introduced into a *ski*⁻ M_1 strain, exacerbates the "*d*isease," that is, growth at 20°C or 37°C becomes defective in addition to the 8°C growth defect seen in such strains even without [D].
[KIL-d]	A mutant form of [KIL-k_1] in which haploid strains unstably maintain and incompletely express M_1, but diploid strains stably maintain and normally express M_1.
[KIL-s]	Deletion mutants of M_1 that prevent the replication of (suppress) the parent M_1 dsRNA. These mutants are analogous to defective interfering particles of animal viruses.
Chromosomal genes	
MAK	*ma*intenance of [*K*IL-k_1]. *MAK* genes comprise at least 32 chromosomal genes necessary to maintain M_1 dsRNA. Mutants carrying the recessive alleles, *mak*, are K^-R^- M-o. At least some of the genes are also required for the maintenance of M_2. *MAK3, MAK10,* and *PET18* are needed by all forms of L-A. *MAK27* is needed by L-A-E.
clo	A complex chromosomal defect resulting in loss of L-B or L-C.

Table 2 (Continued)

SKI	super*k*iller. Mutants carrying the recessive alleles of *ski2, ski3, ski4, ski6, ski7,* or *ski8* produce more killer toxin and have increased copy number of M_1, M_2, L-A, and L-(BC). *ski*$^-$ mutations eliminate the need for some of the *MAK* genes and prevent the exclusion of M_2 by L-A-HN in *mkt* strains. The *ski5* mutant has none of these traits and lacks a cell surface protease that normally degrades the toxin.
MKT	*m*aintenance of [KIL-k_2] in the presence of [NEX]. Strains having a recessive allele, *mkt* (about 80% of laboratory strains), cannot maintain [KIL-k_2] at 30°C if [NEX] (L-A-HN) is present.
MKS	*mkt* suppressor. Like most *ski*$^-$ mutations, mutations in *mks1, mks2,* or *MKS50* prevent exclusion of M_2 by L-A-HN in *mkt1* hosts, but do not show other characteristics of *ski*$^-$ mutations.
KEX	*k*iller *ex*pression. Two chromosomal genes needed to process the toxin precursor. Mutants are K^-R^+. The *kex* mutants also do not properly process the α-hormone precursor, and many other secreted proteins are affected. The *KEX2* product is a protease that cleaves on the $COOH^-$ side of a pair of basic residues.
REX	*r*esistance *ex*pression. One chromosomal gene needed to express M_1-determined resistance to the toxin.
KRE	*k*iller *re*sistant. Three chromosomal genes needed for normal toxin action on sensitive cells. *KRE1* affects β(1-6)glucan, the normal toxin cell wall receptor. *KRE2* also affects toxin binding, but *KRE3* does not.
SEC	*sec*retion. Chromosomal genes for general protein secretion; mutants in these genes are conditionally lethal and also affect killer toxin secretion.

Such viruses probably exist, but their slow spread should make them rare and thus only rarely an object of study. The easy availability of cells whose membranes are exposed to the external environment gives the viruses of vertebrates (and many other metazoans) another easy route for transmission.

Fungi, however, generally have thick cell walls, making penetration difficult for a virus. Perhaps more important, fungi frequently undergo hyphal fusion or anastomosis. Even when two fungal strains are unable to fuse hyphae completely, the "incompatibility reaction" involves a transient hyphal fusion that could allow a virus particle to pass from one individual to another. In saccharomyces, sporulation and mating are probably frequent events in nature. Perhaps the "mycoviruses" have simply evolved to take advantage of the easiest mode of transmission available to them. In some ways, the extracellular route of transmission, involving as it does special requirements for methods of leaving and re-entering cells, is less convenient.

Our choice of the viruses of *S. cerevisiae* assumes that, like other processes in yeast, virus replication is like the intracellular part of virus replication in animal cells: The main difference may be the ease with which these processes may be studied, in particular the ease with which classical genetic studies may be carried out in yeast. This has allowed the identification of over 40 chromosomal genes involved in dsRNA replication as well as a number of cytoplasmic genes affecting this process in various ways. The genetic studies, whose results we will summarize, are serving to identify the cellular components involved, to bring out interactions and relationships among the components, to facilitate the cloning of the various genes involved, and to study the enzymology of the dsRNA replication process.

We will not review our cloning work here, but we have cloned and are studying the *SKI3, SKI8, MAK11, MAK16, MAK18,* and *MKT1* genes (T. Icho, R. B. Wickner, S. S. Sommer, H.-S. Lee, D. Kaback, and T. A. Koh, unpublished data).

The enzymology of dsRNA replication began with the finding that L-A-, L-BC-, and M-containing VLPs synthesize the respective (+) strands (Herring and Bevan, 1977; Bruenn et al., 1980: Welsh and Leibowitz, 1980; Welsh et al., 1980). Recently, (-) strand synthesis has been observed as well for both L-A (Fujimura et al., 1986) and S (a deletion derivative of M_1) (Nemeroff and Bruenn, 1986). We are also beginning to be able to assign the role of chromosomal genes in the enzymatic process (Fujimura and Wickner, 1986,1987).

CHROMOSOMAL GENES AFFECTING dsRNA REPLICATION IN *S. cerevisiae*

While the structure and biochemistry of dsRNA viruses of mammals, plants, insects, etc., can all be studied as easily as those of yeast, only in the yeast system has an extensive study been made of the chromosomal genes affecting dsRNA replication. This is, of course, largely a result of the ease with which complex yeast genetic studies are carried out.

There are over 30 chromosomal genes, called *MAK* genes (*mai*ntenance of *k*iller), whose products are necessary for replication of M_1 dsRNA (Wickner and Leibowitz, 1976,1979; Wickner, 1978,1979). These genes were easy to define using the killer toxin secretion phenotype imparted to cells carrying M dsRNA. They are scattered over the yeast genome. Mapping these genes led to the discovery that *MAK8* is the same as *TCM1*, the gene for ribosomal protein L3 (Wickner et al., 1982) and that *MAK1* is the structural gene for DNA topoisomerase I (Thrash et al., 1984). Polyamines are also needed for M_1 replication (Cohn et al., 1978). However, the role of ribosomal protein L3, DNA topoisomerase I, and polyamines in M replication remains obscure.

Three of the genes originally defined as needed for M dsRNA replication, *MAK3, MAK10,* and *PET18*, were subsequently found to be necessary for L-A replication (Sommer and Wickner, 1982a; Wickner and Toh-e, 1982). Indeed, it was the study of the dependence of the three cytoplasmic genes [HOK], [NEX], and [EXL] (see Table 2) on these three chromosomal genes for their maintenance that led us to the discovery that L dsRNA was comprised of two families (the L-A family and the L-BC family) and that [HOK], [NEX], and [EXL] were all located, in various combinations, on various forms of L-A (Wickner, 1980; Sommer and Wickner, 1982a,1982b; Wickner and Toh-e, 1984). *PET18* has recently been shown to consist of two genes, *MAK31* and *MAK32*, both of which are needed for cell growth and L-A and M dsRNA replication at high temperatures, but not at low temperatures (Toh-e and Sahashi, 1985).

A chance observation, that L-BC congregated 2 + : 2 - in a certain cross, led to the discovery of the *clo*⁻ defect (Wesolowski and Wickner, 1984). This nuclear defect results in the loss of only L-BC. In order to prove that not even one copy per cell of L-BC remains, dsRNA was carefully purified from a large mass of a *clo*⁻ strain. It was found that there were two new species of dsRNA present, unrelated to L-A, L-BC, or M, and which were named T and W (signifying nothing in particular) (Wesolowski and Wickner, 1984). While T and W are present at rather low copy number in cells grown at 20-30°C, they are amplified ten- to twentyfold when cells are grown at 37°C—just the opposite of the heat-curing seen with L-A and M dsRNAs.

The *MKT1* and *MKT2* genes were defined by recessive alleles present in many laboratory strains, making them unable to maintain M_2 in the presence of [NEX], a genetic activity of a certain form of L-A (L-A-HN) (Wickner, 1980,1983).

The *MAK*, *CLO*, and *MKT* genes are all promoters of dsRNA replication. They all encode factors with a positive effect. In contrast, the *SKI* genes, *SKI2, SKI3, SKI4, SKI6, SKI7,* and *SKI8,* all encode factors (presumably proteins) that have a negative effect on the replication of L-A, L-BC, and M dsRNAs (Toh-e et al., 1978; Ball et al., 1984; Ridley et al., 1984). They seem to have no effect on T or W copy number. These genes were first recognized by their *superkiller* mutant phenotype (Toh-e et al., 1978). The *ski*⁻ mutations were subsequently found to suppress many of the *mak*⁻ mutations (Toh-e and Wickner, 1980) as well as *mkt1* (Ridley et al., 1984) (Table 3).

Wild type (SKI^+) killer strains (carrying L-A and M dsRNAs) show no growth defects in comparison with isogenic strains lacking M or lacking both L-A and M. In contrast, ski^- hosts, that are therefore derepressed for L-A and M replication, are cold-sensitive for growth and rapidly die at 8°C if and only if they also carry M dsRNA (Ridley et al., 1984). The copy number of M dsRNA is about 10 times higher in a ski^- strain than in a SKI^+ strain (Table 4). This results in failure of cells to survive at low temperature. Elimination of M without eliminating L-A leaves the cells with far more dsRNA (Ball et al., 1984) (Table 4), but these ski^- L-A M-o cells are not detectably cold-sensitive for growth (Ridley et al., 1984). Thus the growth defect is not due to the total load of dsRNA. If the toxin and immunity precursor protein is eliminated by a deletion (S mutants), the cold sensitivity is unaltered. Thus, the cold sensitivity is not due to overproduction of the toxin or immunity substance (Ridley et al., 1984). It seems that the cold sensitivity is due to the high copy number of the M replicon. We suggest (Figure 1) that the essential role of the *SKI* gene is to prevent the M replicon from overutilizing some *MAK* gene product that is critical for cell growth at low temperatures (Ridley et al., 1984).

Other chromosomal genes involved in dsRNA replication, *mks1*, *mks2*, *MKS50* (Ridley et al., 1984), and *KRB1* (Wickner and Liebowitz, 1977), have been defined as suppressors of *mkt1* and *mak7-1*.

CYTOPLASMIC (NON-MENDELIAN) GENES AFFECTING dsRNA REPLICATION

[KIL-sd_1] (*ski-dependent*) is a mutant cytoplasm found in certain ski^- strains that results in the M_1 in this cytoplasm being dependent for its replication on the ski^- mutation (Toh-e and Wickner, 1979). It was puzzling because after loss or curing of M_1 from [KIL-sd_1] strains and introduction of a wild type L-A-HN + M_1 killer cytoplasm, the [KIL-sd_1] phenotype was gradually selected out. The explanation for this phenomenon gradually became clear as it was found that [HOK], a cytoplasmic gene located on L-A, could complement the defect in the [KIL-sd_1] cytoplasm and that [KIL-sd_1] strains had a normal M_1 but an unusual variant of L-A (called L-A-E) that lacked [HOK]. It was the combination of L-A-E + M_1 that could not be maintained in a SKI^+ strain but was completely stable in a ski^- host (Ridley et al., 1984). This combination gave a lower copy number M_1 in the ski^- host than L-A-HN + M_1 (the usual combination) and so the cell was less cold-sensitive (Ridley et al., 1984). This provided, in retrospect, the explanation

Table 3 Chromosomal Genes: Specificity and Interactions[a]

dsRNA	*mak3*, *mak10*, *pet18*	*mkt1*, *mkt2*	*mak11* group	*mak18* group	*mak16*	*clo*	*ski2*, *ski3*, *ski4*, *ski6*, *ski7*, *ski8*
M_1	-	+	-	-	-	+	↑
M_2	-	-	-	-	-	+	↑
L-A	-	+	+	+	+	+	↑
L-BC	+	+	+	+	+	-	↑
T	+	+	+	+	+	+	→
W	+	+	+	+	+	+	→
Suppressed by *ski*⁻	-	+	+	+	-		
Suppressed by [B]	-		+	-	-		

[a]In the upper part of the table, "+" means the mutant retains the indicated dsRNA, while "-" means the mutant loses the indicated dsRNA. In the lower part of the table, "+" means suppression of the *mak*⁻ mutation is observed, resulting in retention of M dsRNA In the last column, ↑ indicates increased level of that dsRNA species and → indicates same level as found in wild type (SKI^+) strains.

Table 4 Control of M and L-A Copy Number

	dsRNA content		
	L-A-HN	L-A-HN + M_1	
Host genotype	L-A-HN	L-A-HN	M_1
SKI$^+$	40[a]	5	1
ski$^-$	150	≅1	10

[a]Relative copy numbers, based on data of Ball et al. (1984). "1" in this table is the copy number of M_1 in a wild type killer (typically about 180 molecules/cell).

for the selection of the [KIL-sd_1] strains. It also suggests that the role of L-A in M replication and maintenance is to antagonize the repressive effects of the *SKI* gene products. Since L-A encodes the major coat protein of the VLPs, encapsidating both itself and M_1, it is natural to assume that L-A accomplishes this protection by providing this coat protein. However, other possibilities exist, as has been discussed elsewhere (Wickner, 1986).

[HOK] (*h*elper *o*f *k*iller) is thus the function that L-A provides to M to allow it to escape the repressive effects of the *SKI* gene products. The form of L-A found in K_1 strains (L-A-HN) can provide this [HOK] function to M_2 strains, and the form of L-A found in the one K_2 strain studied (L-A-H) can likewise provide this helper function to M_1 dsRNA (Hannig et al., 1985). Thus, the notion that Ls are specific for Ms does not seem to hold, at least not in *MKT*$^+$ strains (see [NEX] below).

[EXL] (as it is now understood) is the ability of some L-A variants (called L-A-E) to exclude either M_1 or M_2 if its L-A is L-A-H (Wickner, 1980; Hannig et al., 1985). L-A-E does this by lowering the copy number of L-A-H; but L-A-E does not exclude L-A-H (Hannig et al., 1985). How L-A-E lowers the L-A-H copy number is not yet clear, but the consequence is that L-A-H, which carries the [HOK] needed by M, is present in such low numbers that it cannot help M. The strain becomes much like a *SKI*$^+$ [KIL-sd_1] strain, in that the resulting combination of L-A-E + M is not stable in the *SKI*$^+$ state. L-A-E does encode a coat protein of about the same size as that encoded by L-A-HN or L-A-H, but L-A-E cannot antagonize *SKI* products to help M replicate, perhaps because of the nature or amount of coat protein provided or perhaps for some other reason.

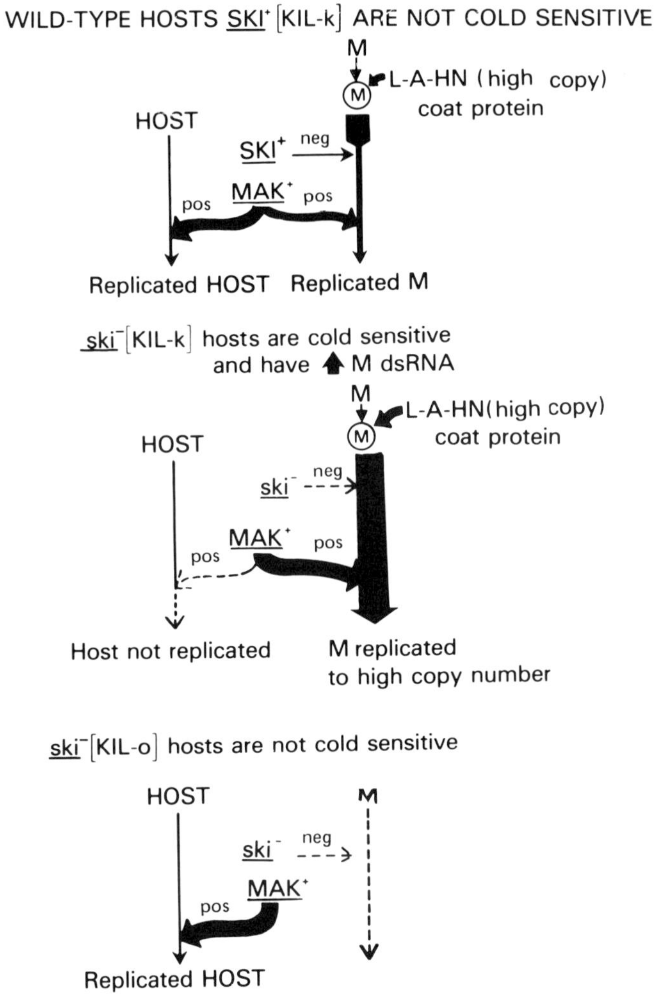

Figure 1 Model to explain the cold sensitivity of growth and its relation to *MAK* genes, *SKI* genes, M dsRNA , and L-A dsRNA The model assumes that the product of a *MAK* gene needed for both low temperature growth and M replication is diverted away from its essential cellular function by high M copy number. The M copy number is controlled, in turn, by *SKI* products and L-A dsRNA. Broken lines indicate an absent or defective product or process. Pos is positive control and neg is negative control or influence of the indicated gene product on the indicated process. (*Source*: Ridley et al., 1984.)

[NEX] was originally defined as a cytoplasmic element making M_2 *non-excludable* by [EXL] (Wickner, 1980). As we now (partially) understand, [NEX] is located on certain L-A variants, called L-A-HN (N = NEX), whose copy number is, unlike that of L-A-H, not lowered by L-A-E (E = EXL) (Sommer and Wickner, 1982a; Wickner, 1983; Hannig et al., 1985). Thus these L-As continue to provide [HOK] function to the M in the cell, and M is thus *non-excludable* by [EXL] (L-A-E).

L-A-HN has a second action. In hosts with a recessive allele of *mkt1* or *mkt2* (defined by this phenomenon), L-A-HN exlcudes M_2 at 30°C, but not at 20°C (Wickner, 1980,1983). This is the one case where, in a sense, an L-A discriminates between M_1 and M_2 because L-A-HN does not exclude M_1 in an *mkt1* or *mkt2* strain. L-A-H does not exclude M_1 or M_2 under any circumstance we have studied. The various exclusion phenomena are summarized in Table 5.

[KIL-b] was first discovered as a cytoplasmic element suppressing the usual requirement of M_1 for *MAK11* for its replication (Toh-e and Wickner, 1980). Further study has shown that, contrary to expectation, [KIL-b] is not a property of M (H. Uemura and R. B. Wickner, submitted), but we do not know where [B] (= *b*ypass), as we now call it, is located. [B] also results in M_1 not requiring several other *MAK* gene products (Toh-e and Wickner, 1980).

[D] (= *d*isease) is a non-Mendelian factor that accentuates the cytopathology that M dsRNA produces in *ski*⁻ strains (Esteban and Wickner, 1987). While *ski*⁻ M_1 strains usually grow normally at 20°C, 30°C, or 37°C and are only defective at 8°C, if [D] is present they show no growth at all at 37°C, grow very slowly at 20°C, and even at 30°C are slower than the wild type or than *ski*⁻ M_1 strains not

Table 5 Exclusion Phenomena

Excluding molecule	Excluded molecule	Host requirements	Comments
S_1	M_1	None known	Analogous to defective interfering viruses of animal cells
M_1	M_2	None known	
L-A-E ([EXL])	M_1 or M_2	L-A-H (not L-A-HN) must be present	L-A-E lowers L-A-H copy number so that [HOK] function is inadequate to maintain M_2
L-A-HN	M_2	*mkt*⁻ or *mkt2*⁻	Only at 30°C or above

carrying [D]. In order to study [D], we have isolated chromosomal mutants (*mad*⁻ = *ma*intenance of [*D*]) that cannot replicate or maintain [D]. We have also found that [D] can be heat-cured. We find that [D] is not a property of M, L-A, L-BC, or W dsRNAs. It is also not located on rho, the mitochondrial genome, or on 2μ DNA. The *mad*⁻ mutants do not lose L-A, M, L-BC, or W, so that [D] is not located on a molecule necessary for the replication of one of these molecules (Esteban and Wickner, 1987).

Many strains that lack [D] activity have a cytoplasmic genetic element that interferes with [D] action or excludes [D]: We cannot yet tell which. We call this element [DIN] ([*D*] *in*terference), and we imagine that [DIN] and [D] are on similar molecules (like L-A-E and L-A-H) (Esteban and Wickner, 1987). Our current efforts are directed at locating [B] and [D] on cytoplasmically-inherited molecules. Since *mad*⁻ mutants lose [D] but do not lose [B] from the same strains, [B] and [D] must be on different molecules.

dsRNA REPLICATION: IN VITRO STUDIES

The yeast dsRNAs replicate in a manner that is coordinated with cell growth. The copy number of each dsRNA remains stable in a given strain. While one study claims that dsRNA is synthesized during specific periods in the cell cycle (Zakian et al., 1981), the results of Newman et al. (1981) and the unpublished results of Bostian and Tipper (cited in Tipper and Bostian, 1984) disagree, indicating that dsRNA is replicated throughout the cell cycle.

In vivo studies indicate that L-A and M dsRNA replicate by a conservative mechanism (Newman et al., 1981; Sclafani and Fangman, 1984) in a sequential manner (Newman and McLaughlan, 1984); that is, for a given dsRNA molecule, first the (+) (message) strand is made and then the (-) strand is synthesized. This is the same as the situation in reovirus replication (Zarbl and Millward, 1983).

Herring and Bevan (1977) first described the RNA polymerase reaction of isolated L-containing VLPs producing single-stranded L RNA, which subsequent studies have shown is (+)-strand RNA (Bruenn et al., 1980). We have shown that this reaction is conservative in the case of L-A and L-BC (+)-strand RNA synthesis and that the (+)-strand single-stranded RNA (ssRNA) is completely extruded from the VLPs (Esteban and Wickner, 1986; Fujimura et al., 1986).

We have also found that a less dense class of VLPs containing only (+)-stranded L-A synthesizes (-) strands in vitro to form L-A dsRNA (Fujimura et al., 1986) (Figures 2 and 3). The VLPs catalyzing this reaction are only found in log-phase cells, suggesting that these particles are specific for replicative synthesis.

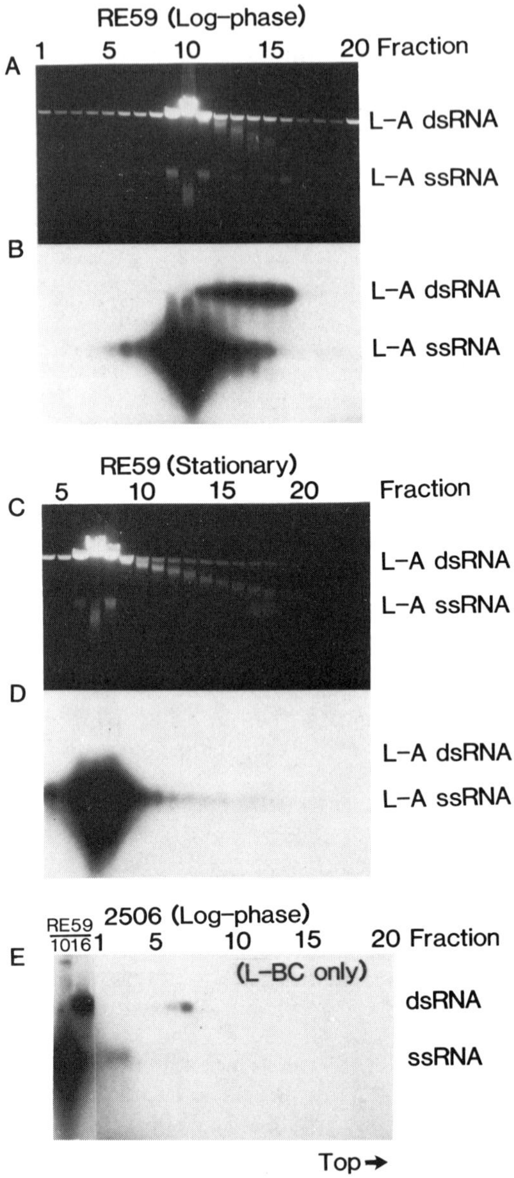

Figure 2 L-A VLPs isolated by differential centrifugation were analyzed on CsCl gradients (Fujimura et al., 1986). The CsCl gradient fractions were dialyzed, and then aliquots were incubated in RNA polymerase reaction mixtures where the gradient fraction supplies the enzyme and template. After phenol extraction, the RNA polymerase

The density of these particles is that predicted for particles identical to those containing L-A dsRNA but carrying only (+) single strands (Fujimura et al., 1986). That this is indeed the structure of these particles is confirmed by several lines of evidence: (1) The L-A dsRNA-containing VLPs have the same major coat protein as these (+) ssRNA-containing particles (Fujimura and Wickner, 1987). (1) After the (+) ssRNA-containing particles synthesize (-) strands in vitro to form L-A dsRNA, they become able to synthesize (+) ssRNA, which they extrude from the VLPs (Fujimura et al., 1986; Fujimura and Wickner, 1987). This indicates that these particles have both (+) and (-) strand-synthesizing RNA polymerases and suggests that the structure of the template determines which activity is observed. (3) In vitro synthesis of (-) strands to form L-A dsRNA results in a density shift of the particles to a density identical to the L-A dsRNA-containing particles isolated from cells (Fujimura and Wickner, 1987). We propose the L-A VLP cycle shown in Figure 4 (Fujimura et al., 1986; Fujimura and Wickner, 1987) to explain these results. We have, as yet, not demonstrated the encapsidation step in vitro, but the existence of the L-A (+) strand particles (Fujimura et al., 1986) indicates that the (+) strands extruded from the "mature" particles must associate with either coat protein(s) or a preformed coat.

To define the roles of chromosomal genes in this process, we used *mak10*^{ts} and *pet18*^{ts} mutants. These strains are temperature-sensitive (ts) for maintenance of L-A dsRNA. We grew *mak10*^{ts} L-A and *pet18*^{ts} L-A strains at the permissive temperature and isolated L-A VLPs by differential centrifugation. These were then analyzed on CsCl equilibrium gradients. We found that the L-A VLPs from both *mak10*^{ts} and *pet18*^{ts} mutants were structurally unstable compared to L-A VLPs from wild type strains (Fujimura et al., 1986; Fujimura and Wickner, 1986; Fujimura and Wickner, 1987). This effect was most dramatic in the case of the L-A dsRNA-containing particles. These particles all broke down on the CsCl gradient (Fujimura and Wickner, 1987).

---

products were analyzed on agarose gels. (A) and (C) show ethidium staining of the gels, while (B) and (D) show autoradiograms of the products. (A) and (B) are the results using VLPs from log phase cells, while (C) and (D) show the results using stationary phase cells. Both log and stationary phase cells have L-A VLPs synthesizing (+) strand L-A ssRNA, but only log phase cells have L-A VLPs that make L-A dsRNA. The newly synthesized strand in these VLPs is the (-) strand (Fujimura et al., 1986). Similar results were obtained for L-BC VLPs (E) (Fujimura et al., 1986).

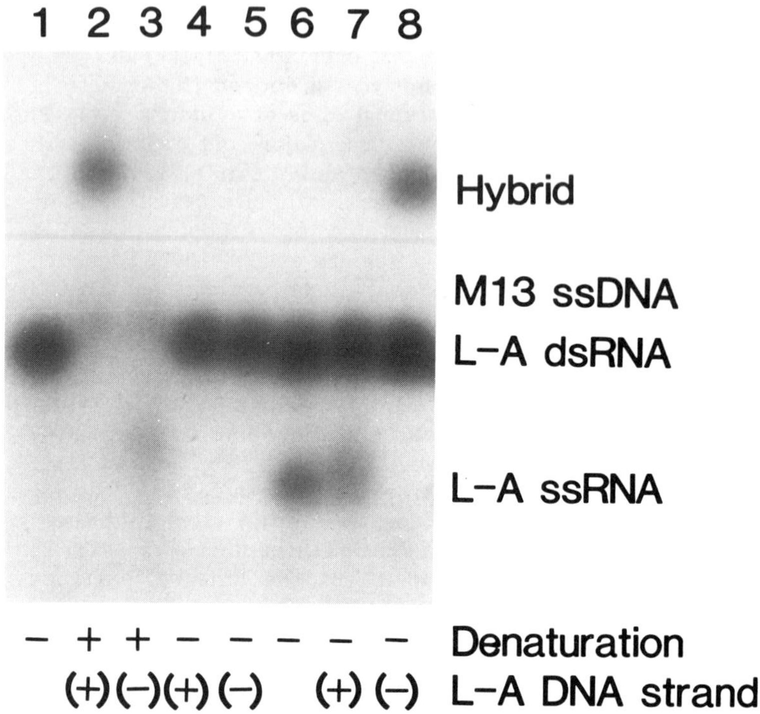

Figure 3 Hybridization of RNA polymerase reaction products of dsRNA-synthesizing particles with each strand of an L-A dsRNA fragment cloned in phage M13. dsRNA-synthesizing particles from the *mak10*^{ts} strain were incubated with an RNA polymerase reaction mixture ([α-^{32}P]UTP used was at 50 Ci/mmol) at 30°C for 45 minutes (a condition under which only dsRNA is produced, lanes 1 to 5) or for 135 minutes (dsRNA and ssRNA produced, lanes 6 to 8), and the products were extracted with phenol. After heat denaturation, or without heat denaturation, the extracted products were hybridized with 2 μg of M13 8pl [(+) strand fragment of L-A dsRNA] or M13 8w3 [(-) strand fragment of L-A dsRNA] (Sclafani and Fangman, 1984), and the hybrids formed were analyzed in an agarose gel. An autoradiogram of the gel is shown. Lanes 1 and 6: RNA polymerase products as control. Lanes 2 and 3: the dsRNA product was heat denatured and hybridized with M13 8pl [(+) strand] and M13 8w3 [(-) strand], respectively. Lanes 4 and 5: without heat denaturation, the dsRNA product was hybridized with M13 8pl [(+) strand] and M13 8w3 [(-) strand], respectively. Lanes 7 and 8: without heat denaturation, both ssRNA and dsRNA products were hybridized with M13 8pl [(+) strand] and M13 8w3 [(-) strand], respectively. (*Source*: Fujimura and Wickner, 1987.)

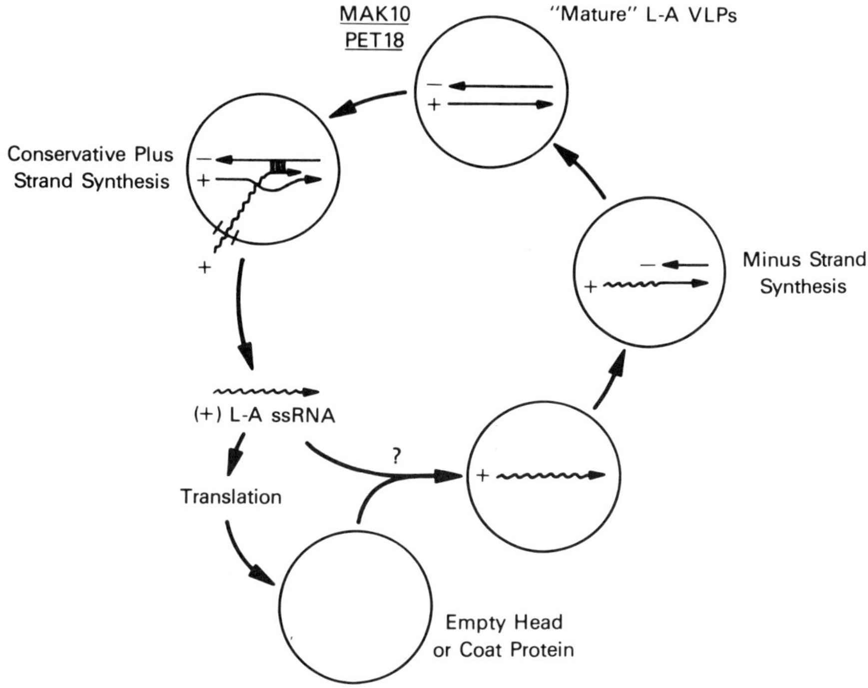

Figure 4 Model of the L-A VLP replication cycle. The conservatively synthesized (+) ssRNA is used to synthesize coat protein and is encapsidated by this coat protein, and perhaps by other host-supplied proteins, such as *MAK10* or *PET18* (*MAK31* plus *MAK32*) to form new particles. Those particles first make a (-) strand on the (+) strand template to form L-A dsRNA. Then the particles use the L-A dsRNA template to again make (+) ssRNA. (*Source*: Fujimura and Wickner, 1987.)

It is possible that either the *MAK10* or the *PET18* gene products or both are present in the L-A VLPs. Alternatively, one or more of these gene products may modify some component of the VLP to make a stable structure. As mentioned above, the *PET18* locus includes two genes, *MAK31* and *MAK32*, both of which are necessary to maintain L-A in vivo and both of which are completely deleted in the $pet18^{ts}$ mutants (Toh-e and Sahashi, 1985). Whether it is *MAK31* or *MAK32* or both whose absence is giving us the observed effects is not yet known.

M_1 dsRNA-containing VLPs can be separated by density into two varieties (Esteban and Wickner, 1986) (Figure 5). Based on their

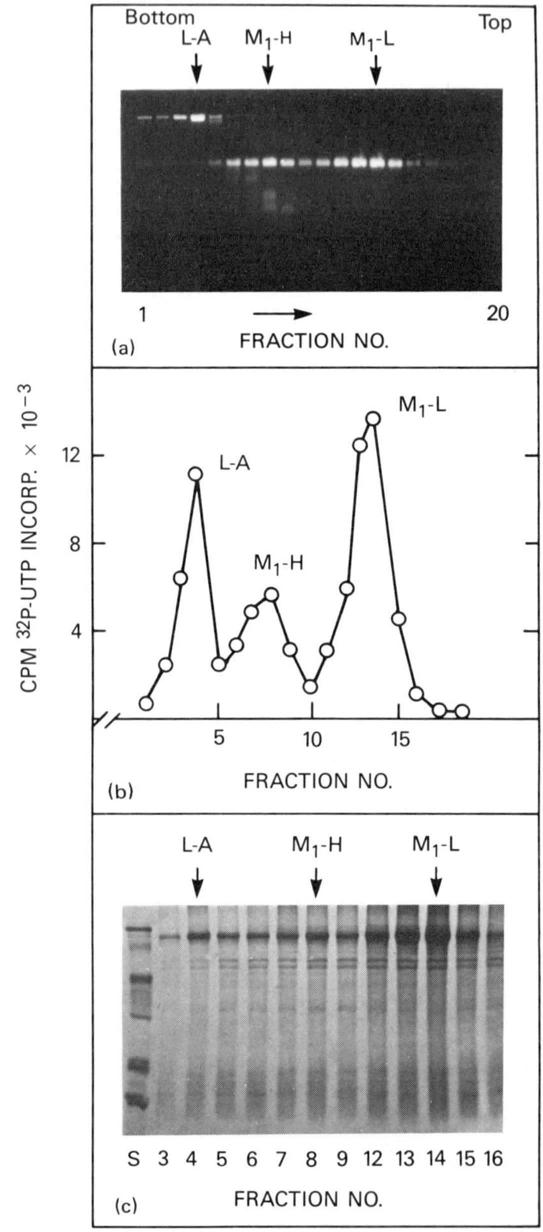

Figure 5 (a) Agarose gel electrophoresis of L-A and M_1 dsRNAs from VLPs purified by CsCl gradient equilibrium sedimentation. (b) RNA polymerase activity in the same fractions. (c) Polyacrylamide gel elec-

hydrodynamic properties, these VLPs have one M_1 dsRNA per particle (M_1-L VLPs) or two M_1 dsRNAs per particle (M_1-H VLPs) (Esteban and Wickner, 1986). L-A dsRNA-containing VLPs always have just one L-A dsRNA molecule per particle. Both M_1-L and M_1-H VLPs have the same major coat protein and the same size by electron microscopy, and both have an RNA polymerase activity producing (+)-strand transcripts.

In the case of M_1-H VLPs, all of the (+)-strand transcript is extruded from the VLPs, making it sensitive to RNase. M_1-L VLPs, in contrast, retain 20-60% of their in vitro transcript inside the VLP, where it can then be converted to M_1 dsRNA by synthesis of a (-) strand (Esteban and Wickner, 1986). The in vitro synthesis of M_1 dsRNA in VLPs has also been observed (Esteban and Wickner, 1986). To explain the existence of M_1-H and M_1-L particles, one must recall that the major coat protein of M-containing VLPs is encoded by L-A (Hopper et al., 1977; Sommer and Wickner, 1982) and thus must be designed to hold 4.5 kbase of dsRNA the size of L-A. When an L-A (+) strand transcript is synthesized, there is no room for it inside the head so it is extruded. The same is true for M_1-H VLPs carrying $2 \times 1.8 = 3.6$ kbase of dsRNA almost a headfull. But M_1-L VLPs have only 1.8 kbase of dsRNA, so M_1 (+)-strand single-strand transcripts are often retained within the VLPs, where they are converted to a second molecule of M_1 dsRNA by (-)-strand synthesis. These ideas are diagrammed in Figure 6.

A similar idea may apply in the ustilago system, where Bozarth and his co-workers (1981) have shown that, whereas the large dsRNAs are encapsidated at one dsRNA molecule per particle, the M dsRNAs are present in one, two, or even three molecules per particle. While the comparable in vitro transcription results have not been reported, it is reasonable to suggest that an analogous headfull mechanism might apply.

---

trophoresis of proteins from the same VLPs. To one aliquot of each fraction (10-20 μl), 0.5 ml of 10% TCA were added. After a 10-min incubation in ice, samples were spun down and pellets washed with 1 ml of 0.1% TCA. The precipitates were dissolved in electrophoresis loading buffer that contained 0.1% sodium dodecyl sulfate and β-MET, and boiled for 5 min. Electrophoresis was carried out in 10% acrylamide gel with 0.1% sodium dodecyl sulfate for 3 hs, and the gel was stained with Commassie Brilliant Blue. Molecular weight standards included phosphorylase B (92,500), bovine serum albumin (66,200), ovalbumin (45,000), carbonic anhydrase (31,000), soybean trypsin inhibitor (21,500), and lysozyme (14,400). The locations of L-A, M_1-H, and M_1-L VLPs in the gradient are indicated by arrows (peak fractions). The major band is the major capsid protein. (*Source*: Esteban and Wickner, 1986.)

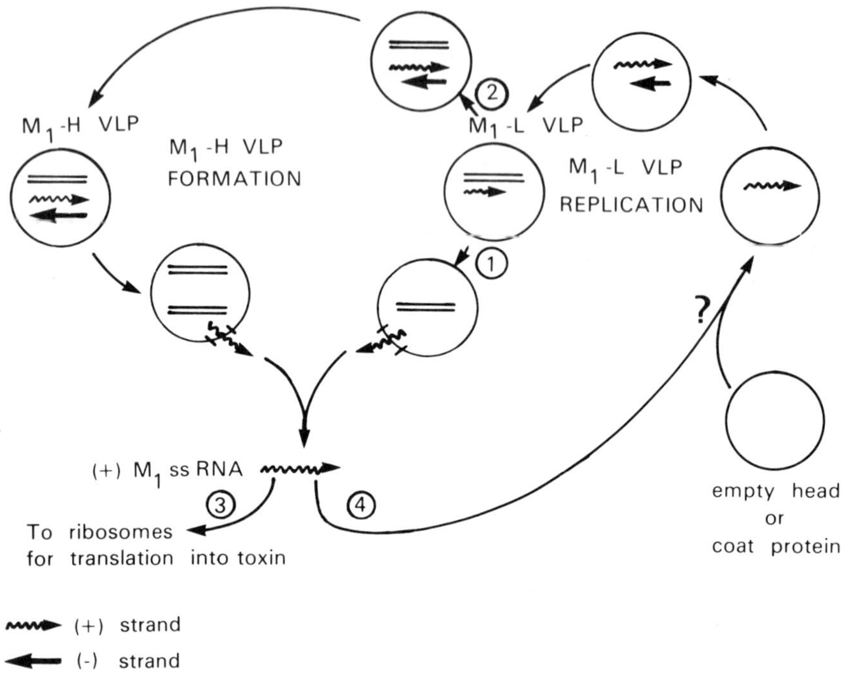

Figure 6 A model of the VLP cycle for M_1 dsRNA transcription and replication. The various types of VLPs have the same coat proteins but differ in RNA content. M_1-L VLPs, that contain one molecule of M_1 dsRNA per particle and have RNA polymerase activity, synthesize (+) M_1 ssRNA, which is used in three different ways: (1) Most is extruded and translated to make toxin [number (3) in the model]. (2) Some is extruded and somehow packaged in a new coat [(4) in the model]. (3) Some of the (+) M_1 ssRNA is retained within the particle [(2) in the model], where it serves as template for (-) strand synthesis to form M_1-H VLPs. M_1-H VLPs are essentially full heads and so their (+) strand transcript is all extruded from the particles and is used for translation and packaging. The particles formed by association of (+) strands with a new coat can then synthesize (-) strands to make new M_1-L particles. (*Source*: Esteban and Wickner, 1986.)

CONCLUSIONS AND PERSPECTIVES

While we feel, subjectively, that the saccharomyces dsRNA virus field is moving rapidly, it is clear that we are still almost at the starting point. Our goal is to understand the role of the many host components, both for the host and for the viruses, to understand the mechanisms by which the dsRNAs replicate, and that replication is controlled by the host. We have observed a myriad of different interactions among the various components of this system. While these observations each teach us something about the system, we are a long way from understanding their bases.

We hope that a thorough understanding of this (relatively simple) system may be of use in the study of viruses infecting mammalian cells. While we have a long way to go to reach the level of understanding that has been achieved for bacteriophage replication, this system is one of the few virus systems being pursued in conjunction with a detailed genetic analysis of the role of the host.

REFERENCES

Ball, S. G., Tirtiaux, C., and Wickner, R. B. (1984). Genetic control of L-A and L-(BC) dsRNA copy number in killer systems of *Saccharomyces cerevisiae*. *Genetics 107*:199-217.

Bevan, E. A., Herring, A. J., and Mitchell, D. J. (1973). Preliminary characterization of two species of dsRNA in yeast and their relationship to the "killer" character. *Nature 245*:81-86.

Boone, C., Bussey, H., Greene, D., Thomas, D. Y., and Vernet, T. (1986). Yeast killer toxin: Site-directed mutations implicate the precursor protein as the immunity component. *Cell 46*:105-113.

Bostian, K. A., Hopper, J. E., Rogers, D. T., and Tipper, D. J. (1980). Translational analysis of the killer-associated virus-like particle dsRNA genome of *S. cerevisiae*: M dsRNA encodes toxin. *Cell 19*:403-414.

Bozarth, R. F., Koltin, Y., Weissman, M. B., Parker, R. L., Dalton, R. E., and Steinlauf, R. (1981). The molecular weight and packaging of dsRNAs in the mycovirus from *Ustilago maydis* killer strains. *Virology 113*:492-502.

Bruenn, J., Bobek, L., Brennan, J., and Held, W. (1980). Yeast viral RNA polymerase is a transcriptase. *Nucleic Acids Res. 8*: 2985-2997.

Cohn, M. S., Tabor, C. W., Tabor, H., and Wickner, R. B. (1978). Spermidine or spermine requirement for killer double-stranded RNA plasmid replication in yeast. *J. Biol. Chem. 253*:5225-5227.

El-Sherbeini, M., Tipper, D. J., Mitchell, D. J., and Bostian, K. A. (1984). Virus-like particle capsid proteins encoded by different L dsRNAs in *Saccharomyces cerevisiae*: Their roles in maintenance of M dsRNA killer plasmids. *Mol. Cell. Biol. 4*:2818-2827.

Esteban, R., and Wickner, R. B. (1986). Three different M_1 RNA-containing virus-like particle types in *Saccharomyces cerevisiae*: *In vitro* M_1 double-stranded RNA synthesis. *Mol. Cell. Biol. 6*: 1552-1561.

Esteban, R., and Wickner, R. B. (1987). A new non-Mendelian genetic element of yeast that increases cytopathology produced by M_1 double-stranded RNA in *ski* strains. *Genetics 117*:394-408.

Fujimura, T., and Wickner, R. B. (1986). Thermolabile L-A virus-like particles from *pet18* mutants of *Saccharomyces cerevisiae*. *Mol. Cell. Biol. 6*:404-410.

Fujimura, T., and Wickner, R. B. (1987). The L-A double-stranded RNA virus-like particle replication cycle in yeast: Particle maturation *in vitro* and the effects of *mak10* and *pet18* mutations. *Mol. Cell. Biol. 7*:420-426.

Fujimura, T., Esteban, R., and Wickner, R. B. (1986). *In vitro* L-A double-stranded RNA synthesis in virus-like particles from *Saccharomyces cerevisiae*. *Proc. Nat. Acad. Sci. USA 83*:4433-4437.

Hannig, E. M., Leibowitz, M. J., and Wickner, R. B. (1985). On the mechanism of exclusion of M_2 double-stranded RNA by L-A-E double-stranded RNA in *Saccharomyces cerevisiae*. *Yeast 1*:57-65.

Herring, A. J., and Bevan, E. A. (1974). Virus-like particles associated with the double-stranded RNA species found in killer and sensitive strains of the yeast *Saccharomyces cerevisiae*. *J. Gen. Virol. 22*:387-394.

Herring, A. J., and Bevan, E. A. (1977). Yeast virus-like particles possess a capsid-associated single-stranded RNA polymerase. *Nature 268*:464-466.

Hopper, J., Bostian, K. A., Rowe, L. B., and Tipper, D. J. (1977). Translation of the L species dsRNA genome of the killer-associated virus-like particles of *Saccharomyces cerevisiae*. *J. Biol. Chem. 252*:9010-9017.

Nemeroff, M. E., and Bruenn, J. A. (1986). Conservative replication and transcription of *Saccharomyces cerevisiae* viral double-stranded RNA *in vitro*. *J. Virol. 57*:754-758.

Newman, A. M., and McLaughlin, C. S. (1984). Sequential replication of L dsRNA. Abstracts of the 12th International Conference on Yeast Genetics and Molecular Biology, Edinburgh, p. 80.

Newman, A. M., Elliott, S. G., McLaughlin, C. S., Sutherland, P. A., and Warner, R. C. (1981). Replication of dsRNA of the virus-like particles in *Saccharomyces cerevisiae*. *J. Virol. 38*:263-271.

Ridley, S. P., Sommer, S. S., and Wickner, R. B. (1984). Superkiller mutations in *Saccharomyces cerevisiae* suppress exclusion of M_2 double-stranded RNA by L-A-HN and confer cold sensitivity in the presence of M and L-A-HN. *Mol. Cell. Biol. 4*:761-770.

Sclafani, R. A., and Fangman, W. L. (1984). Conservative replication of dsRNA in *Saccharomyces cerevisiae* by displacement of progeny single strands. *Mol. Cell. Biol. 4*:1618-1626.

Sommer, S. S., and Wickner, R. B. (1982a). Yeast L dsRNA consists of at least three distinct RNAs; evidence that the non-Mendelian genes [HOK], [NEX] and [EXL] are on one of these dsRNAs. *Cell 31*:429-441.

Sommer, S. S., and Wickner, R. B. (1982b). Co-curing of plasmids affecting killer double-stranded RNAs of *Saccharomyces cerevisiae*: [HOK], [NEX], and the abundance of *L* are related and further evidence that M_1 requires *L*. *J. Bacteriol. 150*:545-551.

Thrash, C., Voelkel, K., Di Nardo, S., and Sternglanz, R. (1984). Identification of *Saccharomyces cerevisiae* mutants deficient in DNA topoisomerase I activity. *J. Biol. Chem. 259*:1375-1377.

Tipper, D. J., and Bostian, K. A. (1984). Double-stranded RNA killer systems in yeast. *Microbiol. Rev. 48*:125-156.

Toh-e, A., and Sahashi, Y. (1985). The *PET18* locus of *Saccharomyces cerevisiae*: A complex locus containing multiple genes. *Yeast 1*:159-172.

Toh-e, A., and Wickner, R. B. (1979). A mutant killer plasmid whose replication depends on a chromosomal "superkiller" mutation. *Genetics 91*:673-682.

Toh-e, A., and Wickner, R. B. (1980). "Superkiller" mutations suppress chromosomal mutations affecting double-stranded RNA killer plasmid replciation in *Saccharomyces cerevisiae*. *Proc. Nat. Acad. Sci. USA 77*:527-530.

Toh-e, A., Guerry, P., and Wickner, R. B. (1978). Chromosomal superkiller mutants of *Saccharomyces cerevisiae*. *J. Bacteriol. 136*:1002-1007.

Vodkin, M., Katterman, F., and Fink, G. R. (1974). Yeast killer mutants with altered dsRNA. *J. Bacteriol. 117*:681-868.

Welsh, D. J., and Leibowitz, M. J. (1980). Transcription of killer virion double-stranded RNA *in vitro*. *Nucleic Acids Res. 8*:2365-2375.

Welsh, D. J., and Leibowitz, M. J. (1982). Localization of genes on dsRNA killer virus of yeast. *Proc. Nat. Acad. Sci. USA 79*: 786-789.

Welsh, D. J., Leibowitz, M. J., and Wickner, R. B. (1980). Virion DNA-independent RNA polymerase from *Saccharomyces cerevisiae*. *Nucleic Acids Res. 8*:2349-2363.

Wesolowski, M., and Wickner, R. B. (1984). Two new double-stranded RNA molecules showing non-Mendelian inheritance and heat inducibility in *Saccharomyces cerevisiae*. *Mol. Cell Biol. 4*:181-187.

Wickner, R. B. (1978). Twenty-six chromosomal genes needed to maintain the killer double-stranded RNA plasmid of *Saccharomyces cerevisiae*. *Genetics 88*:419-425.

Wickner, R. B. (1979). Mapping of chromosomal genes of *Saccharomyces cerevisiae* using an improved genetic mapping method. *Genetics 92*:803-821.

Wickner, R. B. (1980). Plasmids controlling exclusion of the K_2 killer double-stranded RNA of yeast. *Cell 21*:217-226.

Wickner, R. B. (1983). Killer systems in *Saccharomyces cerevisiae*: Three distinct modes of exclusion of M_2 double-stranded RNA by three species of double-stranded RNA, M_1, L-A-E, and L-A-HN. *Mol. Cell Biol. 3*:654-661.

Wickner, R. B. (1986). Double-stranded RNA replication in yeast: The killer system. *Annu. Rev. Biochem. 55*:373-395.

Wickner, R. B., and Leibowitz, M. J. (1976). Chromosomal genes essential for replication of a double-stranded RNA plasmid of *Saccharomyces cerevisiae*: The killer character of yeast. *J. Mol. Biol. 105*:427-443.

Wickner, R. B., and Leibowitz, M. J. (1977). Dominant chromosomal mutation bypassing chromosomal genes needed for killer RNA plasmid replication in yeast. *Genetics 87*:453-469.

Wickner, R. B., and Leibowitz, M. J. (1979). *mak* Mutants of yeast: Mapping and characterization. *J. Bacteriol. 140*:154-160.

Wickner, R. B., and Toh-e, A. (1982). [HOK], a new yeast non-Mendelian trait, enables a replication-defective kilelr plasmid to be maintained. *Genetics 100*:159-174.

Wickner, R. B., Ridley, S. P., Fried, H. M., and Ball, S. G. (1982). Ribosomal protein L3 is involved in replication or maintenance of the killer double-stranded RNA genome of *Saccharomyces cerevisiae*. *Proc. Nat. Acad. Sci. USA 79*:4706-4708.

Zakian, V. A., Wagner, D. W., and Fangman, W. L. (1981). Yeast L double-stranded RNA is synthesized during G1 phase but not the S phase of the cell cycle. *Mol. Cell. Biol.* *1*:673-679.

Zarbl, H., and Millward, S. (1983). The reovirus multiplication cycle. In *The Reoviridae*, W. K. Joklik (Ed.), Plenum Press, New York, pp. 107-196.

6

Structure, Transcription, and Replication of Killer Virus dsRNAs

J. A. Bruenn, M. E. Nemeroff, M. Lee, D. F. Pietras, J. J. Dowhanick, and L. J. Field*
State University of New York at Buffalo, Buffalo, New York

INTRODUCTION

The mycoviruses are double-stranded RNA (dsRNA) viruses whose segments are separately encapsidated. They are not, in general, infectious to intact cells, although they can be infectious to protoplasts (Stanway and Buck, 1984; El-Sherbeini and Bostian, 1987). The *Saccharomyces cerevisiae* viruses (ScV) have one large dsRNA species (L) of about 4.8 kilobase-pairs (kbase) encoding the major capsid polypeptide (Bostian et al., 1980) and, in some cases, a second dsRNA (M) of about 1.9 kbase encoding a secreted killer toxin that kills sensitive cells (Bostian et al., 1984). ScV viral particles have only a single major capsid polypeptide (Reilly et al., 1984), in contrast to the five or six polypeptides in reovirus cores. There are several

Present affiliation: Cold Spring Harbor Laboratory, Cold Spring Harbor, New York

killer subtypes. The L and M dsRNAs associated with killer type one are designated L_1 and M_1. (L_1 is also known as L_{1A} or as L_A.)

In addition to killer and sensitive strains of yeast, there are suppressive sensitive strains. Normally, sensitive haploid strains mated to killer haploids produce killer diploids and killer haploid meiotic products from those diploids. Suppressive sensitive strains produce sensitive diploids when crossed to killer haploids, and sensitive haploids from those crosses. Suppressive sensitive mutants have L_1 and smaller dsRNAs (S) of 0.6 to 1.6 kbase that displace M in these crosses. S dsRNAs are encapsidated in the same capsid polypeptide as L_1 and M_1 (Kane et al., 1979) and are derived from M_1 by internal deletion, sometimes followed by tandem duplication (Bruenn and Kane, 1978; Fried and Fink, 1978; Kane et al., 1979; Bruenn and Brennan, 1980; Thiele et al., 1984a). M is replaced by S due to the faster replication of the latter, and the S dsRNAs are dependent on the same host gene products as M for their maintenance. In some respects then, S dsRNAs are analogous to the defective-interfering (DI) RNAs of animal viruses (Guild and Strollar, 1977; Holland et al., 1979; Kennedy, 1976; Nomoto et al., 1979; O'Hara et al., 1984).

The ScV system is attractive for molecular virology because of the simplicity of the particle structure, the extensive genetics of the host and its interaction with the virus, the stable persistence characteristic of the system, and the existence of defective interfering particles. To answer questions about the transcription, replication, and packaging of the yeast viral dsRNAs, we have cloned and sequenced cDNAs representing all of a 793-basepair defective interfering (DI) yeast viral dsRNA, S14. This is the first yeast viral dsRNA entirely sequenced. Comparison of several DI dsRNAs—S14, S3, and S733—suggests that a very large region derived from the 3' end of the M_1 plus strand is conserved. This region has no extensive open reading frames on either strand but has several repeats of an 11-base consensus sequence also present in the corresponding region of M_2, the type 2 killer toxin encoding dsRNA.

Transcription of the S dsRNAs appears to be similar to that of L and M in that only the viral plus strand appears in the single-stranded products (Thiele et al., 1984a). We find that transcription of S14 in vitro results in equimolar labeling of all regions of the transcript, which has the 5' nucleotide ppGp. The L transcript also has the 5' nucleotide ppGp. Some viral particles are also capable of synthesizing the viral minus strand in vitro. Synthesis of both the viral minus strand and the viral plus strand occurs conservatively. This work, summarized here, is available in several publications (Nemeroff and Bruenn, 1986; Lee et al., 1986; Nemeroff and Bruenn, 1987), in which the materials and methods used are described.

RESULTS

Structure of DI dsRNAs

Features of the Sequence of S14

The 793-basepair sequence of S14 was completed by a combination of RNA and cDNA sequence analysis. All regions have now been sequenced on both strands. The deletion break-point of S14 is at position 253 of M_1, 21 basepairs 3' from the S3 break-point (Figure 1). Reinitiation occurred 540 basepairs from the 3' end of the M_1 (+) strand in S14 but 550 basepairs from the 3' end in S3, since S3 has an additional 10 basepairs (233-243 of S3) not present in S14 or in the 5' region of M_1 (Thiele et al., 1984a). In both cases, the conserved region of the M_1 (+) strand ends with a C residue and resumes at a C residue. The sequence data would predict S14-M_1 heteroduplexes consisting of double-stranded regions of 540 basepairs from one end (measured by electron microscopy as 531 ± 45) and 253 basepairs from the other (measured as 228 ± 35).

The 3' 540-basepair region of S14 (and thus the homologous region of M_1) has no significant open reading frames on either strand, as shown by the computer plot of Figure 2. The only extensive open reading frame in S14 predicted to be used is reading frame 2 on the (+) strand, which encodes the amino-terminal fragment of the toxin prepolypeptide (Bostian et al., 1984). Reading frames 5 and 6 on the (-) strand have a marginal probability of usage and are not more than 60 amino acids long. In addition, there is no free (-) strand synthesized either in vitro (Nemeroff and Bruenn, 1986) or in vivo (Sclafani and Fangman, 1984). Thus, the previous claim that the (-) strand of 3' end of the (+) strand of M_1 encoded a polypeptide of about 175 amino acids (Welsh and Leibowitz, 1982), appears to be erroneous.

Some of the sequence differences between S14 and M_1 or S3 appear to be due to errors in direct RNA sequencing of the latter RNAs (Lee et al. ,1986), but at least two differences are strain differences. The G at position 115 in S14 is an A in M_1 and S3. This G occurs in the third position in an alanine codon and is thus a silent change. The C at position 788 in S14 is an A in S3 and in some M_1s that have been sequenced (Thiele et al., 1984a; Hannig et al., 1984).

Comparison to Other S dsRNAs

We can assign the large T_1 oligonucleotides of S14 (Kane et al., 1979) to portions of this sequence by scanning the sequence for predicted large T_1 oligonucleotides. We have matched these predicted T_1 products with the known T_1 oligonucleotides by comparing their predicted pancreatic RNase products to those determined and by comparing the predicted and measured mobilities of the T_1 oligonucleotides in the

```
              10        20        30        40        50        60
S14     GAAAAATAAAGAAATGACGAAGCCAACCCAAGTATTAGTTAGATCCGTCAGTATATTATT
S3
M1
              70        80        90       100       110       120
S14     TTTCATCACATTACTACACCTAGTCGTAGCGCTGAACGATGTGGCCGGTCCTGCGGAAAC
S3                                                            a
M1                                                            a
             130       140       150       160       170       180
S14     AGCACCAGTGTCATTACTACCTCGTGAAGCGCCGTGGTATGACAAGATCTGGGAAGTAAA
S3
M1
             190       200       210       220       230       240
S14     AGATTGGCTATTACAGCGTGCCACAGATGGCAATTGGGGCAAGTCGATCACCTGGGGTTC
S3                                  ***  c
M1
             250       260       270       280       290       300
S14     ATTCGTAGCGAGCCTCACCTTGAGTCTAACTGGTGGCACGCAGCATATCTCACCCTGAGA
S3                              *

             310       320       330       340       350       360
S14     CTAACTGGCGGCAGGCGACCGGTGAGCATACAGCATGCCCCACTCGATTCGAGACGCGAT
S3

             370       380       390       400       410       420
S14     TCGCGCTCGTAGGTATCGAGCGGCTACGTTGAGCTATTATGGCAGTGACATGCGATTCGC

             430       440       450       460       470       480
S14     GCACTGCCAAGATCAGCTCAGCAAAGTTAAGACCAGTATCGGATATGGT*AGACTACTACA
S3                                                       c   tc

             490       500       510       520       530       540
S14     ATTCGCACAGGTATGAGATTCTCAGTCTAGTGTATGGATGAGTAGTTGAGCCAATGAATC
S3                c
M1
             550       560       570       580       590       600
S14     TAGGGTTTAAATTACTATGCATTGACGTATAACAGGTACAAGCGTAGATAATACTTACTA
S3
M1                                a                                g
             610       620       630       640       650       660
S14     GGCCCCAGCCGGTACACCCTGTATTGAATAAATACGACTATTTGGCCAGGTCTGGACGGG
S3
M1      a
             670       680       690       700       710       720
S14     GCAGTCGAATTACTAGGTTGAGCACACACACGTGAATCACACAACATAACAGTGTAGGAA
S3
M1
             730       740       750       760       770       780
S14     CATAATGTGCCATTCGTAGTCTGAGACGCCGCTAGCCTGGTTTAATGCAACAGCATAGAA
S3
M1
             790
S14     GAAACACCCATCA
S3             a
M1             a
```

Figure 1 Sequence of S14. The sequence shown is that of the cDNA, since most of the sequence was performed on cDNA rather than on RNA. This is the (+) strand. The (+) strand of the dsRNA begins with a pppG and terminates with an A_{OH}. The region of the sequence from the 5' end of the M_1 plus strand is in boldface. The boxed sequences are the five repeats of a rotationally symmetrical consensus

two-dimensional gel system used to separate them. In this way, we have located all but one of the previously reported large T_1 oligonucleotides of S14 as well as many of the large T_1 products of M_1. The large M_1-derived T_1 oligonucleotides from the region deleted in S14 are missing from S14, as well as from S3 and S733, the two other DI dsRNAs so far characterized. Since S733 is also missing 3 of the large T_1 oligonucleotides from the 5' region of the M_1 (+) strand present in S3 and S14, it must beemissing between 86 and 220 basepairs from the 5' region of the (+) strand of M_1 that is present in S14 and S3. This is consistent with the size of S733 (about 600 basepairs). S733 retains at least the 5' 44 bases of the M_1 plus strand. Remarkably, although these three DI RNAs preserve only a small region of the 5' end of the M_1 (+) strand, only 44-253 bases, they retain a very large region from the 3' end of the M_1 (+) strand: 540 in S14, 550 in S3, and at least 333 bases in S733. This region has no large open reading frames in either the (+) strand or in the (-) strand (Figure 2). Its preservation in every DI RNA sug-gests that it is important for maintenance of M_1 derived dsRNAs. A survey of this sequence found 8 repeats of an 11-basepair consensus sequence, (U/C)GCGAUUCGC(A/G). Five of these, showing at least four of five possible identities in a rotational transformation, are boxed in the S14 cDNA sequence (Figure 1).

Initiation of Transcription

Previous results indicated that initiation of viral transcription might not occur in viral particles in vitro (Bruenn et al., 1980) but might require preexisting "pause products" as primers (Brennan et al., 1981). The pause products (2-25 bases long) are uncapped, and the viral transcriptase neither requires nor utilizes S-adenosyl-methionine (Bruenn et al., 1980; Brennan et al., 1981; Welsh et al., 1980). We now show that transcription is initiated in vitro and that the product of transcription has a 5' terminal ppGp.

---

sequence (see text). Differences between the S14, M_1, and S3 sequences are indicated by small letters (mismatches) or asterisks (deletions). The 5' M_1 sequence was determined from the cDNA (Bostian et al., 1984; Skipper et al., 1984) and the S3 and 3' M_1 sequences were determined by direct RNA sequencing (Thiele et al., 1984a; Hannig et al., 1984). The regions of S14 that have been sequenced in M_1 are bases 1-253 and 503-793. The regions in S14 that have been sequenced in S3 are 1-232, 252-327, and 468-793. (From Lee et al., 1986).

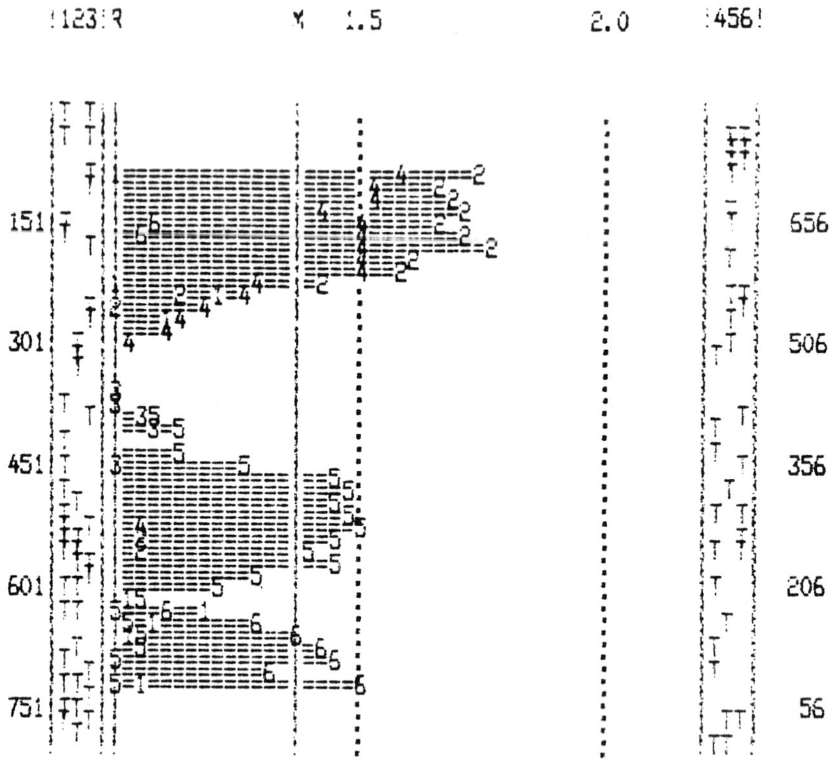

Figure 2 Reading frame analysis of S14. In this plot, the base numbers of the (+) strand are shown at the left margin and the base numbers of the (-) strand at the right margin. The horizontal bars are topped by numbers referring to the reading frame. The reading frames of the (+) strand are indicated at the left and the reading frames of the (-) strand at the right. T indicates a termination codon. The probability of actual usage of the reading frame in question for a region of 50 amino acids around the position of a horizontal bar is proportional to the length of that horizontal bar. These probabilities were calculated by the IBI-Pustell programs from a codon usage table constructed from 19 yeast gene sequences.

Analysis of Large T_1 Products of Transcript

If transcription in vitro is due to completion of already initiated RNAs, the specific radioactivity of those regions of the transcript from the 3' end of completed strands will be higher than that of those regions from the 5' end. This hypothesis was tested by isolation of large T_1

oligonucleotides from the labeled transcript made in vitro. We have deduced the locations of all but one of the previously reported large T_1 oligonucleotides of S14 (Kane et al., 1979). The large T_1 oligonucleotides predicted to arise from the (-) strand were indeed missing from the T_1 fingerprint of the S14 transcript, while the T_1 oligonucleotides predicted to arise from the (+) strand (spots 1, 6, 9, 10, 12, 13, 16, and 45) were present in the transcript. They were identified by their characteristic migration in the two-dimensional gel system and by their pancreatic RNase products (data not shown).

The relative molar amounts of the T_1 oligonucleotides are shown in Table 1, in which correction has been made for the number of A residues in each oligonucleotide, since labeling was with [^{32}P]ATP. All the T_1 oligonucleotides are present at the same molar amount, within experimental error. We conclude that synthesis of transcript occurs in vitro throughout the region tested (bases 52-793) at about the same efficiency. Transcription in vitro does not appear to be due to elongation of previously initiated strands, although the use of very short primers cannot be excluded by these data.

Analysis of Transcript by Hybridization to cDNA Clones

There are no large T_1 oligonucleotides in the first 52 bases of the transcript, so we were unable to establish whether synthesis of this region was also occurring in vitro by analysis of T_1 products. Since the transcriptase pause products from viral particles containing L_1 (the capsid polypeptide encoding dsRNA) are smaller than 50 bases (Brennan et al., 1981) and since these products are present in vast

Table 1 Analysis of T_1 Oligonucleotides of S14 Transcript

T_1 oligonucleotide[a]	Bases	Relative molar amounts
1	52-83	1.0
9 + 16	284-297, 472-485	1.5
12	546-559	1.2
10 + 45	131-144, 588-601	0.79
6	695-712	1.0
13	781-793	1.3

[a]A total of 1.8×10^6 cpm was analyzed. Oligonucleotides 9 and 16, and 10 and 45 were not resolved in this experiment and were counted together.

molar excess in the particles prior to in vitro transcription (Brennan et al., 1981; L. Bobek and J. Bruenn, unpublished data), they might serve as primers for transcription. There do not appear to be such pause products for the ScV-S14 transcript. If the pause products did serve as primers, the first 50 bases of the transcript should be present in much lower molar amount among the labeled products of in vitro transcription.

We tested this prediction by hybridization of transcript synthesized in vitro to M13 cDNA subclones of S14 from every region of the RNA, with the (-) strand present in the phage. Unhybridized regions were removed by treatment with pancreatic RNAse in 6 × SSC. The quantitation of this data (Table 2) shows that the maximum relative molarity of each region of the S14 transcript is the same, within experimental error. We conclude that every region of the transcript, including the 5' 45 bases, is labeled with approximately equal efficiency.

Determination of the 5' End of the Transcript

Our hybridization results do not eliminate the possibility that transcription initiates with small (less than 20-base) primers. However, if such primers were used, the 5' end of the transcript would not be labeled in vitro. We therefore isolated the 5' termini of full-sized transcript labeled in vitro by complete hydrolysis with NaOH followed by high-voltage DEAE paper electrophoresis. The 5' terminal nucleotide, a multiphosphorylated G, is labeled in vitro by both [^{32}P]GTP and ATP but not by UTP, as expected from the sequence of S14 (Table 3). This nucleotide was identified as ppGp by P1 and bacterial alkaline phosphatase digestions. The molar recovery of the 5' nucleotide was the same regardless of whether labeling occurred for 30 min or for 3 hr in vitro. We conclude that transcription does initiate in vitro, and that the 5' terminal nucleotide of the S14 and L transcripts is ppGp.

Mechanism of Transcription and Replication

There are two radioisotopic methods to distinguish conservative from semiconservative viral RNA synthesis by virions in vitro. One method uses in vivo [^{32}P]-labeled virions. If semiconservative (displacement) synthesis occurs, then labeled viral (-) strands should appear in the transcription products in the absence of labeled precursors. This method gave an estimate of a maximum of 2% of particles involved in semiconservative synthesis (Bruenn et al., 1980). More recent experiments gave an estimate of a maximum of 0.7% of the particles involved in semiconservative transcription (J. D. Reilly and J. Bruenn, unpublished data, 1984). These results are less definitive than those obtained by a second method. This second method is to determine the nature of the labeled dsRNA that results from RNA synthesis in vitro by un-

Table 2 Hybridization Analysis of S14 Transcript

Bases	Relative molar amount of transcript[a] (maximum)
1-45	0.53
57-169	0.68
165-229	1.0
224-347	1.0
376-668	0.65
612-793	0.82

[a]The maximum relative molar amount is the relative efficiency of hybridization of S14 transcript labeled in vitro to the M13 cDNA clone homologous to the bases shown as corrected for the number of A residues (label was in [^{32}P]ATP). Unhybridized regions were removed by pancreatic RNAse digestion in 6 × SSC. Controls with the M13 vector DNA or with a plus strand insert gave no hybridization above background.

Table 3 Analysis of the 5' Terminus of ScV Transcripts

RNA	Label	Molar fraction of 5' nucleotide[a] (observed/expected)
S14	ATP	0.56
S14	GTP	0.62
S14	GTP	0.71
S14	UTP	None detected
L	GTP	0.70

[a]The total number of counts analyzed varies from 0.3 to 10.2 × 10^6 cpm. (*Source*: Nemeroff and Bruenn, 1987.)

labeled viral particles in the presence of labeled precursors. Semi-conservative transcription would result in label incorporated into the viral (+) strand in template dsRNA. Conservative replication in "sub-viral" particles containing only the viral (+) strand would result in labeling of viral (-) strand. The ScV RNA polymerase activities are known to result in labeling of dsRNA to the extent of 1-3% of the total RNA synthesized (Herring and Bevan, 1977). We have characterized this labeled dsRNA.

ScV particles from the slower sedimenting portion of the sucrose gradient peak, enriched for ScV-S particles, were incubated under appropriate conditions with three nonradioactive NTPs and one [^{32}P]NTP (1 mCi at 619 Ci/mM). The isolated RNA was purified by CF11 chromatography and agarose gel electrophoresis. Both the S14 dsRNA (excised from the agarose gel) and the crude single-stranded fraction from CF11 were used as probes in separate Southerns with the M13 clones carrying the TaqI fragment (bases 376-668) in opposite orientations. The single-stranded RNA is primarily transcript, since it hybridizes to the TaqI fragment with the sequence of the (-) strand and to all other M13 cDNA clones with the S14 (-) strand (see above). A small amount (less than 4%) of the single-stranded RNA is (-) strand, since there is some detectable hybridization to the (+) strand as well. There is no detectable hybridization to M13 without an insert (Bruenn et al., 1985). In contrast, the S14 dsRNA of this experiment hybridizes mainly (92%) to the (+) strand, and is consequently (-) strand (Figure 3). The DNA on the filter is in molar excess to labeled RNA. These results are quantitated in Table 4: 96% of the single-stranded RNA is transcript and 92% of the label in dsRNA is (-) strand.

We confirmed, by a strand-separation experiment, the synthesis mainly of (-) strand in the dsRNA labeled by in vitro synthesis in this experiment. The strands of S14 are approximately equally labeled with ^{32}pCp by the reaction catalyzed by RNA ligase in vitro. In 5% PAGE of denatured S14, the faster-migrating strand is the (+) strand and the slower-migrating strand the (-) strand, as identified by hybridization to the M13 subclones and by direct RNA sequencing of ^{32}pCp labeled separated strands (Lee et al., 1986). The minus strand of S14 is labeled in dsRNA by the ScV polymerase activity in vitro in this experiment (Figure 4). No labeled (+) strand is apparent, although there are some degradation products present in both dsRNA preparations.

DISCUSSION

Only a small portion (44-82 basepairs) of the 5' region of the positive strand is common to all the DI dsRNAs, sequenced while a large region from the 3' end is common. This 3' region includes five repeats of a

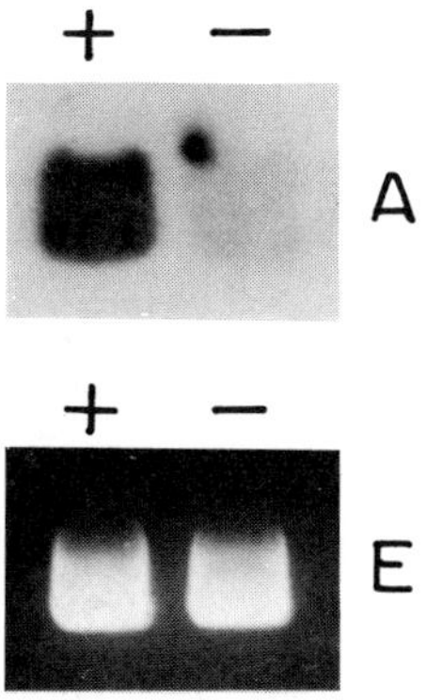

Figure 3 Southern of S14 cDNA probed with S14 dsRNA labeled by synthesis in vitro. The TaqI fragment (bases 376-668) is cloned in M13 mp9 with the (+) strand in the phage (+) or with the minus (-) in the phage (-). Phage DNA (10 μg) in each lane was probed with 2×10^5 cpm of denatured S14 dsRNA labeled by ScV particles in vitro. (A) Autoradiograph of Southern; (E) ethidium bromide stain of gel used for the Southern. (*Source*: Nemeroff and Bruenn, 1986.)

consensus sequence with dyad symmetry, (U/C)GCGAUUCGC(A/G), which is not present in the 5' region of M_1 (Bostian et al., 1984; Skipper et al., 1984) not in any of the sequenced regions of L_1 (Bruenn et al., 1985; Thiele et al., 1984b; J. Bruenn, unpublished, 1987). This suggests that this region may be important for recognition by proteins required for maintenance of M_1 but not for L_1. There are many such host factors (Wickner, 1983). The M dsRNA of ScV subtype 2 (M_2) is also dependent on some of these same gene products (Wickner, 1983), so we might expect M_2 to have repeats of this same sequence in the 3' region of its (+) strand. This sequence is present twice in the 209 basepairs from the 3' region of the M_2 (+) strand that have been

Table 4 Hybridization Analysis of S14 Products Synthesized In Vitro

	cpm above background hybridized to	
	Plus strand	Minus strand
ds S14	785	72
S14 transcript	117	2577

(*Source*: From Nemeroff and Bruenn, 1986.)

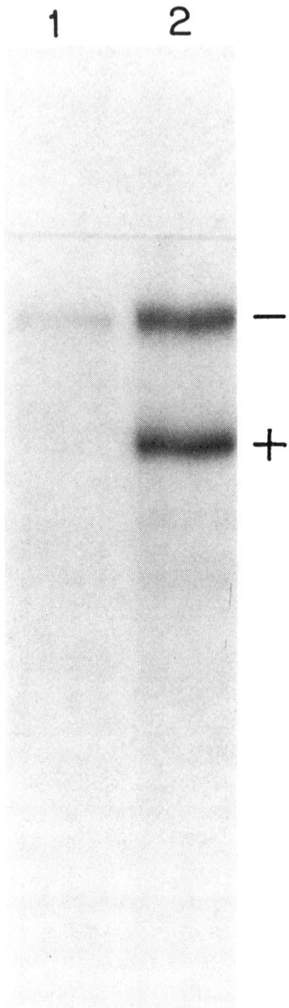

Figure 4 Strand separation of S14 dsRNA labeled by in vitro synthesis. 5% PAGE of denatured dsRNA. Lane 1 has denatured S14 dsRNA labeled by in vitro synthesis. Lane 2 has denatured S14 dsRNA labeled with ^{32}pCp by T4 RNA ligase. (+) strand and (-) strand are indicated. Direction of migration is top to bottom. (*Source*: Nemeroff and Bruenn, 1986.)

determined (Hannig and Leibowitz, 1985) at positions 49-59 and 142-152 from the 3' end of the (+) strand.

We have shown, by three independent techniques, that the ScV-S14 transcript synthesized in vitro is labeled at a constant specific activity throughout and that both the 5' terminus (ppGp) and the 3' terminus are labeled in vitro. Previous results (Bruenn et al., 1980) showed that $\gamma[^{32}P]GTP$ does not label the transcript, although the mature viral dsRNA has 5' pppGp on one or both strands (Bruenn and Keitz, 1976; Bruenn and Kane, 1978) and the L_1 pause products have a 5' multiphosphorylated G (Brennan et al., 1981). This result is now explained by absence of the γ-phosphate on the 5' terminal ppGp of the transcript synthesized in vitro. The identification of ppGp as the 5' terminus of the S14 and L transcripts synthesized in vitro is not inconsistent with previous results on the 5' termini of in vivo labeled ScV dsRNAs (Bruenn and Keitz, 1976; Bruenn and Kane, 1978). The viral dsRNAs remain encapsidated at all times, while the transcripts are extruded from the viral particles. Thus, the ScV phosphotransferase (Georgopoulos and Leibowitz, 1987) might remove the gamma phosphate from the 5' terminus of the transcript as it is extruded, or the viral particles may have other means for distinguishing (+) strands destined for translation from those destined for replication. It will be interesting to see if the viral (-) strand synthesized in vitro and present only in dsRNA duplexes (Nemeroff and Bruenn, 1986) has a pppGp or a ppGp 5' terminus.

The label appearing in ScV dsRNAs during incubation of particles with labeled precursors in vitro is almost entirely (-) strand. From these results, we calculate that the maximum fraction of particles that could be involved in semiconservative transcription is 4×10^{-5}, as opposed to the previous estimate of 2×10^{-2}. Thus, both transcription and replication in ScV appear to be conservative, like those processes in reovirus.

The small amount of labeled (+) strand present in the S14 dsRNA isolated after the in vitro RNA polymerase reaction may be due to contamination with transcript, since 98.7% of the incorporation is into transcript. There are no extensive inverted repeats in S14 (Lee et al., 1986), so we are not observing hybridization of (-) strand to itself. Strand separation gel electrophoresis confirms that the vast majority of the labeled dsRNA in this experiment is (-) strand. There are more interesting interpretations of this labeled (+) strand, however. In some experiments with faster sedimenting ScV particles, a greater proportion of the label in dsRNA is in the (+) strand. There may therefore be particles in which both (+) and (-) strands are synthesized, creating particles with two molecules of S14 dsRNA, as has been observed for M_1-containing particles (T. Fujimura, R. Esteban, and R. B. Wickner, personal communication, 1986).

ACKNOWLEDGMENTS

We thank D. Castanza and B. Keitz for technical assistance; J. Stamos for artwork; D. Kowalski for nuclease P1; K. Bostian and F. Sherman for synthetic oligonucleotides; NIH (grants GM22200 and BRSG S07 RR07066) and the Center for Applied Molecular Biology and Immunology of SUNY for support.

REFERENCES

Bostian, K. A., Hopper, J. E., Rogers, D. T., and Tipper, D. J. (1980). Translational analysis of the killer-associated virus-like particle dsRNA genome of *S. crevisiae*: M dsRNA encodes toxin. *Cell 19*:403-414.

Bostian, K. A., Elliott, Q., Bussey, H., Burn, V., Smith, A., and Tipper, D. J. (1984). Sequence of the preprotoxin dsRNA gene of type 1 killer yeast: Multiple processing events produce a two-component toxin. *Cell 36*:741-451.

Brennan, V. E., Bobek, L. A., and Bruenn, J. A. (1981). Yeast dsRNA viral transcriptase pause products: Identification of the transcript strand. *Nucleic Acids Res. 9*:5049-5059.

Bruenn, J., and Brennan, V. (1980). Yeast viral double-stranded RNAs have heterogeneous 3' termini. *Cell 19*:923-933.

Bruenn, J., and Kane, W. (1978). Relatedness of the double-stranded RNAs present in yeast-like particles. *J. Virol. 26*:762-772.

Bruenn, J., and Keitz, B. (1976). The 5' ends of yeast killer factor RNAs are pppGp. *Nucleic Acids Res. 3*:2427-2436.

Bruenn, J., Bobek, L., Brennan, V., and Held, W. (1980). Yeast viral RNA polymerase is a transcriptase. *Nucleic Acids Res. 8*: 2985-2997.

Bruenn, J., Madura, K., Siegel, A., Miner, Z., and Lee, M. (1985). Long internal inverted repeat in a yeast viral dsRNA. *Nucleic Acids Res. 13*:1575-1591.

El-Sherbeini, M., and Bostian, K. A. (1987). Viruses in fungi: Infection of yeast with the K1 and K2 killer viruses. *Proc. Nat. Acad. Sci. USA 84*:4293-4297.

Fried, H. M., and Fink, G. R. (1978). Electron microscopic heteroduplex analysis of "killer" double-stranded RNA species from yeast. *Proc. Nat. Acad. Sci. USA 75*:4224-4228.

Georgopoulos, D. E., and Leibowitz, M. J. (1987). Nucleotide phosphotransferase, nucleotide kinase and inorganic pyrophosphatase activities of killer virions of yeast. *Yeast 3*:117-129.

Guild, G., and Stollar, V. (1977). Defective interfering particles of Sindbis virus. V. Sequence relationships between SV_{SPD} 42S RNA and intracellular defective viral RNAs. *Virology 77*:175-188.

Hannig, E. M., and Leibowitz, M. J. (1985). Structure and expression of the M_2 genomic segment of a type 2 killer virus of yeast. *Nucleic Acids Res. 13*:4379-4400.

Hannig, E. M., Thiele, D. J., and Leibowtiz, M. J. (1984). *Saccharomyces cerevisiae* killer virus transcripts contain template-coded polyadenylate tracts. *Mol. Cell. Biol. 4*:92-100.

Herring, A. J., and Bevan, E. A. (1977). Yeast virus-like particles possess a capsid-associated single-stranded RNA polymerase. *Nature 268*:464-466.

Holland, J. J., Grauban, E. A., Jones, C. L., and Semler, B. L. (1979). Evolution of multiple genome mutations during long-term persistent infection by vesicular stomatitis virus. *Cell 16*:495-504.

Kane, W. P., Pietras, D., and Bruenn, J. (1970). Evolution of defective interfering double-stranded RNAs in yeast. *J. Virol. 32*: 692-696.

Kennedy, S. I. T. (1976). Sequence relationships between the genome and the intracellular RNA species of standard and defective interfering Semliki forest virus. *J. Molec. Biol. 108*:491-511.

Lee, M., Pietras, D. F., Nemeroff, M., Corstanje, B., Field, L., and Bruenn, J. A. (1986). Conserved regions in defective interfering viral double-stranded RNAs from a yeast virus. *J. Virol. 58*:402-407.

Nemeroff, M. E., and Bruenn, J. A. (1986). Conservative replication and transcription of yeast viral double-stranded RNAs in vitro. *J. Virol. 57*:754-758.

Nemeroff, M. E., and Bruenn, J. A. (1987). Initiation by the yeast viral transcriptase in vitro. *J. Biol. Chem. 262*:6785-6787.

Nomoto, A., Jacobson, A., Lee, Y., Dunn, J., and Wimmer, E. (1979). Defective interfering particles of polio virus. Mapping of the deletion and evidence that the deletions in the genomes of DI (I), (II), and (III) are located in the same region. *J. Mol. Biol. 128*:179-196.

O'Hara, P., Nichol, S., Horodyski, F., and Holland, J. (1984). Vesicular stomatitis virus defective interfering particles can contain extensive genomic sequence rearrangements and base substitutions. *Cell 36*:915-924.

Reilly, J. D., Bruenn, J., and Held, W. (1984). The capsid polypeptides of the yeast viruses. *Biochem. Biophys. Res. Commun.* *121*:619-625.

Sclafani, R. A., and Fangman, W. L. (1984). Conservative replication of yeast double-stranded RNA by displacement of progeny single strands. *Mol. Cell. Biol.* *4*:1618-1626.

Skipper, N., Thomas, D. Y., and Lau, P. (1984). Cloning and sequencing of the preprotoxin-coding region of the yeast M1 double-stranded RNA. *EMBO J.* *3*:107-111.

Stanway, C. A., and Buck, K. W. (1984). Infection of protoplasts of the wheat take-all fungus, *Gaeumannomyces graminis* var. *tritici*, with double-stranded RNA viruses. *J. Gen. Virol.* *65*:2061-2065.

Thiele, D. J., Hannig, E. M., and Leibowitz, M. J. (1984a). Genome structure and expression of a defective interfering mutant of the killer virus of yeast. *Virology* *137*:20-31.

Thiele, D. J., Hannig, E. M., and Leibowitz, M. J. (1984b). Multiple L double-stranded RNA species of *Saccharomyces cerevisiae*: Evidence for separate encapsidation. *Mol. Cell. Biol.* *4*:92-100.

Welsh, J. D., and Leibowitz, M. J. (1982). Localization of genes for the double-stranded RNA killer virus of yeast. *Proc. Nat. Acad. Sci. USA* *79*:786-789.

Welsh, J. D., Leibowitz, M. J., and Wickner, R. B. (1980). Virion DNA-independent RNA polymerase from *Saccharomyces cerevisiae*. *Nucleic Acids Res.* *8*:2349-2363.

Wickner, R. B. (1983). Genetic control of replication of the double-stranded RNA segments of the killer systems in *Saccharomyces cerevisiae*. *Arch. Biochem. Biophys.* *222*:1-11.

7

Transcription and Translation of the Yeast Killer Virus Genome

Michael J. Leibowitz, Iffat Hussain, and Teresa L. Williams
University of Medicine and Dentistry of New Jersey—Robert Wood Johnson Medical School, Piscataway, New Jersey

INTRODUCTION: KILLER VIRUS OF YEAST

Killer virus of yeast is a cytoplasmically-inherited virus conferring upon persistently infected cells of *Saccharomyces cerevisiae*, denoted killers, the ability to secrete a toxin that is lethal to uninfected yeast cells, but to which the killers themselves are resistant (reviewed by Bruenn, 1980; Bussey, 1981; Tipper and Bostian, 1984; Wickner, 1981, 1986). The viral genome consists of two segments of double-stranded (ds) RNA that are separately encapsidated in similar icosahedral virions (Bostian et al., 1980b). The M segment of dsRNA (approximately 1830 basepairs) encodes a polypeptide precursor (preprotoxin) to both toxin and resistance substance (Bostian et al., 1983b). The L-A dsRNA segment (approximately 4980 basepairs) encodes the major capsid protein in which both genome segments are encapsidated (Hopper et al., 1977; Sommer and Wickner, 1982). Most non-killer yeast strains harbor L-A dsRNA-containing virions, although a few totally lack dsRNA. Killer strains all harbor both M and L-A dsRNA; M dsRNA is never found in a strain not also harboring L-A dsRNA, which apparently functions as a "helper" for replication of the dsRNA viral genome segment. Multiple types of L-A dsRNA differing in their helper function for M dsRNA are known, although the molecular basis of their varying helper activities is not fully understood (reviewed by Wickner, 1986).

Multiple types of killer virus exist in yeast, each of which confers upon infected cells the ability to produce a toxin to which only cells infected with a homologous virus are resistant (Young and Yagiu, 1978). The two types which have been most extensively studied are the type 1 and type 2 killer viruses, each of which consists of a specific M dsRNA, denoted M_1 or M_2, respectively, which exists with an L-A dsRNA helper. Different types of L-A dsRNA may occur in different killer strains, but there is no evidence that a specific L-A dsRNA species is required for the replication of each type of M dsRNA (Hannig et al., 1985; Wickner, 1986). However, the L-A dsRNA from the two types of killer have been denoted L_1 and L_2 or L_{A1} and L_{A2} by some workers. The M_1 and M_2 dsRNA species are incompatible with each other. When both are introduced into the same cell by a mating event, the M_1 dsRNA outcompetes the M_2 species, which is "excluded" and thus is absent in progeny cells (Wickner, 1983). This may reflect competition between M_1 and M_2 for a factor or site required for replication.

The sequences of extensive regions of M_1, M_2, and L-A dsRNA molecules have been determined by direct structural analysis (Thiele and Leibowitz, 1982; Thiele et al., 1984a,b; Hannig and Leibowitz, 1985) and by analysis of cDNA clones (Skipper et al., 1984; Bostian et al., 1984; Bruenn et al., 1985; Georgopoulos et al., 1986). Although M_1 and M_2 dsRNA lack extensive sequence homology, they both contain an internal AU-rich region consisting of a 130-200-nucleotide-long polyA tract on the positive strand and a corresponding polyU tract on the negative strand (Fried and Fink, 1978; Hannig et al., 1984; Hannig and Leibowitz, 1985; Hannig et al., 1986). In addition both contain similar hairpin structures at the 5'-terminus of isolated positive strands, in regions which also include the initiating AUG of the first long open reading frame present (Hannig et al., 1984; Leibowitz et al., 1983; Hannig and Leibowitz, 1985). No similar structure exists on the 5' terminus of L-A dsRNA (Thiele et al., 1984a).

KILLER VIRUS GENE EXPRESSION

There is no DNA in killer or non-killer yeasts homologous to the viral dsRNA segments (Wickner and Leibowitz, 1977; Hastie et al., 1978). Therefore, nucleic acid hybridization methods can readily be used to study the in vivo and in vitro expression of the viral genome. Single-stranded positive polarity transcripts of the viral genome segments are present in RNA extracted from infected yeast cells. Full-length and subgenomic in vivo transcripts of M_1 dsRNA bind tightly to oligo(dT)-cellulose, indicating the presence of a polyA tract (Bostian et al., 1983a). Full-length transcripts derived from the genomic dsRNA might function biologically as templates for negative strand synthesis or as translational messages, while subgenomic transcripts, if not degradative artifacts, may function as messenger RNA. Killer virus replication is reviewed in Chapter 5.

Virions from killer yeast cells containing M_1 or M_2 and L-A dsRNA copurify with an RNA polymerase catalyzing the synthesis of full-length positive polarity transcripts of the genomic RNAs; non-killer cells har-boring L-A dsRNA contain similar virions with an enzyme catalyzing transcription of L-A (Herring and Bevan, 1977; Welsh et al., 1980; Welsh and Leibowitz, 1980; Bruenn et al., 1980). Most studies of the viral transcriptase utilized virions purified from infected cells grown into the stationary phase of growth, since these cells contain relatively little RNA polymerase activity catalyzing nucleotide incorporation into dsRNA (Nemeroff and Bruenn, 1986; Williams and Leibowitz, 1987). As discussed in Chapter 5, virions in a distinct class containing the dsRNA-synthesizing replicational RNA polymerase are more plentiful in extracts of actively growing cells (Esteban and Wickner, 1986; Fujimura et al., 1986). For the studies described in this chapter, virions from stationary-phase cells were utilized to maximize the activity of the transcriptase relative to the replicative RNA polymerase. There is no evidence that distinct proteins act catalytically in the transcriptional and replicative RNA polymerase activities found in different types of virions. Virions showing different activities differ in their RNA content, and "replicative" virions can become "transcriptional" after synthesis of their "headfull" of dsRNA (Esteban and Wickner, 1986; Fujimura et al., 1986). The polypeptides active in catalysis by the viral RNA polymerase have not been identified, nor is it known where the host or viral genes encoding them are located.

The transcriptase activity present in virions purified from stationary-phase yeast cells is experimentally useful due to its ability to catalyze the synthesis of large amounts of transcript, which can be purified to homogeneity. The full-length m transcript derived from M_1 dsRNA can be purified to near homogeneity by chromatography on oligo(dT)-cellulose. Although the m transcript lacks a 3'-terminal polyA tract, it binds very tightly to oligo(dT)-cellulose or polyU-Sepharose due to its internal polyA tract, which is encoded by the M_1 dsRNA segment (Hannig et al., 1984; Hannig et al., 1986); m transcript produced in vitro from M_2 dsRNA has similar properties (Hannig and Leibowitz, 1985). The sequence of the m in vitro transcript derived from M_1 dsRNA has been directly determined to be identical with the genomic positive strand at its 5' terminus, while its 3' terminus is identical except for the very last base, which is a mixture of A and G residues on the transcript but is nearly all A on the genomic positive strand (Hannig et al., 1984). This 3' terminal residue has no complement on the negative strand, since all yeast viral dsRNA molecules have a single 3'-terminal unpaired nucleotide which is nearly all A on both strands (Thiele and Leibowitz, 1982; Thiele et al., 1984a,b).

The in vitro m transcript derived from M_1 dsRNA can be translated by the reticulocyte cell-free translation system (Welsh and Leibowitz, 1982), to yield a 32 kDa polypeptide, denoted M-p32 or preprotoxin. Denatured M_1 dsRNA can be translated in various systems to yield an apparently identical polypeptide (Bostian et al., 1980a; Welsh and Leibowitz, 1982). In killer cells, preprotoxin is glycosylated and processed to yield toxin and resistance substance (Bostian et al., 1984) in a series of reactions requiring multiple host cell gene products as reviewed in Chapters 8 and 9 of this book. Preprotoxin can be synthesized in a yeast translation system in response to in vitro m transcript or denatured M_1 dsRNA (Hussain and Leibowitz, 1986). This provides the opportunity to study the translation of a homogeneous messenger RNA synthesized in vitro in a homologous translation system. This system is being used to study the mechanism of translation initiation and the regulatory signals that may control gene expression at this step.

VIRAL TRANSCRIPTION

Initiation in Vitro

Virions purified from stationary-phase killer yeast cells contain mainly transcriptional RNA polymerase activity, which catalyzes incorporation of nucleoside triphosphates into single-stranded RNA (Welsh et al., 1980; Bruenn et al., 1980; Williams and Leibowitz, 1987), with only 1-3% of incorporated nucleotides being incorporated into dsRNA. Although it is known that genomic positive strands and the m and l transcripts all have 5' terminal G residues (Bruenn and Keitz, 1976; Bruenn et al., 1980; Hannig et al., 1984; Thiele et al., 1984a; Hannig and Leibowitz, 1985), radioactivity from $[\gamma\text{-}^{32}P]GTP$ is not incorporated into m and l transcripts by such virions (Georgopoulos and Leibowitz, 1987). Since the high yield and long time course observed for transcript synthesis in vitro are consistent with chain initiation occurring in this reaction (Welsh et al., 1980; Welsh and Leibowitz, 1980), the failure to observe incorporation of radioactivity from $[\gamma\text{-}^{32}P]GTP$ is surprising. The failure to observe this incorporation is probably due to the presence of nucleoside triphosphate phosphotransferase activity in transcriptionally active virions purified from stationary-phase yeast cells. This activity copurified with virions containing M_1, M_2, or L-A dsRNA, as well as with more slowly sedimenting particles lacking RNA polymerase activity and dsRNA. No activity is found in a "mock" virion preparation purified from isogenic yeast cells lacking M and L-A dsRNA, indicating that the phosphotransferase activity does not reside in cosedimenting particles unrelated to the viruses. Under the conditions at which the RNA polymerase reaction is run, the virion-associated nucleoside triphosphate phosphotransferase activity catalyzes the rapid transfer of ^{32}P from [γ-^{32}P]GTP (or any other ribonucleoside triphosphate donor) to the other ribonucleoside triphosphates (or deoxyribonucleoside triphosphates) present in the reaction (Georgopoulos and Leibowitz, 1987). In the reactions in which we unsuccessfully attempted to detect initiation by incorporation of [γ-^{32}P]GTP into RNA, no detectable radioactive GTP remained after incubation (Hannig, 1985). Thus, nucleoside triphosphate phosphotransferase activity explains our failure to detect transcription initiation in vitro. This result still leaves unanswered the question of whether transcription catalyzed by purified virions can initiate new RNA chains in vitro.

In order to answer this question, a nucleoside triphosphate analog was needed which was not active as a substrate for nucleoside triphosphate phosphotransferase yet could still function as a substrate for RNA synthesis in the viral transcriptase reaction. In isolated yeast nuclei, the γ-thio derivatives of ribonucleoside triphosphates resist action of phosphotransferases and kinases, yet serve as substrates for RNA polymerase (Ide, 1981). Therefore, we tested γ-thio-GTP and γ-thio-ATP as both donors and acceptors in the killer-virus-associated nucleoside triphosphate phosphotransferase reaction. The phosphotransferase reactions were run for 1.5 hr at 30°C in the presence of 50 mM Tris-HCl (pH 7.5), 10 mM 2-mercaptoethanol, 100 mM NaCl, 10 mM $MgCl_2$, 0.1 mM EDTA, and virions purified as described (Thiele et al., 1984a), with 20 μM radioactive nucleoside triphosphate donor and 0.5 mM nonradioactive nucleoside triphosphate acceptor. The reactions were stopped by spotting samples (1 μl) on polyethyleneimine (PEI)-cellulose plates (MN 300, Machery-Nagel, Brinkmann Instruments) and nucleoside triphosphates were fractionated by ascending thin layer chromatography. After development of the chromatograms, standard nucleotides were identified utilizing light absorbance at 254 nm, radioactive spots were localized by autoradiography, and each was cut out and quantitated by liquid scintillation spectrometry. Transfer of radioactivity from [^{35}S]-γ-thio-ATP or [^{35}S]-γ-thio-GTP to various nucleoside triphosphate acceptors occurred at 2-5% of the rate of transfer of ^{32}P from [γ-^{32}P]ATP or [γ-^{32}P]GTP to the same acceptors (Table 1). Similarly when nonradioactive γ-thio-GTP was utilized as acceptor for transferred ^{32}P from [γ-^{32}P]ATP, only a low level of transfer was observed and much of that was presumably due to GDP contamination known to be present in this thio-derivative preparation (Table 1). This GDP could act as an acceptor for transfer catalyzed by nucleoside diphosphate kinase activity also copurifying with the virions (Georgopoulos and Leibowitz, 1987).

When [^{35}S]-γ-thio-ATP or [^{35}S]-γ-thio-GTP were used as substrates in the usual RNA polymerase assay utilizing type 1 killer virions (Welsh et al., 1980), the results could not be quantitated by TCA-precipitable radioactivity due to high 0-time blanks. In these reactions,

Table 1 Relative Inactivity of γ-Thio-Nucleoside Triphosphate Substrates for Viral Nucleoside Triphosphate Phosphotransferase

Donor[a]	Acceptor	Product[b]	Isotope transfer (pmoles)	Percentage of control	Solvent[c]
[γ-^{32}P]ATP	GTP	[γ-^{32}P]GTP	8.5[d]	[100]	I
[γ-^{35}S]thio-ATP	GTP	[γ-^{35}S]thio-GTP	0.4	4.7	I
[γ-^{35}S]thio-GTP	ATP	[γ-^{35}S]thio-ATP	0.2	2.3	II
[γ-^{35}S]thio-GTP	CTP	[γ-^{35}S]thio-CTP	0.2	2.3	II
[γ-^{35}S]thio-GTP	UTP	[γ-^{35}S]thio-UTP	0.3	3.5	II
[γ-^{32}P]ATP	γ-thio GTP	[γ-^{32}P]GTP	1.1	12.9[e]	II

[a]Nonradioactive γ-thionucleoside triphosphates were from Boehringer Mannheim; [γ-^{35}S]thio-ATP (S.A. 650 Ci/mmole) was from Amersham; [γ-^{35}S]thio-GTP (S.A. 1022 Ci/mmole) was from New England Nuclear.
[b]In each assay, performed as described in the text, radioactive products were identified by thin layer chromatography on PEI-cellulose plates developed in the indicated solvent and localized by autoradiography and comparison with standards. Subsequently, spots were cut out and quantitated by liquid scintillation spectrometry. No standards were available for γ-thio-CTP and γ-thio-UTP, which were identified by Rf values slightly lower than for CTP and UTP standards.
[c]Solvent I was 0.75 M NaH_2PO_4 (pH 3.5); solvent II was 1 M ammonium bicarbonate.
[d]In assays of transfer of ^{32}P from [γ-^{32}P]ATP or [γ-^{32}P]GTP to each of the other ribonucleoside triphosphates under similar conditions (Georgopoulos and Leibowitz, 1987), the total range of phosphate transferred was 6.2 pmoles (ATP to UTP) to 12.6 pmoles (GTP to CTP).
[e]The relatively high acceptor activity of γ-thio-GTP is probably spurious, since this acceptor is contaminated at the 10-20% level by GDP (Boehringer-Mannheim product specifications), which can function as an acceptor for phosphate from a ribonucleoside triphosphate donor in a reaction catalyzed by nucleoside diphosphate kinase activity also present in the virions (Georgopoulos and Leibowitz, 1987).

0.2 μM [^{35}S]-γ-thio-ATP or [^{35}S]-γ-thio-GTP were incubated in the presence of 20 μM ATP or GTP, respectively, with the usual polymerase assay components. However, after such reactions (2 hr, 30°C), the RNA made in the presence of [^{35}S]-γ-thio-GTP was recovered by phenol extraction for further analysis. The m transcript isolated by affinity chromatography on oligo(dT)-cellulose (Hannig et al., 1984) was radioactive, as indicated by the presence of ^{35}S, which comigrated with m upon electrophoresis through a 1.5% agarose gel. When a similar reaction was run utilizing the same concentration of [^{35}S]-γ-thio-ATP as substrate, much less radioactivity was detected in the m transcript. This result suggests that specific initiation of m transcript occurs in vitro with γ-thio-GTP. Proof of this hypothesis will require sequence analysis to confirm that the incorporated ^{35}S is indeed on the 5' terminal nucleotide of the m transcript.

In contrast to the failure to detect incorporation of radioactivity from [γ-^{32}P]GTP in viral transcription in vitro, Fujimura et al., (1986) observed incorporation of radioactivity from this substrate into L-A dsRNA by replicative particles isolated from logarithmically growing cells of a *ski2* mutant, which overproduces dsRNA. This result suggests that there may be less nucleoside triphosphate phosphotransferase activity relative to RNA polymerase activity in these virions than in the stationary-phase virions we have isolated from wild-type strains. If such a difference exists, it might reflect differences in the genotypes of the strains used, the growth phase, or the virion purification method.

The ability to incorporate specifically modified substrate nucleotides into the 5'-terminal location of viral transcripts opens up the prospect of experimentally altering this site, which is believed to be the initial site of ribosome interaction with messenger RNA in translation (Kozak, 1983).

Conservative Transcription Mechanism

Transcription of the viral dsRNA genome to yield full-length copies might occur semiconservatively, with the newly synthesized strand displacing the genomic positive strand, which is then released from the virion, or conservatively, with both genomic strands persisting and only newly synthesized strands being released as transcript. In the in vitro transcription reactions only 1-3% of the incorporated nucleotides are in genomic dsRNA (Welsh et al., 1980; Bruenn et al., 1980), supporting the conservative transcription model, but not excluding semiconservative transcription if only 1-3% of the virions in the preparation are enzymatically active. In the case of in vitro transcription of L-A dsRNA by isolated "transcriptional" virions from logarithmically growing cells, this latter possibility was excluded based upon the failure of transcription to yield incorporation of radioactive nucleotides into L-A dsRNA over the time course required for the growth of l_A

transcript chains to full length (Fujimura et al., 1986). A small population of particles rapidly synthesizing transcript semiconservatively would be expected to incorporate nucleotides in dsRNA before the appearance of full-length single-stranded RNA product.

Another prediction of the conservative model of transcription is that genomic dsRNA strands should not be displaced as transcript during the transcriptase reaction. Radioactive virions were purified from stationary-phase cells grown in the presence of [^{32}P]phosphate, which contain genomic M_1 and L-A dsRNA with equal amounts of radioactivity in both strands as determined by electrophoretic strand separation (Thiele and Leibowitz, 1982; Thiele et al., 1984a). When these virions were used to catalyze transcription in the presence of [5,6-^{3}H]UTP (ICN Radiochemicals) only 1.4% of the ^{3}H was incorporated into dsRNA, while no detectable ^{32}P appeared in transcript single-stranded RNA (Williams and Leibowitz, 1987). Utilizing nonradioactive virions in reactions run in the presence of [α-^{32}P]UTP, the low levels of incorporation into dsRNA were further characterized by strand separation of this isolated minor product. It was found that both strands of M_1 dsRNA contained equal amounts of radioactivity, but the isotope was only incorporated into the negative strand of L-A dsRNA. This result confirms that at least for L-A dsRNA, replication is not semiconservative, since the low level of nucleotide incorporation into L-A dsRNA is not in the positive strand. Furthermore, it demonstrates that low levels of dsRNA synthesis occur in the virions used in these experiments by the same mechanism seen in the more plentiful replicative virions isolated from logarithmically growing cells (Esteban and Wickner, 1986). Similar results have been obtained with virions containing the S14 deletion mutant of M_1 dsRNA as discussed in Chapter 6 of this book.

TRANSLATION OF VIRAL MESSENGER RNA

Model of Translational Initiation

In prokaryotic cells, translation of messenger RNA initiates at both 5' proximal and internal AUG initiation codons which are preceded by a highly conserved sequence complementary to the 3' terminal region of the 16S RNA of the small ribosomal subunit. Messenger RNA-ribosomal RNA (rRNA) interaction based upon base pairing at these complementary sites appears to be the mechanism by which prokaryotic ribosomes recognize initiation sites on messenger RNA at which they bind (Shine and Dalgarno, 1974; Kozak, 1983). In contrast, eukaryotic ribosomes tend to initiate translation at the 5'-proximal AUG of monocistronic messenger RNA molecules, on which internal initiation of translation tends not to occur (Kozak, 1983). Eukaryotic ribosomes appear to associate specifically with 5' terminal "cap" structures on messenger

RNA, after which they migrate down the message while "scanning" for initiating AUG codons which are defined by their context of flanking sequences, with optimal initiation occurring with AUGs flanked by a purine at position-3 and G at position +4, where the A of the AUG is defined as position +1 (Kozak, 1983). Although the potential role of complementarity between messenger RNA and eukaryotic rRNA in translation initiation has been the subject of speculation based upon the presence of such sequences in some messages, it is clear that no equivalent of the ubiquitous Shine-Dalgarno interaction is universal between messenger RNA and rRNA molecules in eukaryotes (Kozak, 1983), and messenger RNA-rRNA hybridization is not the basis of ribosome binding in translation initiation.

In eukaryotic cells, the relative stability of messenger RNA has led to the speculation that translational control may play a more important role in the regulation of gene expression than is the case in bacteria. Although the exact mechanisms by which translational regulation occurs are not understood on a molecular level, there is evidence that expression of different genes can be controlled at the translational level (Fink, 1986; Hershey et al., 1986). The initiation step of translation is a reasonable point at which regulation of translation might occur (Hershey et al., 1986).

In order to allow differential regulation of expression of different classes of messenger RNA molecules, some specific aspects of their translation initiation signals may differ. It is known that "antisense" DNA hybridizing to messenger RNA can inhibit its translation, in a process called hybrid-arrested translation (Paterson et al., 1977). Similarly, "antisense" RNA can specifically inhibit translation (Singer et al., 1963; Pestka et al., 1984), and in *Escherichia coli* it appears that natural antisense RNA complementary to the 5' terminus of the *ompF* messenger inhibits its translation (Mizuno et al., 1984). This inhibition of translation by antisense RNA is mechanistically similar to the translation inhibitory effect of secondary structure of messenger RNA involving its 5' proximal regions, as has been documented for altered genes for iso-1-cytochrome *c* (*cyc1*) in yeast (Baim et al., 1985). The potential importance of messenger RNA secondary structure in regulating viral gene expression was pointed out by Joklik (1981), who suggested that the complementarity of the 3' termini of some reovirus messenger RNA molecules with their 5'-proximal translation initiation regions might result in intramolecular base-pairing, which might inhibit their translation (Li et al., 1980). It has been proposed (Joklik, 1981) that sites of complementarity to 18S rRNA in the translation initiation regions of such self-complementary reovirus messenger RNA molecules may represent sites at which intermolecular base-pairing with rRNA may destabilize messenger RNA intramolecular bonding which otherwise would inhibit translation, although these structures do not totally explain the differences in translational effectiveness between different reovirus messenger RNAs (Joklik, 1985). Thus, in the case of some messenger RNAs in eukaryotic cells, hybridization to complementary regions of rRNA may represent a mechanism of destabilizing inhibitory messenger RNA secondary structure (Leibowitz et al., 1983).

In the case of the M_1 (Leibowitz et al., 1983; Hannig et al., 1984) and M_2 dsRNA molecules (Hannig and Leibowitz, 1985), self-complementary sequences near the 5' termini cause formation of extensive 5'-terminal stem-and-loop structures in isolated positive strands or m transcripts formed in vitro, whose existence is confirmed by nuclease sensitivity studies. These secondary structures involve the regions surrounding the AUGs initiating the first long open reading frame on these molecules, which in the case of M_1 is known to encode the preprotoxin (M-p32) polypeptide (Bostian et al., 1984). No similar secondary structure is predicted around the 5' proximal AUG of the positive strand or l transcript derived from L-A dsRNA, and this lack of a 5' terminal hairpin is confirmed by nuclease sensitivity studies (Thiele et al., 1984a). The regions surrounding the 5' terminal structures of the transcripts of M_1 and M_2 dsRNA also contain regions complementary to the 3' termini of yeast 18S and 5.8S rRNA molecules (Leibowitz et al., 1983; Hannig and Leibowitz, 1985). As can be seen in Figure 1, rRNA hybridization to these complementary sites in the m transcript derived from M_1 dsRNA could weaken the stability of the 5'-terminal hairpin of the transcript, which might facilitate its function as a translational template. This type of rRNA-messenger RNA interaction of 5.8S rRNA involves the 3'-terminal portion of that molecule, which is also capable of stably base pairing with the 5' terminus of 28S rRNA of the large ribosomal subunit (Olsen and Sogin, 1982). The binding between mouse 5.8S and 28S rRNAs in ribosomes occurs in two conformational states which differ in stability (Peters et al., 1982) suggesting the possible availability of the 5.8S 3' terminus for other types of interactions.

The model proposed for rRNA base pairing resulting in destabilization of messenger RNA secondary structure is potentially applicable to the transcripts of M_1 and M_2 dsRNA segments of the killer virus of yeast, some reovirus transcripts, and perhaps to a population of messages sharing the features of 5' proximal secondary structure and sites of rRNA complementarity. Figure 2 shows the complementarity of mammalian 18S and 5.8S rRNA for the l_2 messenger RNA of reovirus serotype 3. The ability of specific RNA hybridization to messenger RNA to stimulate its translational template activity can be viewed as an alternative regulatory mechanism to that proposed for "antisense" RNA, which in general seems to function as an inhibitor of translation (Paterson et al., 1977; Pestka et al., 1984; Mizuno et al., 1984). If rRNA does, indeed, function in translation in the proposed manner, then it is possible that other RNA or DNA molecules hybridizing at

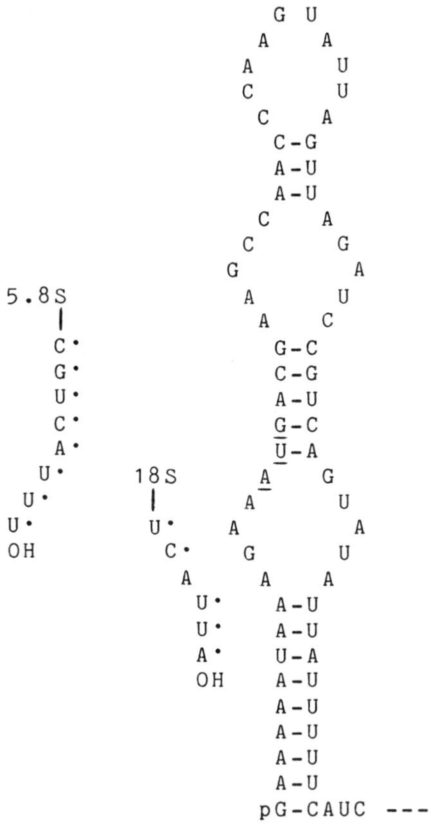

Figure 1 Secondary structure and rRNA complementarity of the 5' terminus of m transcript derived from M_1 dsRNA. The initiating AUG of the preprotoxin open reading frame at positions 14-16 is underlined. The 3'-terminal sequences of yeast 5.8S and 18S rRNA molecules are indicated, with dots indicating sites of complementarity with m, as previously described (Leibowitz et al., 1983). The hairpin structure has a free energy of -13 kcal, and the binding of 5.8S and 18S rRNAs have free energies of -14.0 kcal and -3.2 kcal, respectively.

specific regions of a messenger RNA might also activate its template activity, in a process we denote "hybrid-stimulated translation." This paper presents data indicating that hybrid-stimulated translation can occur in vitro in response to specific synthetic oligodeoxyribonucleotides. The possible role of naturally occurring RNA molecules capable of hybridizing to messenger RNA in translational regulation (Maxwell

```
l2  3'      HO- C U A C U U A A U C C G C G C G A G U ---
                        • • •       • • •
l2  5'   m7GpppG  C U A A A U G G C G C G A U G G C G A --
                        • • • • •               • • •
18S 3'             HO-A U U A C U A G G A A G G C G U C --
                              • • • • • •
5.8S 3'                   HO- U C G CUG C G A G ---
```

Figure 2 Proposed secondary structure and rRNA binding sites of l_2 messenger RNA of reovirus serotype 3. The reovirus sequence is described by Antczak et al. (1982). The mouse 5.8S rRNA sequence and molecular interactions are reviewed by Walker and Pace (1983); rabbit 18S rRNA sequence is from Lockard et al. (1982). In the proposed structure the free energies, according to the rules of Tinoco et al. (1973), are -9.2 kcal for 5'-3' binding by the message, -7.2 kcal for 18S rRNA binding, and -13.4 kcal for 5.8S rRNA binding. The initiating AUG is underlined. Dots indicate potential sites of complementarity to the 5' terminal region of the messenger RNA.

and Martin, 1986) raises the possibility of a variety of in vivo correlates of this phenomenon.

Translation of Viral Transcripts in a Yeast Cell-Free System

As was previously observed in other cell-free translation systems, the m transcript of M_1 dsRNA produced in vitro serves as a template for synthesis of preprotoxin in a message-dependent cell-free translation system derived from yeast cells (Hussain and Leibowitz, 1986). The extraction method used produces a reproducible translation system which is stable for over a year frozen at -70°C. The development of the yeast system is necessary to test the possible role of base pairing of 18S and 5.8S rRNA molecules with m transcript in its translation.

The preprotoxin synthesized in vitro in response to the m transcript has been identified by its 32 kDa size and its immunological reactivity to rabbit antibody against purified killer toxin. The system is also active with rabbit globin and Sindbis virus coat protein messenger RNAs. Thus, the same system translates these two capped messages and the apparently uncapped m transcript. However, the translation of all three messenger RNAs in the yeast system is inhibited by cap analogs, indicating that like other eukaryotic translation systems, the yeast system prefers to initiate at capped 5' termini (Hussain and Leibowitz, 1986). The failure to detect cap structures on the 5' termini of in vitro killer virus transcripts (Bruenn et al., 1980; Hannig et al., 1984; Thiele et al., 1984a) leaves open the possibility that capping may occur on similar transcripts in vivo. If viral transcription occurs in the cytoplasm of infected cells, however, this capping must occur in the cytoplasm rather than in the nucleus, unless the transcripts are transported into the nucleus or obtain their caps in a reaction distinct from the usual nuclear mechanism in yeast and other eukaryotes for capping cellular messenger RNAs (Itoh et al., 1983). Other cytoplasmically replicating viruses, such as influenza, obtain caps for viral messages by scavenging them from host messenger RNA molecules (Robertson et al., 1980).

Initiation Complex Formation

The first step in protein synthesis is the formation of an initiation complex in which the ribosome associates with the messenger RNA in the appropriate configuration to begin its translation. Initiation complex formation is a complex series of reactions representing the result of the action of multiple initiation factors that promote the association of messenger RNA, ribosomal subunits, and the initiator methionyl-tRNA, as recently reviewed (Moldave, 1985). In the usual translation reactions, the initiation complex is relatively short-lived, since elongation soon begins, with the translocation of the ribosome toward the 3' end of the messenger RNA. However, in the presence of antibiotics inhibiting elongation, initiation complexes may be formed in vitro that are stable enough to be isolated and studied. This has been done, for example, for the initiation complexes formed in the wheat germ cell-free translation system in the presence of sparsomycin, in reactions utilizing reovirus messenger RNAs as templates (Kozak, 1982). In these reactions, nuclease protection studies reveal the ribosomes to be bound to the region surrounding the AUG at which translation initiates.

In order to study initiation of translation in the yeast system, similar experiments were performed. As is shown in Figure 3, the association of radioactive m transcript synthesized in vitro with monomeric ribosomes was detected only in the presence of the elongation inhibitor anisomycin. This result indicates that such complexes can be isolated, and may be useful for structural studies to determine the physical nature of the ribosome-messenger RNA interaction. The possible role of RNA-RNA base pairing in this complex may be tested by means of chemical cross-linking, as previously done in the wheat germ system (Nakashima et al., 1980).

Hybrid-Arrested and Hybrid-Stimulated Translation

As discussed above, antisense RNA or DNA hybridizing near the initiation codon of a messenger RNA inhibits its translation in various systems in vivo and in vitro, in a process known as hybrid-arrested trans-

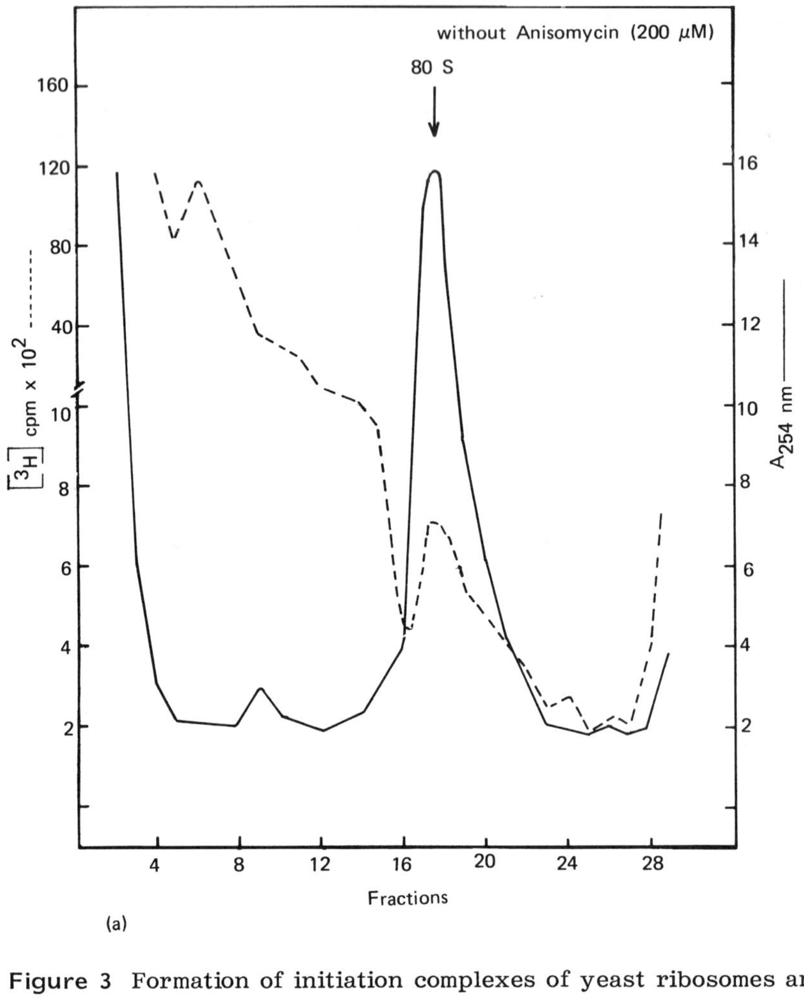

Figure 3 Formation of initiation complexes of yeast ribosomes and m transcript in vitro. Translation reactions were run as described (Hussain and Leibowtiz, 1986) for 15 min at 20°C in the presence of [^{3}H]m transcript synthesized in the presence of [5,6-^{3}H]UTP and purified by oligo(dT)-cellulose chromatography (Hannig et al., 1984) and utilizing a yeast extract not pretreated with micrococcal nuclease. After the reaction, a sample (150 μl) was layered on a 10-30% glycerol gradient (12 ml) in 20 mM Tris-HCl (pH 7.4), 75 mM KCl, 5 mM magnesium acetate, 1 mM dithiothreitol, and RNasin ribonuclease inhibitor (200 U/ml). Gradients were subjected to ultracentrifugation for 2.5 hr at 40,000 rpm at 4°C in a Beckman SW41 rotor. Fractions were collected from the top in an ISCO gradient collector, and absorbance at 254 nm

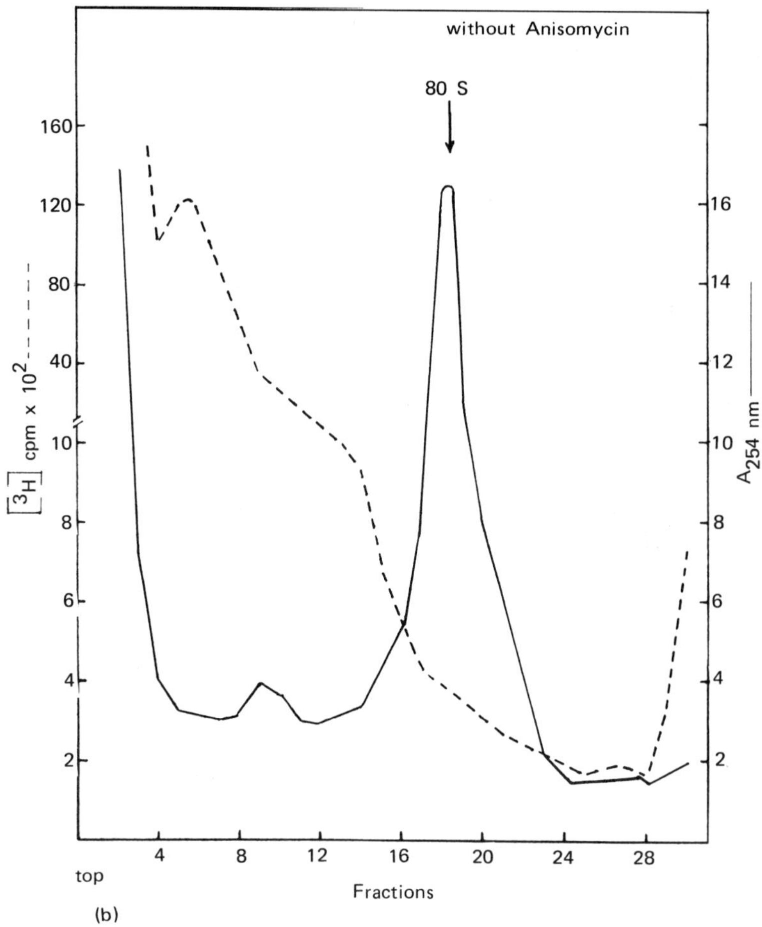

Figure 3 (Continued)

and radioactivity were determined for each fraction. (a) Shows the gradient of a reaction run in the presence of 200 μM anisomycin while (b) shows the gradient of a reaction run in the absence of anisomycin (Pfizer Pharmaceuticals). The 80S peak fractions of radioactivity in (a) were pooled, treated with 1% sodium laurylsarcosinate and proteinase K (100 μg/ml, 10 min, 37°C), and RNA was isolated by phenol: chloroform:isoamyl alcohol (25:24:1) extraction; chloroform:isoamyl alcohol extraction (24:1); and ethanol precipitation. Upon electrophoresis on a denaturing 1% agarose 6.7% formaldehyde gel (Lehrach et al., 1977), the major RNA species recovered comigrated with authentic m transcript.

lation. The distal side of the hairpin on the 5' terminus of the killer virus m transcript presumably acts in a similar manner to inhibit translation initiation on this template. In this system and others, rRNA hybridization to short regions of messenger RNA secondary structure is postulated to destabilize this structure and thus make the message more accessible as a translational template. Therefore, synthetic oligodeoxyribonucleotides were prepared, using an Applied Biosystems DNA Synthesizer (380B), with sequences capable of hybridizing to various regions of the hairpin on the m transcript derived from M_1 dsRNA. The effect of annealing these oligomers to the m transcript was determined in the yeast-cell-free translation system.

Kozak (1986) has observed that in the preproinsulin gene, secondary structure 5' to the initiating AUG seems to be more inhibitory to translation than distal secondary structures, implying that scanning ribosomes prior to initiation are more susceptible to displacement and inhibition due to messenger RNA base pairing than are actively translating ribosomes after initiation has occurred. It seems likely that oligonucleotides hybridizing anywhere on a messenger RNA might inhibit translation if binding to the messenger RNA were tight enough. However, while oligonucleotides binding 5' to the initiating AUG would be expected to inhibit translation in a dose-dependent manner, at critical ratios to messenger RNA, oligonucleotides hybridizing to distal parts of the 5' terminal hairpin of m transcript might stimulate its translation by destabilizing its secondary structure while not preventing ribosome translocation in the elongation reaction.

In order to test this model, the four DNA octamers shown in Figure 4 were synthesized on a 380B DNA Synthesizer (Applied Biosystems, Inc.), using cyanoethyl phosphoramidite chemistry. Oligonucleotide 90 is complementary to eight bases surrounding the initiating AUG; number 91 has a sequence complementary to number 90 and thus is complementary to five of eight bases on the 3' side of the hairpin; number 133 has the sequence complementary to eight bases on the 3' side of the hairpin; and number 92 has the reverse sequence of number 90 and should not base-pair with the m transcript.

The m transcript was prehybridized with each of the octamers and was subsequently translated in the yeast-cell-free system. As is shown in Figure 5 and Table 2, at oligomer:transcript molar ratio of 1000, only oligomer 90, which hybridized near the initiating AUG, inhibits synthesis of preprotoxin. However, as can be seen in Figure 6 and Table 3, at 10 to 20-fold molar excess, hybridized oligomers 91 or 133 resulted in 2 to 3-fold stimulation of translation of this transcript. This stimulation was less at lower or higher ratios of oligomers:m transcript. We suggest that low levels of oligomer were inadequate to destabilize the hairpin of m, while very high levels begin to impede ribosome translocation past the region of hybridization. This stimulatory effect is sequence-specific, since varying doses of oligomer 90 show a

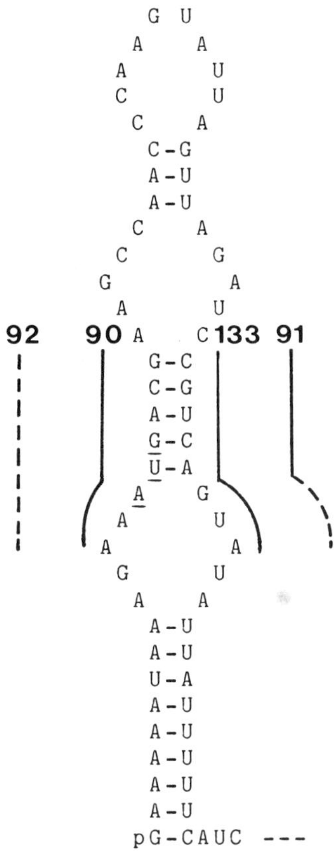

Figure 4 Oligomers tested for effects of hybridization on translation. The indicated four octomers were synthesized, with solid lines indicating sequences complementary to the hairpin structure of the m transcript, and dotted lines indicating noncomplementary sequences. The sequences (5'-3') are oligomer 90, CGTCATTT; oligomer 91, AAATGACG; oligomer 133, TACTGACG; and oligomer 92, TTTACTGC.

dose-dependent inhibition but no translation stimulation at low doses, while oligomer 92 fails to inhibit or stimulate at any dose tested.

Further experiments are now in progress testing the effects of longer oligomers as well as oligomers hybridizing to different sites. Also, physical evidence of the hybridization of the oligonucleotides at the predicted sites will be required to prove this model. However, the data are consistent with the model that the secondary structure at the 5' terminus of the m transcript inhibits translation, and destabilization of this structure improves translational template activity. This

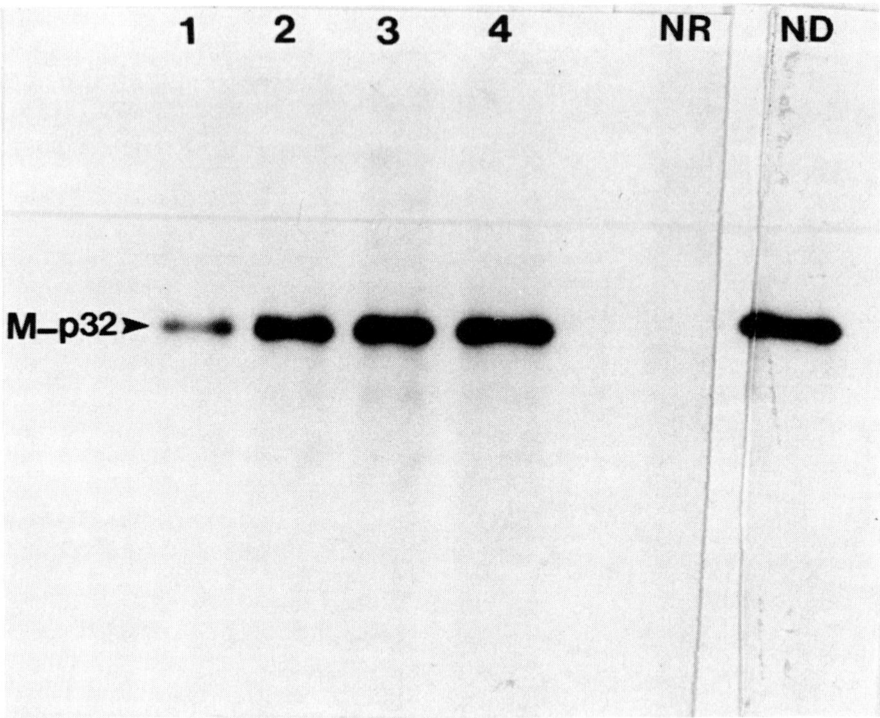

Figure 5 Effect of oligomers on translation of m transcript. Prior to translation, m transcript was hybridized for 1 hr at 37°C with a 1000-fold molar excess of the indicated octamer, in the presence of 1.6 mM magnesium acetate, 288 mM potassium acetate, 39.5 mM HEPES-KOH (pH 7.4), and oligo(dT)-cellulose-purified m transcript (10 μg/ml). Each sample was lyophilized, redissolved in water, and added to a yeast cell-free translation system (micrococcal nuclease-treated) run as described in the presence of [^{35}S]methionine (Hussain and Leibowitz, 1986). After 2-hr reactions at 20°C, reactions were terminated and proteins were separated on a 12.5% polyacrylamide gel run in the presence of SDS. Fluorography of this gel as described in Hussain and Leibowitz (1986) revealed that preprotoxin (M-p32) was the major product synthesized in reactions containing m transcript. Lane 1 included m hybridized with oligomer 90; lane 2, m with oligomer 133; lane 3, m with oligomer 91; and lane 4, m with oligomer 92. Lane NR contains proteins synthesized without added transcript or oligomer, and lane ND had proteins synthesized in response to m transcript mock-hybridized without oligomer.

Table 2 Inhibition of Translation of m Transcript by an Oligomer Hybridizing at the Site of Translation Initiation

Oligomer No.	Preprotoxin synthesis (cpm)[a]
90	539
133	1094
91	1520
92	1323
None	1226

[a]The bands of preprotoxin seen in Figure 5 were cut out and quantitated by liquid scintillation spectrometry. In a control reaction without added RNA, the corresponding gel fragment contained 300 cpm of ^{35}S.

example of hybrid-stimulated translation may occur on other translational templates, and represents an additional mechanism by which antisense nucleic acids may regulate gene expression in vitro and in vivo. The possible role of hybridization of 3'-terminal regions of 5.8S and 18S rRNA molecules with the 5'-proximal region of m transcript in initiation may represent a special example of this mechanism, which may occur in a subclass of messenger RNA molecules. The existence of distinct initiation mechanisms for translation of various messages offers a potential model for the differential regulation of expression of different genes at the translational level.

Cap Function in Killer Virus Translation

Although killer virus transcripts produced in vitro by the virion-associated RNA polymerase reaction have no detectable cap structures (Bruenn et al., 1980), most yeast messages do contain caps on their 5' termini (Sripati et al., 1976). Furthermore, the yeast cell-free translation system apparently prefers capped messages, since cap analogs inhibit translation (Hussain and Leibowitz, 1986). Since as shown above, [γ-^{35}S]thio-GTP seems to function as an initiating triphosphate in viral transcription in vitro, we tested whether m^{7}G5'ppp5'G (Pharmacia P-L Biochemicals) might not also serve as a primer for initiating capped RNA synthesis, as has been done in many DNA-dependent RNA polymerase reactions (for example, see Contreras et al., 1982). Transcription reactions utilizing twice-gradient-purified virions enriched in M dsRNA-containing particles (Williams and Leibowitz, 1987) were run as described (Welsh et al., 1980) in the presence of 20 μM

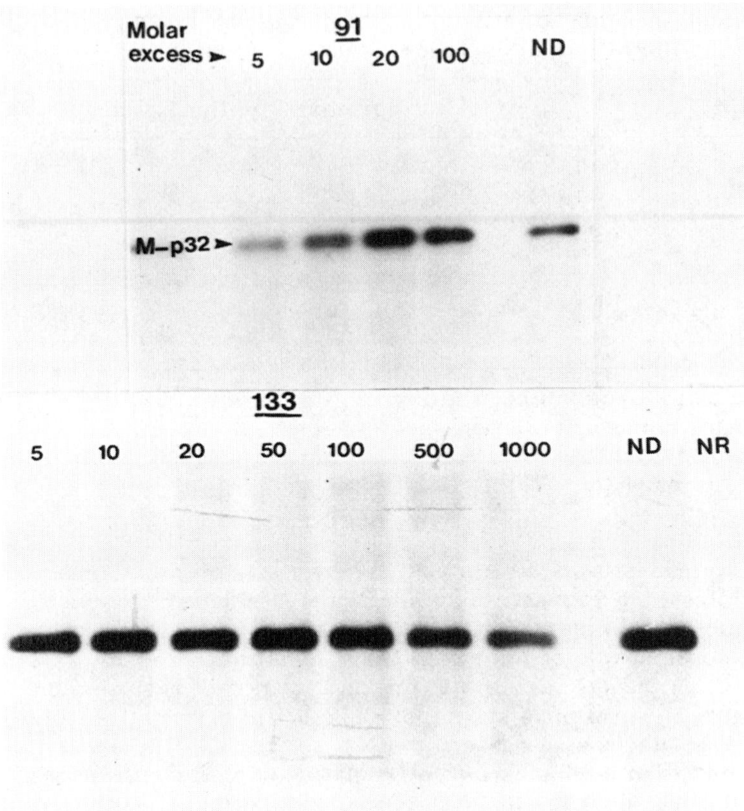

Figure 6 Hybrid-stimulated translation of m transcript. Varying doses of oligomers 91 and 133 were used as described in Figure 5, with the numbers on top of each lane indicating the molar ratio of oligomer to m transcript in each hybridization reaction. Lanes marked ND show proteins synthesized in response to m transcript mock-hybridized without oligomer, and lane NR contains proteins synthesized without added transcript or oligomer. The amount of m transcript used in the experiment with oligomer 133 was one-fifth that used with oligomer 91; different exposures are shown.

[5,6^{3}H]UTP with the addition of 500 μM m^{7}G5'ppp5'G. Reactions run with and without m^{7}G5'ppp5'G yielded the same amount of transcript product, which was purified by oligo(dT)-cellulose chromatography (Hannig et al., 1984). However, when both purified m transcripts were tested in equal amounts for translational template activity in the yeast system (Hussain and Leibowitz, 1986), the amount of preprotoxin

Table 3 Hybrid-stimulated Translation of m Transcript

Experiment	Oligomer	Molar excess[a]	Preprotoxin synthesis (cpm)[b]
I	91	5	1527
	91	10	2134
	91	20	2477
	91	100	1920
	None	—	1076
II	133	5	822
	133	10	1560
	133	20	1020
	133	50	765
	133	100	691
	133	500	588
	133	1000	477
	None	—	507

[a]The molar ratio of oligomer: m transcript is indicated.
[b]The bands of preprotoxin seen in Figure 6 were cut out and quantitated by liquid scintillation spectrometry. In a control reaction without added RNA, the corresponding gel fragment contained 155 cpm of ^{35}S.

made, quantitated as in Tables 2 and 3, was 14-fold greater with template synthesized in the presence of m^{7}G5'ppp5'G than in its absence (I. Hussain and M. J. Leibowitz, 1986, unpublished data). This result suggests that such caps may act as primers for viral transcription in vitro and supports the notion that capped messages are preferred by yeast ribosomes. Structural characterization of these transcripts will be required to confirm this claim.

It is unknown if there is any in vivo correlate of the putative cap-primed transcription observed in vitro in the killer virus system. However, the viral genome is believed to be transcribed in the yeast cytoplasm and the known RNA capping activity of yeast is nuclear (Itoh et al., 1983), raising the question of whether a specific mechanism exists for capping viral messenger RNA in vivo. Specific viral mechanisms for capping cytoplasmic messages, such as the "scavenging" of preformed 5' termini of host-capped messages for attachment to messages produced by influenza virus (Robertson et al., 1980) are a precedent for this type of mechanism.

CONCLUSION

The cytoplasmically-inherited killer virus of yeast has proved useful as a model system for the study of viral gene expression in eukaryotic cells. Virions isolated from infected cells catalyze transcription of the viral genome in vitro, with a conservative transcription mechanism. Data presented here indicates that transcription initiates in vitro, and suggests that both γ-thio-GTP and $m^7G5'ppp5'G$ may act as primers for transcription initiation.

The availability of large quantities of full-length homogeneous viral transcript synthesized in vitro has facilitated study of translation in the yeast cell-free system. Based upon the primary and secondary structure of the m transcript, we have proposed that translation initiation of this message may involve base-pairing of the 3' termini of 5.8S and 18S rRNAs with the initiation region of the m transcript, resulting in destabilization of its secondary structure required for efficient translation to produce preprotoxin. The ability to isolate translation initiation complexes formed in vitro in this system offers the prospect of physically characterizing the RNA-RNA interactions occurring in complex assembly.

The ability of oligonucleotides complementary to distal portions of the 5' proximal hairpin of m transcript to stimulate template activity indicates that destabilization of this secondary structure by nucleic acid hybridization can stimulate translation. Hybrid-stimulated translation represents an additional mechanism by which antisense nucleic acids can alter gene expression. This process is functionally analogous to the mechanism by which "translational attenuator-mediated" regulation increases gene expression of the inducible "MLS" gene of *Staphlococcus aureus* (Dubnau, 1984; Weisblum, 1985). Effects of rRNA and other nucleic acids on messenger RNA secondary structure may also be a translational regulatory mechanism utilized by eukaryotic cells.

ACKNOWLEDGMENT

This research was supported by Public Health Service grant GM-32413 and grant 17-87 from the Foundation of the University of Medicine and Dentistry of New Jersey. T. L. W. is a predoctoral trainee supported by Public Health Service Institutional Research Service Award CA-09069 from the National Cancer Institute. Oligonucleotides used in this work were synthesized by Ms. Lizabeth Flores in the New Jersey Center for Advanced Biotechnology and Medicine DNA Synthesis Facility Network Laboratory.

REFERENCES

Antczak, J. B., Chmelo, R., Pickup, D. J., and Joklik, W. K. (1982). Sequence at both termini of the 10 genes of reovirus serotype 3 (strain Dearing). *Virology 121*:307-319.

Baim, S. B., Pietras, D. F., Eustice, D. C., and Sherman, F. (1985). A mutation allowing a mRNA secondary structure diminishes translation of yeast iso-1-cytochrome *c*. *Mol. Cell. Biol.* 5:1829-1846.

Bostian, K. A., Hopper, J. E., Rogers, D. T., and Tipper, D. J. (1980a). Translational analysis of the killer-associated virus-like particle dsRNA genome of *S. cerevisiae*: M dsRNA encodes toxin. *Cell 19*:403-414.

Bostian, K. A., Sturgeon, J. A., and Tipper, D. J. (1980b). Encapsidation of yeast killer double-stranded ribonucleic acids: Dependence of M on L. *J. Bacteriol. 143*:463-470.

Bostian, K. A., Burn, V. E., Jayachandran, S., and Tipper, D. J. (1983a). Yeast killer dsRNA plasmids are transcribed *in vivo* to produce full and partial-length plus-stranded RNAs. *Nucleic Acids Res. 11*:1077-1097.

Bostian, K. A., Jayachandran, S., and Tipper, D. J. (1983b). A glycosylated protoxin in killer yeast: Models for its structure and maturation. *Cell 32*:169-180.

Bostian, K. A., Elliott, Q., Bussey, H., Burn, V., Smith, A., and Tipper, D. J. (1984). Sequence of the preprotoxin dsRNA gene of type 1 killer yeast: Multiple processing events produce a two-component toxin. *Cell 36*:741-751.

Bruenn, J. A. (1980). Virus-like particles of yeast. *Annu. Rev. Microbiol. 34*:49-68.

Bruenn, J., and Keitz, B. (1976). The 5' ends of yeast killer factor RNAs are pppGp. *Nucleic Acids Res. 3*:2427-2436.

Bruenn, J., Bobek, L., Brennan, V., and Held, W. (1980). Yeast viral RNA polymerase is a transcriptase. *Nucleic Acids Res. 8*: 2985-2997.

Bruenn, J., Madura, K., Siegel, A., Miner, Z., and Lee, M. (1985). Long internal inverted repeat in a viral dsRNA. *Nucleic Acids Res. 13*:1575-1591.

Bussey, H. (1981). Physiology of killer factor in yeast. *Adv. Microb. Physiol. 22*:93-122.

Contreras, R., Cheroutre, H., Degrave, W., and Fiers, W. (1982). Simple efficient *in vitro* synthesis of capped RNA useful for direct expression of cloned eukaryotic genes. *Nucleic Acids Res. 10*:6353-6362.

Dubnau, D. (1984). Translational attenuation. The regulation of bacterial resistance to the macrolide-lincosamide-streptagramin B antibiotics. *Crit. Revs. Biochem. 16*:103-132.

Esteban, R., and Wickner, R. B. (1986). Three different M_1 dsRNA containing virus-like particle types in *Saccharomyces cerevisiae. In vitro* M_1 double-stranded RNA synthesis. *Mol. Cell. Biol. 6*:1552-1561.

Fink, G. R. (1986). Translational control of transcription in eukaryotes. *Cell 45*:155-156.

Fried, H. M., and Fink, G. R. (1978). Electron microscopic heteroduplex analysis of "killer" double-stranded RNA species from yeast. *Proc. Nat. Acad. Sci. USA 75*:4224-4228.

Fujimura, T., Esteban, R., and Wickner, R. B. (1986). *In vitro* double-stranded RNA synthesis in virus-like particles from *Saccharomyces cerevisiae. Proc. Nat. Acad. Sci. USA 83*:4433-4437.

Georgopoulos, D. E., and Leibowitz, M. J. (1987). Nucleotide phosphotransferase, nucleotide kinase and inorganic pyrophosphatase activities of killer virions of yeast. *Yeast 3*:117-129.

Georgopoulos, D. E., Hannig, E. M., and Leibowitz, M. J. (1986). Sequence of the M_1-2 region of killer virus double-stranded RNA. In *Extrachromosomal Elements in Lower Eukaryotes,* R. B. Wickner, A. Hinnebusch, L. Mets, A. Lambowitz, I. C. Gunsalus, and A. Hollaender (Eds.), Plenum, New York, pp. 203-214.

Hannig, E. M. (1985). *The Expression of Saccharomyces cerevisiae Killer-Associated Double-Stranded Ribonucleic Acids.* Ph.D. thesis, Rutgers University and University of Medicine and Dentistry of New Jersey, 174 pp.

Hannig, E. M., and Leibowitz, M. J. (1985). Structure and expression of the M_2 genomic segment of a type 2 killer virus of yeast. *Nucleic Acids Res. 13*:4379-4400.

Hannig, E. M., Thiele, D. J., and Leibowitz, M. J. (1984). Yeast killer virus transcripts contain template coded polyadenylate tracts. *Mol. Cell. Biol. 4*:101-109.

Hannig, E. M., Leibowitz, M. J., and Wickner, R. B. (1985). On the mechanism of exclusion of M_2 double-stranded RNA by L-A-E double-stranded RNA in *Saccharomyces cerevisiae*. *Yeast 1*:57-65.

Hannig, E. M., Williams, T. L., and Leibowitz, M. J. (1986). The internal polyadenylate tract of yeast killer virus M_1 double-stranded RNA is variable in length. *Virology 152*:149-158.

Hastie, N. D., Brennan, V., and Bruenn, J. A. (1978). No homology between double-stranded RNA and nuclear DNA of yeast. *J. Virol. 28*:1002-1005.

Herring, A. J., and Bevan, E. A. (1977). Yeast virus-like particles possess a capsid associated single-stranded RNA polymerase. *Nature 268*:464-466.

Hershey, J. W. B., Duncan, R., and Mathews, M. B. (1986). Introduction: Mechanisms of translational control. In *Translation Control*, M. B. Mathews (Ed.), Cold Spring Harbor Laboratory, Cold Spring Harbor, New York, pp. 1-18.

Hopper, J. E., Bostian, K. A., Rowe, L. B., and Tipper, D. J. (1977). Translation of the L-species dsRNA genome of the killer-associated virus-like particles of *Saccharomyces cerevisiae*. *J. Biol. Chem. 252*:9010-9017.

Hussain, I., and Leibowitz, M. J. (1986). Translation of homologous and heterologous mRNAs in a yeast cell-free system. *Gene 46*:13-23.

Ide, G. J. (1981). Nucleoside 5'-[γS]triphosphates will initiate transcription in isolated yeast nuclei. *Biochemistry 20*:2633-2638.

Itoh, N., Mizumoto, K., and Kaziro, Y. (1983). Partial purification and characterization of mRNA guanylyl transferase from *Saccharomyces cerevisiae*. *FEBS Lett. 155*:161-166.

Joklik, W. K. (1981). Structure and function of the reovirus genome. *Microbiol. Rev. 45*:483-501.

Joklik, W. K. (1985). Recent progress in reovirus research. *Annu. Rev. Genet. 19*:537-575.

Kozak, M. (1982). Analysis of ribosome binding sites from s1 message of reovirus: Initiation at the first and second AUG codons. *J. Mol. Biol. 156*:807-820.

Kozak, M. (1983). Comparison of initiation of protein synthesis in prokaryotes, eukaryotes, and organelles. *Microbiol. Rev. 47*:1-45.

Kozak, M. (1986). Selection of translational start sites in eukaryotic mRNAs. In *Translational Control*, M. B. Mathews (Ed.), Cold Spring Harbor Laboratory, Cold Spring Harbor, New York, pp. 35-41.

Lehrach, H., Diamond, D., Wozney, J. M., and Boedtker, H. (1977). RNA molecular weight determinations by gel electrophoresis under denaturing conditions, a critical reexamination. *Biochemistry 16*: 4743-4745.

Leibowitz, M. J., Thiele, D. J., and Hannig, E. M. (1983). Structure of separated strands of double-stranded RNA from killer virus of yeast: A translational model. In *Double-Stranded RNA Viruses*, D. H. L. Bishop and R. W. Compans (Eds.), Elsevier, New York, pp. 457-462.

Li, J. K.-K., Scheible, P. P., Keene, J. D., and Joklik, W. K. (1980). The plus strand of reovirus gene S2 is identical with its *in vitro* transcript. *Virology 105*:282-286.

Lockard, R. E., Connaughton, J. F., and Kumar, A. (1982). Nucleotide sequence of the 5'- and 3'-domains for rabbit 18S ribosomal RNA. *Nucleic Acids Res. 10*:3443-3457.

Maxwell, E. S., and Martin, T. E. (1986). Complementarity of sequences in low molecular weight RNAs to regions of messenger and ribosomal RNAs. *Nucleic Acids Res. 14*:5741-5760.

Mizuno, T., Chou, M.-Y., and Inouye, M. (1984). A unique mechanism regulating gene expression: Translational inhibition by a complementary RNA transcript (micRNA). *Proc. Nat. Acad. Sci. USA 81*:1966-1970.

Moldave, K. (1985). Eukaryotic protein synthesis. *Annu. Rev. Biochem. 54*:1109-1149.

Nakashima, K., Darzynkiewicz, E., and Shatkin, A. J. (1980). Proximity of mRNA 5'-region and 18S rRNA in eukaryotic initiation complexes. *Nature 286*:226-230.

Nemeroff, M. E., and Bruenn, J. A. (1986). Conservative replication and transcription of *Saccharomyces cerevisiae* viral double-stranded RNA. *J. Virol. 57*:754-758.

Olsen, G. J., and Sogin, M. L. (1982). Nucleotide sequence of *Dictyostelium discoideum* 5.8S ribosomal ribonucleic acid: Evolutionary and secondary structural implications. *Biochemistry 21*:2335-2343.

Paterson, B. M., Roberts, B. E., and Kuff, E. L. (1977). Structural gene identification and mapping by DNA. mRNA hybrid-arrested cell-free translation. *Proc. Nat. Acad. Sci. USA 74*:4370-4374.

Pestka, S., Daugherty, B. L., Jung, V., Hotta, K., and Pestka, R. K. (1984). Anti-mRNA: Specific inhibition of translation of single mRNA molecules. *Proc. Nat. Acad. Sci. USA 81*:7525-7528.

Peters, M. A., Walter, T. A., and Pace, N. R. (1982). Independent binding sites in mouse 5.8S ribosomal ribonucleic acid for 28S ribosomal ribonucleic acid. *Biochemistry 21*:2329-2335.

Robertson, H. D., Dickson, E., Plotch, S. J., and Krug, R. M. (1980). Identification of the RNA region transferred from a representative primer, β-globin mRNA, to influenza mRNA during *in vitro* translation. *Nucleic Acids Res. 8*:925-942.

Shine, J., and Dalgarno, L. (1974). The 3'-terminal sequence of *Escherichia coli* 16S ribosomal RNA: Complementarity to nonsense triplets and ribosome binding sites. *Proc. Nat. Acad. Sci. USA 71*:1342-1346.

Singer, M. F., Jones, O. W., and Nirenberg, M. W. (1963). The effect of secondary structure on the template activity of polyribonucleotides. *Proc. Nat. Acad. Sci. USA 49*:395-399.

Skipper, N., Thomas, D. Y., and Lau, P. (1984). Cloning and sequencing of the preprotoxin-coding region of the yeast M1 double-stranded RNA. *EMBO J. 3*:107-111.

Sommer, S. S., and Wickner, R. B. (1982). Yeast L dsRNA consists of at least three distinct RNAs; evidence that the non-Mendelian genes [HOK], [NEX], and [EXL] are on one of these dsRNAs. *Cell 31*:429-441.

Sripati, C. E., Groner, Y., and Warner, J. R. (1976). Methylated blocked 5' termini of yeast mRNA. *J. Biol. Chem. 251*:2898-2904.

Thiele, D. J., and Leibowitz, M. J. (1982). Structural and functional analysis of separated strands of killer double-stranded RNA of yeast. *Nucleic Acids Res. 10*:6903-6918.

Thiele, D. J., Hannig, E. M., and Leibowitz, M. J. (1984a). Multiple L dsRNA species of *S. cerevisiae*: Evidence for separate encapsidation. *Mol. Cell. Biol. 4*:92-100.

Thiele, D. J., Hannig, E. M., and Leibowitz, M. J. (1984b). Genome structure and expression of a defective interfering mutant of the killer virus of yeast. *Virology 137*:20-31.

Tinoco, I., Jr., Borer, P. N., Dengler, B., Levine, M. D., Uhlenbeck, O. C., Crothers, D. M., and Gralla, J. (1973). Improved estimation of secondary structure in ribonucleic acids. *Nature New Biol. 246*:40-41.

Tipper, D. J., and Bostian, K. A. (1984). Double-stranded ribonucleic acid killer systems in yeast. *Microbiol. Rev. 48*:125-156.

Walker, T. A., and Pace, N. R. (1983). 5.8S ribosomal RNA. *Cell 33*:320-322.

Weisblum, B. (1985). Inducible resistance to macrolides, lincosamides and streptogramin type B antibiotics: The resistance phenotype, its biological diversity, and structural elements that regulate expression -- A review. *J. Antimicrob. Chemother. 16 (Suppl. A)* 63-90.

Welsh, J. D., and Leibowitz, M. J. (1980). Transcription of killer virion double-stranded RNA in vitro. *Nucleic Acids Res. 8*:2365-2375.

Welsh, J. D., and Leibowitz, M. J. (1982). Localization of genes on the double-stranded RNA killer virus of yeast. *Proc. Nat. Acad. Sci. USA 79*:786-789.

Welsh, J. D., Leibowitz, M. J., and Wickner, R. B. (1980). Virion DNA-independent RNA polymerase from *Saccharomyces cerevisiae*. *Nucleic Acids Res. 8*:2344-2363.

Wickner, R. B. (1981). Killer systems in *Saccharomyces cerevisiae*. In *Molecular Biology of the Yeast Saccharomyces: Life Cycle and Inheritance*, J. N. Strathern, E. W. Jones, and J. R. Broach (Eds.). Cold Spring Harbor Laboratory, Cold Spring Harbor, New York, pp. 415-444.

Wickner, R. B. (1983). Killer systems in *Saccharomyces cerevisiae*: Three distinct modes of exclusion of M_2 double-stranded RNA by three species of double-stranded RNA, M_1, L-A-E and L-A-HN. *Mol. Cell. Biol. 3*:654-661.

Wickner, R. B. (1986). Double-stranded RNA replication in yeast: The killer system. *Annu. Rev. Biochem. 55*:373-395.

Wickner, R. B., and Leibowitz, M. J. (1977). Dominant chromosomal mutation bypassing chromosomal genes needed for killer RNA plasmid replication in yeast. *Genetics 87*:453-469.

Williams, T. L., and Leibowitz, M. J. (1987). Conservative mechanism of the *in vitro* transcription of killer virus of yeast. *Virology 158*:231-234.

Young, T. W., and Yagiu, M. (1978). A comparison of the killer character in different yeasts and its classification. *Antonie van Leeuwenhoek J. Microbiol. Serol. 44*:59-77.

8

Secretion and Action of Yeast K1 Killer Toxin

Howard Bussey, Charles Boone, Aleksandra Dmochowska, Deirdre Greene, Hong Zhu, and Susan J. Lolle
McGill University, Montreal, Quebec, Canada

Thierry Vernet, Daniel Dignard, and David Y. Thomas*
Biotechnology Research Institute, National Research Council of Canada, Montreal, Quebec, Canada

INTRODUCTION

Protein processing and secretion and the action of secreted proteins on target cells are central research themes in much current cell biology. Recently, considerable attention has been devoted to the application of genetic and molecular biological approaches to such questions. The yeast, *Saccharomyces cerevisiae*, is a eukaryotic organism receiving great attention: For example, the processing and secretion of both invertase and the alpha mating pheromone have been the subjects of intensive study (see Schekman, 1985 for a review). The use of viral systems, so powerful in both bacterial and mammalian molecular biology, has proved useful in yeast as well. The yeast killer toxin, processed and secreted along the yeast secretory pathway, and the action of the toxin on sensitive cells, serve as useful "viral windows" into these complex processes. A considerable body of knowledge now exists about killer toxin, and it has been comprehensively reviewed (Tipper and Bostian, 1984; Wickner, 1986). Here we attempt to summarize the system, and to describe recent findings on the processing and action of the protein precursor and its product, the secreted toxin, and place our understanding of them in the wider context of cell biology.

*Present affiliation: McGill University, Montreal, Quebec, Canada

PROCESSING AND SECRETION

Structure of the Preprotoxin Gene

Figure 1 schematically illustrates the structure of a cDNA copy of the M_1 preprotoxin gene and its protein product. As described elsewhere in this book, the open reading frame for this gene resides from bases 14-964 from the 5' end of the plus (+) strand of the M_1 dsRNA found in type 1 killer strains of *S. cerevisiae*. The gene confers both the K1 killer and immunity phenotypes on yeast cells, phenotypes which to a large extent have made easy analysis of these yeast dsRNA viruses possible. This toxin/immunity gene codes for a precursor protein (Bostian et al., 1984; Skipper et al., 1984) containing an N-terminal leader sequence followed by the α-subunit of the secreted toxin, an internal glycosylated γ-peptide and β-subunit of the toxin at the C-terminus of preprotoxin. The precursor, targeted by its signal peptide to the endoplasmic reticulum, enters the yeast secretory pathway, where it is glycosylated and proteolytically processed, and the toxin and the immunity components localized to their respective cellular compartments, the extracellular medium and (probably) the plasma membrane, respectively. The system provides a means of studying the events in this complex sequence from a molecular genetic perspective. A technical advance which has greatly assisted progress has been the cloning and expression of a cDNA copy of the gene (Lolle et al., 1984; Hanes et al., 1986). Such expression vectors containing the gene linked to a yeast promoter, such as the ADH-1, can be transformed into non-killer yeast strains, rendering them immune killers.

The Leader Peptide and Entry into the Secretory Pathway

Work with the temperature sensitive secretory defective *sec* mutants demonstrated that the toxin precursor followed the normal secretory pathway in yeast. No toxin was found to be secreted at the restrictive temperature in the *sec* mutants, although, under these conditions, the precursor was found to accumulate at various stages of the pathway, depending upon the mutation (Bussey et al., 1983; Lolle and Bussey, 1986). Several lines of evidence suggest that the N-terminal leader or some part of it directs the precursor protein to the endoplasmic reticulum by functioning as a coventional signal sequence. When the entire 44-amino-acid residue toxin leader sequence plus the first 8 amino acids of the α-peptide were fused to a bacterial cellulase, the cellulase was directed to the extracellular medium when expressed in yeast. Little cellulase secretion was seen in the absence of the killer toxin leader, indicating a signal-like function for the sequence (Skipper et al., 1985). Homologous killer precursor constructs with the leader region deleted show greatly reduced entry into the secretory pathway, as judged by failure to glycosylate the killer precursor, its cytoplasmic accumulation,

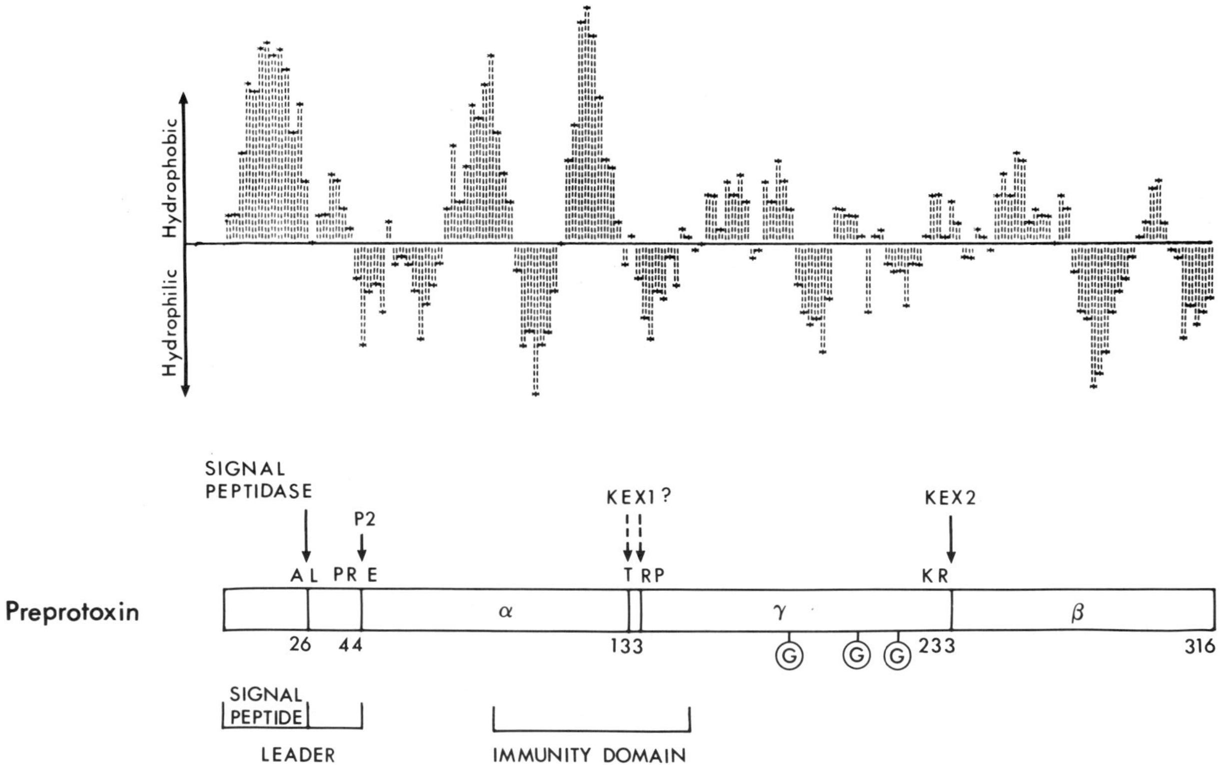

Figure 1 Diagrammatic outline of the structure of the killer toxin-immunity precursor. The joined open boxes represent the four polypeptides making up the precursor, the three circled Gs in the gamma polypeptide show the location of the glycosylated asparagine residues. The sites of precursor cleavage by proteases are indicated by the arrows, with the amino acid residues at the sites given in the single letter code. The numbers represent the relevant amino acid residues in the precursor. The positions of the leader and immunity domains are also shown. The plot above the precursor indicates its hydropathy profile; particularly prominent are the two hydrophobic domains in the alpha polypeptide.

and by greatly reduced secretion of toxin (Lolle and Bussey, 1986). Structural analysis of the leader suggested that nested within the 44 amino acid residues there was an N-terminal 26-amino-acid residue signal sequence (Bostian et al., 1984). Work with deletions of this signal sequence and with an expression system and early *sec* mutants suggests that this signal peptide is necessary for targeting entry to the secretory pathway and that it is cleaved at the endoplasmic reticulum (Lolle and Bussey, 1986). The fate or function of the remaining N terminus, the 27- to 44-amino-acid segment is unknown; presumably it is removed in the Golgi apparatus, where further protoxin protease processing occurs. A similar situation occurs in the mammalian proparathyroid hormone precursor (Potts et al., 1980), where an N-terminal peptide remains following signal cleavage, and is removed in the Golgi apparatus during maturation of the hormone. Removal of the remaining N-terminal fragment may be necessary for activation of these proteins.

Posttranslational Signal Cleavage and Translocation into the Endoplasmic Reticulum of the Killer Precursor

Studies with a killer cDNA expression system have enabled us to dissect some early protoxin precursors during their entry into the endoplasmic reticulum, which suggests that signal cleavage and possibly translocation occur posttranslationally. In this expression system, the overproduction of the toxin precursor from the ADH-1 promoter apparently overloads the entry process, causing precursor buildup. Three such precursor species can be identified and are summarized in Figure 2: a 35 kDa preprotoxin, the primary translation product; a 33 kDa protoxin with the signal removed but unglycosylated; and the 42 kDa glycosylated protoxin. Pulse-chase kinetics, tunicamycin inhibition of glycosylation, endoglycosidase H treatment, and the use of endoplasmic reticulum-blocked *sec* mutants, allow an ordering of these precursors in the entry process, as shown in Figure 2. The 35 and 33 kDa precursors can be detected at the restrictive temperature in the early endoplasmic reticulum-blocked *sec53* mutant, where the 35 kDa species was chased to the one at 33 kDa. At the later endoplasmic reticulum block caused by the mutation in *sec18*, the 35 and 33 kDa species were seen, and were chased to the 42 kDa species at the restrictive temperature. These data strongly suggest that the signal cleavage and the completion of translocation of the killer precursor can occur posttranslationally, as shown in the scheme in Figure 2. Recently, the use of an in vitro yeast translation-endoplasmic reticulum secretion system indicated that posttranslational entry of the yeast α-factor precursor into the endoplasmic reticulum occurred (Hansen et al., 1986; Rothblatt and Meyer, 1986; Waters and Blobel, 1986), and casts doubt on the ubiquity of the conventional view of

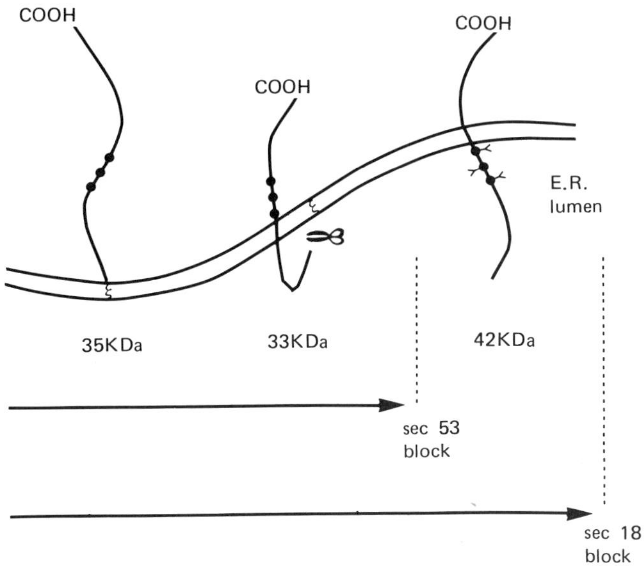

Figure 2 Model of early events in entry of the killer precursor into the yeast endoplasmic reticulum. The precursor is shown entering the lumen of the endoplasmic reticulum via a signal peptide. The proposed block points for the *sec* mutants are indicated by the dotted vertical lines at the ends of the arrows. Signal cleavage and glycosylation are also indicated in the scheme.

contranslational translocation and signal cleavage at the time of endoplasmic reticulum entry.

Processing Proteases

Perusal of Figure 1 indicates that there must be at least four proteolytic cleavages for maturation of the toxin from the precursor. One is caused by the signal peptidase, and a genetic approach has gone some way to defining the other cleavages, though as yet no complete picture has emerged. The starting point in this work was the isolation of mutants that failed to express the killer toxin, while maintaining the dsRNA and expressing the immunity phenotype (Wickner, 1974; Wickner and Leibowitz, 1976). Two nuclear genes were defined in this way, *KEX1* and *KEX2*, and subsequent work showed that mutants harboring defects in these genes were unable to process the toxin precursor. The ordering of the site of processing by these gene products of both the killer and α-factor precursors to the Golgi-secretory vesicle stage (Bussey et al., 1983; Julius et al., 1984b) was consistent with the idea that these genes defined processing proteases, and this has turned out to be the case.

KEX2

Early work (Leibowitz and Wickner, 1976), indicated the pleiotropic nature of the *kex2* mutants, which were sterile as *MAT*-α cells, unable to sporulate when homozygous, and which secreted many unprocessed proteins (Rogers et al., 1979). The finding that *kex2* mutants were *MAT*-α sterile due to their inability to process the α-factor pheromone (Julius et al., 1984a; Julius et al., 1984b) was a key element in initiating a classic series of combined molecular genetic and biochemical studies by the Thorner group, delineating the processing proteases involved in prepro-α pheromone maturation. The *KEX2* gene product, identified as a dibasic endopeptidase, is thought initially to cleave the dibasic amino acid-flanked, repeat elements of the pheromone precursor prior to the mature trimming of the 13-amino-acid pheromone by the *STE13* gene product, a dipeptidyl-aminopeptidase (Julius et al., 1983). In the killer toxin precursor, the *KEX2* gene product almost certainly cuts following the LysArg233 at the γ-β junction, as shown in Figure 1. Evidence to support these findings comes from defining a yeast-membrane-associated, calcium-dependent, sulphydryl, dibasic endoprotease activity absent from *kex2* mutants and increased in amount in yeast transformants containing the *KEX2* gene on a multicopy plasmid. Direct evidence would come from the action of the *KEX2* protein on the killer toxin precursor in vitro. The *KEX2* gene has been sequenced, and although not published, the main features of the deduced gene product are available (Fuller et al., 1985) and are summarized below. It is a large protein containing some 800 amino acid residues. The protein is 91 kDa and the purified protein is 105 kDa, suggesting glycosylation. There is a putative N-terminal signal peptide, followed some 200 residues into the protein by a cysteine-rich region, a negatively charged region which may be the calcium-binding domain, and some 600 residues into the protein a serine- and threonine-rich region that may be glycosylated. Toward the C-terminus of the protein there is a potential membrane-spanning-sequence just prior to amino acid residue 700. The remaining sequence, which may be on the cytoplasmic side of the membrane, has a cysteine residue near the C-terminus, which may be a site for acylation.

Studies on the action of this protease within the secretory pathway, for both the α pheromone and killer precursors, suggest that it is active at the Golgi compartment. It is tempting to speculate that this is its site of localization, though as yet no studies have addressed this important point. The protein has apparent similarities with the calpain class of mammalian proteases. It will be of great interest to see if in multicellular organisms, a homologous protein cleaves the dibasic residues commonly found at the cleavage sites of precursors of hormones and neuropeptides, whose in vivo processing proteases have proved elusive to biochemical approaches. Work on *KEX2* emphasizes again the value of genetic studies on simple systems for the nonlinear solution of biological problems, and satisfyingly justifies the pioneering efforts of the Wickner group in analyzing killer-defective mutants in yeast, which were thought to define unknown genes of general importance in eukaryotic cell biology.

KEX1

To further our understanding of protein processing in yeast we have begun an analysis of the *KEX1* gene. Unlike *KEX2*, this gene is not known to be pleiotropic, as the only phenotype of mutants is failure to process the K1 and K2 killer toxin precursors (R. B. Wickner, cited in Hannig and Leibowitz, 1985). It is our expectation that this gene product has some host cell function, and may well be a processing protease of wide biological significance.

We have cloned the *KEX1* gene by complementation of the *kex1* mutant phenotype, defective killer toxin expression; and shown that it is closely linked to the gene defined by *kex1* mutations on chromosome 7. The DNA sequence of the gene indicated an open reading frame coding for a protein of 729 amino acid residues (A. Dmochowska, unpublished results). The structural features of this protein are shown schematically in Figure 3 and described below. Following a potential N-terminal signal peptide, there is a domain of some 185 amino acid residues with considerable homology with yeast carboxypeptidase Y, a vacuolar serine sulphydryl exopeptidase. The homology extends to the sequence around the conserved serine at the active site, and

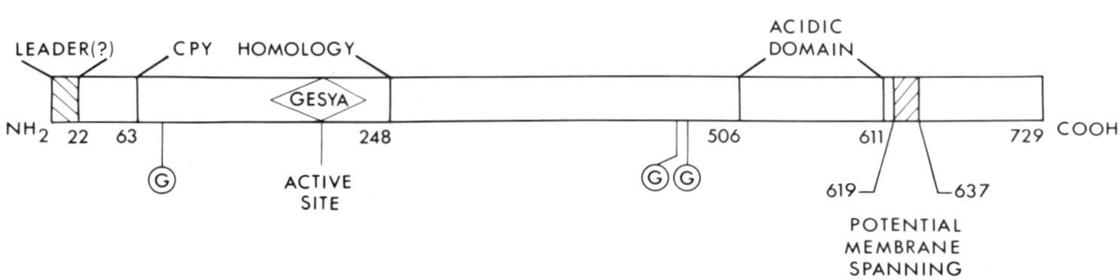

Figure 3 Schematic outline of the structure of the *KEX1* gene product. The potential leader region; the domain with homology to yeast carboxypeptidase Y, including the sequence around the serine at the active site; the acidic domain and the potential membrane spanning region are all indicated, with respect to the numbered amino acid residues in the protein; potential sites of asparagine-linked glycosylation are also shown.

strongly implicates the *KEX1* gene product as a protease. Downstream from the protease domain, there is a region with long stretches of acidic amino acid residues, and at residues 619-637, a hydrophobic, potential membrane-spanning-sequence prior to the C-terminus at residue 729. Three potential asparagine sites for glycosylation occur on the proposed N-terminal, luminal side of the protein. This structure is consistent with a membrane location (Golgi?) for the *KEX1* gene product, but this remains to be determined. The specificity of this protease is also unknown. It was originally suggested from inhibitor studies that the *KEX1* product would be a chymotrypsin-like protease; the protease inhibitors TPCK and ZPCK both inhibit killer toxin secretion by blocking precursor processing (Bussey et al., 1983; Bostian et al., 1984). Such inhibitors, which modify an essential histidyl residue, also abolish carboxypeptidase Y activity (Hayashi et al., 1975), and would also be expected to block the active site of the *KEX1* gene product. These inhibitor studies remain consistent with the *KEX1* protease acting directly on the killer precursor as substrate. A formal possibility remains, however, that the *KEX1* protease activates some proteolytic cascade culminating in some other protease that directly effects killer precursor cleavage.

The actual site of *KEX1*-dependent cleavage in the killer precursor is uncertain, but the P2, ProArg 44, and α-γ junction sites are candidates. Based on the TPCK inhibition data discussed above and the approximate size of the α-peptide, it was conjectured that Trp 130 was a probable site at the α-γ junction of preprotoxin (Bostian et al., 1984). The C-terminus of the α-peptide has not yet been determined and there are other potential cleavage sites in this region, including an ArgPro 135 site. The possibility that the *KEX1* protease also cleaves at the P2, ProArg 44, site of the leader remains open. In mammalian systems a "proline-directed" arginine cleavage protease has been postulated from a compilation of cleavage site data (Schwartz, 1986), and *KEX1* may code for the equivalent activity in yeast. A P2 deleted killer expression vector, with the precursor leader deleted and replaced with an acid phosphatase signal sequence, remains *KEX1*-dependent for killer toxin expression. This indicates that the P2 site is either not a *KEX1* site or not the only *KEX1* site, and implicates some other cleavage site, probably in the α-γ region, as *KEX1*-dependent.

The sequence homology of *KEX1* with carboxypeptidase Y is novel, since no other protease with this active site has so far been found. While carboxypeptidase Y is an exopeptidase, there is no reason to suppose that the *KEX1* protease need also be. The conserved active site in these two proteases conserves the charge-relay mechanism of peptide bond hydrolysis but not necessarily their substrate specificity (see Breddam, 1986, for a review). In yeast, carboxypeptidase Y is localized to the vacuole and requires activation by the vacuolar protease A, whose gene, PrA, is now known to be allelic to the *PEP4* locus, (Woolford et al., 1986; Ammerer et al., 1986). Activation requires the removal of an N-terminal prosequence from the carboxypeptidase Y precursor. This activation, and consequent *PEP4*-dependence, is not found for the *KEX1* protease, and no comparable prosequence is apparent in the determined *KEX1* sequence. Carboxypeptidase Y is found associated in cellular extracts with a 25 kDa polypeptide protease inhibitor (Matern et al., 1974). This inhibitor must be removed for in vitro assay of the protease, and may well serve as another means to regulate carboxypeptidase Y activity in vivo. Removal of such a specific protease inhibitor, which is likely bound to the protease at the active site, could also be a basis of activation of the *KEX1* protease. There is some genetic evidence compatible with this idea; *kex1* suppressors exist (isolated and characterized by Steven S. Sommer), which show an allele-specific partial suppression of *kex1* mutants. These *kos* mutants (*k*ex *o*ne *s*uppressors), could have defective protease inhibitors, resulting in increased *KEX1* protease function in the slightly leaky *kex1* mutants.

We remain ignorant of the normal cellular function of the *KEX1* protease. Disruption of the gene by large insertions or deletions in the genomic copy leads to failure to secrete killer toxin, although such mutant cells remain viable as haploids and diploids and retain the ability to mate in both mating types (A. Dmochowska, unpublished data).

KILLER TOXIN AND IMMUNITY ACTION

The finding that both the immunity and killing phenotypes are conferred by the cDNA expression plasmids containing the toxin precursor gene allowed a genetic approach toward defining the domains of the precursor involved in these processes.

Toxin Action

Previous work using genetic and biochemical approaches has outlined the action of the K1 killer toxin on sensitive yeast cells and this is summarized here, followed by some recent work. The secreted toxin binds initially to sensitive yeast cells by binding to a cell wall receptor containing a (1,6)-β-D-glucan moiety (Bussey et al., 1979; Hutchins and Bussey, 1983). Mutants in two nuclear genes, *kre1* and *kre2*, reduce toxin binding to the wall receptors and result in toxin resistance in the absence of the M_1 dsRNA (Al-Aidroos and Bussey, 1978). Both K1 and K2 toxins appear to use the *KRE1*-dependent wall receptor, but *kre2* mutants retain partial sensitivity to K2 toxin (Rogers and Bevan, 1978). Following wall binding, there are some energy-dependent processes culminating in lethal physiological changes consistent with the formation of toxin-dependent plasma membrane channels as the primary cause. In cells growing actively at pH 4.6, there is a rapid inhibition of net proton pumping from the cells and inhibition of potassium and amino acid uptake (de la Pena et al., 1980; de la Pena et al., 1981). With acidification of the cells, leakage of potassium ions and small metabolites follows (Skipper and Bussey, 1977). Evidence that killer toxins can form membrane channels comes from the in vitro black-lipid bilayer studies of Kagan (1983). This work was done with a toxin from the yeast *Pichia kluyveri*, having physiological action on yeast similar to the K1 toxin (Middelbeek et al., 1980). The *P. kluyveri* toxin formed conductance channels in the black-lipid bilayer; these channels were relatively nonspecific for ions, and were voltage-independent. Similar conductance channels were seen in bilayers treated with K1 killer toxin (B. Kagan, 1983, personal communication).

The in vivo processes leading to lethal membrane damage are poorly understood. There is a mutant, *kre3*, that binds toxin normally to the wall receptor and defines a candidate for a membrane function; others may exist. Heuristically, we have suggested (Bussey, 1981) that a plasma membrane protein receptor is necessary to mediate toxin action at the membrane level. Immunity, retained in spheroplasts and wall-receptor-independent, is also thought to be mediated at a plasma membrane receptor level (see below) where the toxin precursor is proposed to occupy this receptor, preventing toxin action (Boone et al., 1986). Operationally it is possible to divide the toxin action into two functional steps, wall binding and a "catch-all" post-wall-binding step, the killing of yeast spheroplasts. We have begun a mutational analysis of the toxin gene to identify domains essential for each of these functional steps.

Toxin Structure

Based on denaturing gel electrophoretic analyses, the toxin molecule is composed of a disulfide-linked α-γ dimer (Bostian et al., 1984; Vernet et al., 1986). To determine the nature of the native toxin we examined the toxin on nondenaturing gels and by crosslinking studies. The results of the former are shown in Figure 4, where it can be seen that the toxin is capable of forming multimers of the basic α-β dimer. A ladder of such multimers up to octamers can be seen under the conditions employed. It remains to be determined if these multimers are necessary for toxin action. In diphtheria toxin, multimeric species are also seen, but in that case they do not participate in the intoxication process (Carroll et al., 1986). Several modes of action are possible with a dimeric toxin-like killer; one mode of action would postulate that wall binding and channel formation are partitioned to these two polypeptide chains. For example, the α-peptide contains two hydrophobic, potentially membrane spanning, domains, and would be a likely candidate for the channel domain, the β-peptide being involved in wall

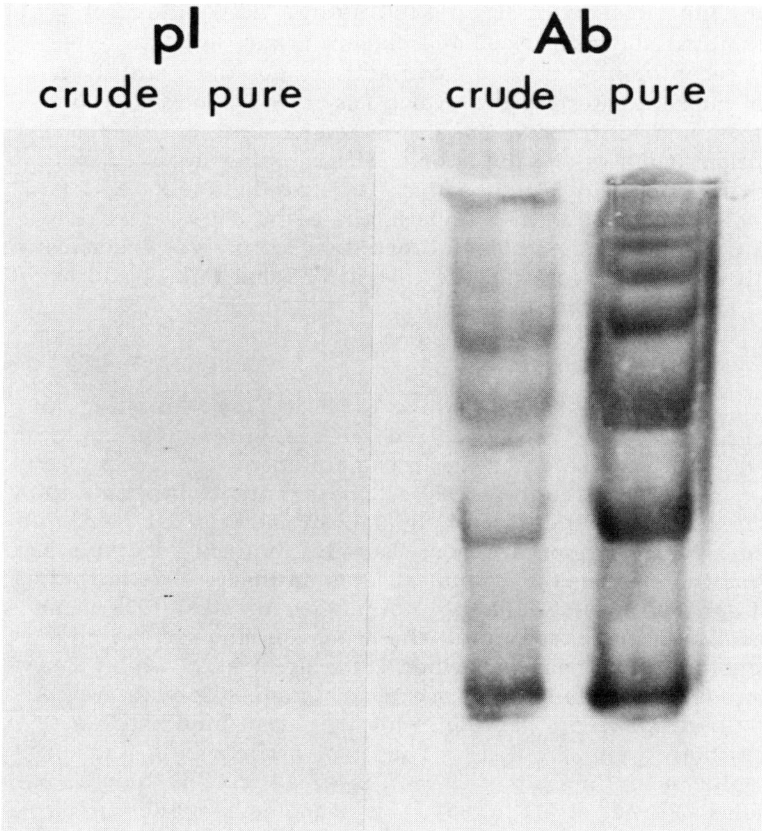

Figure 4 Multimeric structure of killer toxin shown by nondenaturing polyacrylamide gel electrophoresis. Crude or purified toxin was electrophoresed on a 15% acrylamide gel, using the Laemmli system without SDS, and proteins were transferred from the gels to nitrocellulose filters and immunoblotted with antitoxin (Ab) or preimmune antiserum (pI) in a Western blotting procedure. Bands immunospecific to toxin antibody were visualized histochemically using a Vectastain kit. The band with the greatest mobility represents the α-β dimer. The toxin oligomers form a ladder of decreasing mobility above.

receptor binding (Bostian et al., 1984; Bussey, 1984). Several toxins, such as the abrin and ricin class and cholera toxin, use this binary strategy of receptor binding and toxin activity domains, and it forms the basis of current attempts to develop tumor-cell-specific immunotoxins (Olsnes and Pihl, 1973; Mekalamos et al., 1983). Other mechanisms of action, such as the dimer or mutimers being involved in both steps of toxin action, are also possible. One prediction of the two-domain model would be that mutations in the α- or β-coding regions would affect wall binding or channel formation uniquely. Preliminary results with mutants directed to the α and β chains follow, and argue against the simple two-domain model.

Mutants Defective in Toxin Action

Toxins produced by mutants containing three in-frame insertions in the α coding region have been analyzed, all are defective in β-glucan binding, and two are also defective in the ability to kill spheroplasts (Z. Hong, unpublished results, 1986). These mutants implicate the α-peptide in both wall binding and channel formation. Data for mutations in the β coding region are more limited: One mutant with a small deletion near the carboxyl terminus secretes an inactive toxin that is able to kill spheroplasts (Z. Hong, unpublished results, 1986), and a second set of missense mutants at the amino terminus of β with a similar phenotype have been described (Sturley et al., 1986). Based on this limited data, mutants with membrane channel defects are confined to the hydrophobic α peptide, while β-glucan binding defects are found in both α and β chains. The involvement of the β chain in cell wall β-glucan binding has also been inferred from methionine oxidation studies (Vernet et al., 1986), this amino acid residue being found only in the β chain of the toxin. The formal possibility remains that only one peptide chain binds the glucan, but the other, to which it is bound by disulfide bonds, can nonlocally perturb it.

Molecular Basis of Immunity

Yeast strains harboring killer dsRNA virions are immune to the toxin they produce, and the immunity is virion-specific; for example, the K1 immunity system is not immune to the K2-produced toxin system and vice versa. The immune phenotype conferred by the K1 cDNA killer precursor expression vector on yeast transformants indicates that immunity is caused in some way by this precursor molecule or a component of it. This result had been anticipated, and from examination of the structure of the precursor the intersitial γ glycopeptide was suggested to be a candidate for the immunity component (Bostian et al., 1980; Bostian et al., 1984; Bussey et al., 1982; Bostian et al., 1983). To directly test this and alternative models of immunity action, site-directed mutations were made throughout the precursor gene to identify the region coding for the immunity domain (Boone et al., 1986). Mutations were made by site-directed mutagenesis, subcloned in a cDNA expression vector, and mutant plasmids transformed into a sensitive recipient yeast strain and the immunity phenotype scored. A clear result came from these studies: The immunity domain was determined to lie in a region spanning the α-γ part of the precursor (see Figure 1). Mutations perturbing immunity action, measured by plate or toxin survival tests, were confined to this region. The immunity domain, so defined, ruled out the idea that the γ glycoprotein alone was the immunity protein. The simplest tenable model consistent with the mutational analysis was that the precursor itself acted to confer toxin immunity. One plausible way in which it might do so was to form a complex with a plasma membrane receptor necessary for toxin action on sensitive cells. The precursor would then act as a competitive inhibitor of toxin action by binding to the receptor via the same site as the α-peptide of the mature toxin, possibly with the additional γ region increasing the affinity of the precursor for the receptor and forming a more stable complex than with toxin alone. In growing cells, one could envisage the precursor/receptor immunity complex forming during the progression through the secretory pathway following synthesis of the components, so that receptors would be occupied by precursors prior to active toxin processing late in the pathway. In this way, no unoccupied receptors would reach the plasma membrane to be available to interact with mature toxin in the growth medium.

Several phenotypes of killer mutants are consistent with this scheme. Many mutants have altered precursors which fail to be processed and which secrete no killer toxin, yet retain full immunity. The same phenotype is found in the *kex* mutants defective in necessary processing proteases. These mutants indicate that there is no requirement for precursor processing for immunity action. In addition, immunity-defective mutants in the α subunit of toxin that were able to secrete toxin made an inactive toxin, consistent with the hypothesis that there is a domain common to both toxin and immunity action.

A direct prediction of the model is that there will be a receptor, common to both precursor and mature toxin, and this can be examined experimentally, taking a combined genetic and biochemical approach. Candidates for the receptor would be the nuclear gene products of toxin resistance genes not involved in the cell wall binding steps such as *kre3* (see above) or some other toxin resistant mutants in other genes recently isolated by S. Sommer. The predicted receptor mutants should also perturb immunity function in an appropriate non-killer background. The proposed immunity model suggests an interesting new property of a precursor, that of modulating the function of its product on the producing cell. Such a property has implications for precursor regulation in the autocrine action of secreted proteins in multicellular organisms.

PROSPECTS FOR FURTHER WORK

While an outline of the secretion and action of the killer toxin is now available, the cellular processes illuminated by it remain far from being fully understood. By making available (in concert with work on other secreted proteins) mutants and genes involved in the yeast secretory pathway, we can expect further progress in defining events in this pathway, including entry steps, protein processing, and problems of cellular localization. The recent advent of homologous in vitro yeast translation/secretion/processing systems should be a powerful adjunct to the molecular approach. In terms of toxin action and the apparently interrelated one of immunity, many nuclear genes have been defined, and doubtless more remain, but little is known of their role in cellular function. These genes suggest an approach to various areas of cell surface biology: For example, the *kre1* mutant leads to defective (1,6)-β-D-glucan synthesis, and may offer an opening into the molecular biology of the cell wall—a complex organelle and the "final frontier" of the yeast cell.

ACKNOWLEDGMENTS

We thank Steven S. Sommer for providing unpublished information on mutants. We acknowledge support from NSERC operating and strategic grants for part of this work. Issued as National Research Council of Canada publication No. 26302.

REFERENCES

Al-Aidroos, K., and Bussey, H. (1978). Chromosomal mutants of *Saccharomyces cerevisiae* affecting the cell wall binding site for killer factor. *Can. J. Microbiol. 24*:228-237.

Ammerer, G., Hunter, C. P., Rothman, J. H., Saari, G. C., Valls, L. A., and Stevens, T. H. (1986). PEP4 gene of *Saccharomyces cerevisiae* encodes proteinase A, a vacuolar enzyme required for processing of vacuolar precursors. *Mol. Cell. Biol. 6*:2490-2499.

Boone, C., Bussey, H., Greene, D., Thomas, D. Y., and Vernet, T. (1986). Yeast killer toxin: Site-directed mutations implicate the precursor protein as the immunity component. *Cell 46*:105-113.

Bostian, K. A., Hopper, J. E., Rogers, D. T., and Tipper, D. J. (1980). Translational analysis of the killer-associated virus-like particle dsRNA genome of *S. cerevisiae*: M-dsRNA encodes toxin. *Cell 19*:403-414.

Bostian, K. A., Jayachandran, S., and Tipper, D. J. (1983). A glycosylated protoxin in killer yeast: Models for its structure and maturation. *Cell 32*:169-180.

Bostian, K. A., Elliott, Q. A., Bussey, H., Burn, V., Smith, A., and Tipper, D. J. (1984). Sequence of the preprotoxin dsRNA gene of type 1 killer yeast: Multiple processing events produce a two-component toxin. *Cell 36*:741-751.

Breddam, K. (1986). Serine carboxypeptidases. A review. *Carlsberg Res. Commun. 51*:83-128.

Bussey, H. (1981). Physiology of killer factor in yeast. *Adv. Microb. Physiol. 22*:93-112.

Bussey, H. (1984). Yeast killer toxin and an immunity protein are processed from a composite precursor. *Microbiol. Sci. 1*:62-66.

Bussey, H., Saville, D., Hutchins, K., and Palfree, R. G. E. (1979). Binding of yeast toxin to a cell wall receptor on sensitive *Saccharomyces cerevisiae*. *J. Bacteriol. 140*:888-892.

Bussey, H., Sacks, W., Galley, D., and Saville, D. (1982). Yeast killer plasmid mutations affecting toxin secretion and activity and toxin immunity function. *Mol. Cell. Biol. 2*:346-354.

Bussey, H., Saville, D., Greene, D., Tipper, D. J., and Bostian, K. A. (1983). Secretion of yeast killer toxin: Processing of the glycosylated precursor. *Mol. Cell. Biol. 3*:1362-1370.

Carroll, S. F., Barbieri, J. T., and Collier, R. J. (1986). Dimeric form of diphtheria toxin: Purification and characterication. *Biochemistry 25*:2425-2430.

de la Pena, P., Barros, F., Gascon, S., Ramos, S., and Lazo, P. S. (1980). Primary effects of yeast killer toxin. *Biochem. Biophys. Res. Commun. 96*:544-550.

de la Pena, P., Barros, F., Gascon, S., Lazo, P. S., and Ramos, S. (1981). Effect of yeast killer toxin on sensitive cells of *Saccharomyces cerevisiae*. *J. Biol. Chem. 256*:10420-10425.

Fuller, R. A., Brake, A. J., Julius, D. J., and Thorner, J. (1985). The KEX2 gene product required for processing of prepro-α-factor is a calpain-like endopeptidase specific for cleaving at pairs of basic residues. In *Protein Transport and Secretion*, M. J. Gething (Ed.), Cold Spring Harbor Laboratory, New York, pp. 97-102.

Hanes, S. D., Burn, V. E., Sturley, S. L., Tipper, D. J., and Bostian, K. A. (1986). Expression of a cDNA derived from the yeast killer preprotoxin gene: Implications for processing and immunity. *Proc. Nat. Acad. Sci. USA 83*:1675-1679.

Hannig, E. M., and Leibowitz, M. J. (1985). Structure and expression of the M2 genomic segment of a type 2 killer virus of yeast. *Nucleic Acids Res. 13*:4379-4400.

Hansen, W., Garcia, P. D., and Walter, P. (1986). In vitro protein translocation across the yeast endoplasmic reticulum: ATP dependent posttranslational translocation of the prepro-α-factor. *Cell 45*:397-406.

Hayashi, R., Bai, Y., and Hata, T. (1975). Evidence for an essential histidine in carboxypeptidase Y. Reaction with the chloromethyl ketone derivative of benzyloxocarboxyl-L-phenylalanine. *J. Biol. Chem. 250*:5221-5226.

Hutchins, K., and Bussey, H. (1983). Cell wall receptor for yeast killer toxin: Involvement of a (1→6)-beta-D-glucan. *J. Bacteriol. 154*:161-169.

Julius, D., Blair, L., Brake, A., Sprague, G., and Thorner, J. (1983). Yeast alpha-factor is processed from a larger precursor polypeptide: The essential role of a membrane-bound dipeptidyl amino peptidase. *Cell 32*:839-852.

Julius, D., Brake, A., Blair, L., Kunisawa, R., and Thorner, J. (1984a). Isolation of the putative structural gene for the lysine-arginine-cleaving endopeptidase required for processing of yeast prepro-α-factor. *Cell 37*:1075-1089.

Julius, D., Schekman, R., and Thorner, J. (1984b). Glycosylation and processing of prepro-α-factor through the yeast secretory pathway. *Cell 36*:309-318.

Kagan, B. L. (1983). Mode of action of yeast killer toxins: Channel formation in lipid bilayer membranes. *Nature 302*:709-710.

Leibowitz, M. J., and Wickner, R. B. (1976). A chromosomal gene required for killer plasmid expression, mating and sporulation in *Saccharomyces cerevisiae*. *Proc. Nat. Acad. Sci. USA 73*:2061-2065.

Lolle, S., and Bussey, H. (1986). *In vivo* evidence for post-translational translocation and signal cleavage of the killer preprotoxin of *Saccharomyces cerevisiae*. *Mol. Cell. Biol. 6*:4274-4280.

Lolle, S., Skipper, N., Bussey, H., and Thomas, D. Y. (1984). The expression of cDNA clones of yeast M1 double-stranded RNA in yeast confers both killer and immunity phenotypes. *EMBO J. 3*: 1383-1387.

Matern, H., Hoffman, M., and Holzer, H. (1974). Isolation and characterization of the carboxypeptidase Y inhibitor from yeast. *Proc. Nat. Acad. Sci. USA 71*: 4874-4878.

Mekalamos, J. J., Schwartz, D. L., Pearson, G. D. N., Harford, N., Groyne, F., and deWilde, M. (1983). Cholera toxin genes: Nucleotide sequence, delection analysis and vaccine development. *Nature 306*: 551-557.

Middelbeek, E. J., Stumm, C., and Vogels, G. D. (1980). Effects of *Pichia kluyveri* killer toxin on sensitive cells. *Antonie van Leeuwenhoek J. Microbiol. Serol. 46*: 205-220.

Olsnes, S., and Pihl, A. (1973). Isolation and properties of abrin: A toxic protein inhibiting protein synthesis. Evidence for different biological functions of its two constituent-peptide chains. *Eur. J. Biochem. 35*: 179-185.

Potts, J. T., Kronenberg, H. M., Habener, J. F., and Rich, A. (1980). Biosynthesis of parathyroid hormone. *Ann. N.Y. Acad. Sci. 343*: 38-55.

Rogers, D., and Bevan, E. A. (1978). Group classification of killer yeasts based on cross-reactions between strains of different species and origin. *J. Gen. Microbiol. 105*: 199-202.

Rogers, D. T., Saville, D., and Bussey, H. (1979). *Saccharomyces cerevisiae* killer expression mutant *kex2* has altered secretory proteins and glycoproteins. *Biochem. Biophys. Res. Commun. 90*: 187-193.

Rothblatt, J. A., and Meyer, D. I. (1986). Secretion in yeast: Reconstitution of the translocation and glycosylation of α-factor and invertase in a homologous cell-free system. *Cell 44*: 619-628.

Schekman, R. (1985). Protein localization and membrane traffic in yeast. *Annu. Rev. Cell Biol. 1*: 115-143.

Schwartz, T. W. (1986). The processing of peptide precursors. "Proline-directed arginyl cleavage" and other monobasic processing mechanisms. *FEBS Lett. 200*: 1-10.

Skipper, N., and Bussey, H. (1977). Mode of action of yeast toxins: Energy requirement for *Saccharomyces cerevisiae* killer toxin. *J. Bacteriol. 129*: 668-677.

Skipper, N., Thomas, D. Y., and Lau, P. C. K. (1984). Cloning and sequencing of the preprotoxin-coding region of the yeast M1 double-stranded RNA. *EMBO J. 3*: 107-111.

Skipper, N., Sutherland, M., Davies, R. W., Kilburn, D., Miller, R. C., Warren, A., and Wong, R. (1985). Secretion of a bacterial cellulase by yeast. *Science 230*:958-960.

Sturley, S., Tipper, D. J., Elliott, Q., LeVitre, J., and Bostian, K. A. (1986). Mutations in the cDNA copy of yeast killer M1-dsRNA defining the immunity determinant. In *Extrachromosomal Elements in Lower Eukaryotes, Meeting Abstracts*, page 80, University of Illinois.

Tipper, D. J., and Bostian, K. A. (1984). Double-stranded ribonucleic acid killer systems in yeast. *Microbiol. Rev. 48*:125-156.

Vernet, T., Boone, C., Greene, D., Thomas, D. Y., and Bussey, H. (1986). Yeast killer toxin: Site-directed mutations affecting toxin and immunity domains of the precursor gene. In *Natural Anti-Microbial Systems, Part I: Antimicrobial Systems in Plants and Animals*, FEMS Symposium No. 35, Chapter 16, A. K. Charnley, M. E. Rhodes-Roberts, R. M. Cooper, G. W. Gould, and R. G. Board (Eds.), Bath University Press, pp. 222-231.

Waters, M. G., and Blobel, G. (1986). Secretory protein translocation in a yeast cell-free system can occur posttranslationally and requires ATP hydrolysis. *J. Cell Biol. 102*:1543-1550.

Wickner, R. B. (1974). Chromosomal and nonchromosomal mutations affecting the "killer character" of *Saccharomyces cerevisiae*. *Genetics 76*:423-432.

Wickner, R. B. (1986). Double-stranded RNA replication in yeast: The killer system. *Annu. Rev. Biochem. 55*:373-395.

Wickner, R. B., and Leibowitz, M. J. (1976). Two chromosomal genes required for killing expression in killer strains of *Saccharomyces cerevisiae*. *Genetics 82*:429-442.

Woolford, L. B. D., Daniels, L. B., Park, F. J., Jones, E. W., Van Arsdell, J. N., and Innis, M. A. (1986). The PEP4 gene encodes an aspartyl protease implicated in the posttranslational regulation of *Saccharomyces cerevisiae* vacuolar hydrolases. *Mol. Cell. Biol. 6*:2500-2510.

9

Acquisition and Expression of the Killer Character in Yeast

S. L. Sturley,* M. El-Sherbeini, S.-H. Kho, J. L. LeVitre,† and K. A. Bostian‡
Brown University, Providence, Rhode Island

INTRODUCTION

Since its discovery in the early 1960s (Bevan and Makower, 1963), the killer dsRNA gene system of *Saccharomyces cerevisiae* has rapidly become the best understood dsRNA genome: a 4.7 kilobase-pair L_1-dsRNA, present in the cell cytoplasm and encapsidated in a 40 nm icosahedral virus-like particle (ScV-L_1 VLP), and a similarly encapsidated, 1.9 kilobase M_1 dsRNA (Hopper et al., 1977; Bostian et al., 1980). The M_1 dsRNA directs the production of a dimeric, exocellular toxin and specific immunity to this toxin. The mycotoxin consists of two dissimilar 9.5 and 9.0 kDa disulfide-linked, nonglycosylated subunits, denoted α and β, respectively. Toxin-producing cells kill sensitive strains by the disruption of cytoplasmic membrane function (Bussey, 1981).

---

*Present affiliation: University of Wisconsin, Madison, Wisconsin
†Present affiliation: Collaborative Genetics, Inc., Waltham, Massachusetts
‡Present affiliation: Merck Sharp & Dohme Research Laboratories, Rahway, New Jersey

Extensive genetic analysis of this phenomenon (for comprehensive reviews, see Wickner, 1981; Mitchell and Bevan, 1983; and Tipper and Bostian, 1984) has revealed a complex array of interactions between the viral and host genomes. These include host-encoded functions concerned with the maintenance (e.g., the *MAK* genes, Wickner, 1978) or expression (e.g., the *KEX* genes, Wickner and Leibowitz, 1976) of the killer virus, as well as viral genes involved in plasmid incompatibility (e.g., *EXL*, Wickner, 1980). Rapid and sophisticated advances in molecular genetic techniques in *S. cerevisiae* make it likely that an even more detailed knowledge of this system will emerge. However, until recently, two major stumbling blocks have impeded this rapid progress.

First, the relative intractability of the multiple-copy dsRNA system in mutational analysis compared to DNA-encoded traits has rendered the assignment of structure-function relationships to the viral genome virtually impossible. There is a paucity of naturally occurring killer plasmid mutations and, despite attempts to introduce mutations in the dsRNA (Bussey et al., 1982), relatively few novel phenotypes have arisen. Furthermore, recombination within each RNA genomic segment has not been observed. To overcome this, Hanes et al. (1986); Lolle et al. (1984); and Boone et al. (1986) produced and expressed a cDNA copy of the toxin encoding M_1 dsRNA gene. Our approach has been to regulate the expression of wild-type and mutant alleles of the gene by their fusion to the phosphate-regulated promoter for the repressible yeast acid phosphatase structural gene, *PHO5* (Sturley et al., 1986).

Second, experiments designed to study the interactions of different forms of the virus in various genetic backgrounds have been hampered by the necessity to construct appropriate strains whose isogenic natures are difficult to confirm. The type 1 killer virus is maintained in nature by vertical transmission and, to date, no infectious cycle has been demonstrated. We have uniquely overcome this problem by developing infection protocols based upon the in vitro permeation of the host cell wall (El-Sherbeini and Bostian, 1987).

Both of these approaches have led to new avenues of research. The mutagenic analysis of the preprotoxin cDNA described here and elsewhere (Sturley et al., 1986) has begun to elucidate mechanisms of killing and immunity and general aspects of protein processing, and the infection experiments have provided an ideal opportunity to study host-viral interactions. Moreover, it is now possible, as we will see, to study the interactions of these viruses with non-saccharomyces yeasts.

STRUCTURE OF THE PREPROTOXIN GENE

The toxin encoding M_1 dsRNA is separated into two regions of approximately 1.0 and 0.6 kilobases, by an adenine-uracil rich denaturation "bubble" of approximately 0.2 kilobases (Figure 1(a), Fried and Fink,

1978; Bostian et al., 1983b). The smaller region and the denaturation bubble possess no substantial open reading frames (ORFs) and are probably involved in maintenance of the virus and its encapsidation by products of the L genome. Naturally occurring defective interfering particles are deletion derivatives of the full length M genome lacking the majority of the preprotoxin coding region. The remaining sequences are sufficient for the maintenance of the M dsRNA. In particular, the presence of five repeats of a consensus sequence, (T/C)GCGATTCGC(A/G), may play a significant role in these mechanisms (Lee et al., 1986). Plasmid mutations (e.g., [Kil-b]; Tyagi et al., 1984; Wickner, 1976), that alter the patterns of maintenance of the M genome, may map to these non-coding regions.

We have constructed a DNA version of the 1.0 kilobase preprotoxin encoding segment of the M_1 dsRNA by cDNA cloning. Comparison of the N-terminal sequences of the secreted toxin subunits with the nucleotide squence of the cDNA clone has enabled us to locate them within a 316 amino-acid ORF (preprotoxin or M_1-P1)with the linear domain structure shown in Figure 1 (Bostian et al., 1984; Skipper et al., 1984). The secreted α subunit of preprotoxin (86 residues) is preceded by a 44-amino-acid N-terminal segment called δ. The α subunit and the C-terminal domain of preprotoxin, β (83 amino acids), are separated by γ, a segment of approximately 103 amino acids, carrying all three N-glycosyl substituents of protoxin. The γ region has been postulated to be the immunity determinant (Bostian et al., 1984). We have shown (Hanes et al., 1986) that the preprotoxin cDNA, lacking most of δ, confers both the killing and immunity phenotypes when expressed from the *PHO5* promoter and secretion leader in an M_1 dsRNA-free yeast strain. To investigate the maturation and functions of preprotoxin, we have reconstructed the 5' terminal sequence of the preprotoxin cDNA and fused it to the *PHO5* promoter and transcriptional terminator [Figure 1(b)]. This was performed by a series of in vitro DNA manipulations (Sturley et al., 1986) involving the annealing of eight overlapping phosphorylated single-stranded oligonucleotides, corresponding to the 5' sequence of the preprotoxin gene, to form a double-stranded molecule with sequences from the ATG of preprotoxin to the *Pst*I site at base 115. This was then fused to the *Pst*1 site of the preprotoxin cDNA and to an *Eco*RI site downstream of the *PHO5* promoter in plasmid p1A1-R1 (Parent et al., 1985). One such plasmid, pSH6, was shown to possess the predicted gene fusion, *PTX1-316* (PHO5-*T*o*X*in 1). The "allelic" designation, 316, refers to the position of the last amino acid prior to translation termination. This nomenclature is also used for mutations in the *PHO5*-preprotoxin gene, where the number indicates the position of the first alteration from the wild-type sequence (see below). The complete *PTX1* gene [Figure 1(b)] was constructed by the fusion of a synthetic 120-mer, containing the *PHO5* transcription terminator, to the *Bgl1* site immediately downstream of the preprotoxin translational stop codon in *PTX1-316*.

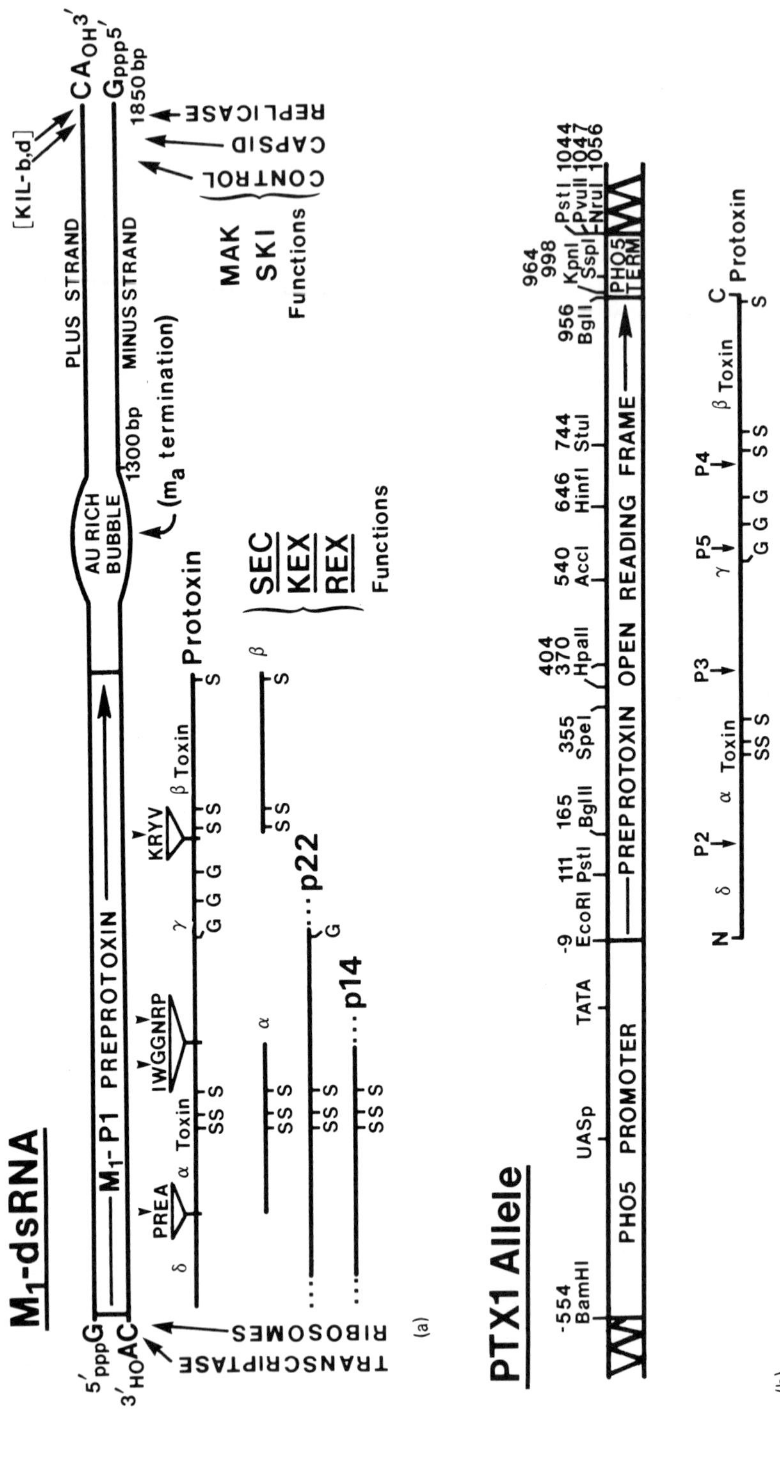
M1-dsRNA
PLUS STRAND
MINUS STRAND
AU RICH BUBBLE
1300 bp
1850 bp
(ma termination)
[KIL-b,d]
M1-P1 PREPROTOXIN
Protoxin
Toxin
KRYV
IWGGNRP
PREA
p22
p14
TRANSCRIPTASE
RIBOSOMES
CONTROL
CAPSID
REPLICASE
MAK
SKI
Functions
SEC
KEX
REX
Functions
(a)
PTX1 Allele
PHO5 PROMOTER
PREPROTOXIN OPEN READING FRAME
PHO5 TERM
-554 BamHI
UASp
TATA
-9 EcoRI
111 PstI
165 BglII
355 SpeI
404
370 HpaII
540 AccI
646 HinfI
744 StuI
956 BglI
964
998 KpnI SspI
PstI 1044
PvuII 1047
NruI 1056
P2
P3
P5
P4
Toxin
Protoxin
(b)

Figure 1 (a) The structure of the M_1-dsRNA, preprotoxin encoding gene. The AU-rich denaturation bubble is the probable site of transcription termination, yielding m_a as its product. The sequences 3' to the AU-rich bubble are preserved in defective interfering particles and probably are essential for efficient replication and maintenance of the dsRNA. *MAK* and *SKI* functions are therefore shown to act at this point and plasmid mutations (KIL-b,d) involved in these processes are also presumed to map to this region. The precursor and mature products (α, β, p22, p14) of the preprotoxin ORF (bases 14 to 964 of the (+) strand) are also shown to scale. *SEC*, *KEX*, and *REX* indicate the host functions involved in the maturation process. G indicates the point of attachment of a single mannose based glycosylation unit. S indicates the position of cysteine residues, potentially involved in the formation of sulphydryl bridges between the mature secreted subunits. The amino acids surrounding putative processing sites are denoted by the single-letter code.

(b) The *PTX1* allele: A regulated cDNA copy of M_1 dsRNA. This consists of a gene fusion between the *PHO5* promoter and terminator and the complete preprotoxin cDNA in a standard 2μ-based yeast vector (vector sequences indicated by the wavy line). Unique restriction endonuclease sites are as shown. P2-P5 are probable proteolytic processing sites as described in the text. Transformants bearing this construct express phosphate-regulated killing and immunity.

When these plasmids were used to transform cells of the M_1 dsRNA-free, non-killer yeast strain, GG100-14D (α *his3 ura3 trp 1-1 pho5 pho3)*, they expressed phosphate-regulated killing and immunity phenotypes. Levels of phenotypic expression were marginally higher than those obtained with strain K12-1, a wild-type killer harboring M_1 dsRNA. Killing and immunity phenotypes expressed by the *PTX1-316* allele were also two- to fourfdld higher than the levels expressed by pSH-GB1, in which the majority of δ was replaced with the *PHO5* secretion leader (Hanes et al., 1986). Northern blot hybridization analysis of total RNA from these transformants confirmed the regulated expression of the *PTX1-316* gene. In vitro translation of total RNA from derepressed cells harboring *PTX1-316* produced a 35 kDa protein, immunoprecipitable with antitoxin IgG, with the same mobility as the immunoprecipitable translation product of wild-type M_1 dsRNA transcripts.

MATURATION AND SECRETION OF TOXIN

Immunoprecipitation of killer cell extracts with antitoxin IgG identifies an intracellular 43 kDa precursor, protoxin, from which α and β are derived (Bostian et al., 1983a; Bussey et al., 1983). Tunicamycin treatment of killer cells and in vitro endoglycosidase H treatment of protoxin (Bussey et al., 1983; Bostian et al., 1983a) demonstrate that protoxin differs from the 35 kDa preprotoxin molecule by the incoporation of three mannose-based $(GlcNAc)_2(Man)_a$-$(Glu)_3$ glycosylation units. The maturation of preprotoxin in wild-type killer cells is governed by the *SEC* (Novick et al., 1980), and *KEX* genes and may also require the action of enzymes sensitive to the chymotrypsin inhibitor, tosyl-L-phenylalanylchloromethylketone (TPCK) (Tipper and Bostian, 1984; Hanes et al., 1986). Lesions in these genes result in a failure to secrete toxin and intracellular accumulation of protoxin (Bussey et al., 1983). In addition, *kex1* and *kex2* mutants retain immunity but are unable to kill sensitive strains (Wickner and Leibowitz, 1976). The *KEX2* gene product is an endopeptidase which cleaves on the carboxyl side of paired basic residues, especially Lys-Arg (Julius et al., 1984). The γ-β junction (P4, residue 234, LysArgTyrVal) is a likely *KEX2* cleavage site, although there is a similar sequence (P5, LysArg-SerAsp) within the γ domain, at residue 188. The δ-α junction, (P2, residue 44, ProArgGluAla) is a good candidate for a trypsin-like endopeptidase, while the sequence around the α-γ boundary (P3, residue 129, IleTrpGlyGlyAsnArgPro) possesses a potential chymotrypsin-like cleavage site, with breakage of the Trp-Gly bond the most likely event. However, the existence of "proline directed arginyl cleavage" (Schwartz, 1986) in a variety of systems raises the possibility that one endopeptidase (probably the *KEX1* gene product) with activity against ProArg (cleavage towards the C-terminus) and ArgPro (cleavage towards the N-terminus) residues, may catalyze cleavage at both

P2 and P3. An additional step in the processing of preprotoxin may be mediated by the *REX1* gene product (Wickner, 1974). Strains bearing mutant alleles of this gene appear to secrete toxin at normal levels but are partially defective in immunity. It remains to be confirmed whether this is due to defective production of the immunity determinant or some other phenomenon unrelated to the maturation of killer toxin.

The δ domain has many of the features of a conventional secretion signal peptide, including a hydrophobic-membrane-spanning peptide sequence between residues 12 and 27 (Bostian et al., 1984) and a potential leader endopeptidase cleavage site at residue 26 (Perlman and Halvorson, 1983) which appears to be cleaved in vitro in the presence of dog pancreas membranes (Bostian et al., 1983a). However, the evidence for internal cleavage of δ in vivo is somewhat conflicting; for example, the effects of TPCK on preprotoxin maturation suggest that cleavage may not occur, at least until after glycosylation (Hanes et al., 1986). If δ persists in protoxin, it may also be of importance in providing a membrane anchor, possibly facilitating correct or differential endopeptidase processing in the Golgi or secretory vesicle.

Expression of the *PTX1* allele in *S. cerevisiae* has provided new insight into the processing of protoxin. The *PTX1-316* construct directed the secretion of α and β toxin subunits at high levels in cells grown under conditions of phosphate limitation. The predominant toxin-related protein in membrane fractions of derepressed *PTX1-316* transformant cells, revealed by immunoblotting with antitoxin IgG, was a 43 kDa species, identical in mobility to protoxin extracted from wild-type killer cells [K12-1, Figure 2, lane (a)]. In addition, a less abundant 35 kDa preprotoxin species was also discernible in the same preparations, presumably equivalent to M_1-P1. Two minor protein species of approximately 38 and 40 kDa are glycosylation intermediates containing 1 and 2 core units as indicated by their change in mobility upon hydrolysis with endoglycosidase H (not shown). Four other proteins were also consistently present: Two species of 9.5 and 9.0 kDa either represent intracellular forms of the secreted α and β toxin subunits or toxin eluted from cell wall fragments present in the membrane preparation. In these extracts, α was much less abundant than β. Since the same disparity did not occur between secreted subunits, this is unlikely to reflect differing affinities for the toxin antibody. It may result from selective binding of β to the cell wall. Two other, previously uncharacterized antigenic species are termed p22 and p14, based on their respective mobilities. The possible functions of these proteins are discussed below. Their presence implies cleavage of protoxin at a previously undiscovered processing site. While p14 has the mobility of an uncleaved δ-α complex, the size and degree of glycosylation of p22 [1 $(GlcNAc)_2(Man)_9$-$(Glu)_3$ unit, as indicated by endoglycosidase H treatment] are best explained by a processing event within the γ domain, indicated as P5 in Figure 3. Furthermore, p22 also reacts with an anti-γ IgG raised against

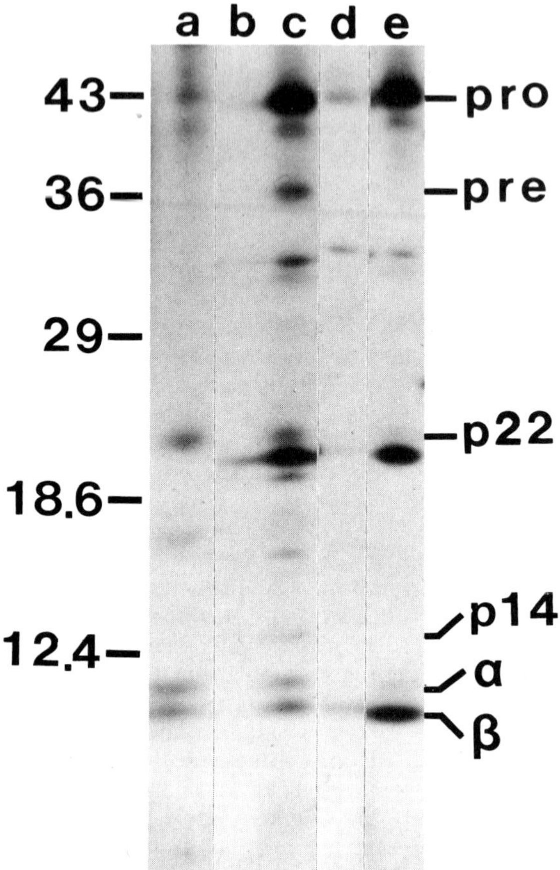

Figure 2 Expression of the *PTX1* allele: Processing and maturation. Proteins were extracted from the appropriate yeast strain in the presence of Triton-X-100 by published procedures (Bostian et al., 1983), fractionated by SDS-polyacrylamide electrophoresis, transferred to nitrocellulose, and incubated with antitoxin IgG. Antigen-antibody complexes were detected using iodinated protein A (Burnette, 1981). Lane (a) is a protein extract from a wild-type, M_1-dsRNA containing strain, K12-1. Lanes (b) and (c) are protein extracts from cells bearing the *PTX1* allele, grown under repressed and derepressed conditions respectively. Lanes (d) and (e) are M_1 dsRNA, containing *kex1* and *rex1* mutant strains, respectively. The "pro" and "pre" refer to the positions of the killer protoxin and preprotoxin species respectively; p22, p14, α and β are described in the text. The species not labeled represent nonspecific reaction with antibody.

a β-galactosidase-γ protein fusion (S. L. Sturley, J. Levitre, and K. A. Bostian, 1986, unpublished observations).

We have further investigated the origin of p22 by immunoblotting of proteins extracted from killer strains harboring wild type and mutant alleles of genes putatively involved in processing. The p22 species is present in wild type, M_1 dsRNA killer cell extracts [Figure 2, lane (a)], indicating that it is not an artifact of the *PHO5* regulated expression of *PTX1-316.* although levels of p22 appear to be reduced in cell extracts of a *kex1* mutant [Figure 2, lane (d)], darker exposures indicate it to be present at wild-type levels, relative to the amount of protoxin in the same cell extract. It is absent from *rex1* mutant cell extracts, although protoxin accumulates at a high level, as do the intracellular forms of α and β [Figure 2, lane (e)]. The data suggests a model for the production of p22, α, and β as shown in Figure 3(b). This invokes differential processing of a common pool of protoxin into p22 (mediated by *REX1*) and separate maturation of the precursor into α and β under the influence of the *KEX1* and *KEX2* endopeptidases. There is a potential cleavage site at residue 188 (*KEX2*-like, LysArgSerAsp) and an alternate endopeptidase cleavage site at residue 180 (AlaAsnAla), with cleavage after the second alanine residue. Cleavage at either site would produce a 22 kDa protein, with the physical properties of p22. The implications of this model with regard to the expression of immunity are discussed in greater detail in the next section.

MECHANISM OF KILLING AND IMMUNITY

The action of toxin on sensitive cells appears to be a two-stage process (Figure 4), initially involving binding of toxin to a 1,6-β-D-glucan cell wall receptor (Hutchins and Bussey, 1983), probably encoded by the *KRE1* and *KRE2* genes. Mutations at either of these loci confer toxin resistance, reduce cell wall binding of toxin, and modify the 1,6-β-D-glucan content of the cell wall. However, *kre1* and *kre2* mutant spheroplasts are sensitive to toxin, indicating that the actual killing process occurs at the plasma membrane level. Mutations at a third locus, *KRE3*, confer toxin resistance to spheroplasts, consistent with the involvement in the killing process of a plasma-membrane-located receptor (Al-Aidroos and Bussey, 1978). Killing occurs from a rapidly induced increase in the permeability of the cytoplasmic membrane to protons, and the associated leakage of metabolic pools (de la Pena et al., 1981). Immunity probably also occurs at the plasma membrane, since spheroplasts of toxin-producing cells retain immunity. An attractive model involves the alteration or masking of the *KRE3* receptor by the immunity determinant, rendering it inaccessible to killer toxin.

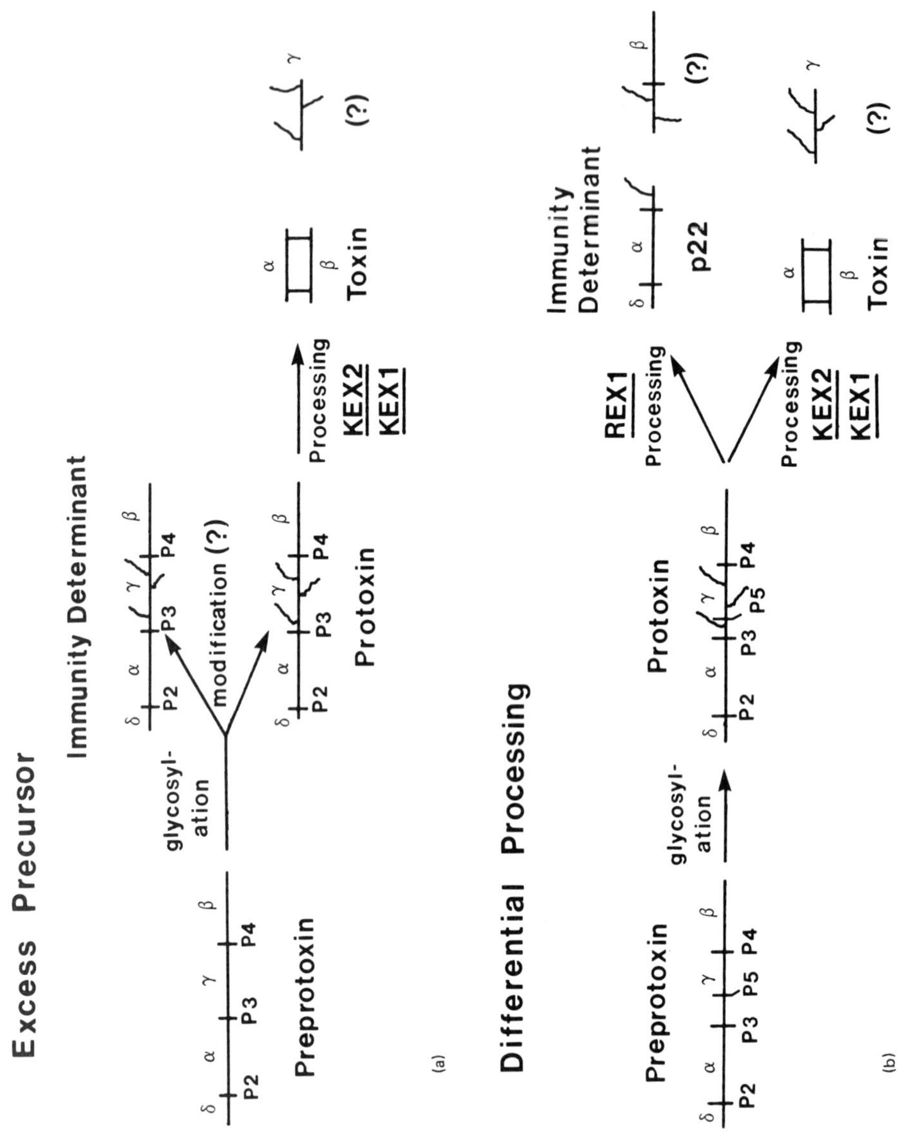
Excess Precursor
Preprotoxin
δ α γ β
P2 P3 P4
glycosyl-
ation
modification (?)
Immunity Determinant
Protoxin
Processing
KEX2
KEX1
Toxin
(?)
(a)
Differential Processing
Preprotoxin
P2 P3 P5 P4
Protoxin
REX1
Processing
Immunity
Determinant
p22
(?)
Toxin
(b)

Hydropathicity analysis (Kyte and Doolitle, 1982) of the amino acid sequence of preprotoxin predicts the α domain to possess two highly hydrophobic regions (residues 72-91 and 112-127) separated by an hydrophilic region containing acidic amino acids and all of the cysteine residues found in α (Bostian et al., 1984). It has been postulated (Tipper and Bostian, 1984), that insertion of this region of α into a membrane could create the cation channels responsible for cell death. The γ domain could act as the immunity determinant by interacting with the same receptor during secretion. However, the absence of apparent peptide homology between α and γ is inconsistent with such a model (Tipper and Bostian, 1984). The C-terminal domain of preprotoxin β lacks potential membrane-spanning regions. However, by analogy to the abrin and ricin class of toxins (Olsnes and Pihl, 1973), this subunit has been theorized to function as a ligand for the 1,6-β-D-glucan cell wall receptor (Bostian et al., 1984).

An initial test of the role of γ in immunity was to express this domain in the absence of the other protoxin regions. In this instance, expression of γ alone might confer immunity. We therefore constructed an in-frame fusion of the γ region, together with 16 preceding residues of α and 11 succeeding residues of β, to the *PHO5* promoter-leader construct used in pSH-GB1. No discernible phenotype resulted from expression of this construct in strain GG100-14D. The γ domain is presumably a processing product of protoxin but its fate has not been determined. It is not detected with antitoxin (α and β) IgG, and antibodies generated against a β-galactosidase-γ protein fusion have so far proved unsuccessful in detecting species other than protoxin and p22.

Figure 3 Models for the processing of type 1 killer preprotoxin and the expression of immunity. Two are proposed, involving different fates for the glycosylated precursor. Both suggest the binding via a common region of α to a membrane receptor (*KRE3* ?). Thus, the receptor is altered so that it can no longer interact with α delivered from outside of the now-immune cell. Excess precursor (a) requires the initial binding of protoxin (possibly modified) to the receptor to confer immunity, with the subsequent processing of unbound protoxin to produce active α and β. Differential processing (b) predicts two fates for an intracellular pool of protoxin, *REX1*-mediated cleavage to give an immunity determinant (p22 ?) and an alternate series of proteolytic events (performed by the *KEX1* and *KEX2* gene products) to produce the secreted toxin subunits. Likely processing sites are indicated as P2-P5, using the nomenclature of Bostian et al. (1984). P1 (not shown) is the predicted leader peptidase recognition site within δ. P5 is the alternate cleavage site at residue 188. Glycosylation is denoted by the appendages to the γ domain.

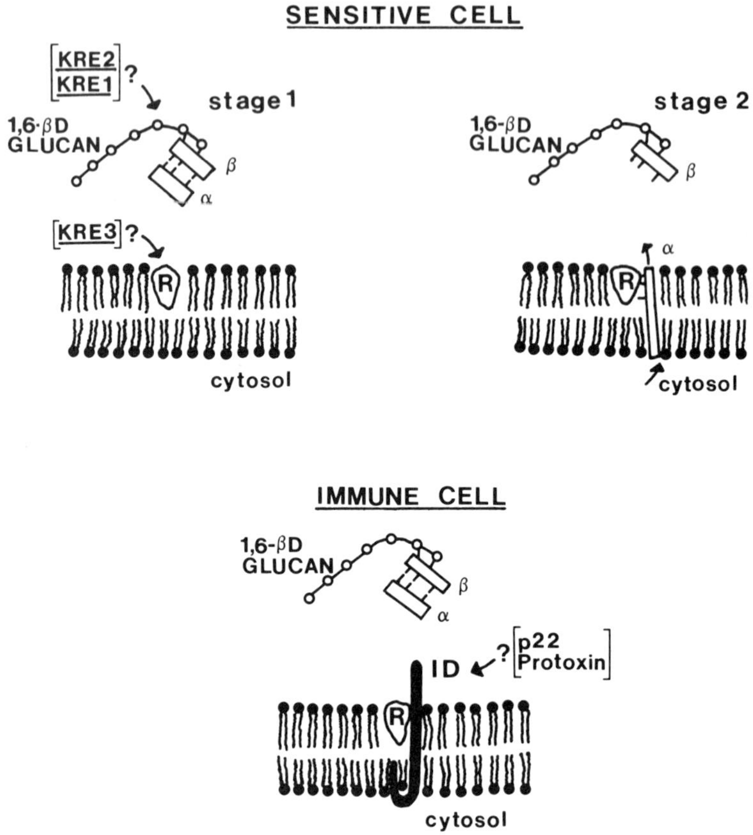

Figure 4 Mechanisms of killing and immunity by type 1 killer toxin. Killing is envisaged as a two-stage process, initially involving the binding of the α-β dimer to a 1,6-β-D glucan cell wall receptor (encoded by the *KRE1* and *KRE2* genes) via the β subunit. In the second lethal step, α binds to the plasma membrane receptor (*KRE3*), resulting in an increase in permeability and death of the cell. Immunity would also occur at this stage, by binding of the immunity determinant (p22 or protoxin) to the same receptor, making it inaccessible to the α subunit, so producing an immune cell.

In order to further characterize the functional domains of preprotoxin, a series of mutations were introduced into *PTX1-316* (plasmid pSH6) and the phenotypes studied under derepressed conditions in strain GG100-14D (Sturley et al., 1986). Mutations were of several types. Insertion of the TERM linker, CTAGTCTAGACTAG, generated amber termination codons in all reading frames, producing truncated protoxins (*PTX1-116, -177,* and *-240*). Use of a host strain carrying the tyrosine-inserting *SUP7* mutation allowed analysis of expression of the "read-through" products of these mutants, resulting in the production of insertion mutations (*PTX1-116*^{sup}, *-177*^{sup}, and *-240*^{sup}). A C-terminal deletion of the β region was also constructed (*PTX1-239*) and termination close to the δ-α boundary occurred as a result of a construction error (*PTX1-34*). Finally, oligonucleotide mutagenesis was used to create a missense mutation (*PTX1-234*). In each case, RNA blot hybridizations and in vitro translation of in vivo transcripts demonstrated the phosphate-repressible production of functional transcripts producing antitoxin IgG precipitable proteins of the predicted length. Immunoblotting of in vivo synthesized proteins revealed the presence of these proteins or their fully glycosylated counterparts. The gene products of the mutant *PTX1* alleles and their phosphate-regulated killer ($K^{+/-}$) and immunity ($R^{+/-}$) phenotypes are shown in Figure 5. In addition to assaying killer toxin against intact cells, the sensitivity of spheroplasts ($P^{+/-}$) to the secreted proteins from these strains was also tested.

The Active Site of α Is Closely Involved with the Expression of Immunity

The immunity conferred by SH-GB eliminates a role for δ in immunity determination. The same conclusion can be drawn from the inability of the *PTX1-34* allele (in which translation terminates close to the δ-α boundary) to confer immunity. Similarly, it is apparent from the immune phenotypes conferred by alleles lacking the complete β region (*PTX1-239, -240,* and *-240*^{sup}), that this domain is also not required for the expression of immunity. However, since *PTX1-116* and *-116*^{sup}, which carry a termination and disruption of the second hydrophobic region of α respectively, do not express immunity, it seems likely that this region of protoxin is inherently involved in the process. The partial immunity conferred by the *PTX1-177* and *-177*^{sup} alleles (defective in the N-terminal portion of γ) defines the region upstream from residue 177 as being the likely determinant of immunity, although the remainder of protoxin may facilitate accurate processing or localization of the immunity determinant. Two new models for immunity, involving either differential processing or accumulation of protoxin, are presented in Figure 3. Both models postulate that two forms of the "active site" of α can interact with the hypothetical *KRE3* plasma membrane receptor, through the same residues but with different results. Mature α, delivered from outside the cell, kills cells as a consequence of this interaction. In contrast, α as a component of protoxin, or in partially or differentially processed protoxin, would interact with the receptor during secretion, either leading to its loss, for example by shunting

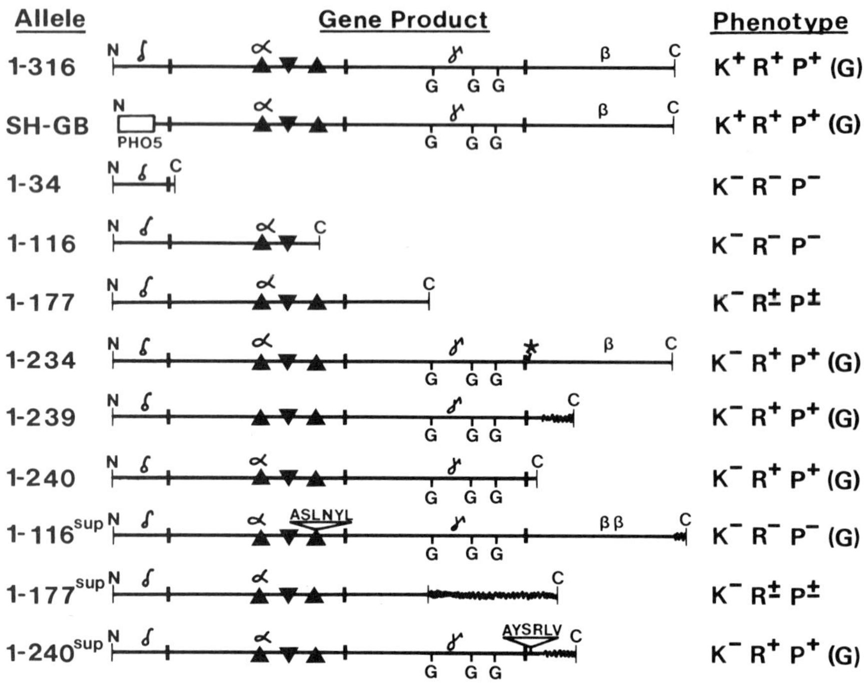

Figure 5 Mutagenic analysis of the *PTX1* allele. Mutations in the *PTX1* gene were produced as described in the text. Under conditions of phosphate limitation, the mutant alleles express the products shown here with the corresponding phenotypes. Killer and immunity ($K^{+/-}$, $R^{+/-}$) assays were performed by standard methods. The sensitivity of M_1-dsRNA free spheroplasts ($P^{+/-}$) to the secreted proteins of derepressed strains carrying these mutant alleles was determined as described by Sturley et al. (1986). G indicates the position of glycosylation; (G) indicates the presence of glycosylation on proteins produced by each mutant as indicated by endoglycosidase-H treatment. The upright and inverted triangles signify the regions of high and low hydropathicity, respectively, within α, believed to delineate the membrane-breaching active site of this domain. Allele 1-316 is the wild type allele (43 kDa product). SH-GB possesses the *PHO5* leader in place of the δ domain of protoxin (40 kDa product). Alleles 1-34, 1-116, 1-177, and 1-240 all possess insertions encoding amber termination at the points shown. They produce polypeptides of 5, 13, 19, and 34 kDa respectively. Allele 1-234 has a missense mutation at positions 234 and 235 (indicated by *) replacing TyrVal with AspPro. It produces a wild-type sized polypeptide. Allele 1-239 has a slight C-terminal extension over the 1-240 gene product (36 kDa) derived from vector sequences (indicated by a wavy line). Alleles $1\text{-}116^{sup}$, $1\text{-}177^{sup}$, and $1\text{-}240^{sup}$ are the read-through products of alleles 1-116, 1-177, 1-240 in a *SUP7* amber suppressing host strain. This results in the in-frame insertion of the amino acids shown or in the case of $1\text{-}177^{sup}$, a 77 residue missense C-terminal extension.

of the complex to the vacuole, or to masking as a consequence of tight binding. Uncleaved protoxin [Figure 3(a)] could perform this function early in the secretion pathway, before access to the processing enzymes, presumed to act in a Golgi compartment (Tipper and Bostian, 1984). Alternatively, processing at residue 182 or 188 within γ [Figure 3(b)] could preempt the other processing events (α-γ and γ-β) to produce p22, which may no longer be a substrate for cleavage between α and γ.

To address the problem of whether p22 or protoxin is the immunity determinant, we have investigated the antitoxin immunoreactive protein species in mutant strains defective in the expression of killing (*kex1* mutants) or immunity (*rex1* mutants). Both appear to be processing mutants since they accumulate high levels of protoxin within the cell (as previously reported for *kex1* and *kex2* strains by Bussey et al., 1983). If p22 is the immunity determinant, a *rex1* mutant strain carrying M_1 dsRNA would be expected to lack this protein species. Similarly, a *kex1* mutant would lack the α and β subunits of killer toxin but possess normal levels of p22. In general, this is the case [Figure 2, lanes (d) and (e)]. However, the *kex* and *rex* mutants used here were leaky and expressed low levels of killing and immunity respectively. Thus our *rex1* strain (R12a) was not truly suicidal and did possess a very low level of p22. Similarly, *kex1-1* mutants display partial killing of sensitive strains and possessed low levels of both intracellular and extracellular α and β subunits. Darker exposures of the autoradiogram shown in Figure 2 show *kex1-1* mutant cell extracts to have wild-type levels of p22, relative to the concentration of protoxin in the same preparation. Furthermore, overexpression of protoxin from *PTX1-316* in *kex1-1* strains indicated the presence of similar levels of p22 to those found in wild-type cells transformed with the same construct (not shown). We are currently isolating strains bearing tighter mutant alleles of these genes to verify these observations.

α Is the Active Killing Agent of Toxin, While β May Recognize the Cell Surface Receptor

We hypothesized that the α domain of toxin attacks the membrane and that its delivery to this site in intact cells is dependent on the interaction of the β component of the *PTX1-239* or *PTX1-240* alleles, deleted for most of β, produce fully glycosylated truncated polypeptides lacking all of the β domain, except the first six residues. Normal processing produced the intracellular α toxin component. However, the secreted form may have been labile or inefficiently secreted since it was not detected. Although intact cells were not killed by derepressed culture supernatants, spheroplasts of a sensitive strain did not survive such treatment. This is consistent with α acting as the lethal toxin component and β bringing α in contact with its site of action at the plasma membrane by binding to the cell wall. Cells harboring *PTX1-234* accumulated normal levels of a fully glycosylated protoxin of normal size and processed it with secretion of near normal quantities of α and β subunits, identified by size and immunigenicity. The only difference from wild type should be the replacement of TyrVal at the N-terminus of β by AspPro. This was compatible with processing by the *KEX2* and other proteases, and with toxin export. However, this toxin was inactive against whole cells, although secreted proteins did kill spheroplasts as effectively as wild-type toxin. Moreover, very little of the modified β can be eluted from the cell walls of *PTX1-234* transformants, suggesting that the defect in β function may be binding to the cell wall receptor and that the active site of β is at the N-terminus (S. L. Sturley, D. J. Tipper, and K. A. Bostian, 1986, unpublished data).

DEPENDENCE OF M ON L

Another aspect of toxin and immunity expression under investigation in this laboratory is the dependence of M_1 maintenance on L dsRNA. M_1 dsRNA is found only in cells possessing L dsRNA, which confers no recognizable phenotype other than support of M. The identification of the major 88 kDa capsid protein VL_{1A}-P1 (ScV-P1) as the gene product of L, and the demonstration that VL_{1A}-P1 packages both L and M dsRNA, explains part of this dependency relationship (Bostian et al., 1980; El-Sherbeini et al., 1984). L may also provide part or all of the polymerase or replicase activity required for the maintenance of both the L and M genomes.

Several laboratories have demonstrated the coexistence of multiple species of L in a variety of *S. cerevisiae* strains (Sommer and Wickner, 1982; Field et al., 1982; El-Sherbeini et al., 1983). By examination of several yeast strains for L dsRNAs and their in vivo and in vitro translation products, we have identified four unique species of L-sized dsRNA: L_{1A}, L_{1BC}, L_{2A}, and L_{2BC} (Table 1, Figure 6). L_{1A} encodes VL_{1A}-P1. L_{1BC} is contained in VLPs with a single unique 82 kDa capsid protein, VL_{1BC}-P1, encoded by L_{1BC}. In strains containing both L_{1A} and L_{1BC} [Figure 6, lane (c)], M_1 is encapsidated only in VL_{1A}-P1. Although L_{1A} and L_{2A} share some sequence homology, they encode distinctly different capsid proteins (El-Sherbeini et al., 1984; M. El-Sherbeini, 1985, unpublished observations). Some of these variants of L dsRNA have been correlated with an ability to exclude or maintain M_2 (expression of the type 2 killer phenotype) and M_1 dsRNAs in different genetic backgrounds. For example, the [HOK] phenotype, characterized by its ability to maintain replication defective mutant forms of M dsRNA, has been correlated with L_{1A}

Table 1 Genomes and Gene Products Within Different Mycoviral Systems

dsRNA genome (kilobases)	In vivo products (kDa)	In vitro products (kDa)
K1 system		
L_{1A} (4.7)	VL_{1A}-P1 (88)	L_{1A}-P1 (88)
		L_{1A}-P2 (76)
L_{1BC} (4.7)	VL_{1BC}-P1 (82)	L_{1BC}-P1 (82)
		L_{1BC}P2 (78)
M_1 (1.9)	Protoxin (42)	M_1-P1 (34.8)
K2 system		
L_{2A} (4.7)	VL_{2A}-P1 (84)	L_{2A}-P1 (84)
L_{2B} (4.7)	VL_{2B}-P1 (83)	L_{2B}-P1 (83)
M_2 (1.7)	Protoxin (ND)[a]	M_2-P1 (42)
L_y system		
L_y (4.9)	VL_y-P1 (83)	L_y-P1 (83)
	VL_y-P2 (77)	L_y-P2 (77)
		L_y-P3 (74)
		L_y-P4 (68)
L_{Pc} system		
L_{Pc1} (3.6)[b]	Vl_{Pc}-P1 (130)	L_{Pc1}-P1 (130)
L_{Pc2} (3.3)[b]		L_{Pc2}-P2 (120)
L_{Pc3} (3.2)[b]	Vl_{Pc}-P3 (92)	L_{Pc3}-P3 (92)
L_{Pc4} (2.9)[b]		ND

[a]ND, not determined
[b]Sizes calculated from data of Edmondson et al., 1986.
Source: Adapted from El-Sherbeini et al., 1986.

(Sommer and Wickner, 1982; El-Sherbeini et al., 1983). Similarly, another form of L_{1A} dsRNA is correlated with the [EXL] encoded ability to exclude M_2 dsRNA in crosses involving these dsRNA species.

We have also compared and contrasted the dsRNA genomes of other fungi to the yeast killer system. The dsRNA systems of *Yarrowia lipolytica* and *Penicillium chrysogenum* are described in Table 1.

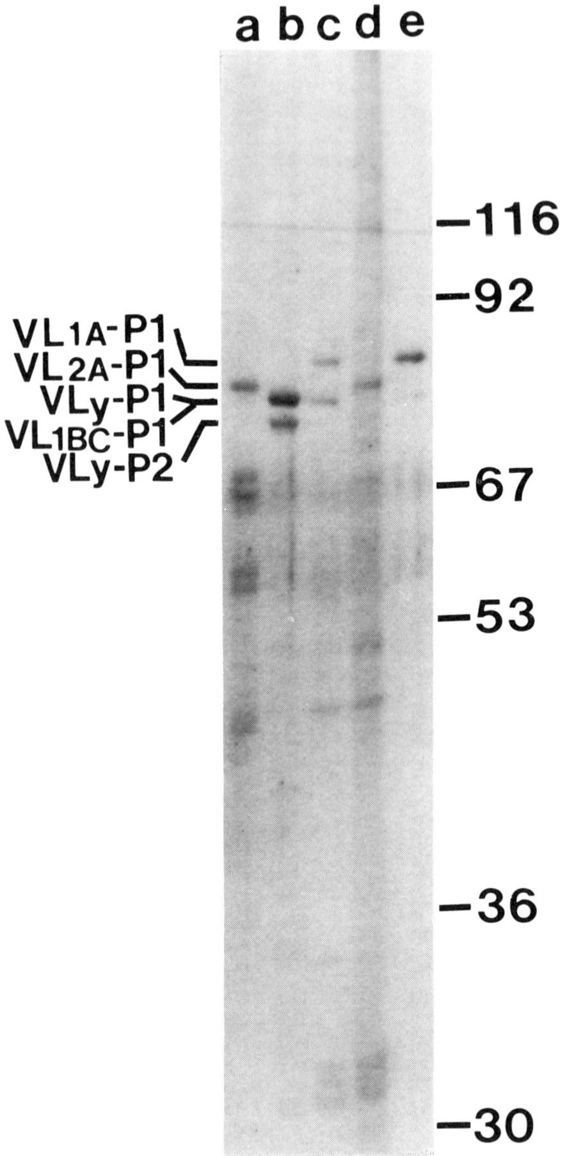

Figure 6 SDS-polyacrylamide gel electrophoresis of viral proteins encoded by different L dsRNA species. Virus-like particles were prepared from *Y. lipolytica* strain CBS6124-2 and from *S. cerevisiae* strains with known L dsRNA content. VLP samples from sucrose density gradients were denatured, subjected to electrophoresis on a 10% SDS-poly-

Strains of *Y. lipolytica*, a dimorphic, hydrocarbon-utilizing yeast, were only recently shown to possess VLPs with a species of dsRNA the same approximate size as L dsRNA in *S. cerevisiae* (Groves et al., 1983; Trenton et al., 1985). It has been suggested that *Y. lipolytica* also produces a killer toxin (Groves et al., 1983), although this has not yet been confirmed. We have demonstrated that the L-dsRNA of *Y. lipolytica* (L_y-dsRNA) encodes its own capsid polypeptide (El-Sherbeini et al., 1987). The major in vivo VLP protein (VL_y-P1, 83 kDa), is identical by peptide mapping to the major in vitro translation product of L_y- dsRNA (L_y-P1), and essentially identical to a second, less-abundant virus-associated protein VL_y-P2 of 77 kDa [Figure 6, lane (b)]. Like *S. cerevisiae*, the two capsid proteins may also derive from multiple species of dsRNA. VL_y-P1 and VL_y-P2 are distinct from the capsid polypeptides produced by either L_{1A} or L_{1BC}-dsRNA of *S. cerevisiae*, and L_y-dsRNA shares very little or no sequence homology with either L_{1A}, L_{1BC}, or L_2-dsRNAs, thus establishing that *Y. lipolytica* VLPs constitute a distinctly different dsRNA viral system from the K1 or K2 killer systems in *S. cerevisiae* (El-Sherbeini et al., 1987).

The *P. chrysogenum* system is more complex than those of *S. cerevisiae* and *Y. lipolytica.* PcV particles consist of three to four dsRNAs, designated L_{Pc1} to L_{Pc4}, separately encapsidated in identical protein coats (Wood and Bozarth, 1972; Edmondson et al., 1984). Comparison of the in vivo and in vitro translation products of the dsRNAs (Table 1) show that L_{Pc1} encodes the major capsid protein of the PcV virus (M. El-Sherbeini, J. E. Hopper, R. F. Bozarth, and K. A. Bostian, 1986, unpublished data).

These studies have demonstrated that each L species (L_{1A}, L_{1BC}, L_{2A}, L_{2B}, L_y, and L_{Pc1}) encodes a unique capsid protein. This has been confirmed by RNA hybridization analysis using probes made from the individual dsRNAs, implying origins of diversity. These results also correlate unique mycoviral particles with different host species.

Further comparative study of these and other dsRNA systems will provide additional insight into the origins and dynamics of host-virus interactions. One complexity, however, in studying the dependency relationships between different capsid encoding dsRNAs and M, is the

---

acrylamide slab gel, and stained with Coomassie blue. The position of migration of standard protein markers of known molecular weight are indicated to the right, consisting of β-galactosidase (116 kDa), phosphorylase A (92 kDa), bovine serum albumin (67 kDa), glutamate dehydrogenase (53 kDa), and carbonic anhydrase (30 kDa). Lanes are: (a) strain 482 (L_{2A}, L_{2B} M_2); (b) strain CBS6124.2 (L_y); (c) strain K11.A (L_{1A}, L_{1BC}); (d) strain Y110 (L_{2A}, L_{2B}, M_2); and (e) strain K23.A (L_{1A}, L_{1BC}, M_1). (*Source*: From El-Sherbeini et al., 1986.)

inability to introduce and mix different dsRNAs in different hosts, without cytoplasmic or nuclear exchange. To overcome this limitation, we have developed an efficient method for the extracellular transmission of viral particles into a variety of fungal organisms. Preliminary experiments demonstrate that this approach will be a valuable tool to study viral-host and viral-viral interactions. The infection protocol provides an efficient alternative approach to previous techniques of cell mating or cell fusion between strains, allowing the capability of following the kinetic properties of viral propagation, particularly with regard to amplification, copy number control, interference, and exclusion.

EXTRACELLULAR TRANSMISSION OF THE ScV VIRUS

It has generally been believed that mycoviruses, including the yeast killer system, are incapable of extracellular infection, although reports of protoplast uptake of VLPs have been reported in other fungal organisms (Stanway and Buck, 1984). However, we have modified existing protocols in our laboratory for DNA transformation and VLP preparation, and used them successfully to infect yeast cells with extracellular VLPs (El-Sherbeini and Bostian, 1987).

Infection of Spheroplasts and Competent Cells

Virus particles prepared from appropriate K1 and K2 killer strains were filter sterilized and used to inoculate spheroplasts of a standard sensitive uracil requiring (*ura3-50*) host strain, GG100-14D (containing only L_{1A} and L_{1BC} dsRNA and the appropriate capsid proteins, as shown by polyacrylamide gel electrophoresis). The inoculation was done in the presence of a *URA3*-containing, 2-μm based plasmid (YEp24, Parent et al., 1985) and selection was made for *URA3*-regenerated spheroplasts (Hinnen et al., 1978). Uracil prototrophs were assayed for the killer phenotype by standard methods. With the K1 killer virus inoculum, 67% of the transformants assayed had also been infected with ScV-M_1. The infected clones were highly stable killers with similar M_1 dsRNA copy numbers to the donor strain (K7) and the same genotype as the host strain, with the exception of the uracil marker. As expected, the ScV-M_1 infected clones showed the same profile of capsid polypeptides as the recipient host strain. All possessed VL_{1A}-P1 and VL_{1BC}-P1.

In a similar experiment, the viruses of the type 2 killer system were used to inoculate the same host strain, yielding slightly different results. Only 4% of the transformed colonies were infected with ScV-M_2. The newly infected clones showed the expected phenotype of GG100-14D transformed with YEp24. However, the infected transformants varied in the stability of their killer phenotype. Two of the original infected clones were passed through several rounds of subculturing in complete medium, after which 34% and 86% of the colonies tested had lost the killer phenotype. Those subclones that had retained the K2 virus after this process were stable killers with M_2 dsRNA copy numbers equivalent to the original K2 donor. Interestingly, the ScV-M_2 infected clones contained VL_{1A} (the L_{1A} capsid endogenous to the host) but lacked VL_{2A} (the capsid protein of L_{2A}, from the K2 donor). As first suggested by Hannig et al. (1985), L_{1A} is apparently capable of fulfilling the capsid requirement for maintenance of M_2. The observed variation in M_2 maintenance suggests that M_2 is subject to exclusion in early generations due to poor replication or segregation of virus at initially low copy number.

The work was also extended to demonstrate the ability of yeast cells to incorporate the killer virus without removal of the cell wall, by stimulating uptake with lithium acetate. Cells of strain GG100-14D were rendered competent by the lithium acetate procedure of Ito et al. (1983) and inoculated with K1 killer virus in the presence of the plasmid YEp24. As before, selection was made for *URA3* transformants. Of the uracil prototrophs tested, 3% had been infected with ScV-M_1.

Infection of Naturally Mating Cells

Having established that the yeast killer virus can be transmitted under circumstances involving removal (by spheroplasting) or permeabilization (lithium acetate treatment) of the cell wall, we decided to test infectivity under more natural conditions, such as during mating. Two haploid yeast strains, devoid of the killer virus, with complementing nutritional markers, were subjected to mating in liquid medium in the presence of the killer virus. Selection of zygotes was made by plating appropriate dilutions on minimal medium. Of 50 randomly isolated diploids, 3 had acquired the killer phenotype. All three killer clones sporulated when plated onto sporulation medium, unlike the parental haploids. Tetrad analysis of the three killer diploids upon microdissection revealed a 4:0 segregation for the killer:sensitive phenotype and a 2:2 segregation for all the unlinked nutritional markers present in the parental strains, confirming that the killer diploids were derived from mating of two sensitive parents.

The establishment of more natural conditions for viral infection could be taken as evidence that extracellular transmission is a productive route for viral inheritance in the wild, provided that the conditions used also occur in nature. Virus particles are found in vivo at levels as high as several thousand per cell, providing a feasible pool for infection. However, cell lysis or liberation of viral particles would have to occur in the same habitat, stimulating cell surface transport. Several natural cell surface alterations, in addition to those responsible for infection of mating cells, such as the budding process occurring during cytokinesis, mechanical damage, or conditions of nutritional stress stimulating membrane transport, may provide additional avenues for viral penetration or escape.

CROSS-SPECIES TRANSFER OF MYCOVIRUSES

A wide variety of mycoviral systems are known to exist in nature, and studying the relationships and properties of these viruses will be greatly facilitated by the infection protocols described above. The development of transformation protocols for fungi such as *Y. lipolytica* (Davidow et al., 1985) and *Schizosaccharomyces pombe* (Beach and Nurse, 1981) provides a unique opportunity to study the interactions of totally unrelated viruses in a variety of hosts. Our current research along these lines is twofold. First, we are studying the interactions of different mycoviruses with each other and with different hosts. To this end, we have designed experiments to infect two different hosts with the same mycovirus and to bring together two different mycoviruses in the same genetic background. Second, we are testing the ability of different fungal hosts to express the M_1 dsRNA preprotoxin gene, as well as the ability of *S. cerevisiae* hosts to express other killer toxins. This will help define host contributions to the mechanisms and requirements for toxin secretion and the expression of immunity.

Most fungal hosts naturally harbor mycoviruses that are not associated with any phenotypic effect, e.g., the well-defined mycoviruses of *Aspergillus* and *Penicillium* and the more recently discovered mycoviruses of *Y. lipolytica*. In many instances, such as in *S. cerevisae* and *Ustilage maydis*, mycoviruses are associated with distinct phenotypes, i.e., the ability to kill sensitive fungi. The killer phenotype of *U. maydis*, the smut fungus of maize (for which a transformation system exists), is also associated with a segmented, cytoplasmic dsRNA genome, individually encapsidated in VLPs. The genome consists of at least one representative of each size class of dsRNA: heavy (H1-4), medium (M1-3), and light (L1). However, there is only one major capsid polypeptide, encoded by one or more of the H genomes (Koltin, 1977). An M species is required for the expression of the killer phenotype together with a representative of the H and L species for maintenance of the virus (Koltin and Kandel, 1978). In a few fungi, such as *S. pombe*, dsRNA mycoviruses are not readily found in natural isolates. In addition to the contrasting mycoviral composition of these fungal organisms, the hosts themselves have distinct morphological characteristics, unique replicative and mating cycles, and different degrees of differentiation [Figure 7(a)].

Our immediate objective has been to demonstrate that cross-species transfer of mycoviruses is possible. This has been accomplished bilaterally for *S. cerevisiae* and *Y. lipolytica* and their L genomes [Figure 7(b)]. Using the LiAc infection protocol, purified *Y. lipolytica* virus (YlV) was introduced into *S. cerevisiae* strain GG100-14D. Analysis of the capsid polypeptide profiles of the infected cells [Figure 7(b), lane (e)] and dsRNA content by RNA blot hybridization (not shown), reveal the presence of L_y dsRNA at a concentration similar to that found in the natural host. The newly acquired dsRNA genome was stably maintained over numerous generations of cell growth. Thus, YlV is more than adequately maintained by the presumed *MAK* functions present in GG100-14D, to the extent that in certain circumstances L_y excludes the endogenous L_{1A} component [Figure 7(b), lane (b)] of the new host [Figure 7(b), lane (e)], leaving L_{1BC} at approximately its original concentration. The loss of L_{1A} by this route in an M_1 dsRNA containing strain may lead to the concomitant loss of the M_1 genome and the killer character, unless Ly dsRNA is sufficient for M_1 replication and packaging, since the remaining L_{1BC} species is incapable of packaging the M_1 dsRNA genome.

In very recent experiments, L_{1A} dsRNA from *S. cerevisiae* strain K7 was successfully introduced into a dsRNA free *Y. lipolytica* strain. Interestingly, the normally high and stable concentration of L_{1A} seen in *S. cerevisiae* is also exhibited in *Y. lipolytica* (M. El-Sherbeini, S.-H. Kho, and K. A. Bostian, 1987, unpublished data), a concentration much higher than the normal YlV virus. Superinfection of the new strain with the ScV-M_1 virus should enable us to examine preprotoxin gene expression in the new host.

Several natural barriers to the successful expression of preprotoxin, however, can be envisaged. First, the host may be inadequate in its maintenance attributes, in which case ScV-M_1 may fail to replicate. Alternatively, the virus may replicate normally, but functional transcripts may not be produced. Standard RNA blot hybridization and in vitro translation analysis would distinguish these possibilities. Furthermore, if the gene product is made but requires processing, as for killer toxin encoded by ScV-M_1, the absence of maturation enzymes would preclude the expression of the expected phenotype. The expression of a cDNA copy of the gene in question from an appropriate promoter would determine if this was the situation. At the same time, the inability to maintain ScV-M_1, to express protoxin or immunity, or to secrete toxin in different organisms would raise interesting questions concerning the natural dissemination and evolution of these mycoviral systems. Do *S. pombe* strains, for example, lack endogenous dsRNA virus by chance, because they lack the appropriate genes to maintain or express the virus, or do they have a natural defense system not found in the other fungi considered here? In *S. cerevisiae*, for example, it appears that several of the *SKI* gene products protect

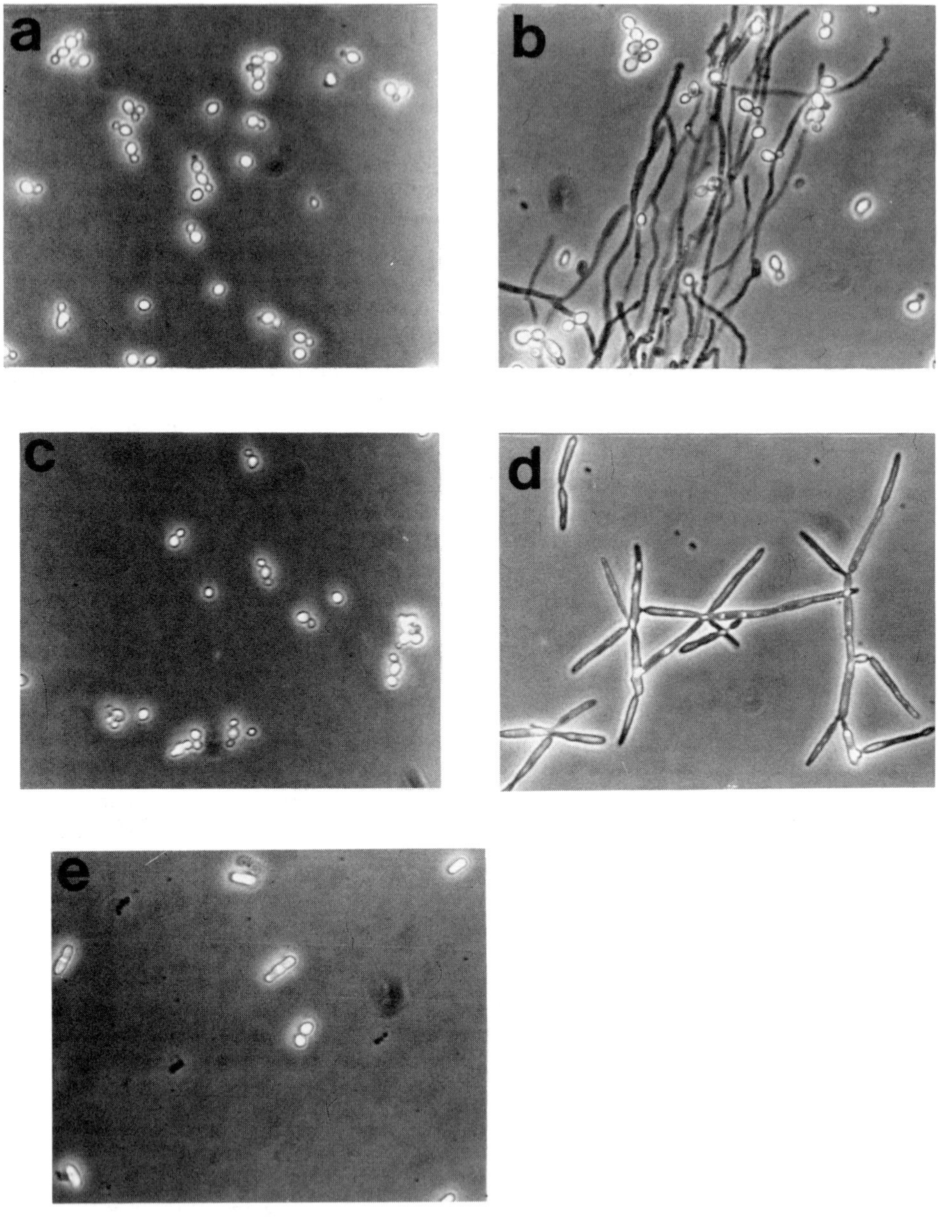

Figure 7 Cross-species transfer of viral genomes. Left: Photomicrograph of various host fungal strains and infected progeny. Original magnification X320, with phase contrast of wet-mounted preparations

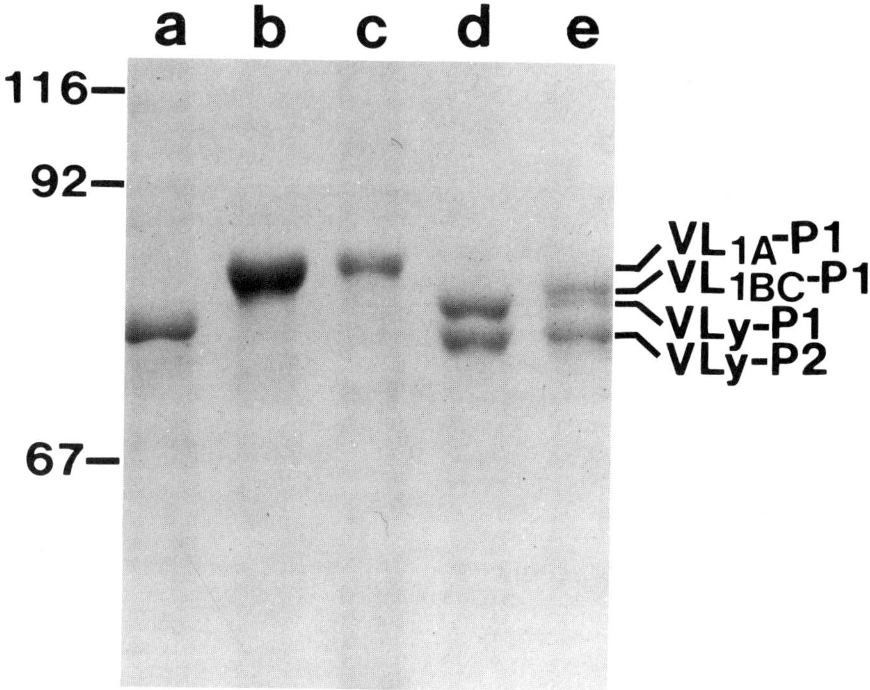

Figure 7 (Continued) showing: (a) *S. cerevisiae* strain GG100-14D. (b) *Y. lipolytica*. (c) *S. cerevisiae* GG100-14D, infected with the mycovirus from *Y. lipolytica*. (d) *U. maydis*. (e) *S. pombe*. Right: Capsid protein profiles of viruses isolated from various fungal strains. Capsid polypeptides were prepared and analyzed by PAGE as described previously (El-Sherbeini et al., 1984). Lanes are: (a) *S. cerevisiae* strain 110, a type 2 killer strain (containing VL_{2A}-P1); (b) *S. cerevisiae* strain K7, a type 1 killer (containing VL_{1A}-P1 and VL_{1BC}-P1) and (c) the host *S. cerevisiae* strain GG100-14D (also containing VL_{1A}-P1 and VL_{1BC}-P1). Lane (d) is the capsid profile of *Y. lipolytica* (VL_Y-P1 and P2) and lane (e) shows the capsid polypeptides present in a strain of GG100-14D infected with the *Y. lipolytica* virus [see micrograph (c)] consisting of VL_{1BC}, VL_Y-P1, and VL_Y-P2.

the host from the cytopathic effects of elevated copy numbers of the dsRNA genomes.

CONCLUSIONS

Pursuing the lines of investigation described above will allow us to determine what factors are essential for the maintenance and expression of the killer phenotype in yeast and whether these factors are ubiquitous among strains and fungal species in nature. Also, the marked stability of the dsRNAs and high levels of toxin secretion has focused considerable attention for the use of this system in the expression and secretion of heterologous proteins. With further development of our infection protocols, the killer mycoviruses might ultimately form the basis of a broad host range vector system for fungi in general.

ACKNOWLEDGMENTS

This work was supported in part by HHS grants GM32496 and GM36441 from the National Institute of General Medical Sciences to K. A. B. We thank D. Beach, L. Davidow, and D. J. Tipper for yeast strains and plasmids; S. Hanes and S. Parent for comments on the manuscript; and Cecilia Barry for technical assistance.

REFERENCES

Al-Aidroos, K., and Bussey, H. (1978). Chromosomal mutants of *Saccharomyces cerevisiae* affecting the cell wall binding site for killer factor. *Can. J. Microbiol. 24*:228-237.

Beach, D., and Nurse, P. (1981). High-frequency transformation of the fission yeast *Schizosaccharomyces pombe*. *Nature 290*:140-142.

Bevan, E. A., and Makower, M. (1963). The physiological basis of the killer character in yeast, in *Genetics Today*, XIth Int. Congr. Genet., Vol. 1 (Geerts, S. J., Ed.). Pergamon, Oxford, England, pp. 202-203.

Boone, C., Bussey, H., Greene, D., Thomas, D. Y., and Vernet, T. (1986). Yeast killer toxin: Site-directed mutations implicate the precursor protein as the immunity component. *Cell 46*:105-113.

Bostian, K. A., Sturgeon, J. A., and Tipper, D. J. (1980). Encapsidation of yeast killer double-stranded ribonucleic acids: Dependence of M on L. *J. Bacteriol. 143*:463-470.

Bostian, K. A., Jayachandran, S., and Tipper, D. J. (1983a). A glycosylated protoxin in killer yeast: Models for its structure and maturation. *Cell 32*:169-180.

Bostian, K. A., Burn, V. E., Jayachandran, S., and Tipper, D. J. (1983b). Yeast killer dsRNA plasmids are transcribed *in vivo* to produce full and partial-length plus-stranded RNAs. *Nucl. Acids Res. 11*:1077-1097.

Bostian, K., Elliot, Q., Bussey, H., Burn, V., Smith, A., and Tipper, D. J. (1984). Sequence of the preprotoxin dsRNA gene of type 1 killer yeast: Multiple processing events produce a two component toxin. *Cell 36*:741-751.

Burnette, W. N. (1981). "Western blotting": Electrophoretic transfer of proteins from sodium dodecyl sulfate-polyacrylamide gels to unmodified nitrocellulose and radiographic detection with antibody and radioiodinated protein A. *Analyt. Biochem. 112*:195-203.

Bussey, H. (1981). Physiology of killer factor in yeast. *Adv. Microb. Physiol. 22*:93-112.

Bussey, H., Sacks, W., Galley, D., and Saville, D. (1982). Yeast killer plasmid mutations affecting toxin secretion and activity and toxin immunity function. *Mol. Cell. Biol. 2*:346-354.

Bussey, H., Saville, D., Greene, D., Tipper, D. J., and Bostian, K. A. (1983). Secretion of yeast killer toxin: Processing of the glycosylated precursor. *Mol. Cell. Biol. 3*:1362-1370.

Davidow, L. S., Apostolakos, D., O'Donnell, M. M., Proctor, A. R., Ogrydziak, D. M., Wing, R. A., Stasko, I., and DeZeeuw, J. R. (1985). Integrative transformation of the yeast *Yarrowia lipolytica*. *Curr. Genet. 10*:39-48.

de la Pena, P., Barros, F., Gascon, S., Lazo, P. S., and Ramos, S. (1981). Effect of yeast killer toxin on sensitive cells of *Saccharomyces cerevisiae*. *J. Biol. Chem. 256*:10420-10425.

Edmondson, S. P., Lang, D., and Gray, D. M. (1984). Evidence for sequence heterogeneity among the dsRNA segments of *Penicillium chrysogenum* mycovirus. *J. Gen. Virol. 65*:1591-1599.

El-Sherbeini, M., and Bostian, K. A. (1987). Viruses in fungi: Infection of yeast with the K1 killer virus. *Proc. Nat. Acad. Sci. USA 84*:4293-4297.

El-Sherbeini, M., Bevan, E. A., and Mitchell, D. J. (1983). Two biochemically and genetically different forms of L dsRNA of *Saccharomyces cerevisiae* exist: One form, L2, is correlated with the [HOK] plasmid. *Curr. Genet. 7*:63-68.

El-Sherbeini, M., Tipper, D. J., Mitchell, D. J., and Bostian, K. A. (1984). Virus-like particle capsid protein encoded by different L double stranded RNAs of *Saccharomyces cerevisiae*: Their roles in maintenance of M double stranded killer plasmids. *Mol. Cell. Biol. 4*:2818-2827.

El-Sherbeini, M., Levitre, J., Bostian, K. A., and Mitchell, D. J. (1987). Genc-protein assignments within the yeast *Yarrowia lipolytica* dsRNA viral genome. *Curr. Genet. 11*:483-490.

Field, L. J., Bobek, L., Brennan, V., Reilly, J. D., and Bruenn, J. (1982). There are at least two viral double-stranded RNAs of the same size: An explanation for viral exclusion. *Cell 31*:193-200.

Fried, H. M., and Fink G. (1978). Electron microscopic heteroduplex analysis of killer double stranded RNA species from yeast. *Proc. Nat. Acad. Sci. USA 75*:4224-4228.

Groves, D. P., Clare, J. J., and Oliver, S. G. (1983). Isolation and characterisation of a double stranded RNA virus like particle from the yeast *Yarrowia lipolytica*. *Curr. Genet. 7*:185-190.

Hanes, S. D., Burn, V. E., Sturley, S. L., Tipper, D. J., and Bostian, K. A. (1986). Expression of a cDNA derived from the yeast killer preprotoxin gene: Implications for processing and immunity. *Proc. Nat. Acad. Sci. USA 83*:1675-1679.

Hannig, E., Leibowitz, M. J., and Wickner, R. B. (1985). On the mechanism of exclusion of M2 double-stranded RNA by L-A-E double-stranded RNA in *Saccharomyces cerevisiae*. *Yeast 1*:57-65.

Hinnen, A. J., Hicks, J. B., and Fink, G. R. (1978). Transformation of yeast. *Proc. Nat. Acad. Sci. USA 75*:1929-1933.

Hopper, J. E., Bostian, K. A., Rowe, L. B., and Tipper, D. J. (1977). Translation of the L-species dsRNA genome of the killer-associated virus-like particles of *Saccharomyces cerevisiae*. *J. Biol. Chem. 252*:9010-9017.

Hutchins, K., and Bussey, H. (1983). Cell wall receptor for yeast killer toxin: Involvement of 1,6-β-D-glucan. *J. Bacteriol. 154*:161-169.

Ito, H., Fukuda, Y., Murata, K., and Kimura, A. (1983). Transformation of intact yeast cells treated with alkali cations. *J. bacteirol. 153*:163-168.

Julius, D., Brake, A., Blair, L., Kunisawa, R., and Thorner, J. (1984). Isolation of the putative structural gene for the lysine-arginine-cleaving endopeptidase required for processing of yeast prepro-α-factor. *Cell 37*:1075-1089.

Koltin, Y. (1977). Virus-like particles in *Ustilago maydis*: Mutants with partial genomes. *Genetics 86*:527-534.

Koltin, Y., and Kandel, J. S. (1978). Killer phenomenon in *Ustilago maydis*: The organization of the viral genome. *Genetics 88*:267-276.

Kyte, J., and Doolitle, R. F. (1962). A simple method for displaying the hydropathic character of a protein. *J. Mol. Biol. 157*:105-132.

Lee, M., Pietras, D. F., Nemeroff, M. E., Corstanje, B. J., Field, L. J., and Bruenn, J. A. (1986). Conserved regions in yeast defective-interfering viral double-stranded RNAs. *J. Virol. 58*: 402-407.

Lolle, S., Skipper, N., Bussey, H., and Thomas, D. Y. (1984). The expression of cDNA clones of yeast M1 double-stranded RNA in yeast confers both killer and immunity phenotypes. *EMBO J. 3*: 1383-1387.

Mitchell, D. J., and Bevan, A. E. (1983). ScV "killer" viruses in yeast, in *Yeast Genetics, Fundamental and Applied Aspects*, J. F. T. Spencer, D. M. Spencer, and A. R. W. Smith (Eds.). Springer-Verlag, New York, pp. 371-419.

Novick, P., Field, C., and Schekman, R. (1980). Identification of 23 complementation groups required for post-translational events in the yeast secretory pathway. *Cell 21*:205-215.

Olsnes, S., and Pihl, A. (1973). Isolation and properties of Abrin, a toxic protein inhibiting protein synthesis. Evidence for differential function of its two constituent peptide side chains. *Eur. J. Biochem. 25*:179-185.

Parent, S. A., Fenimore, C., and Bostian, K. A. (1985). Vector systems for the expression, analysis and cloning of DNA sequences in *S. cerevisiae*. *Yeast 1*:83-138.

Perlman, D., and Halvorson, H. O. (1983). A putative signal peptidase recognition site and sequence in eukaryotic and prokaryotic signal peptides. *J. Mol. Biol. 167*:391-409.

Schwartz, T. W. (1986). The processing of peptide precursors. *FEBS Lett. 200*:1-10.

Skipper, N., Thomas, D. Y., and Lau, P. C. K. (1984). Cloning and sequencing of the preprotoxin-coding region of the yeast M_1-dsRNA. *EMBO J. 3*:107-111.

Sommer, S. S., and Wickner, R. B. (1982). Yeast L dsRNA consists of at least three distinct RNAs: Evidence that the non-Mendelian genes [HOK], [NEX] and [EXL] are on one of these dsRNAs. *Cell 31*:429-441.

Stanway, C. A., and Buck, K. W. (1984). Infection of protoplasts of the wheat take-all fungus, *Gaeumannomyces graminis* var. *tritici*, with double-stranded RNA viruses. *J. Gen. Virol. 65*:2061-2065.

Sturley, S. L., Elliot, Q., LeVitre, J., Tipper, D. J., and Bostian, K. A. (1986). Mapping of functional domains within the *S. cerevisiae* type 1 killer preprotoxin. *EMBO J. 5*:3381-3389.

Tipper, D. J., and Bostian, K. A. (1984). Double-stranded ribonucleic acid killer systems in yeast. *Microbiol. Rev. 48*:125-156.

Treton, B. Y., Le Dall, M., Heslot, H. (1985). Virus-like particles from the yeast *Yarrowia lipolytica*. *Curr. Genet. 9*:279-284.

Tyagi, A. K., Wickner, R. B., Tabor, C. W., and Tabor, H. (1984). Specificity of polyamine requirements for the replication and maintenance of different double-stranded RNA plasmids in *Saccharomyces cerevisiae*. *Proc. Nat. Acad. Sci. USA 81*:1149-1153.

Wickner, R. B. (1974). Chromosomal and nonchromosomal mutations affecting the "killer character" of *Saccharomyces cerevisiae*. *Genetics 76*:423-432.

Wickner, R. B. (1976). Mutants of the killer plasmid of *Saccharomyces cerevisiae* dependent on chromosomal diploids for expression and maintenance. *Genetics 82*:273-285.

Wickner, R. B. (1978). Twenty-six chromosomal genes needed to maintain the killer double-stranded RNA plasmid of *Saccharomyces cerevisiae*. *Genetics 88*:441-452.

Wickner, R. B. (1980). Plasmids controlling exclusion of the K2 killer double-stranded RNA plasmid of yeast. *Cell 21*:217-226.

Wickner, R. B. (1981). Killer systems in *Saccharomyces cerevisiae*, in *Molecular Biology of the Yeast Saccharomyces: Life Cycle and Inheritance*, J. N. Strathern, E. W. Jones, and J. R. Broach (Eds.). Cold Spring Harbor Press, Cold Spring Harbor, New York, pp. 415-444.

Wickner, R. B., and Leibowitz, M. J. (1976). Two chromosomal genes required for killer expression in killer strains of *Saccharomyces cerevisiae*. *Genetics 82*:429-442.

Wood, H. A., and Bozarth, R. F. (1972). Properties of virus-like particles of *Penicillium chrysogenum*: One double-stranded RNA molecule per particle. *Viology 47*:604-609.

10

The Killer System of *Ustilago maydis*: Secreted Polypeptides Encoded by Viruses

Yigal Koltin
Tel Aviv University, Ramat Aviv, Israel

Ustilago maydis AS AN EXPERIMENTAL SYSTEM

Ustilago maydis is the causative agent of corn smut. The fungus is a heterothallic Basidiomycete (Hanna, 1929) that induces in corn plants the development of neoplastic tissue in which the sexual stage of the smut develops (Figure 1). The various phases of ploidy (haplophase, diplophase, and dikaryophase) are readily distinguishable, and all can be manipulated in synthetic medium under laboratory conditions. Diplophase, as the sexual stage, is the only phase in the life cycle that requires plant tissue for development and will not proceed through meiosis in synthetic medium.

The haploid phase consists of a unicellular, yeastlike growth. The oblong cells reproduce by budding. The cells are uninucleate and can be maintained in a vegetative state indefinitely (Figure 2). Sexual reproduction is dependent on an interaction between two haploids, compatible mates. Compatibility is determined by a single gene with two alleles, *a1* and *a2* (Rowell and DeVay, 1954; Rowell, 1955). Following an interaction between compatible mates, sexual development is initiated by the formation of mycelium and dikaryotic cells that are typical of the Basidiomycetes (Day and Anagnostakis, 1971a,b). Al-

Figure 1 Neoplastic development in corn as a result of infection by *Ustilago maydis* (left); uninfected plant (right).

though this stage can be formed in synthetic medium it is unstable and will proceed in its sexual development only in the plant tissue of the host. The development in the host tissue is regulated by a second gene with multiple alleles, designated *b* (Puhalla, 1970; Day et al., 1971; Silva, 1972). Thus far, some 24 alleles of *b* have been identified. This gene is unlinked to *a* (Rowell, 1955).

Within the plant tissue the mycelial form spreads and induces the formation of neoplastic growth that is seen as galls on leaves, stems, and in corncobs. The compatible haploid nuclei fuse to form the zygote and this phase is noteworthy in its accumulation of melanin. The zygotes are referred to as brandspores or teliospores; this is the phase that endures from one growth period to the next. The brandspores can be germinated and the diploid nuclei undergo meiosis as a germ tube is formed. Within the germ tube, known as the promycelium, septa are formed, separating the cytoplasm between each of the haploid nuclei. Form each cell of the promycelium a spore is formed. Thus, a complete tetrad of meiotic products can be obtained from each brandspore. The spores divide further by a mitotic division. Each of the haploid spores can form a yeastlike haploid colony (Figure 3).

Occasionally, irregular meiosis occurs and the homologous chromosomes do not migrate to opposite poles. This irregularity does not affect directly the progression of the development of other phases subsequent to meiosis and thus spores are formed that remain unreduced.

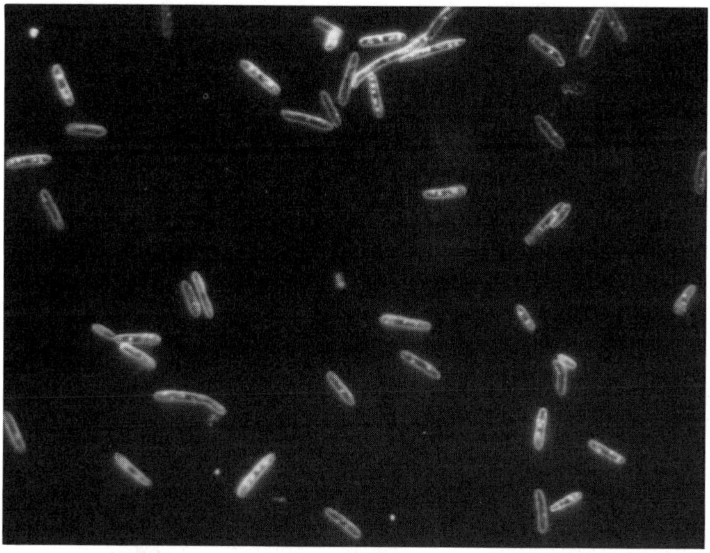

(a)

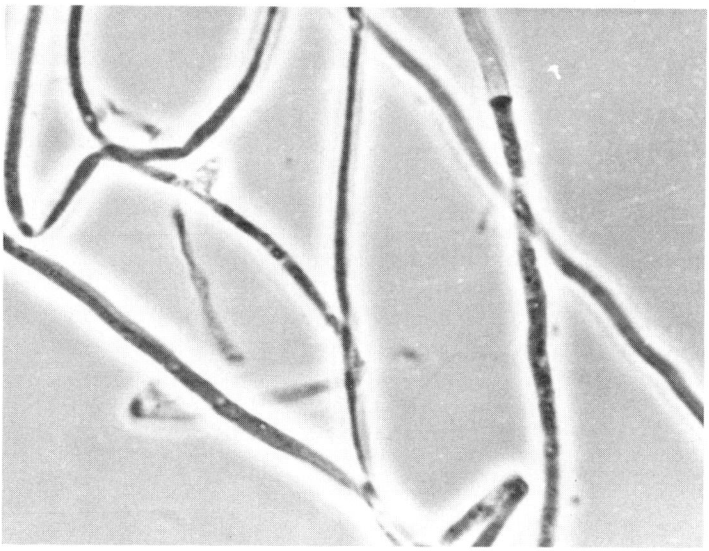

(b)

Figure 2 Cell morphology of *Ustilago maydis*. (a) Haploid cells. (b) Heterokaryotic mycelium.

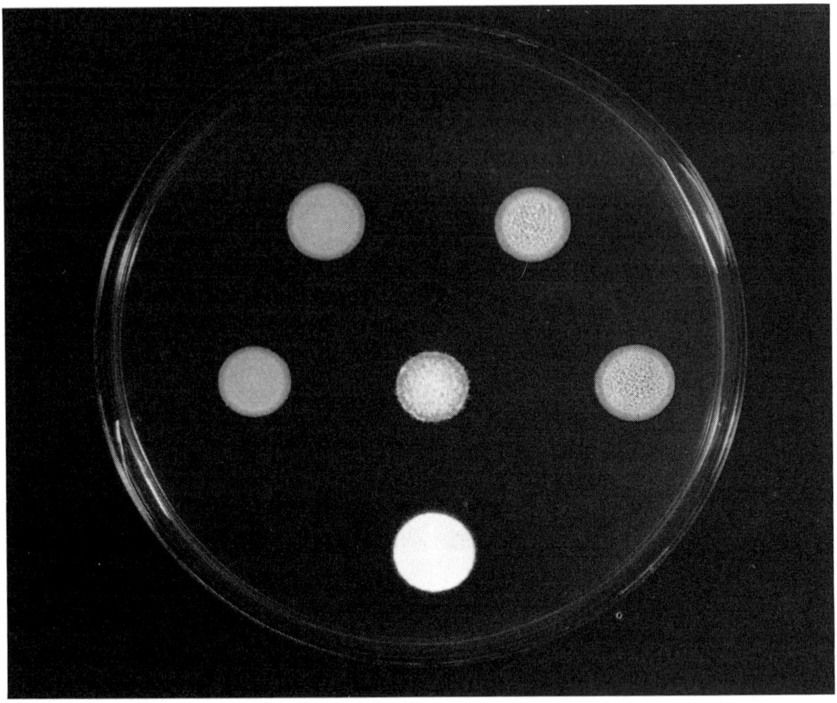

Figure 3 Morphological differentiation between haploid colonies (upper and middle rows) and a diploid colony (bottom). A heterokaryon formed between compatible mates is similar to the diploid colony.

The morphology of such diploid colonies is very distinct from the yeast-like haploid colonies seen in Figure 3 since they develop mycelia (Day and Anagnostakis, 1971b). Furthermore, cells from such colonies can effectively infect the plant without an earlier interaction with a compatible mate and are therefore called solopathogenic (Christensen, 1931; Holliday, 1961a,b; Puhalla, 1970). Only after the brandspores undergo meiosis do the constituents of the diploid become apparent among the haploid progeny.

Genetics

Ustilago maydis as a genetically amenable system has a number of advantages. As mentioned above, its life cycle can be controlled, the organism is heterothallic, individual tetrads can be isolated, and each state of ploidy can be obtained for experimentation. The disadvantages are that the sexual stage is obtained only in plant tissue, and that

10-14 days are required to complete a sexual cycle. In addition, the tetrads are borne externally (Figure 4) and therefore only unordered tetrads are obtained. Random spore samples may also be used for analysis of segregation. Various investigators encountered difficulties recovering all 4 meiotic products even when tetrads were isolated after each meiotic spore replicated 3-4 times. Nonetheless, the haploid phase is convenient for genetic analysis and the induction of mutations, as shown initially by Perkins (1949). The yeastlike character of this phase permits the use of replica plating even for testing mating types (Figure 5). The stable diploid phase that can be maintained in a vegetative state indefinitely and induced to undergo meiosis offers excellent material for the study of interallelic and intergenic interactions, gene conversion, and mitotic recombination (Holliday, 1964,1965,1967; Esposito and Holliday, 1964; Puhalla, 1970). The heterokaryon offers the means to study internuclear and intranuclear complementation in conjunction with the study of interallelic and intergenic interactions in the diploid phase.

In addition, the heterokaryotic phase offers an exceptional situation in which cytoplasmic inheritance can be studied. The instability

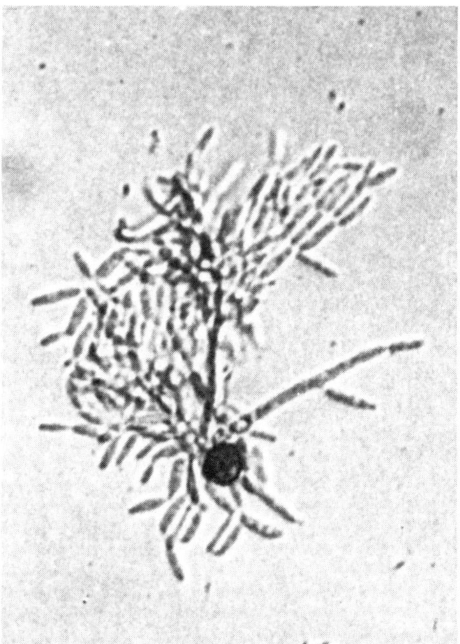

Figure 4 Germinated teliospore. The individual yeastlike cells are the haploid products of meiosis that occurred in the teliospore.

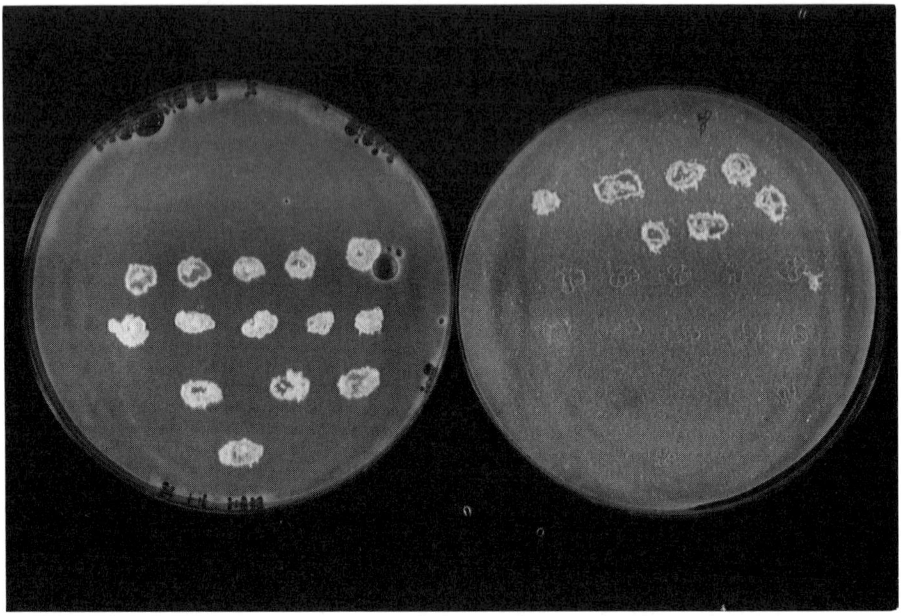

Figure 5 Replica plating of haploid colonies on tester lawns to determine their mating type. If tested strains are compatible (differing in both *a* and *b* loci), heterokaryotic mycelium develops.

of the heterokaryon on synthetic medium is an essential component for the performance of tests to determine whether specific phenotypes are transmissable via cytoplasmic mixing of the two mates, known also as cytoduction. Since heterokaryon formation is dependent on an interaction between mates with different incompatibility genes, one can ensure the formation of a heterokaryon between two different mates. The haploid phase is distinguishable morphologically from the heterokaryotic phase. Therefore, the formation of the heterokaryon is easily discernible and the dissociation of the heterokaryon and the recovery of its haploid parental components are clearly recognized. The transmission of a specific phenotype merely by cytoplasmic mixing is a clear indication for the relation between a cytoplasmic element and the examined phenotype.

Ustilago maydis is used in genetic studies involving other basic genetic phenomena such as meiotic recombination and repair mechanisms (Unrau, 1975; Pukilla, 1978; Holliday and Dickson, 1977). Currently, efforts are directed toward the development of useful shuttle vectors and an efficient transformation system. An initial successful effort had been reported (Banks, 1983a,b) using a *Saccharomyces cerevisiae*-

Escherichia coli shuttle vector with the aminoglycoside phosphotransferase gene as the selector and the yeast 2 μm sequences as the replicator. The constructed plasmid did not contain any sequences from *U. maydis* and the efficiency of transformation was low. Recently, two efficient transformation systems were developed (S. Leong, personal communication; Lorz, personal communication). Cloning of the genes may offer a shortcut for mapping the genome using nucleic acid hybridization after chromosome separation by electrophoresis (Schwartz and Cantor, 1984; Carle and Olson, 1984). The chromosome number in *U. maydis* is still unknown; the earlier estimates of two chromosomes (Ainsworth and Sampson, 1950) have been superseded by most recent findings by electrophoretic separation, that suggest the presence of 11 chromosomes (S. Leong, personal communication). In a short time, chromosome separation and gene cloning should establish the genetic map of this organism.

Biochemistry

The biochemistry of *U. maydis* has hardly been explored thus far but the biology of the organism does not impose any difficulty for biochemical studies. The yeastlike phase can be grown in defined medium. On solid medium it forms discrete colonies and in liquid medium the organism grows well in shake cultures. The cell density that is obtained is ca. 10^8 cells per ml in complete medium and ca. 10^7 cells per ml in minimal medium. The generation time is ca. 2.5 hr in complete medium and 3-4 hr in minimal medium. Thus, in terms of cell density, *S. cerevisiae* and *U. maydis* reach the same density under similar conditions. Although special media were devised for *U. maydis* (Stevens, 1974) the standard media used for *S. cerevisiae* (Sherman et al., 1982) are also used for studies with *U. maydis*. Aeration is essential since the organism does not tolerate microaerophilic conditions.

Extensive proteolytic activity as is found in *S. cerevisiae* is not observed in *U. maydis*. Also, only a few proteins are secreted. Some extracellular nuclease activity is present (Holliday and Halliwell, 1968) but it has not affected the recovery of high molecular weight nuclear DNA nor the extraction of dsRNA. The cell walls of *U. maydis* are accessible to degradation by the enzymes derived from *Trichoderma harzianum* (Novozyme 234) using the protocol devised for *S. cerevisiae* (Sherman et al., 1982), including the osmotic stabilization of the protoplasts with sorbitol (Figure 6). The regeneration of the protoplasts is well over 90%.

In addition to the biochemical studies related to the killer system of *U. maydis*, remarkable progress has recently been made using this organism in the identification and purification of enzymes and proteins related to recombination and DNA replication (Holloman et al., 1981; Rowe et al., 1981; Kmiec and Holloman, 1982,1983,1984; Kmiec et al.,

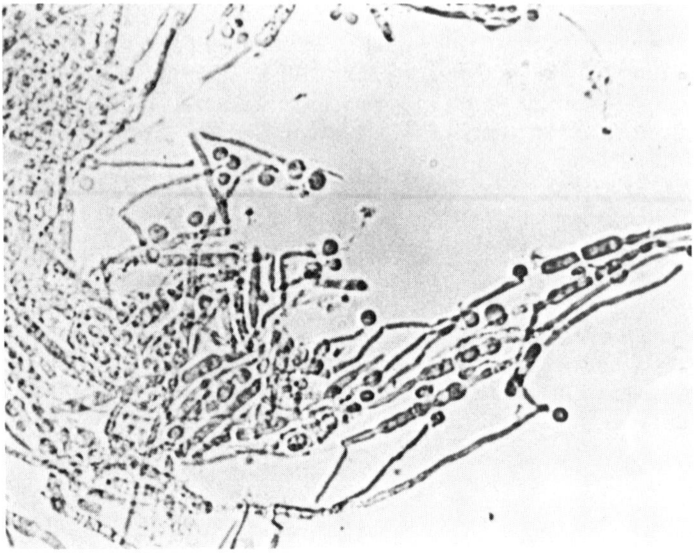

(a)

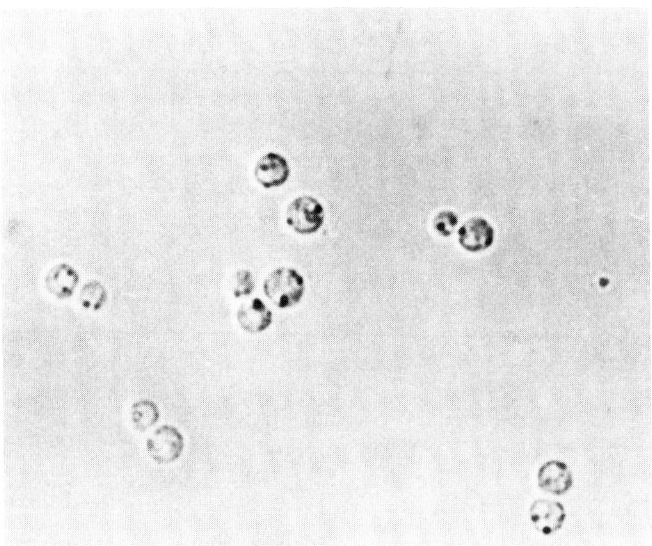

(b)

Figure 6 Generation of spheroplasts. (a) Initial stages indicated by the release of spheroplasts from tips of the elongated cells. (b) Spheroplasts stabilized in 1 M sorbitol.

Banks, 1973; Rusche et al., 1980; Holloman, 1975; Kmiec et al., 1985). Other biochemical studies with *U. maydis* have included the study of ergosterol biosynthesis (Henry and Sisler, 1981; Georgopolus and Sisler, 1970).

INTERSTRAIN INHIBITION

Detection and Incidence

Interstrain inhibition in ustilago was detected for the first time during a study of compatibility reactions in solid medium in an effort to simplify the identification of incompatibility alleles and to estimate their frequency (Puhalla, 1968). In some compatible interactions using specific strains a clear inhibition was noticed and the typical mycelium did not develop. The survey included strains from various parts of the United States and Canada. The inhibition was noticed only in interactions with four strains. Later tests with strains from Poland (Silva, 1972) and from Mexico (Y. Koltin and P. R. Day, unpublished data) and the examination of a local population in Connecticut (Day, 1981) clearly indicated that the incidence of strains causing interstrain inhibition is ca. 1%.

There are no indications for any linkage between the inhibitory effect and a specific mating type nor are there any indications that would suggest some relationship between the degree of virulence of the pathogen and the inhibition phenomenon. Furthermore, although few populations have been screened for the effect, it does not appear to occur with a specific geographical distribution. Recently, such strains have been detected in China (Lu, personal communication). It appears that every survey will lead to the detection of such strains in nature, providing the sample size is sufficiently large.

Specificity of Inhibition

The population of *U. maydis* can be divided on the basis of the inhibitory phenomenon to two types, strains that elicit the inhibitory effect and strains that do not elicit such an effect. The former will be referred to henceforth as "killer" strains and the latter as nonkiller strains.

Among the nonkillers, the response to killer strains was not uniform. Some were sensitive to all killer strains, some to none, and most were sensitive to some killer strains but not to all. These results suggested that the inhibitory effect is not exclusively of one type and that the inhibitory phenomenon reflects a group of different effects with different specificities similar to those found in the bacterial colicins (Konisky, 1982). This was further supported by tests of the response of each killer strain to the other killer strains. Mutual sensitivity was detected among the killer strains (Figure 7) and the pattern

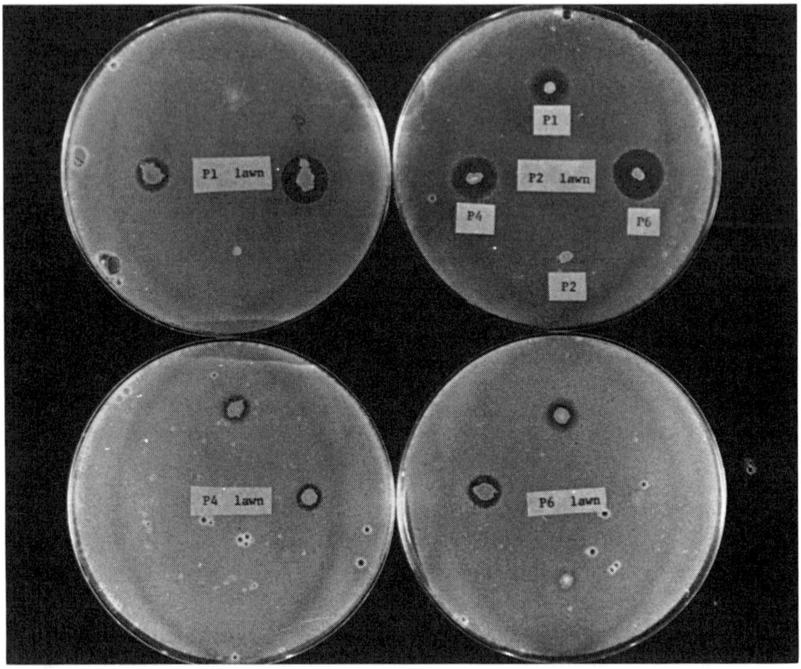

Figure 7 Mutual sensitivity of "killer" strains. Upper right, 3 killer strains (P1, P4, and P6) displaying the toxic activity on a sensitive lawn (P2 is the sensitive strain used as a lawn). Tested strains are plated in all the plates in the same order, but P1 is used as a lawn in upper-left-hand plate, P4 is used as a lawn in the botton-left-hand plate, and P6 in the bottom right. Notice mutual sensitivity among the killer strains. (*Source*: From Koltin and Day, 1976a.)

that emerged suggested the occurrence of three killer specificities, designated P1, P4, and P6. It is unclear whether the three specificities reflect the entire spectrum found in nature; only additional surveys can provide a clear picture of the extent of this phenomenon. The three known specificities are from the study of Puhalla (1968), and no new specificities were identified later by Day (1981) in a survey of a local population. The resistance to the killer effect may suggest that either *U. maydis* evolved under the selection pressure of such a phenomenon or that the inhibitory factors are similar to other agents that impose a selection pressure on populations of this organism.

Among the native killer strains, each one expressed either specificity P1, P4, or P6. Efforts to expand the expression of a single cell to more than one killer type revealed that certain combinations can be derived experimentally, but all three pairwise combinations are not found. P1 and P4 can be expressed in a stable form by a single cell (Figure 8) (R. Levine and Y. Koltin, unpublished data), whereas P1 and P6, or P4 and P6 cannot be maintained in a stable form by a single cell (Wigderson and Koltin, 1982). In vegetative cells derived from a heterokaryon synthesized by mating two compatible auxotrophic strains on minimal medium, the exclusion of P6 by both P1 and P4 is clearly noticed. The exclusion is nearly complete after 9-10 generations. Similar results have been noticed in progeny from crosses between killer strains with different specificities. Only from a cross of P1 × P4 were progeny displaying both specificities obtained. In all other combinations, exclusion relations led to the loss of P6 from a cross of P4 × P6 and in a cross of P1 × P6. In earlier studies (Koltin and Day, 1976a) it appeared that P1 and P6 mutually exclude each other, yet in later studies, as a result of some change in the strains or the factors associated with the elements encoding the inhibitory function, these relations were modified. All the currently available killer strains display unilateral exclusion in which P6 is always excluded. The molecular basis for the exclusion relations will be discussed later.

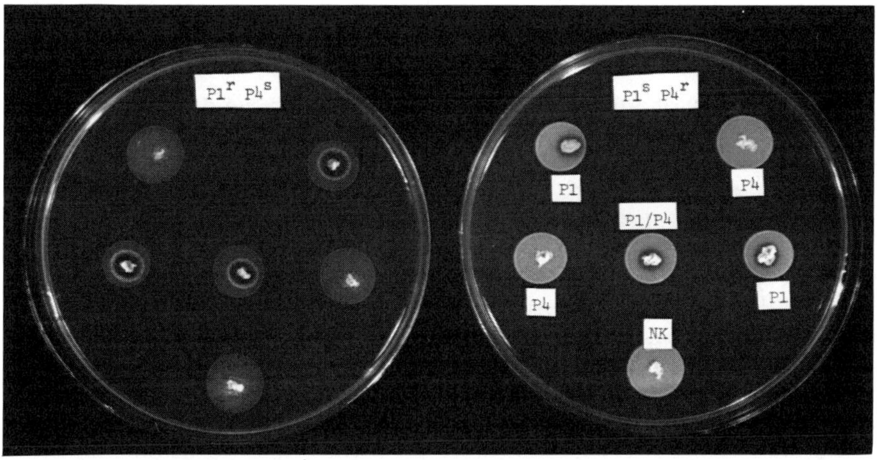

Figure 8 Dual killer activity of P1 and P4 in a colony derived from a single cell (middle). The lawns in the left plate are all from a strain resistant to P1 and sensitive to P4. The lawn in the plate on the right are all from a strain sensitive to P1 and resistant to P4. In the upper row killer strains (left to right) P1 and P4 spotted, in the middle row killer strains (left to right) P4, a strain with dual specificity (center), and P1 killer strain. (*Source*: From Wigderson and Koltin, 1982.)

Beyond the intraspecies specificity, the killer phenomenon appears to be highly species-specific within the order Ustilaginales (Koltin and Day, 1975). A wide spectrum of fungi and bacteria has been tested to date for sensitivity to the killer substance. An emphasis was placed on the examination of microorganisms that compete for the same host, in an effort to examine the biological significance of the killer phenomenon as an antibiotic effect with some selective advantage. The result obtained clearly indicate that none of the bacterial specimens tested are sensitive. Of the fungal species tested, including various representatives of filamentous forms from the Ascomycetes and Basidiomycetes, are all insensitive. Furthermore, the yeasts *S. cerevisiae* and *Candida albicans* are insensitive; likewise many yeastlike forms of the marine Basidiomycetes were all insensitive to all the killer types (Koltin, 1986). The only forms that displayed sensitivity were other Ustilaginales, primarily those that are pathogens of other graminaeous plants such as wheat, oats, and barley. Slight sensitivity was found also in a smut of ornamentals. However, it is clear that the phenomenon is highly specific to closely related species. Ustilago has a high degree of specialization for specific hosts and the sensitive Ustilaginales have no common host. It does not appear that the inhibitory phenomenon is related to competition but rather to a common receptor for the killer substance that is typical of certain species within the Ustilaginales.

Genetics

Genetic analysis of interstrain inhibition included random spore analysis, tetrad analysis, and cytoduction experiments. All the results suggested that the killer phenomenon is related to a cytoplasmically inherited element (Puhalla, 1968; Koltin and Day, 1976a). Crosses between killer strains and nonkiller strains were performed. Such crosses are feasible even between a killer strain and a sensitive strain, since the crosses are performed by injection of cell suspensions in corn seedlings. In the plant tissue, for an unknown reason, the inhibitory effect does not occur against sensitive strains and the mating interaction between killer and sensitive strains is not affected. In random spore analysis, in most crosses, nearly 80-90% of the progeny displayed the killer phenotype. The nonkillers were either resistant to the specific killer type or sensitive to it. These results could be interpreted as the inheritance of a cytoplasmic trait or a polygenic trait. However, tetrad analysis yielded 4:0 tetrads (killer vs. nonkiller progeny) and the most convincing evidence that the killer phenomenon is related to a cytoplasmic trait was obtained by transmission of the killer phenotype via cytoduction to the nonkiller parent.

In the genetic analysis performed initially by Puhalla (1968) with P1, an additional cytoplasmically inherited phenotype was identified. Among the nonkiller progeny from a cross between a killer strain and a sensitive strain, some resistant progeny were shown to transmit the resistance as a cytoplasmic trait. This phenotype was designated P3 and was referred to as cytoplasmic immunity (Koltin and Day, 1976a). Thus, it appeared that a cytoplasmic immunity factor is associated with the killer trait, conferring immunity to the killer cells. In later studies the results with P1 were repeated (Peery et al., 1982) and similar relations were also suggested for P4 and P6 killer subtypes (Koltin and Day, 1976b). However, the latter results have been reexamined recently (T. Peery, A. Finkler, and Y. Koltin, unpublished data) and current data suggests that the immunity element may not exist in P6.

The resistance to the killer effect in naturally occurring nonkiller strains is determined by nuclear genes (Puhalla, 1968; Koltin and Day, 1976a). Resistance to each subtype of killer, P1, P4, and P6 is conferred by a different gene, and the recessive allele of each gene confers resistance. The genes appear unlinked although resistance to P1 and P4 may be loosely linked. All the native killer strains carried the resistance allele. Thus, if cytoplasmic immunity occurs simultaneously with the cytoplasmic killer element, the carriers are protected by the nuclear resistance genes in addition to the immunity factor. The recessive nature of the resistance suggested that these nuclear genes encode a receptor for the killer substance. However, the role of these genes seems to be more complex (Finkler et al., 1984). In cytoplasmic transmission experiments it appears that only P1 strains can transmit the cytoplasmic element related to the inhibitory phenotype to sensitive strains. P4 and P6 are transmitted to nonkiller cells only if they contain the specific nuclear resistance gene effective against the inhibitory substance of the donor. The function of these genes is not merely related to resistance, since killer diploids heterozygous for the nuclear resistance allele were formed. Heterozygotes for the nuclear resistance genes are sensitive to the inhibition if they do not carry the element related to the killer function. Heterozygous diploids expressing the inhibitory function were thought to be immune due to a cytoplasmic immunity factor. Most recent findings, however, indicate that once the diploids are sporulated all the progeny are resistant, suggesting that those diploids were homozygous for the resistance allele, probably due to somatic recombination or the selection of rare resistant cells from the sensitive strain. Furthermore, the dissociation of linkage between the P4 and P6 killer element and immunity could not be demonstrated as it was for P1. Thus, it appears at this time that there is no cytoplasmic element that confers immunity to P6; probably the same is true for P4. The protection of the cells in those cases is conferred only by the nuclear gene. This may be the reason that all native strains contain the nuclear resistance allele corresponding to the expressed killer specificity. The earlier report by Koltin and Day (1976a) that cytoplasmic immunity was detected in P4 and P6 is probably incorrect, due to a confusion of nuclear resistance and immunity along with the low viability in tetrads. This combined effect provided an impression of a 4:0 segregation of the resistance in some tetrads.

DOUBLE-STRANDED RNA VIRUSES

Relation to the Inhibitory Phenotype

Virus particles in *U. maydis* were detected by Wood and Bozarth (1973) in strains used earlier by Day and Anagnostakis (1973) in cytoplasmic mixing experiments. The experiments were performed with P1 strains, P3 strains, and with sensitive strains. The results conformed precisely to the expectations: P1 and P3 strains contained virus particles (Figure 9) and the sensitive strains were devoid of virus particles. Later studies suggested that these results may have been a coincidence or resulted from a low titer of the viruses in these specific sensitive strains, since nonkiller strains and even sensitive strains can contain virus particles (Koltin, 1977; Koltin et al., 1978). Some sensitive strains contain the viral nucleic acids without complete capsids (Koltin and Day, 1976b) but this is not always the case. Viral nucleic acids have been extracted from almost every nonkiller strain tested in the

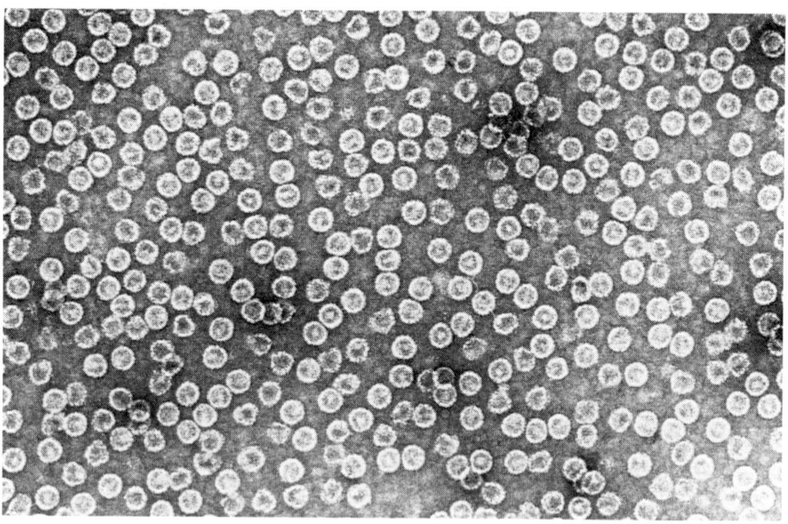

Figure 9 Virus particles of *U. maydis* (41-43 nm in diameter) after CsCl density gradient.

laboratory and from almost every sporidial culture from a local population of *U. maydis* examined by Day (1981).

The particles detected were ca. 41-43 nm in diameter (Figure 9). The particles formed a number of sedimentation components in sucrose gradient and in CsCl density gradients but all the components contained virsues with a capsid of the same size. The nucleic acids extracted from the viral particles were double-stranded RNA based on (1) the melting profile, (2) the sensitivity of the molecules to RNase in reactions performed at low but not at high ionic strength, (3) the insensitivity of the nucleic acids to DNase, and (4) the positive reaction with orcinol and negative reaciton with diphenylamine. The same type of particles were detected also in killer strains of type P4 and P6 (Koltin and Day, 1976a).

Morphologically, the viruses purified from P1, P4, and P6 strains were indistinguishable, and antibodies obtained against one virus reacted with the others. The major capsid protein was identical in the viruses purified from each of the killer types with a molecular weight of 75 kDa. The structure of the viral capsid included 120 molecules of the major protein. Very few other minor proteins were detected in purified capsids (Koltin et al., 1978). The dsRNA extracted was segmented and the most distinct factor typical of each of the killer types was the pattern of segments of dsRNA recovered directly from killer strains and from purified virus particles. The recovery of the same pattern from particles and from whole cells indicated that all the dsRNA molecules are encapsidated.

Three basic patterns of the major dsRNA segments are known to date, each corresponding to a specific killer type: P1 with P6 segments, P4 with 7 segments, and P6 with 5 segments. As described in Figure 10 and Table 1 the segments are divided into three size classes and have been designated Heavy (6.7-2.6 kilobase pairs), Medium (1.7-0.9 kilobase pairs) and Light (0.3 kilobase pairs). Within each class the segments are designated according to the position of the segment in the gel, in a descending order from the heaviest segment to the lightest segment.

Limited sequence data indicate that the 3' terminal residue of most segments consists of a mixture of A-OH or G-OH. In addition, some segments display a certain degree of heterogeniety in the 3' terminal sequences that were analyzed. This heterogeneity suggests inadequate resolution of some segments as a result of an overlap of different segments in the gels. The 3' ends of the segments of the *S. cerevisiae* viruses and the ustilago viruses are within a consensus sequence A/U A/U A/U A/U A/U A/U U/G C A/G-OH, which may reflect a functional similarity. Some of the segments end with a sequence UUUUUCA/G-OH, which is the transcriptase initiation consensus sequence of the *S. cerevisiae* virus (Bruenn and Brennan, 1980; Brennan et al., 1981). This may be also the transcription initiation site for the ustilago viruses.

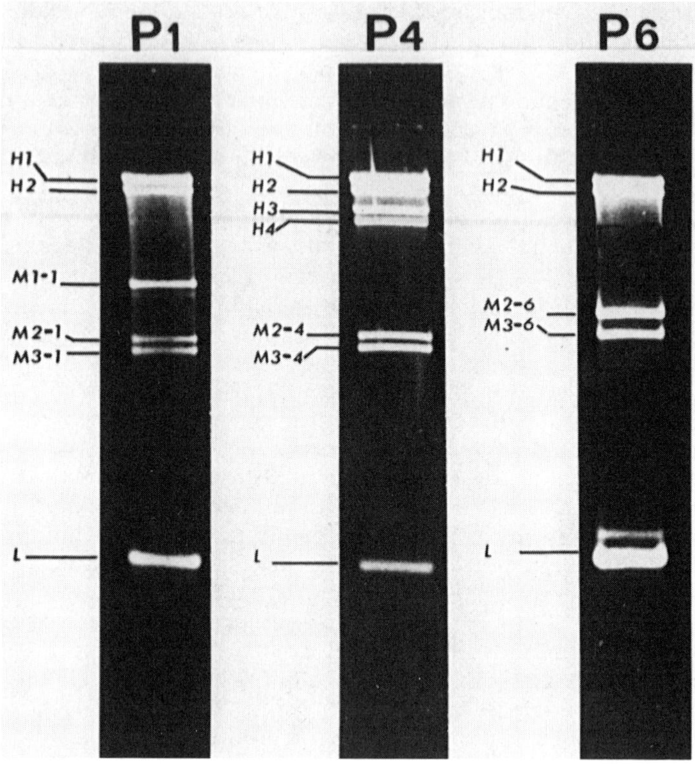

Figure 10 Three dsRNA patterns from different killer strains P1, P4, and P6, in 5% polyacrylamide gel electrophoresis. H, M, and L indicate segment designation. P6 used currently in all studies contains only H1, M2-6, and L.

The patterns described in Figure 10 are those that were identified originally in the native killer strains. However, some instability in the viral genome has been noticed either as the loss of specific segments or the addition of new segments. Some of the modifications of the viral genome are advantageous for experimental purposes since these variations are helpful in mapping viral functions. As an example, the P6 subtype used today in the laboratory includes only three of the original six segments, H1, M2-6, and L. These are sufficient for viral maintenance, coat formation, and the expression of the killer substance. Additional segments heavier than M2 and another heavier than L have also been noticed in P6. The former was found in a UV-induced nonkiller of P6 and the latter as a fraction that is seen clearly

Table 1 Characteristics of the *U. maydis* Virus

	Empty particles	Particles containing segment H1	H2	M3 (a)	M3 (b)	M3 (c)
CsCl (gm/ml^{-3})	1.278	1.418	1.394	1.370	1.341	1.314
$S_{20,w}$ (10^{-13}/sec)	98	172	157	–	–	–
MW dsRNA segment recovered $\times 10^6$ Da	–	4.2	3.1	0.67	0.67	0.67
MW of virus $\times 10^6$ Da	9.8	14.0	13.0	11.81	11.14	10.47

Source: From Bozarth et al. (1981).

in some extractions and not in others (see Figure 10). In addition, some dsRNA bands do not resolve well. The P4 H3 band includes two segments that can be resolved only in 2.5% polyacrylamide gels. In P1 and P4, an additional L segment lighter than the consistent L segment is noticed in some extractions. Also, a segment slightly lighter than M3 of P6 that constitutes less than 5% of the dsRNA extracted from virions can be detected if the dsRNA loaded on the gels is overloaded. Thus, some dynamic alterations of bands should be anticipated as noticed by Sommer and Wickner (1984) in the viruses of *S. cerevisiae* and in other dsRNA rotaviruses in which genome rearrangements are seen during serial passage (Hundley et al., 1985). The detection of additional bands and overlapping bands is made probable either by better resolution among segments of a similar size or by more refined procedures such as curing, end-labeling, and nucleic acid hybridization [that resolved two L segments of the same size in the viruses of *S. cerevisiae* (reviewed by Tipper and Bostian, 1984)]. The bands that do not appear consistently may be similar to the dsRNA molecules that Wesolowski and Wickner (1984) detected recently and can be seen in cells grown at suboptimal conditions or may reflect clonal variation phenomenon as described by Sommer and Wickner (1984) and further elaborated by Hannig et al. (1986) although a polyadenylate tract is not known in the M dsRNA of the ustilago virus. Most of the wild population of *U. maydis* contains dsRNA. The patterns detected seem to be based on the three major subtypes P1, P4, and P6 and most strains contain the heavy segments.

Organization of the Viral Information

The organization of the viral dsRNA is such that each group of segments, H and M, is separately encapsidated in different particles (Bozarth et al., 1981). In purification and characterization of *U. maydis* viruses, contrary to other dsRNA viruses (such as reovirus), a number of density components are observed. If all the dsRNA segments were encapsidated in a single particle, two density components should be obtained in a preparation of such viruses, one corresponding to empty particles and the other to full particles. In purification of *U. maydis* viruses, a number of density components are obtained (Table 1). These density components are a result of the separate encapsidation of dsRNA segments. Results from extraction of the nucleic acids from particles of different densities indicate that different molecules are encapsidated in particles with different densities. The differences in density result from the fact that the same type of capsid encapsidates molecules of different size. The difference between the sedimentation coefficients of empty and full capsids suggests that each particle can contain up to 24% dsRNA. Such particles band at a density of 1.42 g/cm in CsCl. Since the molecular weight of empty capsids is 9.8×10^6 Da and the heaviest particles are 14.0×10^6 Da, these particles must contain only one segment of the H class. The heaviest segment is 4.2×10^6 Da. Combinations such as H1 + H2 would cause the capsids to be heavier than any particles that have ever been detected.

Among virions containing the M segment, a number of density components are detected even among capsids containing only one type of M segment. Thus, particles encapsidating M dsRNA segments can contain up to three molecules in one particle, not exceeding 24% dsRNA per particle as the upper limit. A similar packaging of dsRNA was found in the *S. cerevisiae* virus (Wickner, 1986) and it is suggested that it is related to the mode of replication of the virions. It is unclear whether L is also encapsidated separately although, current data on its origin suggest that these segments are separately encapsidated.

Mapping Viral Functions

The mapping of viral functions is derived by two basic methods. One is based on the correlation between the loss of a specific segment and the loss of a specific function or the essentiality of specific segments. A second method is based on the in vitro translation of isolated segments and the identification of the products as those participating in specific viral functions. Both approaches have been used and some of the viral functions have been mapped (Table 2) (Koltin et al., 1978; Koltin et al., 1980; Peery et al., 1982; Dalton et al., 1985; T. H. Chang, J. A. Bruenn, N. Held, T. Peery, and Y. Koltin, unpublished data).

Table 2 Characteristics of Viral Subtypes: Segment Size and Function

VIRAL SUBTYPE								
P1			P4			P6		
Segment[a]	Size[b]	Function[c]	Segment	Size	Function	Segment	Size	Function
H1	6.1-6.7	Coat	H1	6.1-6.7	Coat	H1	6.1-6.7	Coat
H2	4.5-5.0	?	H2	4.5-5.0	Coat	H2	4.5-5.0	–
M1	1.4-1.7	Toxin	H3	3.0-3.9	Coat	M2	1.0-1.2	Toxin
M	0.9-1.1	?	H4	2.6-3.6	Coat	M3	0.9-1.1	–
M	0.9-1.1	?	M3	0.9-1.1	?	L	0.3	Toxin expression
L	0.3	Immunity	L	0.3	?			

[a]Only the major segments are listed.
[b]Range in kilobase pairs from estimates using PAGE (Bozarth et al., 1981), agarose gel electrophoresis, and electron microscopy (Field et al., 1983).
[c]Deduced from in vitro translation segments M1 of P1; M2 of P4 and P6; H2 of P4, and genetic studies.

Mutants with less than the complete set of segments typical to each killer type can be induced by various agents such as M-methyl-N-nitro-nitrosoguanidine and ultraviolet radiation (Koltin, 1977; Kandel and Koltin, 1978; R. Levine, unpublished data). In addition, some of the nonkiller progeny from crosses between killer strains and nonkiller strains contain only a portion of the segments originally found in the parental killer strain. Using a mutagen, nonkiller cells can be obtained from a killer strain at a frequency of 10^{-2}-10^{-3}. Also, from crosses of killer × killer strains, nonkiller progeny constitute some 10% of the total progeny. Strains in which spontaneous loss of segments occurred were also detected and included in these studies. The variants obtained by these methods were tested for the maintenance of virus particles, killer expression, and cytoplasmic immunity.

One striking feature that emerged from these studies is that some strains have only H segment(s), yet none of the strains contain segments of M or L without H. Two strains had L without M and none had M without L. The relations between M and L will be discussed in more detail later. The occurrence of H alone and the lack of any example of the occurrence of M and L without H suggested an interdependence between M, L, and H. The essential functions for maintenance such as encoding the coat protein, viral transcriptase, and replicase are thought to be encoded by the H segments. Although a virus associated RNA polymerase activity has been characterized (Ben-Zvi et al., 1984), it is unclear whether it is a viral or a host function. As for the major coat protein it was recently (Dalton et al., 1985) produced by the in vitro translation of denatured purified H2 segment of P4. This in vitro translation product was shown to be similar to the viral major coat protein by partial proteolytic digestion and peptide mapping. However, the H2 segments could code for ca. 1500 amino acids or an additional protein of 75,000 Da beyond the major coat protein that is of this size. H1 is 20% larger than H2 and could even code for an additional protein. Thus, the H segments may encode more than one product.

The M segments were assumed to play a role in the expression of the inhibitory function. All killer strains contain M segments and a loss of the function is very often associated with the loss of certain M segments, M2 in P4 and P6. In P1 a loss of M segments was also associated with the loss of the killer function but it involved always the simultaneous loss of M1 and M2. The relationship between the two segments is not known.

The deletion data left unclear the question of whether the dsRNA encodes the inhibitory substance or a signal that activates a host function. Convincing genetic data supporting the contention that the M segments encode the inhibitory function came from the exchange of specific M segments accompanied by a change of killer specificity (Koltin et al., 1978). From a cross of a P4 killer strain with a nonkiller, three types of nonkiller progeny were obtained. Two of the nonkillers were lacking M2 and L of P4 and one was also lacking M3 of P4. Each of the nonkillers was crossed to a P6 killer strain and progeny were obtained expressing the P6 specificity and containing M2 and M3 from P6 and M3 of P4. In the strain that lacked the P4 M2 and M3, new M2 and M3 dsRNA segments from P6 restored the killer function and along with it the change of specificity from P4 to P6. The information that the loss of M2 in strains derived from P4 and P6 killers is associated with the loss of the killer function and genetic data suggesting that M2 is associated with the killer function was further supported by a spontaneous mutant of P6, in which segments H2 and M3 were deleted. This mutant remained an active killer. Thus, the information pointed to M2 as the toxin-encoding segment.

Recently, in vitro translation of purified M1 of P1 and M2 of P4 and P6 was performed. In each case, a protein was encoded (Table 2). The products of the translation reaction of P6 were reacted with antiserum against the P6 toxin (T. H. Chang, J. A. Bruenn, N. Held, T. Peery, and Y. Koltin, unpublished data) and confirmed that the product is the pretoxin molecule. From the size of the translation product, it appears that the product obtained from each killer type is the only protein that can be encoded from this segment unless it contains overlapping reading frames.

The L segment was proposed to be involved in toxin expression without specifying its function (Kandel and Koltin, 1978). This suggestion was based on its ability to restore expression of the inhibitory function in a viral mutant lacking L. Later it was suggested that L confers immunity in cells containing the P1 subtype since nonkiller cells containing L were insensitive and those lacking it were sensitive to inhibition by P1 (Peery et al., 1982). Attempts to derive similar mutants in P4 and P6 failed, and therefore it is not known whether L plays a similar role in P4 and P6. Translation of L from P1 did not yield any detectable product, and available sequence data of the entire segment suggest that it is not informative. There are four initiation codons in this 355 basepair segment and four termination codons (T. H. Chang, J. A. Bruenn, N. Held, T. Peery, and Y. Koltin, unpublished data). However, it is conceivable that L plays a role in transport, as shown for other small RNAs (Walter and Blobel, 1982), and as such may also confer immunity.

RNA Polymerase

As in all other fungal viruses in which the RNA polymerase was investigated (Buck, 1980), in ustilago virus the enzymatic activity studied in P6 copurifies with the virus particles (Ben-Zvi et al., 1984). Conditional mutants of this enzyme have not been recovered thus far, and the enzyme was not purified. Therefore, it is not known whether this activity is encoded by the virus or the host and whether cells free of the dsRNA viruses contain this activity.

The activity associated with the particles is affected by the host. The specific activity of the polymerase reaction is at its maximum in virions purified from late log-early stationary phase cells. The specific activity is low in particles purified from cells in stationary phase since many particles at that stage lack a template. Most of the particles purified from cells at that stage do not contain dsRNA.

The products of the polymerase reaction are single-stranded RNA molecules and therefore the polymerase activity associated with the virions is a transcriptase. The reaction is sensitive to Na^+ ions and levels above 0.02 mM cause inhibition. The reaction requires Mg^{+2}, and its pH optimum is 8.9, somewhat higher than that of the *S. cerevisiae* virus (Welsh et al., 1980; Bruenn et al., 1980). Three classes of transcripts are synthesized corresponding to the three segments, and their size is nearly half the molecular weight of the respective dsRNA segments of P6. Since the dsRNA segments of different classes are separately encapsidated, the synthesis of ssRNA homologous to the three segments suggests that all the viral capsids contain this enzymatic activity. The transcripts are released from the virions. In reactions containing bentonite, the particles adhere to the bentonite. The particles are removed by centrifugation and ssRNA homologous to the dsRNA segments can be recovered. By a comparison of the distribution of labeled precursor in ssRNA and dsRNA after extraction of particles, and the distribution of incorporation in the supernatant and the particles, it appears that as soon as a transcript is completed it is released from the particle. In most experiments, less than 5% of the incoproration is into dsRNA molecules.

In pulse-chase experiments, the smallest transcript was shown to be synthesized after the large and medium transcripts. The molar ratio of these transcripts is 1:5:4 for the large, medium, and small transcripts, respectively. These results suggest some relationship between the medium and small transcripts. From most recent information on the relationship between those two segments it is conceivable that the transcript of L is not synthesized independently but occurs as a processed product of the M message (T. H. Chang, J. A. Bruenn, N. Held, T. Peery, and Y. Koltin, unpublished data). If this is proven to be correct, the molar ratios between the messages of H, M, and L derived from in vitro transcription with virus particles suggest that the viruses contain an RNA-processing enzyme. The dsRNA-dependent RNA polymerase reaction has become a standard procedure to identify the position of the virions in sucrose gradients during purification of the viruses. The reaction can proceed for many hours, producing microgram quantities of transcripts, provided that a coat-associated RNase activity is inhibited. Thus, the transcripts can now be used for translation experiments as well as for cloning of the viral genome via cDNA derived from these transcripts.

Relatedness of the Segments of dsRNA

To resolve the interrelatedness of the dsRNA segments within each virus subtype and between the subtypes, a series of hybridizations were performed (Field et al., 1983). The segments of each subtype were eluted from agarose or polyacrylamide gels and each segment was used as a probe after 3'-end labeling with ^{32}P of the denatured dsRNA or with randomly primed cDNA derived from individual segments. The probes were hybridized with each virus subtype. The only segments that did separate to an extent that permitted their elution individually were H3 and H4 of P4, and M2 and M3 of both P1 and P4. The simplified version of P6 was used, containing only segments H1, M2, and L. A summary of the results is shown in Figure 11.

Within each subtype it is clear that the H segments bear no homology among themselves. Likewise, the M segments that can be seen in P1 seem to lack homology between M2 + M3 to M1. The only homologies found within each subtype were between segments M1 and L of P1, M2 + M3 to L of P4, and M2 and L in P6. Contrary to anticipation of homology among the H segments based on earlier genetic data (Koltin et al., 1978) none was detected. The genetic data, based on a series of virus types containing segments H2, H3, and H4, or H3 and H4, and having normal capsids, predicted that the H segments contain redundant information. The hybridization data suggests that if such redundancy occurs, a strong genetic drift has in fact occurred within these segments. The resolution of this aspect must await the cloning and sequencing of these segments. This will provide significant data on the extent of the drift in dsRNA, for which available information is limited.

The homology detected within each complex between the L segment and one of the M segments is interesting. Heteroduplex analysis shows that L hybridizes to one end of the M segment. It was proposed, based on genetic data, that L is essential for expression of the inhibitory effect (Kandel and Koltin, 1978) and is related to the cytoplasmic immunity of P1 (Peery et al., 1982). The hybridization data clearly shows a relation between L and the segment encoding the inhibitory substance. Recent information based on sequencing data of this segment indicates that L is homologous to one end of the M segment and that that end is a noncoding region of the M segment. Also, one of its 3' termini is different from all other known termini. Whether it is derived independently or as a processed product of either the M dsRNA or an M transcript is not known, although circumstantial evidence suggests the latter (Figure 12).

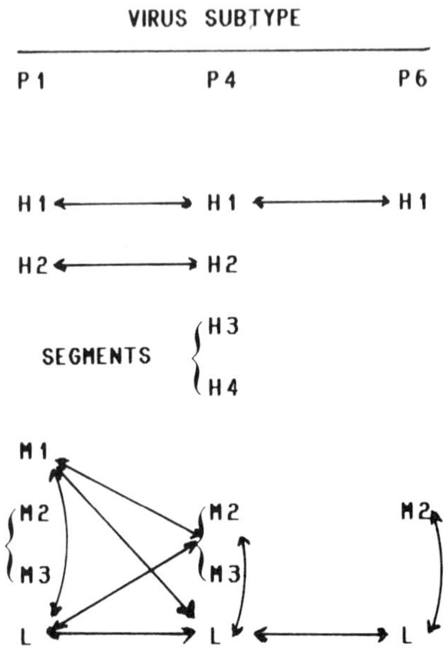

Figure 11 Relatedness as determined by nucleic acid hybridization of the dsRNA segments within and between subtypes. Arrows indicate homology.

Only the H1 segment displays cross hybridization among all three viral subtypes. It appears that L also bears a certain degree of homology, although L of P1 did not hybridize with L of P6, which may reflect a greater relatedness of the L species of P1 and P4 than of either of these with L of P6. In general, some similarity is found among three L segments. Also, the relationship between the M segments suggests that the segments of P1 and P4 that encode the inhibitory substance are similar and both are quite different from M2 of P6, which is the segment associated with the expression and specificity of the killer effect.

The similarity found between M1 of P1 and M2 + M3 of P4 is reflected also by the fact that the only dual killer specificity that was derived successfully in the laboratory was between P1 and P4. This similarity may render them indistinguishable to the replicative mechanism. However, if a simple mechanism of replicative advantage was the determining factor, then M2 of P4, being the smallest segment determining the killer function, should have a selective advantage in cases in which both M1 of P1 and M2 of P4 are contained in one cell. The retention of both would argue against this simple mechanism. The sequence analysis once again should resolve the subtle differences between these

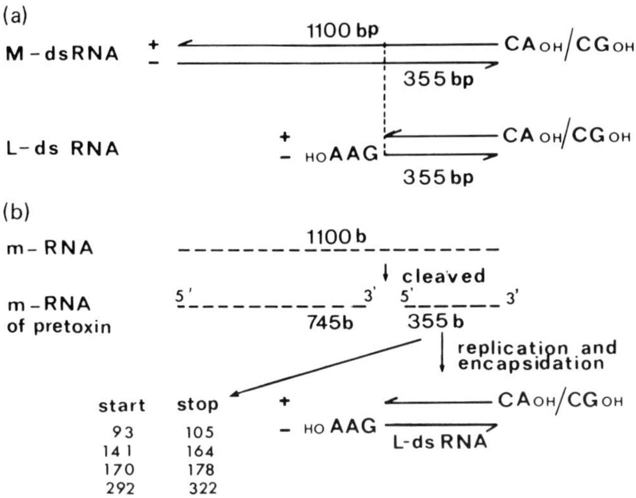

Figure 12 (a) Relation of the L segment of P1 to the toxin coding segment M1-1. (b) Hypothetical scheme for the generation of L from M-1. The positions of the four short open reading frames of L dsRNA are indicated.

segments and should provide more sound data for understanding of the exclusion relationships based on their nucleic acid sequence.

THE VIRUS-ENCODED TOXIN

Structure and Physical Properties

Conditions for maximal production of the inhibitory substance and growth do not coincide. Maximum accumulation of the substance occurs in early stationary phase corresponding to maximum viral transcription activity. Also, maximal growth occurs in pH 4.0, whereas maximal toxin accumulation is obtained at pH 7.0. The latter reflects an actual difference in optimum pH for synthesis of the substance since it is not inactivated at low pH and is not degraded in supernatants of cultures at pH 4.0.

Initial indications on the nature of inhibitory substance were obtained by Hankin and Puhalla (1971) in studies of interstrain inhibition caused by P1 viruses. The material was heat-labile, sensitive to basic conditions, and relatively stable in an acidic environment (pH 2.0 for 24 hr). In neutral conditions, the substance was very stable. The inhibitory substance displayed a strong affinity to Millipore filters. Indications on its proteinaceous nature were inferred from the loss of the inhibitory activity after exposure to proteolytic enzymes. Later, results of repeated tests with the inhibitory substances of P1, P4, and P6 using proteolytic enzymes, as well as DNase and RNase, indicated that only the proteolytic enzymes were effective in degradation of the substances, and all were precipitated with ammonium sulphate (Koltin and Day, 1975). The inhibitory substances encoded by P1, P4, and P6 viral subtypes were designated KP1, KP4, and KP6 (killer protein), respectively.

The three killer proteins migrated in 12.5% polyacrylamide SDS gels (pH 7.0) to the same position, corresponding to a molecular weight of 10 kDa (Kandel and Koltin, 1978). However, by a number of properties such as heat lability, affinity to concanavalin A, migration in nondenaturing gels (7.5% polyacrylamide pH 8.3 and pH 4.3) and the characteristics of the zone of inhibition, it was evident that the killer proteins are distinct and differ in their conformation and the interaction with sensitive cells. The zone of inhibition formed by exposure of a sensitive strain to KP1 and KP4 does not form a sharp boundary, and within the zone of inhibition some viable cells remain. KP6 produces a sharp zone of inhibition and causes a lethal effect within a short time. In growth kinetics the same effect is noticed when sensitive cells are grown in the presence of the killer proteins. The survivors from exposure to KP1 or KP4 are inhibited in their ability to multiply for ca. 48 hr and resume growth with a doubling time about twice the normal rate.

Analytical procedures, including purification of the killer proteins and electrophoresis, suggested that all three types consist of a single polypeptide. However, viral mutants affecting the expression of KP6 were obtained, and among those some had all the dsRNA segments found originally in the parent type. In addition, these mutants secreted a polypeptide that migrated in gel electrophoresis in a manner indistinguishable from the active polypeptide. These mutants were tested for in vitro complementation and found to fall into two complementation groups (Kandel and Koltin, 1978). However, since only one band corresponding to the active polypeptide was noticed in SDS gels, it was assumed that the complementation obtained is intracistronic and that the active substance functions as a homomultimer. Recently, by a modification of the protocol for electrophoresis (18% polyacrylamide-0.09% bisacrylamide) for better resolution of low molecular weight polypeptides in the region in which normally one polypeptide was noticed, two are currently detected with a molecular weight of 10 and 12.5 kDa. Each of the viral mutants in the two complementation groups forms only one of the two complementary polypeptides. Furthermore, a new temperature-sensitive mutant recently isolated was shown to lose the killer activity at the restrictive temperature as a result of poor synthesis of one of the polypeptides. In addition, a protocol based on the purification of the polypeptides by ion-exchange followed by size separation results in clear distinction between the two polypeptides in some fractions. The fractions containing only one polypeptide are devoid of the inhibitory activity. Thus, at least KP6 consists of two polypeptides (Peery et al., 1987), as does the killer toxin of *S. cerevisiae* (Tipper and Bostian, 1984).

The polypeptides of KP6 do not seem to interact in solution. There are no intermolecular S–S bonds, and in native nonreducing gels (15% polyacrylamide, pH 8.3) two distinct bands can be detected in preparations of the active toxin, and only one of these polypeptides in preparations from each of the complementary mutants, with the heavier 12.5 kDa polypeptide migrating at a faster rate than the 10.0 kDa polypeptide under these conditions.

The polypeptides are insensitive to SDS even if heated for 10 min at 80°C in the presence of 1% SDS. However, mercaptoethanol inactivates the toxin. The 10 kDa polypeptide is highly sensitive and is inactivated within 2 min. The 12.5 kDa polypeptide is not affected under these conditions. However, incubation of this larger polypeptide in mercaptoethanol for 2 min at 80°C inactivates it. Therefore, it appears that both of the polypeptides contain intramolecular S–S bonds but no intermolecular S–S bonds. Also, an environment of high ionic strength does not interfere with the activity of the toxin. Therefore, electrostatic bonds between the polypeptides seem unlikely.

Function and Expression

The biological role of the killer proteins is not known and it is doubtful that strains expressing this activity would have a selective advantage, since *Ustilago maydis* is a heterothallic species dependent on heterokaryosis for pathogenecity. On plants, the inhibitory activity is probably neutralized by adsorption to cellulose, as indicated by its high affinity binding to Millipore filters. Therefore, the impression is that the killer activity is either a laboratory phenomenon or a viral function exerting its selective advantage for the virus rather than for the fungal host. The exclusion relations among different viral subtypes were considered as a function that may have resulted from a specific dsRNase activity that could be related to the killer function. Efforts were directed to resolve the type of enzymatic activities associated with the killer activity.

The studies on the mode of action were performed primarily as in vitro reactions with purified KP4 and KP6 and defined substrates. A nucleolytic activity copurified with the toxins (Levine et al., 1979). Protein synthesis was shown to be inhibited by KP4 and KP6 in in vitro reactions using the reticulocyte system (Koltin and Levine, 1979). The inhibition results from the degradation of both the mRNA and the rRNA. The activity was effective also on ssRNA and dsDNA as an endonuclease incising single strands. The activity is similar to the bacterial colicins E2 and E3 (Holland, 1975; Konisky, 1982). The effect of the toxin on intact ribosomes is currently under investigation. Initial results suggest a similarity to the cytotoxic nuclease L-sarcin (Wool, 1984).

It is not known yet whether the catalytic activity is associated with one polypeptide and the recognition function with the second polypeptide of the toxins. Both functions are performed by a single polypeptide of the colicins. It is clear, however, that the two polypeptides act cooperatively in a sequential manner. Both polypeptides are essential for the activity. If cells are exposed to one polypeptide and then washed and exposed to the second polypeptide, the results indicate that the sequential order is critical. Only if sensitive cells are exposed first to the 10 kDa polypeptide and then to the 12.5 kDa polypeptide they are affected. If the order is reversed the cells are not affected. It is conceivable that the 10 kDa polypeptide makes the cells accessible to the second component.

The receptors for the toxin have not yet been identified. Recent studies suggest that the cell walls are the major target for the interaction with the cells. The resistant protoplasts are sensitized as the cell wall regenerates (R. Steinlauf, T. Peery, and Y. Koltin, unpublished data). The recessive alleles for each of the three genes that confer resistance to each of the toxins may serve to identify the receptors for the toxin. In yeast, the cell wall receptor was identified as (1,6)-β-D-glucan (Hutchins and Bussey, 1983). However, a higher degree of specificity may operate to distinguish between the different toxins KP1, KP4, and KP6 if their specificity is based upon receptor binding. The receptor identified in yeast may be a general component of the cell wall typical to the species, and additional components must be identified. The resistance of the protoplasts to the toxin in ustilago suggests that the receptors are localized in the cell wall unless the removal of the cell walls modifies membrane receptors rendering the cells insensitive to the toxin. An additional possible hypothesis is that the cell wall is the site of initial interaction between the two toxin polypeptides and after some modification they jointly interact with the cell via the cell membrane.

CONCLUDING REMARKS

The viruses of ustilago and the virus-encoded polypeptides have been resolved to an extent that emphasizes some properties of the toxins that are of interest as general biological problems: (1) How are the polypeptides, as a toxic complex, processed and expressed by the cell? (2) What is the mechanism that makes these polypeptides such narrow-spectrum antibiotics? (3) What is the mechanism for the cidal and static activity of the polypeptides? As for the virus, how does a complex system of a major virus and a number of satellites coordinate its expression and replication in synchrony with the cell cycle? In addition, what is the selective advantage confered by the dsRNA viruses of the fungi and their nucleic acids that make them so ubiquitous in this group of microorganisms? It is to answer these questions that future efforts will be directed.

REFERENCES

Ainsworth, G. C., and Sampson, K. (1950). *The British Smut Fungi (Ustilaginales)*. Kew Surrey, The Commonwealth Mycological Institute.

Banks, G. R. (1973). Mitochondrial DNA synthesis in permeable cells. *Nature New Biol.* *245*:196-199.

Banks, G. R. (1983a). Transformation of *Ustilago maydis* by a plasmid containing yeast 2 micron DNA. *Curr. Genet.* *7*:73-77.

Banks, G. R. (1983b). Chromosomal DNA sequences from *Ustilago maydis* promote autonomous replication of plasmids in *Saccharomyces cerevisiae*. *Curr. Genet.* *7*:79-84.

Ben-Zvi, B. S., Koltin, Y., Mevarech, M., and Tamarkin, A. (1984). RNA polymerase activity in virions from *Ustilago maydis*. *Molec. Cell. Biol.* *4*:188-194.

Bozarth, R. F., Koltin, Y., Weissman, M. B., Parker, R. L., Dalton, R. F., and Steinlauf, R. (1981). The molecular weight and packaging of dsRNA in the mycovirus from *Ustilago maydis* killer strain. *Virology* *113*:492-502.

Brennan, V. E., Bobek, L. A., and Bruenn, J. A. (1981). Yeast dsRNA viral transcriptase: Identification of the transcript strand. *Nucleic Acids Res.* *9*:5049-5059.

Bruenn, J. A., and Brennan, V. E. (1980). Yeast viral dsRNAs have hetergenous 3' termini. *Cell* *19*:923-933.

Bruenn, J. A., Bobek, L., Brennan, V., and Held, W. (1980). Yeast viral RNA polymerase is transcriptase. *Nucleic Acids Res.* *8*:2985-2997.

Buck, K. W. (1980). Viruses and killer factors of fungi. *Symp. Soc. Gen. Microbiol.* *30*:329-375.

Carle, G. F., and Olson, M. V. (1984). Separation of chromosomal DNA molecules form yeast by orthogonal-field-alteration by electrophoresis. *Nucleic Acids Res.* *12*:5647-5664.

Christensen, J. J. (1931). Studies on the genetics of *Ustilago zeae*. *Phytopath. Z. 4*:129-188.

Dalton, R. E., Podila, G. K., Lurkey, W. H., and Bozarth, R. F. (1985). *In vitro* translation of the major capsid polypeptide from *Ustilago maydis*. *Virus Res. 3*:153-163.

Day, P. R. (1981). Fungal virus populations in corn smut from Connecticut. *Mycologia 73*:379-391.

Day, P. R., and Anagnostakis, S. L. (1971a). Corn smut dikaryon in culture. *Nature New Biol. 231*:19-20.

Day, P. R., and Anagnostakis, S. L. (1971b). Meiotic products form natural infections of *Ustilago maydis*. *Phytopathology 61*: 1020-1021.

Day, P. R., Anagnostakis, S. L., and Puhalla, J. E. (1971). Pathogenicity resulting from mutations at the *b* locus of *Ustilago maydis*. *Proc. Nat. Acad. Sci USA 68*:533-535.

Esposito, R. E., and Holliday, R. (1964). The effect of 5'-fluorodeoxyuridine on genetic replication and mitotic crossing over in synchronized cultures of *Ustilago maydis*. *Genetics 50*:1009-1017.

Field, L. J., Bruenn, J. A., Chang, T. H., Pinchasi, O., and Koltin, Y. (1983). Two *Ustilago maydis* viral dsRNAs of different size code for the same product. *Nucleic Acids Res. 11*:2765-2778.

Finkler, A., Peery, T., and Koltin, Y. (1984). Mitotic transmission of the *Ustilago* virus M dsRNA is dependent on the nuclear resistance genes. *Genetics 107*:s232.

Georgopolus, S. C., and Sisler, H. D. (1970). Gene mutation eliminating Antimycin A-tolerant electron transport in *Ustilago maydis*. *J. Bacteriol. 103*:745-764.

Hankin, L., and Puhalla, J. E. (1971). Nature of a factor causing interstrain lethality in *Ustilago maydis*. *Phytopathology 61*:50-53.

Hanna, W. F. (1929). Studies on the physiology and cytology of *Ustilago zeae* and *Sorosporium reilianum*. *Phytopathology 19*:415-442.

Hannig, E. M., Williams, T. L., and Leibowitz, M. J. (1986). The internal polyadenylate tract of yeast killer virus M1 double-stranded RNA is variable in length. *Virology 152*:149-158.

Henry, M. J., and Sisler, H. D. (1981). Inhibition of ergosterol biosynthesis in *Ustilago maydis* by the fungicide 1-[2,4-(Dichlorophenyl)-4-ethyl-1,3-dioxolan-2-ylmethyl]-1H-1,2,4-triazole. *Pest Sci. 12*:98-102.

Holland, I. B. (1975). Physiology of colicin action. *Adv. Microbiol. Physiol.* *12*:55-139.

Holliday, R. (1961a). The genetics of *Ustilago maydis*. *Genetics* *50*:1009-1017.

Holliday, R. (1961b). Induced mitotic crossing over in *Ustilago maydis*. *Genet. Res.* *2*:231-248.

Holliday, R. (1964). A mechanism of gene conversion in fungi. *Genet. Res.* 5:282-304.

Holliday, R. (1965). Induced mitotic crossing over in relation to genetic replication in synchronously dividing cells of *Ustilago maydis*. *Genet. Res.* *6*:104-120.

Holliday, R. (1967). Altered recombination frequencies in radiation sensitive strains of *Ustilago*. *Genet. Res.* *9*:275-288.

Holliday, R., and Dickson, J. M. (1977). The detection of post-meiotic segregation without tetrad analysis in *Ustilago maydis*. *Molec. Gen. Genet.* *153*:331-335.

Holliday, R., and Halliwell, R. E. (1968). An endonuclease-deficient strain of *Ustilago maydis*. *Genet. Res.* *12*:95-98.

Holloman, W. K. (1975). Protein from *Ustilago* which form an acid-soluble complex with deoxyribonucleic acid. *J. Biol. Chem.* *8*:2993-3000.

Holloman, W. K., Rowe, T. C., and Rusche, J. R. (1981). Studies on nuclease from *Ustilago maydis*. *J. Biol. Chem.* *256*:5087-5094.

Hundley, F., Biryahwaha, B., Gow, M., and Desselberger, U. (1985). Genome rearrangements of bovine Rotavirus after serial passage at high multiplicity of infection. *Virology* *143*:88-103.

Hutchins, K., and Bussey, H. (1983). Cell Wall Receptor for yeast killer toxin: Involvement of (1-6)-β-D-glucan. *J. Bacteriol.* *154*:161-168.

Kandel, J., and Koltin, Y. (1978). Killer phenomenon in *Ustilago maydis*. Comparison of the killer proteins. *Expt. Mycol.* *2*:270-278.

Kmeic, E. B., and Holloman, W. K. (1982). Homologous pairing of DNA molecules promoted by a protein from *Ustilago*. *Cell* *29*:367-374.

Kmeic, E. B., and Holloman, W. K. (1983). Heteroduplex formation and polarity during strand transfer promoted by *Ustilago* rec 1 protein. *Cell* *37*:857-864.

Kmiec, E., and Holloman, W. K. (1984). Synapsis promoted by a protein form *Ustilago* rec 1 protein. *Cell 36*:593-598.

Kmeic, E. B., Kroeger, P. E., Brougham, M. J., and Holloman, W. K. (1983). Topological linkage of circular DNA molecules promoted by *Ustilago* protein and topisomerase. *Cell 34*:919-929.

Kmeic, E. B., Angelides, K. J., and Holloman, W. K. (1985). Left-handed DNA and the synaptic pairing reaction promoted by *Ustilago* Rec 1 protein. *Cell 40*:139-145.

Koltin, Y. (1977). Virus-like particles in *Ustilago maydis*: Mutants with partial genomes. *Genetics 86*:527-534.

Koltin, Y. (1986). The killer systems of *Ustilago maydis*. In *Fungal Virology*, K. W. Buck (Ed.). CRC Press, Boca Raton, Florida, pp. 109-143.

Koltin, Y., and Day, P. R. (1975). Specificity of *Ustilago maydis* killer proteins. *Applied Microbiol. 30*:694-696.

Koltin, Y., and Day, P. R. (1976a). Inheritance of killer phenotype and double stranded RNA in *Ustilago maydis*. *Proc. Nat. Acad. Sci USA 73*:594-598.

Koltin, Y., and Day, P. R. (1976b). Suppression of the killer phenotype in *Ustilago maydis*. *Genetics 82*:629-637.

Koltin, Y., and Levine, R. (1979). Fungal viruses and killer factors—*Ustilago maydis* killer proteins. In *Fungal Viruses*, H. P. Molitoris, M. Hollings and H. A. Wood (Eds.). Springer-Verlag, New York, pp. 120-128.

Koltin, Y., Mayer, I., and Steinlauf, R. (1978). Killer phenomenon in *Ustilago maydis*: Mapping viral functions. *Molec. Gen. Genetics 166*:181-186.

Koltin, Y., Levine, R., and Peery, T. (1980). Assignment of functions to segments of the dsRNA genome of *Ustilago maydis*. *Molec. Gen. Genetics 178*:173-178.

Konisky, J. (1982). Colicins and other bacteriocins with established modes of action. *Annu. Rev. Microbiol. 36*:125-144.

Levine, R., Koltin, Y., and Kandel, J. S. (1979). Nuclease activity associated with the *Ustilago maydis* virus-induced killer proteins. *Nucleic Acids Res. 6*:3717-3731.

Peery, T., Koltin, Y., and Tamarkin, A. (1982). Mapping the immunity of the *Ustilago maydis* P1 virus. *Plasmis 7*:52-58.

Peery, T., Shabat-Brand, T., Steinlauf, R., Koltin, Y., and Bruenn, J. (1987). Virus-encoded toxin of *Ustilago maydis*: Two polypeptides are essential for activity. *Molec. Cell Biol. 7*:470-477.

Perkins, D. D. (1949). Biochemical mutants in the smut fungus *Ustilago maydis*. *Genetics 34*:607-626.

Puhalla, J. E. (1964). The formation of diploid *Ustilago maydis* on agar medium. *Phytopathology 54*:1771-1772.

Puhalla, J. E. (1968). Compatibility reactions on solid medium and interstrain inhibition in *Ustilago maydis*. *Genetics 60*:461-474.

Puhalla, J. E. (1970). Genetic studies of the *b* incompatibility locus of *Ustilago maydis*. *Genet. Res. 16*:229-232.

Pukilla, P. (1978). The recognition of mismatched base pairs in DNA by DNase I from *Ustilago maydis*. *Molec. Gen. Genet. 161*:245-250.

Rowe, T. C., Rusche, J. R., Brougham, M. J., and Holloman, W. K. (1981). Purification and preparation of a topoisomerase from *Ustilago maydis*. *J. Biol. Chem. 256*:10354-10361.

Rowell, J. B. (1955). Functional role of compatibility factors and an *in vitro* test for sexual compatibility with haploid lines of *Ustilago zeae*. *Phytopathology 45*:370.

Rowell, J. B., and DeVay, J. E. (1954). Genetics of *Ustilago zeae* in relation to basic problems of its pathogenecity. *Phytopathology 44*:356-362.

Rusche, J. R., Rowe, T. C., and Holloman, W. K. (1980). Purification and characterization of nuclease beta from *Ustilago maydis*. *J. Biol. Chem. 255*:9117-9123.

Schwartz, D. C., and Cantor, C. R. (1984). Separation of yeast chromosome sized DNAs by pulsed field gradient electrophoresis. *Cell 37*:67-75.

Sherman, F., Fink, G. R., and Hicks, J. B. (1982). *Methods in Yeast Genetics*. Cold Spring Harbor Laboratory, Cold Spring Harbor, New York.

Silva, J. (1972). Alleles at the *b* incompatibility locus in Polish and North American population of *Ustilago maydis* (DC) Corda. *Physiol. Plant Path. 2*:333-337.

Sommer, S. S., and Wickner, R. B. (1984). Double-stranded RNA that encode killer toxins in *Saccharomyces cerevisiae*. Unstable size of M double-stranded RNA and inhibition of M2 replication by M1. *Molec. Cell Biol. 4*:1747-1753.

Stevens, R. B. (1974). Genetic studies with *Ustilago maydis*, in *Mycology Guidebook*. University of Washington, Seattle and London, pp. 506-524.

Tipper, D. C., and Bostian, K. A. (1984). Double-stranded ribonucleic acid killer systems in yeast. *Microbiol. Rev. 48*:125-156.

Tyler, K. L., and Field, B. N. (1985). Reovirus and its replication, in *Virology*, B. N. Field, D. M. Knipe, R. M. Chonock, J. L. Melnick, B. Roizman, and R. E. Slope, (Eds.). Raven Press, New York, pp. 823-862.

Unrau, P. (1975). The excision of pyrimidine dimers form the DNA of mutant and wild type strains of *Ustilago*. *Mut. Res. 29*:53-65.

Walter, P., and Blobel, G. (1982). 1 SL small cytoplasmic RNA is an integral component of the signal recognition particle. *Nature 299*:691-698.

Welsh, J. D., Leibowitz, M. J., and Wickner, R. B. (1980). Virion DNA-independent RNA polymerase from *Saccharomyces cerevisiae*. *Nucleic Acids Res. 8*:2349-2363.

Wesolowski, M., and Wickner, R. B. (1984). New double-stranded RNA molecules showing non-Mendelian inheritance and heat inducibility in *Saccharomyces cerevisiae*. *Molec. Cell Biol. 4*:181-187.

Wickner, R. B. (1986). Double-stranded RNA replication in yeast: The killer system. *Annu. Rev. Biochem. 55*:373-395.

Wigderson, M., and Koltin, Y. (1982). Dual toxin specificaties and the exclusion relations among the *Ustilago* dsRNA viruses. *Curr. Genet. 5*:127-136.

Wood, H. A., and Bozarth, R. F. (1973). Heterokaryon transfer of virus-like particles associated with a cytoplasmically inherited determinant in *Ustilago maydis*. *Phytopathology 63*:1019-1021.

Wool, I. G. (1984). The mechanism of action of the cytotoxic nuclease L-sarcin and its use to analyse ribosome structure. *Trends Biochem. Sci. 9*:14-17.

11

Killer Systems and Pathogenic Yeast

Judith S. Kandel
California State University, Fullerton, California

INTRODUCTION

The killer phenomenon, the secretion by yeast cells of a polypeptide toxin that kills sensitive cells of the same species and sometimes from other genera, was first described in *Saccharomyces cerevisiae* (Bevan and Makower, 1963). Strains within a species may produce different toxins. Killer cells are immune to the action of the toxin that they produce but may be sensitive to toxins produced by other organisms. Cells that are not affected by heterologous toxins are termed resistant.

In *Saccharomyces cerevisiae* the killer phenomenon is associated with the presence of double-stranded RNAs (dsRNAs) in viruslike particles (VLPs). Double-stranded RNAs (designated L, with an approximate molecular weight of 3.4×10^6) are present in all killer strains and in most nonkillers and sensitive cells. Smaller dsRNAs (designated M, with a molecular weight of approximately 1.5×10^6) are also always present in killer strains (Bevan et al., 1973; Vodkin and Fink, 1973). The M dsRNAs code for the production of killer toxin and presumably the immunity factor (Bostian et al., 1980). M dsRNAs are dependent on the presence of L dsRNA for maintenance and replication.

Surveys of yeast strain collections and natural isolates have shown that the killer phenomenon is widespread and some interactions can be found among most genera (Philliskirk and Young, 1975; Stumm et al., 1977; Young and Yagiu, 1978; Kandel and Stern, 1979; Middlebeek et al., 1980b). However, failure to demonstrate certain killer/sensitive relationships may result from the methodology of these surveys. For example, most attempts to identify new killers have utilized indicator strains sensitive to known killer-producing yeast under the conditions that are optimal for the known killer. The production and activity of killer toxins is influenced by conditions such as growth medium, pH, temperature, and agitation. The majority of toxins thus far identified are optimally active at a pH below 5; consequently, most assays have been performed below pH 5.

The killer phenomenon, dsRNA, and VLPs have all been demonstrated in pathogenic fungi. This chapter will review the currently available information and discuss the implications of these findings for the control of fungal infections.

KILLER TOXIN PRODUCTION BY PATHOGENIC YEASTS

Human disease caused by yeasts is usually opportunistic. Most often the organisms belong to the genera candida, torulopsis, or cryptococcus. Toxin production has been described in all three genera. No one, however, has surveyed the yeast forms of the dimorphic fungal pathogens for killer toxin production.

Torulopsis glabrata Toxin

The first detailed report of killer toxin production by a pathogenic yeast was in the opportunist *Torulopsis glabrata* (Bussey and Skipper, 1975). The toxin, called PEST (pool efflux-stimulating toxin), was characterized and compared to a *Saccharomyces cerevisiae* toxin in its ability to kill sensitive strains of *S. cerevisiae*. Further studies on toxins from torulopsis were performed by Kandel and Stern (1979) and Santell and Kandel (1982).

The toxins produced by *T. glabrata* are not genus- or species-specific. They can be distinguished by their killing specificity into several classes. Some can kill only *S. cerevisiae*, some only *S. cerevisiae* and *T. glabrata*, while others have a wider range including certain cryptococci and candida isolates. The activity of the toxin against sensitive cells is potentiated by the polyene drug amphotericin B. Amphotericin B does not make resistant cells susceptible to the activity of toxin. Thus, killing of *S. cerevisiae* is enhanced by amphotericin B, whereas no killing by toxin is observed when the target cell is

C. albicans (Coats and Kandel, 1979). Among 39 clinical isolates tested for killer activity, only 7 were positive. Thus, there is no apparent relationship between killer phenotype and pathogenicity.

A toxin (denoted torulopis toxin) with a killing specificity different from PEST has been partially purified from the culture filtrate of a clinical isolate obtained at autopsy. The temperature optimum for killing by this toxin is approximately 30°C. This differs from *S. cerevisiae* toxins, which have an optimum between 18-22°C (Bussey, 1981). At 22°C, the killing activity of torulopsis toxin is only 50% of its optimum. Both toxins display little activity above pH 5.0; however, the optimum pH reported for most toxins of saccharomyces is between 4.6-4.8 (Young and Yagiu, 1978), whereas for torulopsis toxin it is between pH 4.0-4.4. Temperature, pH, and agitation also are different in their affect on the stability of the toxins. The torulopsis toxin is quite stable. It can be stored indefinitely at 4°C or in the freezer without losing activity. It retains its activity for at least 1 hr at temperatures up to 37°C and for 2 hr at a pH between 3.0 and 7.0. In contrast, most saccharomyces toxins are inactivated above pH 5 or at temperatures above 30°C in the absence of a stabilizing agent such as glycerol (Young and Yagiu, 1978; Palfree and Bussey, 1979).

Bostian et al. (1984) determined the molecular weight of crude and partially purified saccharomyces toxin by SDS-polyacrylamide gel electrophoresis. They showed that the toxin consists of two subunits of similar size linked by disulfide bonds, to form a molecule of approximately 18,000 molecular weight. The molecular weight of partially purified torulopsis toxin determined under nonreducing conditions in 12% SDS-polyacrylamide gels is estimated at 16,000 to 18,000 (Figure 1).

Table 1 compares some properties of torulopsis toxin with toxins from several nonpathogenic yeasts.

Bussey and Skipper (1975) showed that PEST causes sensitive cells to leak potassium, lose AMP, and shut down macromolecular synthesis after a brief lag period. The lethal effect of torulopsis toxin was observed by staining toxin-treated cells with methylene blue and by plating survivors to determine colony-forming units. There is a marked difference in viability measured by these methods. Most cells of saccharomyces that are treated with torulopsis toxin over a 150-min incubation period are smaller and contain fewer buds than control (untreated) cells. However, they exclude methylene blue, indicating that although they are not reproducing they are viable. Under these conditions, colony-forming units decrease throughout the incubation period. At 150 min, only 3% of toxin-treated cells are able to form colonies, while more than 80% still exclude methylene blue. Thus, although toxin binding is rapid, the lethal event is not completed by 150 min. The degree of killing depends upon toxin concentration and growth conditions. Equivalent concentrations of toxin will kill 83% of sensitive cells after 50 min if the incubation is in buffer whereas only

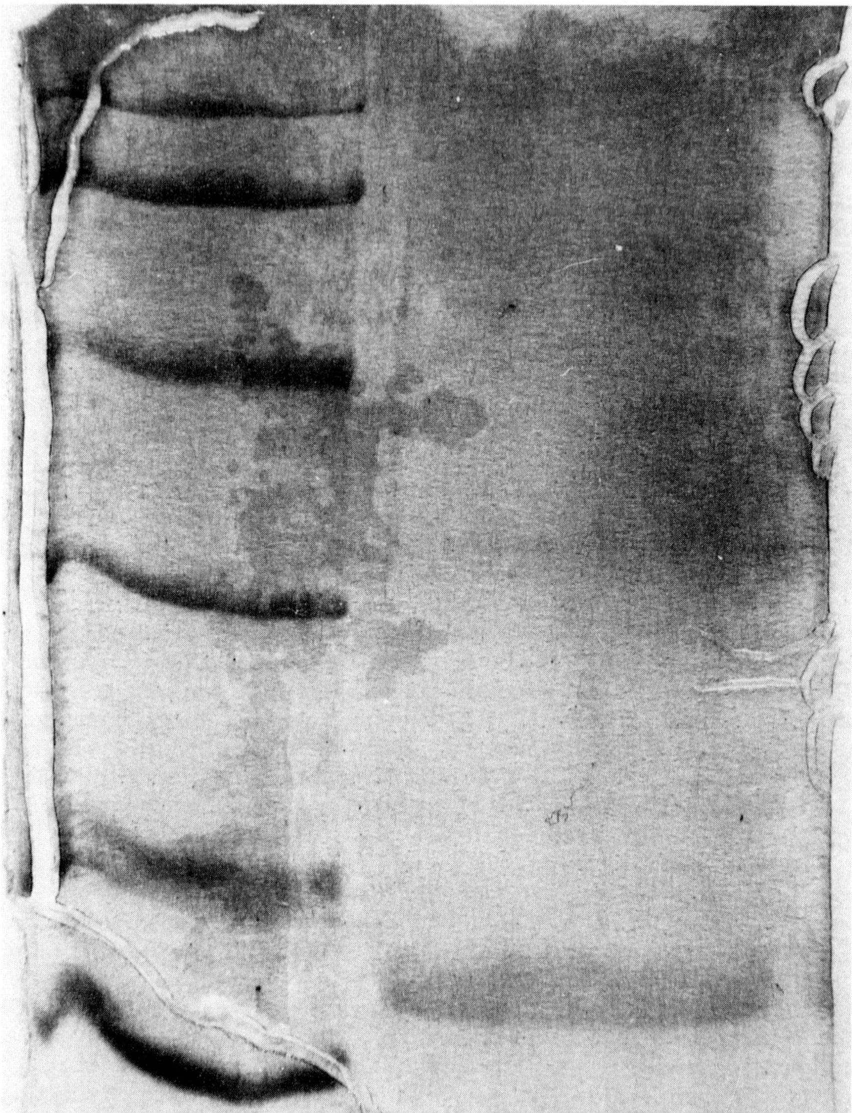

Figure 1 Molecular weight estimation of torulopsis toxin by electrophoresis on 12% SDS-polyacrylamide gel. The gel was stained with silver according to the procedure of Wray et al. (1981). Left, marker proteins; right, partially purified toxin.

Table 1 Comparison of Torulopsis Toxin with Toxins from Nonpathogenic Yeast

Property	*T. glabrata*[a]	*S. cerevisiae*[b]	*K. lactis*[c]	*P. kluyveri*[d]
Genetic template	?	dsRNA	dsDNA	?
Heat cured	No	Yes	Yes	No
pH optimum	4.0-4.4	4.6-4.8	4-8	3-4.5
Temperature optimum	28-30°C	18-22°C	25-30°C	28°C
Action	Lethal	Lethal	Inhibitory	Lethal
Molecular weight	16-18,000	18,500 (Subunits, 9000, 9500)	>100,000 (Subunits, 27,000, 80,000)	19,000
Composition	Protein	Protein	Glycoprotein	Glycoprotein

[a]Santell and Kandel (1982).
[b]Bussey (1981).
[c]Sugisaki et al. (1984).
[d]Middlebeek et al. (1979).

5% are killed in broth. Thus, this torulopsis toxin, like PEST, apparently does not require energy to induce damage (Skipper and Bussey, 1977). Organisms that survive toxin treatment form smaller colonies than controls. This indicates that sublethal doses of toxin either reduce the growth rate or inhibit growth temporarily.

The rescue of sensitive cells of *S. cerevisiae* from the action of a pichia killer toxin was reported by Middlebeek et al. (1980a). Saccharomyces cells can be similarly rescued from the lethal effect of torulopsis toxin early in the incubation period. If cells exposed to toxin for 90 min or less are brought to pH 7.0 and plated on an agar medium at pH 7.0, the toxin is apparently inactivated before it irreversibly damages the cells. Dilution of cells prior to plating at pH 4.4 does not inactivate toxin or reverse its action. The addition of cycloheximide to a concentration of 25 μg/ml into the dilution medium reduces the number of cells rescued by 50%. Thus, protein synthesis is required for rescue, presumably to allow the cell to repair sublethal damage caused by the action of the toxin.

Examination of torulopsis toxin-treated cells in the scanning electron microscope shows surface alterations that include deep pits, cracking, shallow depressions, bleb formation, and bud scar collapse not seen in untreated cells (Figure 2) (Moore et al., 1984). Cell lysis is not observed. This is consistent with the observation by Bussey and Skipper (1975) that macromolecules do not leak from toxin-damaged cells. The amount of morphological damage varies with incubation conditions, but is always less apparent than the decrease in viability. Cells treated with toxin in buffer have more damage than those treated in broth. Cells in buffer may be unable to repair damage as it occurs in the early stages of toxin action.

Toxins from Other Pathogenic Yeasts

Candida

Candida species are the most commonly isolated yeast pathogens of humans. They are frequently associated with diseases in cancer patients, kidney transplant recipients, diabetics, pregnant women, and persons with genetic immune dysfunctions or endocrinopathies. The clinically important species of candida include *Candida albicans*, the most commonly isolated pathogen in the genus, *Candida tropicalis, Candida pseudotropicalis, Candida parapsilosis, Candida krusei, Candida guillermondii* and *Candida rugosa*. Among these pathogenic candida, killing activity has been reported for all but *C. parapsilosis* and *C. rugosa* (Rogers and Bevan, 1978; Kandel and Stern, 1979; Middlebeek et al., 1980b; Morace et al., 1984). The incidence of killers within each species is low. Table 2 lists the toxin-producing candida as well as other species of yeast that show sensitivity to their toxins.

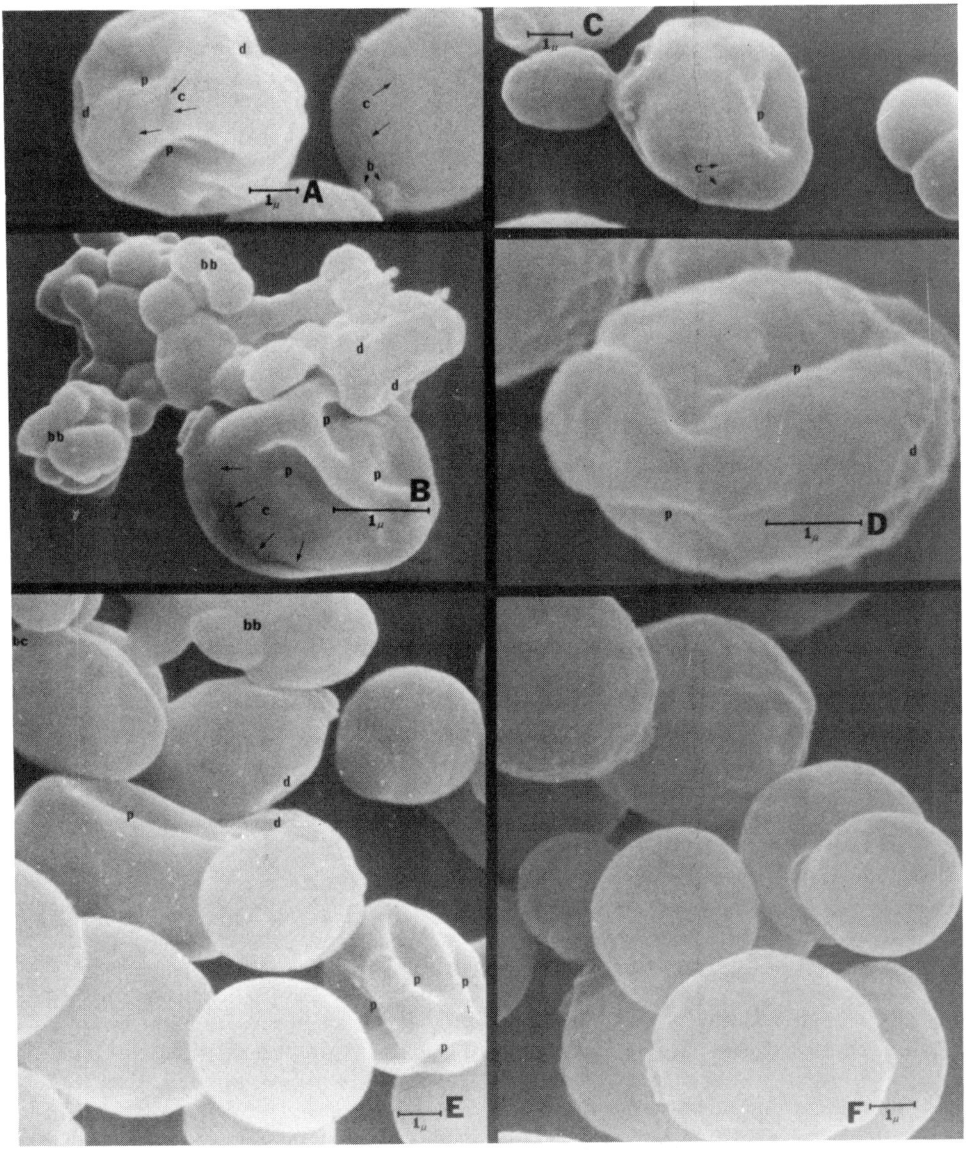

Figure 2 Morphological damage after treatment with torulopsis toxin. *S. cerevisiae* was incubated in buffer pH 4.4 with toxin and examined after (A) 1.5 hr, (C) 3.5 hr, (E) 4.5 hr. (B) Cells were exposed to toxin for 1.5 hr and the pH raised to 7.0 for an additional 3-hr incubation. (D) Cells were exposed to toxin for 3.5 hr and the pH raised to 7.0 for an additional 1 hr incubation. (F) Cells incubated in the absence of toxin for 4.5 hr. Damage includes blebs (b), bubbling (bb), bud scar collapse (bc), cracks (c), depressions (d), and pits (p). (Photographs courtesy of J. Moore.)

Table 2 Killer Activity among Candida Species

Killer[a]	Sensitive[b]	Reference
C. albicans	*S. cerevisiae* *Saccharomyces drosophilarum* *T. glabrata*	Rogers and Bevan (1978)
C. tropicalis	*C. rugosa* *C. krusei* *S. cerevisiae* *T. glabrata*	Kandel and Koltin (1981)
C. pseudotropicalis	*C. rugosa* *S. cerevisiae*	Kandel and Koltin (1981)
C. krusei	*S. cerevisiae*	Middlebeek et al. (1980b)
C. guillermondii	*S. cerevisiae* *C. tropicalis* *C. pseudotropicalis*	J. Kandel (unpublished data) Morace et al. (1984)
Candida valida[c]	*S. cerevisiae* *T. glabrata* *C. parapsilosis* *Kluyteromyces fragilis*	Kandel and Stern (1979) Young and Yagiu (1978)

[a]Not all isolates within the species are killers.
[b]Not all isolates within the species are sensitive.
[c]This organism is nonpathogenic.

The two strains described by Rogers and Bevan (1978) are the only killers reported in *C. albicans*, despite the fact that by far the largest number of pathogenic yeast strains tested has been from this species.

Killers have been identified among several candida species that are not considered pathogenic. For example, Philliskirk and Young (1975) showed that *Candida valida* isolated from draught beer could kill *S. cerevisiae*. Young and Yagiu (1978) later demonstrated a wider killing range for this strain, including *T. glabrata*. Kandel and Stern (1979) extended the range to include *C. parapsilosis*.

The low incidence of killers among pathogenic candida is possibly explained in one of the following ways: (1) The methods used for screening may be inappropriate for candida. The problem may be in the choice of media, incubation conditions, and/or indicator strains. (2) The amount of toxin produced may be very low and undetectable without concentration. Middlebeek et al. (1980b) and Kandel and Stern (1979) showed in other genera that certain culture filtrates have to be concentrated 25- to 100-fold in order to produce a killing zone that can be seen easily. (3) Extracellular proteases may be destroying toxin in the culture medium. Production of an acid protease is very common in clinical isolates of *Candida albicans* and has been reported to be related to virulence (Kwon-Chung et al., 1985). Strains of *S. cerevisiae* that secrete a surface protease exhibit reduced killing activity compared to a mutant with a defective protease (Bussey et al., 1983). (4) The low incidence of killers among candida may be real. It is possible that they cannot maintain cytoplasmic dsRNAs needed for toxin production. Alternatively, cytoplasmic dsRNA may be related to hypovirulence (see below). In this case, dsRNA-containing toxin producers would be at a selective disadvantage in infections and probably would not be seen in clinical isolates.

Cryptococcus

Stumm et al. (1977) first reported killer activity in a soil isolate of cryptococcus that was later identified as *Cryptococcus laurentii*, a pathogen on rare occasions. The toxin is active only against other cryptococci, including one strain of *Cryptococcus neoformans* (Middlebeek et al., 1980c). The mode of action of this toxin against *Cryptococcus albidus* is similar to that of the saccharomyces toxin. Sensitive cells leak intracellular potassium and ATP; they are not killed when oxidative phosphorylation is inhibited by dinitrophenol. The cryptococcus toxin differs from most saccharomyces toxins in other properties. It has an optimum activity at pH 5.0, is stable over a broad pH range, and is not inactivated after 30 min at 45°C. Toxin derived from *S. cerevisiae* strain 28 shares these characteristics (Pfeiffer and Radler, 1982).

Certain clinical isolates of *C. neoformans* inhibit the growth of *T. glabrata* (Kandel, 1980). Clear zones of inhibition surrounding colonies of *C. neoformans* are observed on seeded plates after 24 hr. These zones become overgrown upon further incubation.

SENSITIVITY OF PATHOGENIC YEAST TO HETEROLOGOUS TOXINS

In contrast to the low incidence of killers among pathogenic yeast, there is a high incidence of sensitivity to toxins produced by certain genera. The toxins lethal to pathogenic yeast may have practical applications for epidemiological typing and possibly for design of synthetic derivatives for use as antifungal agents. Candida species (especially *C. albicans*) *T. glabrata*, and *C. neoformans* are the yeast pathogens of greatest clinical concern.

An early report by Wilbur-Murphy and MacMillan (1976) showed that, although *C. albicans* was resistant to a toxin produced by *S. cerevisiae*, its spheroplasts were sensitive. Viability was determined by measurement of respiration rate. Rogers and Bevan (1978) later described killing of two strains of *C. albicans* by several saccharomyces killers and the PEST-producing *T. glabrata*. *C. albicans* is also sensitive to toxins produced by hansenula and pichia (Middlebeek et al., 1980b) and kluyveromyces. Morace et al. (1984) found a high frequency of sensitivity among candida other than *C. albicans* to killers from the genera pichia, hansenula, and kluyveromyces.

Saccharomyces toxin has been shown by several groups to kill *T. glabrata* (Bussey and Skipper, 1976; Kandel and Stern, 1979). *T. glabrata* is also sensitive to toxin-producing hansenula, pichia (Morace et al., 1984) *T. glabrata*, and *C. albicans* (Kandel and Stern, 1979). It is inhibited by certain strains of *C. neoformans*.

Table 3 summarizes the killer-sensitive relationships among killers of different genera and clinical isolates of pathogenic yeast.

Epidemiological Typing

Polonelli et al. (1983) reported the first attempt at epidemiological typing using killers as "myocins," analogous to the use of bacteriocins to type bacteria. They selected nine pichia and hansenula killers to survey 100 *C. albicans* isolates primarily from the vagina, sputum, and throat. Each isolate was sensitive to at least one killer yeast. These nine strains differentiated the *C. albicans* into 25 different types. However, greater than 50% of the isolates were of a single type, sensitive to all nine killers. No correlation was observed between type and clinical site from which the pathogen was collected. Caprilli et al. (1985) expanded the typing assay by adding 12 killers. They found that 78% of their 100 *C. albicans* isolates (from the skin and vaginal tract) were sensitive to all nine of the original killers, whereas 48% were sensitive to all 21 killers. Thus almost half of the isolates were a single biotype. This is not unexpected since the majority of the 21 killers were reported to have activity against 92% to 100% of all *C. albicans* against which they had been tested in screening assays.

In an attempt to improve on the typing assay, J. Kandel and B. Gallivan (unpublished data) compared the original nine hansenula and pichia killers and eight additional killers, most of which displayed activity against less than 80% of *C. albicans* isolates. Unlike the earlier surveys, the clinical isolates were from blood, throat, and respiratory specimens from AIDS patients, as well as vaginal and skin sites. Eight of the nine killers used in the previous studies killed all 100 *C. albicans* and were therefore ineffective in differentiating among isolates. The nine remaining killers differentiated the 100 *C. albicans* into 41 groups. The largest was comprised of 17 organisms. No correlation

Table 3 Killer-Sensitive Relationships Among Yeasts of Different Genera and Pathogenic Yeasts

	Killer activity[a]						
Seeded strain[b]	Saccharomyces	Torulopsis	Candida	Hansenula	Pichia	Cryptococcus	Kluyveromyces
C. albicans	+	-	-	+	+	-	+
C. tropicalis	-	+	+	+	+	-	+
C. pseudotropicalis	-	-	+	+	+	-	+
C. parapsilosis	+	+	+	+	+	-	+
C. krusei	+	-	+	-	+	-	+
C. guillermondii	+	+	+	+	+	-	-
C. rugosa	+	+	+	ND	ND	-	ND
T. glabrata	+	+	+	+	+	+	+
Cr. neoformans	+	+	-	+	+	+	-

[a]+, killing; -, no killing; ND, not done. Not all isolates within a species are killers.
[b]Not all isolates within a species are sensitive.
Sources: Rogers and Bevan (1978); Kandel and Stern (1979); Middlebeek et al. (1980b); Morace et al. (1984).

was observed between biotype and specimen collection site. However, all eight wound isolates were sensitive to at least five killers whereas only two of seventeen blood isolates exhibited such sensitivity.

In another survey, killer toxins were isolated from seven strains of hansenula and one strain of pichia and measured quantities were placed on the surface of seeded agar plates (Polonelli et al., 1985). Each toxin was active against all isolates but the sizes of the zones of inhibition varied. Isolates were classified "0" or "1" for their response to each of the eight toxins. Thus an eight-digit binary code was generated. Fourteen groups were identified among 100 strains tested by this method. The authors did not indicate the distribution of isolates among these groups.

Strain differentiation of Candida other than *C. albicans* and of *C. neoformans* was also performed using killers from six genera (Morace et al., 1984). *C. parapsilosis* isolates were sensitive to killers among candida, hansenula, pichia, rhodotorula, and trichosporon species, but were unaffected by a *S. cerevisiae* killer. *C. pseudotropicalis* were sensitive to killers in the same genera, but to a lesser degree. Sixteen strains of *C. parapsilosis* and 20 strains of *C. pseudotropicalis* were tested using the original nine typing isolates described earlier. The *C. parapsilosis* were sensitive to all killers and therefore fell into a single group. Seven strain types were identified among the *C. pseudotropicalis*. Most of the 22 *C. tropicalis* isolates also responded to the hansenula and pichia killer strains; seven biotypes were identified. Killing of *Cryptococcus neoformans* was greatest by strains of hansenula and trichosporon, but some were killed by pichia and *S. cerevisiae*. Kandel and Stern (1979) had demonstrated several of these relationships and in addition shown that *S. cerevisiae*, *T. glabrata* (including the PEST-producing strain), and *C. valida* killers are active against various isolates of *C. neoformans* and several candida species. It is interesting to note that killers from most hansenula species kill *T. glabrata* strains, including killers. However, none of four *Hansenula anomala* killers, three of which killed 100% of *C. albicans*, had any activity against torulopsis. This toxin may bind to a different surface or target receptor than toxins produced by the other hansenula species.

Antifungal Activity of Hansenula Toxins

Pathogenic yeast can be killed efficiently by several hansenula toxins. Their mode(s) of action against the pathogens is unknown but it is likely from preliminary studies that some of these toxins are similar to the best characterized toxins from saccharomyces, torulopsis, and pichia. Hansenula toxins are most active between pH 4 and 5. The toxin from *Hansenula mrakii* causes sensitive *C. albicans* to shrink and stop budding. Binding of toxin is rapid but cells can be rescued during a reversible phase by incubation at pH 7.0. The toxin has an apparent molecular weight between 15,000 and 25,000. The toxin produced by *H. anomala* is somewhat different. It contains two fractions with differential activity (Sawant et al., 1986). One has a molecular weight between 10,000 and 30,000 and the second has a molecular weight above 30,000. Both fractions kill *C. albicans*, but only the lower molecular weight fraction kills *Saccharomyces ludwigii*.

DOUBLE-STRANDED RNA AND VIRUSES IN PATHOGENIC YEAST

The genetic information responsible for the production of killer toxins has been identified in three cases. In *Saccharomyces cerevisiae* and the corn smut *Ustilago maydis*, the genetic information can be transmitted cytoplasmically through mating. In both cases, a linear plasmid of dsRNA has been isolated and shown by in vitro translation to code for the production of a preprotoxin that is processed into the secreted product (Bostian et al., 1980; Simon et al., 1981). Saccharomyces killers can be made to lose this dsRNA and consequently the killer phenotype by heat (Wickner, 1974) or cycloheximide treatments (Fink and Styles, 1972). In *Kluyveromyces lactis*, two linear dsDNA plasmids are present in killer strains, one of which appears to encode the toxin (Gunge et al., 1981). The DNA plasmids and the killer phenotype can be transferred into nonkiller saccharomyces by spheroplast fusion (Gunge and Sakaguchi, 1981) or transformation (Gunge et al., 1982).

Attempts to demonstrate cytoplasmic inheritance in killer systems from pathogenic yeasts have been hampered by the absence of sexual reproduction. Attempts to isolate dsRNA from these killer strains have also been largely unsuccessful. Similarly, killers from most genera other than saccharomyces can not be cured by heat or cycloheximide treatments.

Torulopsis glabrata and *Candida albicans*

The earliest demonstration of the presence of dsRNA in *T. glabrata* was by immunoprecipitation using partially purified yeast cell homogenates and antiserum to the synthetic dsRNA polyinosinic:polycytidylic acid (Trimmell, 1976). Two of four strains, one killer and one nonkiller, gave positive results. Two of six *C. albicans* isolates also reacted with the antiserum. Extraction of large nucleic acids, molecular weight greater than 3×10^6, from both torulopsis and candida was accomplished by pronase treatment of a phenol-extracted cell homogenate followed by CF-11 chromatogrpahy (Kandel and Koltin, 1981). No apparent differences were observed between the patterns in torulopsis killers and nonkillers. The presence of DNA, however, was not excluded. There has been only one report that *Candida albicans* contain virus-like particles (Mehta et al., 1982). These authors stated that preliminary experiments suggested that these VLPs, ranging in size from 12 to 30 nm, contain dsRNA.

Using a modified phenol extraction and CF-11 chromatography, Wilson and Kandel (1985) detected unique dsRNA patterns in a single *T. glabrata* killer and nonkiller. The toxin-producing strain contained two species of dsRNA with molecular weights similar to the L and M dsRNAs of saccharomyces. In addition it contained two lower-molecular-weight dsRNAs. A nonkiller strain contained only the largest dsRNA. Protoplasts of an adenine-requiring, nystatin-resistant mutant of the killer strain and a methionine-requiring, fluorocytosine-resistant mutant of the nonkiller strain were fused in the presence of polyethylene glycol, and prototrophic killers were selected. One killer with the auxotrophic and drug resistance characteristics of the nonkiller strain was isolated. This fusion-derived killer contained the four dsRNA species found in the parental killer. These preliminary results suggest that the killer determinant in *T. glabrata* is cytoplasmically inherited and may be coded for by dsRNA, although the possibility of another cytoplasmic nucleic acid encoding the killer phenotype cannot be excluded. In this case it is significant that the killer phenotype in torulopsis is not eliminated by cycloheximide or heat treatments. Wickner (1974) cured *S. cerevisiae* of its killer genome by incubation at 37 to 40°C. *T. glabrata* is a human pathogen and is adapted to grow at 37°C. Higher temperatures therefore may be required to cure opportunistic yeast of killer activity. *T. glabrata* is highly resistant to cycloheximide compared to saccharomyces. Thus it is likely that the methods of Fink and Styles (1972) cannot be applied to this species. In saccharomyces, strains resistant to protein synthesis inhibition by cycloheximide are also resistant to curing by that specific agent (Leibowitz, 1982).

Candida tropicalis

The presence of dsRNA in opportunistic yeast other than *T. glabrata* and *C. albicans* has not been demonstrated. Viruslike particles however have been isolated by Nesterova et al. (1973) from a temperature-sensitive strain of *C. tropicalis*. The particles were 80 to 110 nm in diameter and were surrounded by membrane. Strains lacking the particles were obtained by subculture. When strains carrying and lacking VLPs were compared, it was found that strains carrying the VLPs were more sensitive to osmotic pressure and had an altered morphology (Nesterova, 1974). VLPs isolated by differential centrifugation could be transferred to *S. cerevisiae* if added during mating of opposite types

(Nesterova et al., 1974). The VLPs were not maintained in *S. cerevisiae* upon subculture, however.

Histoplasma capsulatum

While no information is available on the presence of killers among the yeast forms of the primary fungal pathogens, there are several reports of isolation of dsRNA and VLPs from these organisms. dsRNA with a molecular weight of 1.5×10^6 is present in viruslike particles in *Histoplasma capsulatum* (Adler and DempWolff, 1980). The concentration of dsRNA per gram of fungal tissue varies from isolate to isolate but is constant for a given strain. The level of dsRNA is inversely related to the virulence of the isolate. Virulence was measured by the spleen weight and colony-forming units per spleen observed after inoculation of mice with a sublethal dose of *H. capsulatum* (Newton et al., 1982; Newton, 1983). Infection with strains containing dsRNA at a concentration of about 0.07 mg/g of fungal tissue resulted in spleen weights between 66 and 134 mg and colony-forming units ranging from 4 to 180 per spleen. In contrast, strains with lower levels of dsRNA, 0.03 mg/g of fungal tissue, produced spleen weights between 255 and 434 mg and colony-forming units per spleen ranging from 439 to 8293. The reduction in virulence in strains containing higher levels of dsRNA is possibly related to the immunomodulatory activity of mycoviral dsRNA. This hypothesis is supported by the observation that exogenously administered dsRNA, isolated from *Penicillium chrysogenum*, reduces the virulence of *H. capsulatum* (Olson and Adler, 1983). In addition, Wright and Adler-Moore (1985) have shown that dsRNA isolated from a nonkiller strain of *S. cerevisiae* enhances both the cellular and humoral immune responses in mice. The adjuvant effects produced by mycoviral dsRNA are more potent than those produced by an equivalent dose of polyinosinic:polycytidylic acid.

Other Yeasts

Recently, dsRNA was isolated form *Blastomyces dermatiditis* (Adler and Stevens, personal communication). No dsRNA was found in a screening survey of *C. neoformans*.

No dsRNA has been found in most non-saccharomyces killer yeasts. If the genetic information for killing in these organisms is dsRNA, it (or any VLP in which it is encapsidated) may be associated with cellular proteins or lipids that interfere with extraction by standard methods. Alternatively, the genetic information for the production of killer toxins may be found on a linear dsDNA plasmid, as is the case in *Kluyveromyces lactis*. Gunge et al. (1981) failed to isolate plasmid DNA from 32 other species examined, however. These included saccharomyces, candida, torulopsis, and hansenula isolates. In addition, the kluyveromyces toxin differs markedly from most other killers: It is a high-molecular-weight glycoprotein, active at pH 6.5, and inhibits adenylate cyclase in *S. cerevisiae* (Sugasaki et al., 1983; Sugasaki et al., 1984). A third possibility is that the information for toxin production is coded by the nuclear genome. Hybridization studies with isolated RNA messages or cDNAs would provide an answer to this hypothesis. These experiments, however, are far from trivial.

CONCLUSIONS

Pathogenic yeast participate in a variety of killer/sensitive relationships. Thus far the incidence of killers among these yeast isolates is low, with the highest numbers found in *T. glabrata* and very few, if any, among *C. albicans*. It is unlikely, therefore, that toxins are important as yeast virulence factors. In addition, the low pH required for activity is found only in selected body sites or in lysosomal granules, indicating a probable lack of selection for the killer phenotype in clinical infections.

The incidence of sensitivity among pathogens is high, particularly to toxins produced by nonpathogens. The failure of individual strains to be killed is probably related to cell surface properties. If spheroplasts of resistant strains are sensitive to these toxins, it might indicate a universal membrane target for antifungal activity. The toxins themselves are proteins that are too large to use in systemic fungal therapy without eliciting an immune response. However, if they can be delivered without encountering the immune system or if active subunits or derivatives can be found, they might prove useful (providing they have no adverse effect on human cells).

The use of killers toxins as "fungicins" for epidemiological typing of *C. albicans* or to differentiate among species of candida has potential. At this time, however, the system still needs refinement to be practical.

ACKNOWLEDGMENTS

I wish to thank Michael Tushla, Mike Davis, and Steve Karl for their assistance in preparation of this manuscript.

REFERENCES

Adler, J. P., and DempWolff, U. (1980). Double-stranded RNA associated with VLPs in *Histoplasma capsulatum*. *Abstr. Annu. Meet. Am. Soc. Microbiol.*, p. 273.

Bevan, E. A., and Makower, M. (1963). The physiological basis of the killer character in yeast. *Proc. XI Int. Congr. Genet. 1*:203.

Bevan, E. A., Herring, A. J., and Mitchell, D. J. (1973). Preliminary characterization of two species of dsRNA in yeast and their relationship to the "killer" character. *Nature 245*:81-86.

Bostian, K. A., Hopper, J. E., Rogers, D. J., and Tipper, D. J. (1980). Translational analysis of the killer-associated virus-like particle dsRNA genome of *S. cerevisiae*: M-dsRNA encodes toxin. *Cell 19*:403-413.

Bostian, K. A., Elliott, Q., Bussey, H., Burn, V., Smith, A., and Tipper, D. J. (1984). Sequence of the preprotoxin dsRNA gene of type 1 killer yeast: Multiple processing events produce a two-component toxin. *Cell 36*:741-751.

Bussey, H. (1981). Physiology of killer factor in yeast. *Adv. Microb. Physiol. 22*:93-112.

Bussey, H., and Skipper, N. (1975). Membrane-mediated killing of *Saccharomyces cerevisiae* by glycoproteins from *Torulopsis glabrata*. *J. Bacteriol. 124*:476-483.

Bussey, H., and Skipper, N. (1976). Killing of *Torulopsis glabrata* by *Saccharomyces cerevisiae* killer factor. *Antimicrob. Agents Chemother. 9*:352-354.

Bussey, H., Steinmetz, O., and Saville, D. (1983). Protein secretion in yeast: Two chromosomal mutants that oversecrete killer toxin in *Saccharomyces cerevisiae*. *Curr. Genet. 7*:449-456.

Caprilli, F., Prignano, G., Latella, C., and Tavarozzi, S. (1985). Amplification of the killer system for differentiation of *Candida albicans* strains. *Mykosen 28*:569-573.

Coats, H. A., and Kandel, J. S. (1979). Potentiation of killer toxins by amphotericin B. *Abstr. Annu. Meet. Am. Soc. Microbiol.*, p. 367.

Fink, G. R., and Styles, C. A. (1972). Curing of a killer factor in *Saccharomyces cerevisiae*. *Proc. Nat. Acad. Sci. USA 69*:2846-2849.

Gunge, N., and Sakaguchi, K. (1980). Intergeneric transfer of deoxyribonucleic killer plasmids, pGKL1 and pGLK2, from *Kluyveromyces lactis* to *Saccharomyces cerevisiae* by cell fusion. *J. Bacteriol. 147*:155-160.

Gunge, N., Tamaru, A., Ozawa, F., and Sakaguchi, K. (1981). Isolation and characterization of linear dexoyribonucleic acid plasmids from *Kluyveromyces lactis* and the plasmid-associated killer character. *J. Bacteriol.* *145*:382-390.

Gunge, N., Murata, K., and Sakaguchi, K. (1982). Transformation of *Saccharomyces cerevisiae* with linear DNA killer plasmids from *Kluyveromyces lactis.* *J. Bacteriol.* *151*:462-464.

Kandel, J. S. (1980). Killer systems in pathogenic yeast. In *Human and Animal Mycology,* Int. Congr. Series No. 480, E. S. Kuttin and G. L. Baum (Eds.). Excerpta Medica, Amsterdam, pp. 259-262.

Kandel, J. S., and Koltin, Y. (1981). Killer phenomenon and double-stranded RNA in Candida and Torulopsis species. *Abstr. Annu. Meet. Am. Soc. Microbiol.*, p. 191.

Kandel, J. S., and Stern, T. A. (1979). Killer phenomenon in pathogenic yeast. *Antimicrob. Agents Chemother.* *15*:568-571.

Kwon-Chung, K. J., Lehman, D., Good, C., and Magee, P. T. (1985). Genetic evidence for role of extracellular proteinase in virulence of *Candida albicans.* *Infect. Immun.* *49*:571-575.

Leibowitz, M. J. (1982). Role of protein synthesis in the replication of the killer virus of yeast. *Curr. Genet.* *5*:161-163.

Mehta, R. J., Nash, C. H., and Bozzola, J. J. (1982). Virus-like particles and lytic plaque formation in lawns of *Candida albicans.* *J. Bacteirol.* *152*:502-505.

Middlebeek, E. J., Hermans, J. M. H., and Stumm, C. (1979). Production, purification, and properties of a *Pichia kluyveri* killer toxin. *Antonie van Leeuwenhoek J. Microbiol. Serol.* *45*:437-450.

Middlebeek, E. J., Crutzen, A. Q. H., and Vogels, G. D. (1980a). Effect of potassium and sodium ions on the killing action of a *Pichia kluyveri* toxin in cells of *Saccharomyces cerevisiae.* *Antimicrob. Agents Chemother.* *18*:519-524.

Middlebeek, E. J., Hermans, J. M. H., Stumm, C., and Muytjens, H. L. (1980b). High incidence of sensitivity to yeast killer toxins among Candida and Torulopsis isolates of human origin. *Antimicrob. Agents Chemother.* *17*:350-354.

Middlebeek, E. J., Peters, J. G. W. H., Stumm, C., and Vogels, G. D. (1980c). Properties of a *Cryptococcus laurentii* killer toxin and conditional killing effect of the toxin on *Cryptococcus albidus.* *FEMS Microbiol. Lett.* *9*:81-84.

Moore, J. C., Adler, J. P., and Kandel, J. S. (1984). Scanning electron microscopy of eucaryotic cells exposed to killer toxin. *Abstr. Annu. Meet. Am. Soc. Microbiol.*, p. 296.

Morace, G., Archibusacci, C., Sestito, M., and Polonelli, L. (1984). Strain differentiation of pathogenic yeasts by the killer system. *Mycopathologia 84*:81-85.

Nesterova, G. F. (1974). On two cases of viral conversion in yeasts. *Mikrobiologiya 43*:919-921.

Nesterova, G. F., Kyarner, Y., and Soom, Y. O. (1973). Virus-like particles in *Candida tropicalis*. *Mikrobiologiya 42*:162-164.

Nesterova, G. F., Karner, J., and Kozlova, T. M. (1974). The possible intercellular transfer of virus-like particles from *Candida tropicalis* to *Saccharomyces cerevisiae*. *Mikrobiologiya 43*:1095-1097.

Newton, K. (1983). Study of the relationship between dsRNA concentration and virulence in strains of *Histoplasma capsulatum*. Masters thesis, California State Polytechnic University, Pomona.

Newton, K., Adler, J. P., and Howard, D. H. (1982). Relationship between dsRNA concentration and virulence in *Histoplasma capsulatum*. *Abstr. Annu. Meet. Am. Soc. Microbiol.*, p. 336.

Olson, J., and Adler, J. P. (1983). Reduced pathogenicity of *Histoplasma capsulatum* in C57B1/6 mice following treatment with mycoviral dsRNA. *Abstr. Annu. Meet. Am. Soc. Microbiol.*, p. 393.

Palfree, R., and Bussey, H. (1979). Yeast killer toxin: Purification and characterization of the protein toxin from *Saccharomyces cerevisiae*. *Eur. J. Biochem. 93*:487-493.

Pfeiffer, P., and Radler, F. (1982). Purification and characterization of extracellular and intracellular killer toxin of *Saccharomyces cerevisiae* strain 28. *J. Gen. Microbiol. 128*:2699-2706.

Philliskirk, G., and Young, T. W. (1975). The occurrence of killer character in yeasts of various genera. *Antonie van Leeuwenhoek J. Microbiol. Serol. 41*:147-151.

Polonelli, L., Archibusacci, C., Sestito, M., and Morace, G. (1983). Killer system: A simple method for differentiating strains of *Candida albicans*. *J. Clin. Microbiol. 17*:774-780.

Polonelli, L., Castagnola, M., Rossetti, D. V., and Morace, G. (1985). Use of killer toxins for computer-aided differentiation of *Candida albicans* strains. *Mycopathologiya 91*:175-179.

Rogers, D., and Bevan, E. A. (1978). Group classification of killer yeasts based on cross-reactions between strains of different species and origin. *J. Gen. Microbiol. 105*:199-202.

Santell, L., and Kandel, J. S. (1982). Characterization of a killer toxin from *Torulopsis glabrata*. *Abstr. Annu. Meet. Am. Soc. Microbiol.*, p. 330.

Sawant, A. D., Abdelal, A. T., and Ahearn, D. G. (1986). Killer toxins of *Hansenula anomala*. *Abstr. Annu. Meet. Am. Soc. Microbiol.*, p. 270.

Skipper, N., and Bussey, H. (1977). Mode of action of yeast toxins: Energy requirement for *Saccharomyces cerevisiae* killer toxin. *J. Bacteriol. 129*:668-677.

Simon, S., Gorecki, M., and Koltin, Y. (1981). In vitro translation of the Ustilago virus dsRNA. *Genetics 97*:S97-98.

Stumm, C., Hermans, J. M. H., Middlebeek, E. J., Croes, A. F., and deVries, G. J. M. L. (1977). Killer-sensitive relationships in yeasts from natural habitats. *Antonie van Leeuwenhoek J. Microbiol. Serol. 43*:125-128.

Sugisaki, Y., Gunge, N., Sakaguchi, K., Yamasaki, M., and Tamura, G. (1983). Killer toxin inhibits adenylate cyclase of sensitive yeast cells. *Nature 304*:464-466.

Sugisaki, Y., Gunge, N., Sakaguchi, K., Yamasaki, M., and Tamura, G. (1984). Characterization of a novel killer toxin encoded by a double-stranded linear DNA plasmid of *Kluyveromyces lactis*. *Eur. J. Biochem. 141*:241-245.

Trimmell, D. (1976). Immunological detection of double-stranded RNA in yeasts. Masters thesis. California State University, Fullerton.

Vodkin, M., and Fink, G. (1973). A nucleic acid associated with a killer strain of yeast. *Proc. Nat. Acad. Sci. USA 70*:1069-1072.

Wickner, R. B. (1974). "Killer character" of *Saccharomyces cerevisiae*: Curing by growth at elevated temperature. *J. Bacteriol. 117*:1356-1357.

Wilbur-Murphy, J., and MacMillan, J. D. (1976). Synergism of yeast killer factor and Oeskovia yeast cell wall-degrading enzyme against *Candida albicans*. *Abstr. Annu. Meet. Am. Soc. Microbiol.*, p. 92.

Wilson, C., and Kandel, J. (1985). Cytoplasmic transfer of the killer phenomenon in *Torulopsis glabrata*. *Abstr. Annu. Meet. Am. Soc. Microbiol.*, p. 241.

Wray, W., Boulikas, T., Wray, V. P., and Hancock, R. (1981). Silver staining of proteins in polyacrylamide gels. *Anal. Biochem. 118*:197-203.

Wright, C. L., and Adler-Moore, J. P. (1985). The adjuvant effects of mycoviral dsRNA and polyinosinic:polycytidylic acid on the murine immune response. *Biochem. Biophys. Res. Commun.* *131*:949-955.

Young, T. W., and Yagiu, M. (1978). A comparison of the killer character in different yeasts and its classification. *Antonie van Leeuwenhoek J. Microbiol. Serol.* *44*:59-77.

12

Kluyveromyces Linear DNA Plasmids

Norio Gunge
Kumamoto Institute of Technology, Kumamoto, Japan

INTRODUCTION

The 2 μm DNA plasmid, the first eukaryotic plasmid, was discovered by Sinclair et al. (1967). This plasmid exists in most laboratory strains of *Saccharomyces cerevisiae* and was regarded as the only plasmid of yeast until several years ago. In screening for new plasmids from various yeast species, we have found that the petite negative yeast *Kluyveromyces lactis* (IFO 1267) harbors two linear DNA plasmids, pGKL1, with 8.9 kilobase pairs, and pGKL2, with 13.4 kilobase pairs, 50 to 100 copies per cell (Gunge et al., 1981). The finding of the linear yeast plasmids was surprising, because, with the exception of the mitochondrial linear plasmids from cytoplasmic male sterile maize (Pring et al., 1977; Kemble and Bedbrook, 1980; Kemble and Thompson, 1982), all prokaryotic and eukaryotic plasmids were thought to be circular. Subsequently, linear plasmids were reported in a variety of eukaryotes: in plants, *Sorghum bicolor* (Pring et al., 1982), *Brassica campestris*, *Brassica napus* (Palmer et al., 1983; Erickson et al., 1985); in fungi, *Ascobolus immersus* (Francou, 1981), *Claviceps purpurea* (Tudzynski et al., 1983), *Morchella conica* (Meinhardt and Esser, 1984), *Rhizoctonia solani* (Hashiba et al., 1984); and in yeast, *Saccharomyces kluy-*

veri (Kitada and Hishinuma, 1987). These findings indicate that linear plasmids may occur rather frequently in eukaryotes. In prokaryotes, a linear plasmid exists in *Streptomyces rochei* (Hayakawa et al., 1979; Hirochika et al., 1984).

Primary interest of the *K. lactis* plasmids is in their linear structure, which raises a question as to how the ends of the plasmids replicate. Since DNA polymerases require a primer to initiate the 5' to 3' synthesis, the 5' ends of linear DNA genomes would remain unsynthesized or gapped after the removal of the primer, unless some specific mechanism provided for their synthesis. In this connection, certain linear DNA viruses such as human adenoviruses and bacillus bacteriophages are known to have specialized structures: inverted terminal repeats and a protein covalently attached to the 5' ends (Rekosh et al., 1977; Mellado and Salas, 1983; Yoshikawa and Ito, 1981). These terminal structures are essentially involved in the viral DNA replication. Similar structures also exist in the yeast linear plasmids. There are inverted repeats of 202 base pairs in pGKL1 and 184 in pGKL2 (Sor et al., 1983; Hishinuma et al., 1984) whose 5' ends are associated with their specific proteins (Kikuchi et al., 1984). The pGKL DNA replication probably depends on protein priming and a strand displacement mechanism. Linear plasmids from various eukaryotes and *Streptomyces rochei*, as far as examined, also have similar adenovirus type terminal structures (Kemble and Thompson, 1982; Erickson et al., 1985; Francou, 1981; Tudzinski et al., 1983; Meinhardt and Esser, 1984; Hirochika et al., 1984).

A second noteworthy character of the linear plasmids is their association with the killer phenomenon (Gunge et al., 1981; Gunge and Sakaguchi, 1981; Wésolowski et al., 1982), making a marked contrast to other yeast plasmids: 2 μm DNA, 2 μm DNA-like plasmids, 3 μm DNA (Gunge, 1983), and the *S. kluyveri* linear plasmid, all of which are cryptic in biological function. In the *K. lactis* killer system, pGKL1 confers on the host cells the ability to secrete a killer toxin, lethal to a variety of yeast species (killer phenotype), and resistance to the toxin (immunity phenotype). Plasmid pGKL2 probably controls the replication and maintenance of the pGKL plasmids (Gunge et al., 1981; Wésolowski et al., 1982; Niwa et al., 1981). *K. lactis* killer toxin consists of three subunits: a glycosylated polypeptide of approximately 100 kDa and two smaller nonglycosylated components of ∿30 and ∿27 kDa (Sugisaki et al., 1984; Stark and Boyd, 1986; T. Fujimura, 1984, unpublished results). The toxin properties are distinct from those of the dsRNA-encoded killer toxin of *S. cerevisiae* in several aspects: ranges of pH and temperature for the toxin activity (Gunge et al., 1981), killing spectrum (Gunge et al., 1981; Gunge and Sakaguchi, 1981), mechanism of the toxin action (Sugisahi et al., 1983), etc.

A third interesting feature of the linear plasmids of *K. lactis* is their replicating ability in heterologous yeasts. The pGKL plasmids are transferable to *S. cerevisiae* (Gunge and Sakaguchi, 1981; Gunge et al., 1982), *Kluyveromyces fragilis*, and *Candida pseudotropicalis* (Sugisaki et al., 1985). This chapter describes the methodology of the plasmid transfer into heterologous yeasts, the behavior and biological problems in a different yeast, and some approaches to construction of pGKL plasmid-based vectors for use in recombinant DNA technology.

PLASMID TRANSFER INTO HETEROLOGOUS YEASTS

Plasmid transfer into heterologous yeasts was attempted in the following three ways: (1) Intergeneric protoplast fusion (Gunge and Sakaguchi, 1981); (2) Transformation with isolated pGKL plasmids (Gunge et al., 1982); (3) Illegitimate mating (Sugisaki et al., 1985).

Procedure (1) is based on the fact that intergeneric protoplast fusion in yeast generally gives rise to a heterokaryon rather than to a nuclear fusant (Ferenczy, 1981): Parental nuclei may coexist in fused protoplasts but dissociate from each other during the course of the protoplast regeneration. In the meantime, extrachromosomal elements or autonomously replicating plasmids may be randomly distributed among the regenerating protoplasts. Experimentally, protoplasts of *K. lactis* killer 2105-1D (*ade1 ade2 leu*, pGKL1 and pGKL2) were mixed with protoplasts of *S. cerevisiae* AH22 (*a leu2 his4 can*^{R} ρ^0) in the presence of 30% polyethylene glycol (PEG) and fused protoplasts were regenerated on yeast nitrogen-base medium containing histidine and leucine. Out of the regenerated colonies, 336 were randomly isolated and tested for the killer phenotype, and 84 killer colonies harboring pGKL1 and pGKL2 plasmids were obtained. They were all ρ^0, and genetically and morphologically the same as *S. cerevisiae* AH22, except for the acquired killer character. Thus, the pGKL killer plasmids could be transferred into *S. cerevisiae* with a frequency of 25% (84/336). The plasmids stably replicated in multiple copies and expressed the killer and immunity phenotypes. Using the protoplast fusion technique, the pGKL plasmids were also transferred from *K. lactis* killer into *Candida pseudotropicalis* (Sugisaki et al., 1985).

Procedure (2) is based upon the conventional yeast transformation technique (Hinnen et al., 1978). Protoplasts (3×10^5 cells) of AH22 were mixed with 1 μg each of pGKL1 and pGKL2 plasmid DNAs in the presence of 30% PEG and regenerated on complete medium. Out of 2000 regenerated colonies, two pGKL killer transformants were isolated.

Procedure (3) depends on a rare mating phenomenon between different yeast species (Wickerham and Burton, 1956). Cells of *K. lactis* killer YS283 (*met*, maltose assimilation positive) were mixed with cells of *K. fragilis* NRRL Y-610 (*MET*, maltose assimilation negative) on complete medium. The mixed culture was incubated overnight at 30°C and spread on minimal medium containing maltose as a carbon source. After 5 days, colonies appeared with a frequency of 10^{-3} to 10^{-4}, and some of them were pGKL killers. The rationale for this approach is based on the possible production of heterokaryons in a mixed culture by rare mating between yeasts. The cell growth that ensues is accompanied by nuclear dissociation and free distribution of the plasmids.

INCOMPATIBILITY WITH MITOCHONDRIAL DNA IN *S. cerevisiae*

Plasmid Instability

In the k1 and k2 killer systems of *S. cerevisiae*, M_1 and M_2 dsRNA plasmids encode their specific killer and immunity phenotypes. Their replication in yeast cells depends upon the presence of the L-A dsRNA plasmid (Wickner, 1981). The M and L-A dsRNA plasmids are the genomes of cytoplasmically inherited viruses. When k1 and k2 killers are crossed, M_2 is excluded by M_1 as a result of their replicative competition (Wickner, 1983), in a process resembling bacterial plasmid incompatibility. Now that the pGKL plasmids are transferable to *S. cerevisiae*, the question arises as to whether or not both pGKL and dsRNA killer systems can coexist in the same cell. To answer this problem, we introduced M_1 and L-A dsRNA plasmids into a *S. cerevisiae* pGKL killer, using the *kar1* cytoduction technique (Conde and Fink, 1976). As a result, the pGKL plasmids were found to be compatible with the dsRNA killer plasmids, but, unexpectedly, pGKL plasmids were incompatible with mitochondrial DNA in *S. cerevisiae* (Gunge and Yamane, 1984). Experimentally, a protoplast fusant pGKL killer of *S. cerevisiae* F102-2 (*a leu2 his4 can*^{R} ρ^0, pGKL1 and pGKL2) was crossed with *S. cerevisiae* JC11 (α his4 kar1 ρ^+, M_1, and L-A dsRNA) and the resultant ρ^+ cytoductants of F102-2 were isolated on a suitable selection medium. The cytoductants were then tested for killing activity, using the pGKL toxin-specific sensitive tester *S. cerevisiae* B511-4C and the k1 toxin-specific sensitive tester *Saccharomyces italicus* IFO 0725. Out of 94 ρ^+ cytoductants tested, 78 were killers against both testers (++ type) and the remaining 16 killed only *S. italicus* IFO 0725 (-+ type) (Table 1). Unexpectedly, the ++ type cytoductant clones were all unstable for the pGKL killer phenotype. When subcloned, they segregated into a mixture of ++ and -+ type clones. These ++ type segregants were again unstable and, when subcloned, they were mixed with -+ type cells (Table 1). The -+ type cells contained M_1 and L-A dsRNA but lacked both of the pGKL plasmids, suggesting that the instability of ++ type killers was due to the frequent loss of pGKL plasmids, rather than a defect in the expression of killer activity from stable plasmids.

Table 1 Stability of pGKL and dsRNA Plasmids in ρ^+ Cytoductants

	No. of clones					% Plasmids	
		Plasmid type[a]					
Strain	Total	++	+-	-+	--	pGKL	dsRNA
F102 cytoductant	94	78	0	16	0	83	100
Subclone 1[b]	36	25	0	9	2	69	94
Subclone 2	54	32	0	22	0	59	100
Subclone 3	47	36	1	10	0	79	98
Subclones 4 to 10	336	151	0	184	1	45	100

[a]++, The presence of pGKL and dsRNA plasmids; +-, the presence of pGKL plasmids; -+, the presence of dsRNA plasmids; --, the absence of pGKL and dsRNA plasmids.
[b]All subclones were derived from single F102 cytoductant colonies of the ++ plasmid type.

Stabilization in ρ^0 Mutants

The plasmid instability observed in the ++ type of cytoductant described above was not due to the presence of dsRNA plasmids, but was caused by incompatibility of the pGKL plasmids with mitochondrial DNA of *S. cerevisiae*. This was shown by the fact that ethidium-bromide-induced ρ^0 (mitochondrial DNA-less) mutants of unstable ++ type clones were stable and consisted of almost all (more than 95%) ++ type cells. The pGKL plasmids thus stably coexisted with dsRNA killer plasmids in ρ^0 mutants, and the killer toxins of both killer systems were simultaneously produced in the same cells.

Incompatibility with mitochondrial DNA was also observed in the following crosses. A stable ρ^0 ++ type cytoductant CD1-2192-335 EB1 was crossed with pGKL-less ρ^+ nonkiller testers and with ρ^0 mutants derived from these testers. The transmission of pGKL plasmids into diploid clones was compared (Table 2). The transmission frequency was extremely low (0-11%) in ρ^+ diploids ($\rho^0 \times \rho^+$ crosses) but high (60-80%) in ρ^0 diploids ($\rho^0 \times \rho^0$ crosses).

Analysis with Various Mitochondrial Mutants

A question arises as to whether the stable pGKL plasmid transmission in ρ^0 diploids is due to the absence of mitochondrial genome (ρ^+) or to the respiration-deficiency of the ρ^0 cells. To answer this question,

Table 2 Effect of mtDNA on Transmission of pGKL Plasmids into Diploids

Cross[a]	Total	No. of clones: Plasmid type[b]				% Plasmids	
		++	+-	-+	--	pGKL	dsRNA
$\rho^0 \times \rho^+$							
CD1-2191-335-EB1 × 0596R-3A	162	0	0	161	1	0	99
CD1-2191-335-EB1 × 705	102	0	0	102	0	0	100
CD1-2191-335-EB1 × 722	104	11	0	93	0	11	100
$\rho^0 \times \rho^0$							
CD1-2191-335-EB1 × 0596R-3A-EB	54	34	0	20	0	63	100
CD1-2191-335-EB1 × 705-EB	61	45	0	16	0	63	100
CD1-2191-335-EB1 × 722-EB	59	47	0	12	0	80	100

[a]All crosses of *S. cerevisiae* were between a ++ ρ^0 cytoductant and either ρ^+ or ρ^0 nonkiller testers (-- ρ^+ or -- ρ^0).
[b]Symbols are as described in Table 1 footnote.

we mated a stable ρ^0 killer to a chromosomal petite mutant (*pet9*) carrying an intact mitochondrial genome. No plasmid transmission was observed, indicating that plasmid transmission in ρ^0 strains was not related to respiration deficiency. The effects of *mit*$^-$ mutations (point or small deletion mutations in mitochondrial respiratory genes) were also studied. In crosses of a stable ρ^0 pGKL-bearing *S. cerevisiae* killer with *mit*$^-$ mutants defective in cytochrome oxidase genes (*oxi1*, *oxi2*, and *oxi3*) or in the cytochrome b gene (*cob*), all diploid progeny were pGKL-less nonkillers. We next examined the effects of ρ^- mutations (mitochondrial large deletion mutations). A stable ρ^0 ++ type killer CD15-2EB18 was crossed with each of three ρ^- mutants, OIP2, CEP2, and CE02, which were derived from the ρ^+ nonkiller strain MH41-7B. As shown in Table 3, the transmission frequency of pGKL plasmids was as high as 70 to 80% in all of these $\rho^0 \times \rho^-$ crosses and was indistinguishable from that of a control $\rho^0 \times \rho^0$ cross, although transmission as low as 3.7% was observed in a cross with the parental ρ^+ strain. Mitochondrial maps of the three mutants, OIP2, CEP2, and CEO2, are shown in Figure 1. Each of the mutants represents a specific large deletion on the mitochondrial genome and, taken together, they cover the whole genome. Therefore we could not find any specific regions responsible for the incompatibility with the pGKL plasmids.

The above results raise another question as to the possibility that mitochondrial protein products or protein synthesis may have been involved in the plasmid instability phenomenon, since mitochondrial protein synthesis exists in mit$^-$ mutants, but not in ρ^0 and ρ^- mutants. This problem could be solved by performing crosses with syn$^-$ mutants (point or small deletion mutants in mitochondrial protein synthesis-related genes), if these were available. Since such mutant strains did not exist in our laboratory, we approached this problem by examining how the pGKL plasmid transmission in ρ^0 stable killer by ρ^+ nonkiller crosses would be affected in the presence of chloramphenicol (3 mg/ml) which specifically inhibits mitochondrial protein synthesis. All of the diploids were nonkillers under these conditions, suggesting that mitochondrial protein synthesis appears not to be required for mitochondrial exclusion of pGKL plasmids in *S. cerevisiae*. The result must be taken with some caution, however, because we could not rule out the possibility that chloramphenicol also blocked the formation of terminal proteins of pGKL plasmids and that this was the reason for the loss of killer plasmids in these experiments.

Another possible reason for the pGKL plasmid exclusion by mitochondrial DNA could be competition between replication of mitochondrial DNA and the plasmid DNAs, with mitochondrial DNA out-competing the plasmids. However, it should be noted that a portion (20 to 30% on the average) of diploid progenies in pGKL $\rho^0 \times \rho^0$ crosses were still pGKL-less (Tables 2 and 3). Thus, it appears that, in addition to the mitochondrial genome, other undefined factors may be involved

Table 3 Transmission of pGKL Plasmids in ρ^0 Killer by ρ^- Nonkiller Crosses

Cross	No. of clones					% Plasmids	
		Plasmid type[a]					
	Total	++	+-	-+	--	pGKL	dsRNA
CD15-2EB18 × MH41-7B(ρ^+)	54	2	0	52	0	3.7	100
CD15-2EB18 × OIP2	54	39	0	15	0	72.2	100
CD15-2EB18 × CEP2	54	44	0	10	0	81.5	100
CD15-2EB18 × CEO2	54	44	0	10	0	81.5	100
CD15-2EB18 × MH41-7B-EB(ρ^0)	47	35	0	12	0	74.5	100

[a]Symbols are as described in the Table 1 footnote.

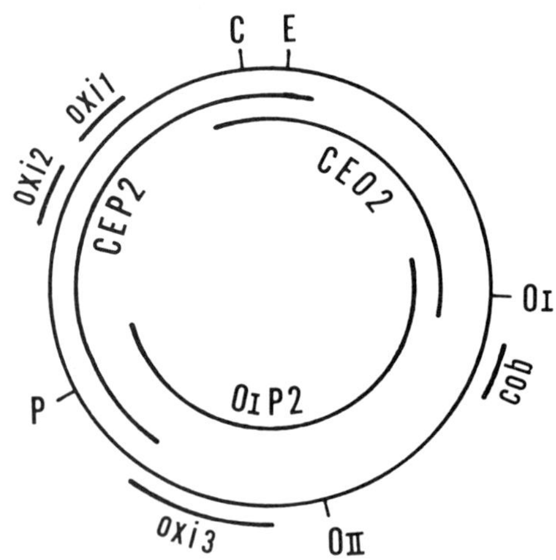

Figure 1 Mitochondrial mapping of ρ^- mutants. C, E, O_I, O_{II}, and P refer to mitochondrial drug resistance genes for chloramphenicol, erythromycin, oligomycin (two loci), and paromomycin, respectively.

in the instability problem. Last, we should also recall that, unlike the situation in *S. cerevisiae*, the pGKL plasmids are stably maintained in the obligatory aerobic petite negative yeast *K. lactis* from which the pGKL plasmids were originally isolated. The phenomenon of mitochondrial incompatibility raises new questions concerning interactions between the organelle DNA and plasmids. This interaction is reminiscent of the instability of the killer virus dsRNA of *S. cerevisiae* in the presence of a functional mitochondrial genome in *mak10* mutants of that organism (Wickner, 1977).

VECTOR CONSTRUCTION

pGKL Plasmid ARS

The linear pGKL plasmids have some characteristic features: (1) high copy numbers in cells (up to 100 per cell); (2) the ability to replicate in heterologous yeasts; (3) the production of a toxin with a large molecular size; and (4) linear structure. These properties may be useful in construction of plasmid vectors with high copy numbers, shuttle vectors between various yeast species, and powerful secretion vectors. Linear vectors of a novel type may also be constructed using the specialized terminal structures, which may act as telomeres.

We first looked for autonomously replicating sequences (ARS) in the pGKL plasmids, which may function as the origin of DNA replication in *S. cerevisiae* and/or the fission yeast *Schizosaccharomyces pombe*, using an integrating vector YIp32 as an ARS screening vector. YIp32 transforms *S. cerevisiae* (*leu2*) and *S. pombe* (*leu1*) to LEU^+ phenotype at a very low frequency (at most, 10 transformants per 1 μg DNA) as a result of chromosomal integration, but if combined with ARS element(s), the transformation efficiency increases 100 to 1000 times (Struhl et al., 1979; Beach et al., 1980). Based on this principle, several restriction fragments of pGKL1 and pGKL2 were found to contain the ARS elements (N. Gunge and C. Yamane, 1984, unpublished results). For example, a hybrid DNA pYGP [Figure 2(a)], constructed by an insertion of a *Pst* I-cut pGKL1 internal fragment (6.4 kilobases) to a *Pst* I site of YIp32, efficiently transformed both of *S. cerevisiae* AH22 (*leu2*) and *S. pombe* JY226 (*leu1*) to their respective wild types by a lithium acetate transformation technique (Ito et al., 1983). Similarly, a *Hind* III internal fragment (5.6 kilobases) of pGKL1, a *Hind* III terminal fragment (0.94 kilobases) of pGKL1, and a *Bam* HI-*Pst* I fragment (5.0 kilobases) of pGKL2 expressed the ARS activity in both the above two yeasts. A hybrid plasmid pYGU1 [Figure 2(a)], constructed by an insertion of the *S. cerevisiae URA3* DNA (1.1 kilobase) to a *Hind* III site of the pGKL1 portion on pYGP, autonomously replicated in *K. lactis* KH101-5B (*uraA*), *S. cerevisiae* YNN27 (*ura3*), and *S. pombe* JY280 (*ura4*). All these pGKL fragments showing the ARS activity were found to contain one or two copies of ARS

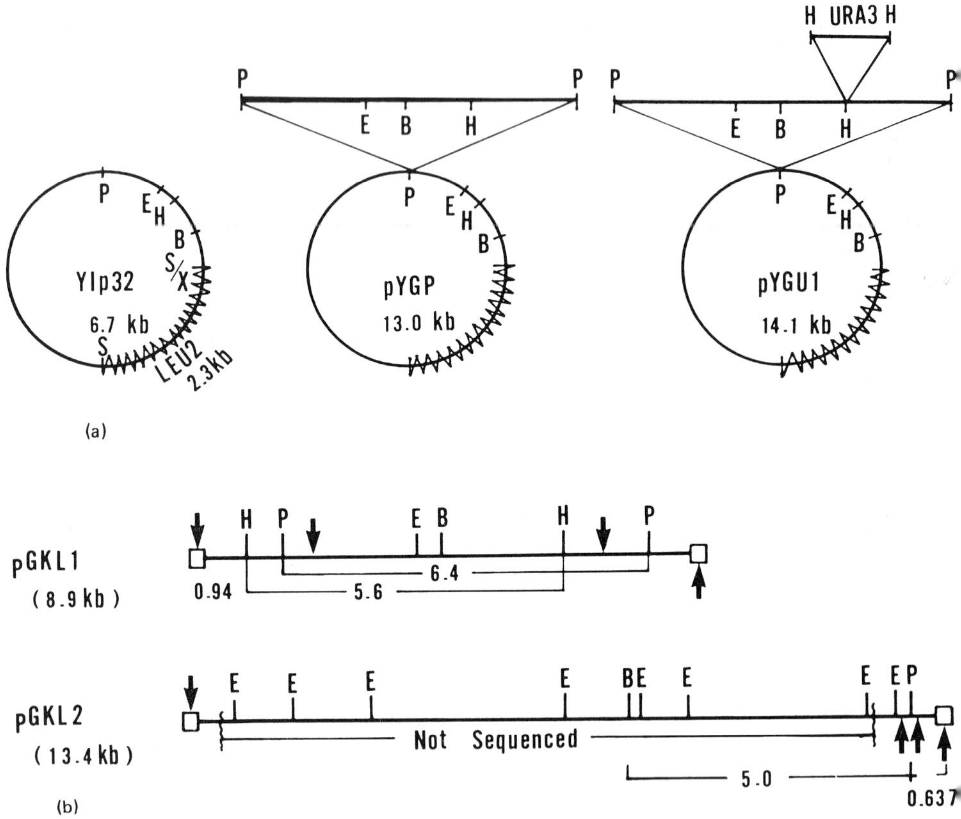

Figure 2 Construction of pGKL-ARS containing hybrid plasmids. (a) Physical maps of an integrating vector YIp32 and the pGKL-ARS containing hybrid plasmids. (b) Physical maps of pGKL1 and pGKL2. B, *Bam* HI; E, *Eco*RI; H, *Hind* III; P, *Pst* I; S, *Sal* I; X, *Xho* I. □: inverted terminal repeats. Arrows indicate positions of ARS consensus sequence.

consensus sequence of A/T TTTAT T/G TT A/T which was reported in *S. cerevisiae* (Broach et al., 1982), as illustrated [Figure 2(b)]. However, whether these ARS elements were commonly used in *S. pombe* and/or *K. lactis* as well as in *S. cerevisiae* is questionable. In fact, *S. cerevisiae ARS1* and the *ori* of 2 μm DNA did not properly function in *K. lactis* and *S. pombe* (Das and Hollenberg, 1982; Gaillardin et al., 1983). Host specificity of ARS activity in *S. cerevisiae* and *S. pombe* has also been reported in an ARS assay (Maundrell et al., 1985). Similarly, we have found that a cloned terminal fragment (637 basepairs) of pGKL2 containing two ARS consensus sequences functioned in *S. cerevisiae*, but not in *K. lactis* (Fujimura et al., 1987).

The mitotic stability of the pGKL-ARS hybrid plasmids constructed above was examined, with the result that all of them were found to be unstable in *S. cerevisiae* and *K. lactis*: less than 1% of transformant cells retained the plasmids after 10 generations of growth in nonselective medium. In *S. pombe*, however, the same hybrid plasmids were much more stable: 30% or more of the transformant cells carried the plasmids under the same condition of growth. In every case, they did not require the presence of pGKL2 for the replication, indicating that the replication mechanism is different from that of natural linear pGKL plasmids. If the problem of host specificity could be resolved, however, the pGKL fragments showing ARS activity in heterologous yeasts would be useful in construction of yeast shuttle vectors of a wide host range.

Circularization of pGKL-Based Vectors

The natural pGKL plasmids probably depend for the replication on the specialized terminal structures. Therefore, if a selection marker could be inserted directly into the internal regions of the plasmids, the resulting hybrid vectors could be expected to replicate as linear molecules. Such an experiment was first attempted by de Louvencourt and his co-workers (1983). They attempted to insert the *S. cerevisiae URA3* DNA into suitable internal restriction sites of pGKL1, but hybrid plasmids recovered from *K. lactis* transformants were all circular. In a similar experiment, we attempted to insert the *S. cerevisiae LEU2* DNA into a unique *Pst* I site of pGKL2 and to transform *S. cerevisiae*. We chose pGKL2 as an element of linear vector to be constructed, because pGKL2 can replicate alone in *S. cerevisiae* (N. Gunge, 1985, unpublished results) as well as in *K. lactis* (Wésolowski et al., 1982; Niwa et al., 1981), unlike pGKL1, which requires pGKL2 for replication. The unique *Pst* I site exists in the right-hand terminal region of pGKL2, located 453 basepairs from the inverted terminal repeat (Figures 2(b) and 3), and no large open reading frame occurs in this region (Hishinuma et al., 1984). In our experiment, as in those discussed previously, the hybrid plasmids recovered from transformants were all circular (Fujimura et al., 1987). The experimental design and results are briefly described below.

pGKL2 plasmid (13.4 kilobases) was digested with *Pst* I into two fragments of 637 and about 12.8 kilobases, and mixed with a 4.0 kilobase *LEU2* DNA obtained by gel-purification of *Pst* I digests of pAJ50. The DNA mixture was treated with T4 DNA ligase and introduced into *S. cerevisiae* (*leu2*) by transformation (Figure 3). Dozens of LEU^+ transformants were obtained on selective medium, all of which harbored

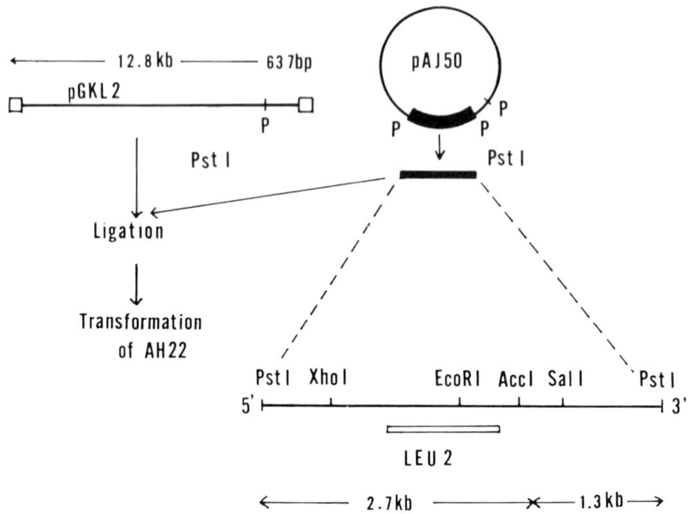

Figure 3 Scheme of insertion of yeast *LEU2* DNA into pGKL2.

free hybrid plasmids hybridizable to both radioactive *LEU2* DNA and pGKL2 probes. Southern blotting analysis indicated, however, that all plasmid DNAs recovered from transformants migrated in agarose gel in the forms of closed circular and open circular plasmids, but not in the form of linear plasmids. Molecular sizes of the hybrid plasmids were smaller than the total of DNAs employed for the hybrid plasmid construction, suggesting the occurrence of some deletion(s) in one or both of the cloned *LEU2* and pGKL2 DNAs. A 3.3 kilobase hybrid plasmid pF3 was chosen and the structure was analyzed in detail, the result of which is given in Figure 4. It consists of a 2.7 kilobase *LEU2* DNA fragment and a 637 basepair *Pst* I fragment of pGKL2. Another 12.8 kilobase *Pst* I fragment of pGKL2 and the remaining 1.3 kilobase *LEU2* DNA fragment were lost. Restriction and sequence analyses revealed that the *Pst* I-cut end of the 637 basepair pGKL2 fragment was ligated to the *Pst* I site in the 5' flanking region of *LEU2* DNA, while the free end of pGKL2 fragment was joined to the position 180 basepairs downstream of *Acc* I site in the 3' flanking region (Figures 3 and 4; Froman et al., 1984). The *LEU2* DNA fragment downstream from the junction point was lost. The plasmid pF3 was present in 50 to 100 copies per cell and its mitotic and meiotic stability was as high as that of 2 μm DNA-*ori* based hybrid plasmids.

The reason why the hybrid *LEU2*-pGKL2 plasmid was circularized is unknown. It should be noted, however, that all hybrid plasmids

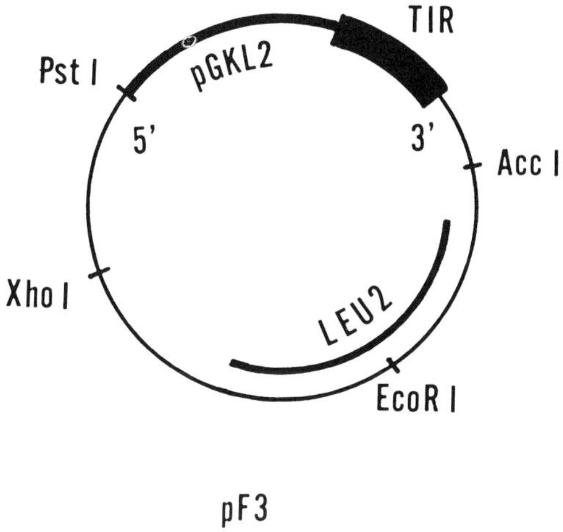

Figure 4 Physical map of a circular hybrid plasmid pF3. TIR indicates an inverted terminal inverted repeat of pGKL2.

constructed in the above experiment were circular. Moreover, the unexpected plasmid circularization was reproducibly observed in two independent laboratories using different host yeast species. These facts strongly suggest that some rational cause(s) must exist for the circularization. One possibility is that the expression of cloned *LEU2* gene is only possible in the nucleus, but not in the cytoplasm where natural pGKL plasmids are located (Gunge et al., 1982). If so, the constructed hybrid plasmids must enter the nucleus so that the *LEU2* gene can be expressed, and within the nucleus, the proper replication of the plasmids may have to depend upon the internal pGKL-ARS element(s) rather than the specialized terminal structure of the pGKL2 plasmid. The plasmid circularization may have been needed for it to act as a substrate for the nuclear replication system. Another possibility is that some pGKL plasmid gene(s) required for the terminal replication of the linear plasmids may have been lost as a result of deletion(s) of certain pGKL sequences or may have been destroyed by the internal insertion of markers. Consequently, the hybrid plasmids became unable to replicate in linear forms, thus selecting for circularization which allows use of the internal pGKL-ARS element(s) for replication. If the above view is correct, the successful construction of pGKL-based linear vectors will require suitable DNA ligation conditions and the selection of proper insertion sites in the pGKL plasmids.

ACKNOWLEDGMENT

The author thanks Dr. H. Fukuhara for his providing strains MH41-7B, OIP2, CEP2, and CEO2, and Chieko Fujimura for her excellent technical assistance.

REFERENCES

Beach, D., Piper, M., and Shall, S. (1980). Isolation of chromosomal origins of replication in yeast. *Nature 284*:185-187.

Broach, J. R., Li, Y.-Y., Feldman, J., Jayaram, M., Abraham, J., Nasmyth, K. A., and Hicks, J. B. (1982). Localization and sequence analysis of yeast origins of DNA replication. *Cold Spring Harbor Lab. Symp. Quant. Biol. 47*:1165-1173.

Conde, J., and Fink, G. R. (1976). A mutant of *Saccharomyces cerevisiae* defective for nuclear fusion. *Proc. Nat. Acad. Sci. USA, 73*:3651-3655.

Das, S., and Hollenberg, C. P. (1982). A high-frequency transformation system for the yeast *Kluyveromyces lactis*. *Curr. Genet. 6*:123-128.

de Louvencourt, L., Fukuhara, H., Heslot, H., and Wésolowski, M. (1983). Transformation of *Kluyveromyces lactis* by killer plasmid DNA. *J. Bacteriol. 154*:737-742.

Erickson, L., Beversdorf, W. D., and Pauls, K. P. (1985). Linear mitochondrial plasmid in *Brassica* has terminal protein. *Curr. Genet. 9*:679-682.

Ferenczy, L. (1981). Microbial protoplast fusion. In *Genetics as a Tool in MIcrobiology*, Glover, S. W. and Hopwood, D. A. (Eds.). Society for General Microbiology Symposium 31. Cambridge University Press, pp. 1-34.

Francou, F. (1981). Isolation and characterization of a linear DNA molecule in the fungus *Ascobolus immersus*. *Mol. Gen. Genet. 184*:440-444.

Froman, B. E., Tait, R. C., and Rodriguez, R. L. (1984). Nucleotide sequence of the 3' terminal region of the *LEU2* gene from *Saccharomyces cerevisiae*. *Gene 31*:257-261.

Fujimura, H., Hishinuma, F., and Gunge, N. (1987). Terminal segment of *Kluyveromyces lactis* linear DNA plasmid pGKL2 supports autonomous replication of hybrid plasmids in *Saccharomyces cerevisiae*. *Curr. Genet. 12*:99-104.

Gaillardin, C., Fournier, P., Budar, F., Kudla, B., Gerbaud, C., and Heslot, H. (1983). Replication and recombination of 2 μm DNA in *Schizosaccharomyces pombe*. *Curr. Genet.* *7*:245-253.

Gunge, N. (1983). Yeast DNA plasmids. *Ann. Rev. Microbiol.* *37*: 253-276.

Gunge, N., and Sakaguchi, K. (1981). Intergeneric transfer of deoxyribonucleic acid killer plasmids, pGKL1 and pGKL2, from *Kluyveromyces lactis* into *Saccharomyces cerevisiae* by cell fusion. *J. Bacteriol.* *147*:155-160.

Gunge, N., and Yamane, C. (1984). Incompatibility of linear DNA killer plasmids pGKL1 and pGKL2 from *Kluyveromyces lactis* with mitochondrial DNA from *Saccharomyces cerevisiae*. *J. Bacteriol.* *159*:533-539.

Gunge, N., Tamaru, A., Ozawa, F., and Sakaguchi, K. (1981). Isolation and characterization of linear deoxyribonucleic acid plasmids from *Kluyveromyces lactis* and the plasmid-associated killer character. *J. Bacteriol.* *145*:382-390.

Gunge, N., Murata, K., and Sakaguchi, K. (1982). Transformation of *Saccharomyces cerevisiae* with linear DNA killer plasmids from *Kluyveromyces lactis*. *J. Bacteriol.* *151*:462-464.

Hashiba, T., Homma, Y., Hyakumachi, M., and Masuda, I. (1984). Isolation of a DNA plasmid in the fungus *Rhizoctonia solani*. *J. Gen. Microbiol.* *130*:2067-2070.

Hayakawa, T., Tanaka, T., Sakaguchi, K., Otake, N., and Yonehara, H. (1979). A linear plasmid like DNA in *Streptomyces* sp. producing lankacidin group antibiotics. *J. Gen. Appl. Microbiol.* *25*:255-260.

Hinnen, A., Hicks, J. B., and Fink, G. R. (1978). Transformation of yeast. *Proc. Nat. Acad. Sci. USA,* *75*:1929-1933.

Hirochika, H., Nakamura, K., and Sakaguchi, K. (1984). A linear DNA plasmid from *Streptomyces rochei* with an inverted terminal repetition of 614 base pairs. *EMBO J.* *3*:761-766.

Hishinuma, F., Nakamura, K., Hirai, K., Nishizawa, R., Gunge, N., and Maeda, T. (1984). Cloning and nucleotide sequence of the linear DNA killer plasmids from yeast. *Nucleic Acids Res.* *12*:7581-7597.

Ito, H., Fukuda, K., Murata, K., and Kimura, A. (1983). Transformation of intact yeast cells treated with alkali cations. *J. Bacteriol.* *153*:163-168.

Kemble, R. J., and Bedbrook, J. R. (1980). Low molecular weight circular and linear DNA in mitochondria from normal and male-sterile *Zea mays* cytoplasm. *Nature 284*:565-566.

Kemble, R. J., and Thompson, D. (1982). S1 and S2, the linear mitochondrial DNAs present in a male sterile line of maize, possess terminally attached proteins. *Nucleic Acids Res. 10*:8181-8190.

Kikuchi, Y., Hirai, K., and Hishinuma, F. (1984). The yeast linear DNA killer plasmids, pGKL1 and pGKL2, possess terminally attached proteins. *Nucleic Acids Res. 12*:5685-5692.

Kitada, K., and Hishinuma, F. (1987). A new linear DNA plasmid isolated from the yeast *Saccharomyces kluyveri. Mol. Gen. Genet. 206*:377-381.

Maundrell, K., Wright, A. P. H., Piper, M., and Shall, S. (1985). Evaluation of heterologous ARS activity in *Saccharomyces cerevisiae* using cloned DNA from *Schizosaccharomyces pombe. Nucleic Acids Res. 13*:3711-3722.

Meinhardt, F., and Esser, K. (1984). Linear extrachromosomal DNA in the morel *Morchella conica. Curr. Genet. 8*:15-18.

Mellado, R. P., and Salas, M. (1983). Initiation of phage φ29 DNA replication by the terminal protein modified at the carboxyl end. *Nucleic Acids Res. 11*:7397-7407.

Niwa, O., Sakaguchi, K., and Gunge, N. (1981). Curing of the killer deoxyribonucleic acid plasmids of *Kluyveromyces lactis. J. Bacteriol. 148*:988-990.

Palmer, J. D., Shields, C. R., Cohen, D. B., and Orton, T. J. (1983). An unusual mitochondrial DNA plasmid in the genus *Brassica. Nature 301*:725-728.

Pring, D. R., Levings, III., C. S., Hu, W. W. L., and Timothy, D. H. (1977). Unique DNA associated with mitochrondria in the "S" type cytoplasm of male-sterile maize. *Proc. Nat. Acad. Sci. USA, 74*:2904-2908.

Pring, D. R., Conde, M. F., Scerz, K. F., and Levings, III., S. S. (1982). Plasmid-like DNAs associated with mitochondria of cytoplasmic male-sterile *Sorghum. Mol. Gen. Genet. 196*:180-184.

Rekosh, D. M. K., Russell, W. C., Bellett, A. J. D., and Robinson, A. J. (1977). Indentification of a protein linked to the ends of adenovirus DNA. *Cell 11*:283-295.

Sinclair, J. H., Stevens, B. J., Sanghavi, P., and Rabinowitz, M. (1967). Mitochondrial satellite and circular DNA filaments in yeast. *Science 156*:1234-1237.

Sor, F., Wésolowski, M., and Fukuhara, H. (1983). Inverted terminal repetitions of the two linear DNA associated with the killer character of the yeast *Kluyveromyces lactis*. *Nucleic Acids Res. 11*:5037-5044.

Stark, M. J. R., and Boyd, A. (1986). The killer toxin of *Kluyveromyces lactis*: Characterization of the toxin subunits and identification of the genes which encode them. *EMBO J. 5*:1995-2002.

Struhl, K., Stinchcomb, D. T., Scherer, S., and Davis, R. W. (1979). High frequency transformation of yeast: Autonomous replication of hybrid DNA molecules. *Proc. Nat. Acad. Sci. USA, 76*:1035-1039.

Sugisaki, Y., Gunge, N., Sakaguchi, K., Yamasaki, M., and Tamura, G. (1983). *Kluyveromyces lactis* killer toxin inhibits adenylate cyclase of sensitive yeast cells. *Nature 304*:464-466.

Sugisaki, Y., Gunge, N., Sakaguchi, K., Yamasaki, M., and Tamura, G. (1984). Characterization of a novel killer toxin encoded by a double-stranded linear DNA plasmid of *Kluyveromyces lactis*. *Eur. J. Biochem. 141*:241-245.

Sugisaki, Y., Gunge, N., Sakaguchi, K., Yamasaki, M., and Tamura, G. (1985). Transfer of DNA killer plasmids from *Kluyveromyces lactis* to *Kluyveromyces fragilis* and *Candida pseudotropicalis*. *J. Bacteriol. 164*:1373-1375.

Tudzynski, P., Duvell, A., and Esser, K. (1983). Extrachromosomal genetics of *Claviceps purpurea*. I. Mitochondrial DNA and mitochondrial plasmids. *Curr. Genet. 7*:145-150.

Wésolowski, M., Algeri, A., Goffrini, P., and Fukuhara, H. (1982). Killer DNA plasmids of the yeast *Kluyveromyces lactis*. I. Mutations affecting the killer phenotype. *Curr. Genet. 5*:191-197.

Wickerham, L. J., and Burton, K. A. (1956). Hybridization studies involving *Saccharomyces lactis* and *Zygosaccharomyces ashybi*. *J. Bacteriol. 71*:290-295.

Wickner, R. B. (1977). Deletion of mitochondrial DNA bypassing a chromosomal gene needed for maintenance of the killer plasmid of yeast. *Genetics 87*:441-452.

Wickner, R. B. (1981). Killer system in *Saccharomyces cerevisiae*. In *The Molecular Biology of the Yeast Saccharomyces: Life Cycle and Inheritance*, Strathern, J. N., Jones, E. W., and Broach,

J. R. (Eds.). Cold Spring Harbor Laboratory, New York, pp. 415-444.

Wickner, R. B. (1983). Killer systems in *Saccharomyces cerevisiae*: Three distinct modes of exclusion of M2 double-stranded RNA by three species of double-stranded RNA, M1, L-A-E, and L-A-HN. *Mol. Cell. Biol.* 3:654-661.

Yoshikawa, H., and Ito, J. (1981). Terminal proteins and short inverted repeats of the small *Bacillus* bacteriophage genomes. *Proc. Nat. Acad. Sci. USA*, *78*:2596-2600.

13

Structural Properties of dsRNA: Genetic Elements Associated with Plant Pathogenic Fungi

James Tartaglia* and Donald L. Nuss
Roche Institute of Molecular Biology, Roche Research Center, Nutley, New Jersey

INTRODUCTION

Double-stranded (ds) RNA is found in plants in the form of genomic segments of dsRNA viruses, replicative forms of single-stranded RNA viruses, and plasmid-like molecules of ill-defined origin and function. Double-stranded RNA genetic elements, in a variety of forms, are also commonly associated with plant pathogenic fungi. Plant pathologists have become increasingly interested in these genetic elements, principally because of recent advances in the use of dsRNAs as diagnostic markers of viral infection in higher plants (recently reviewed in Dodds et al., 1984; Valverde et al., 1986; and Chapter 14, this volume) and accumulating evidence that dsRNA genetic elements can modulate virulence in some plant pathogenic fungi (Castanho et al., 1978; Ghabrial, 1980; Anagnostakis, 1982; Van Alfen, 1982; Finkler et al., 1985). Nevertheless, little information has been presented concerning the structural properties of these molecules. In this regard, the genomic segments of dsRNA viruses, e.g., wound tumor virus (WTV) (Nuss,

**Present affiliation*: Wadsworth Center for Laboratories and Research, New York State Department of Health, Albany, New York

1984; Asamizu et al., 1985) and *Ustilago maydis* viruses (Koltin et al., 1978; Field et al., 1983) are the best characterized. Until recently (Tartaglia et al., 1986; Hiremath et al., 1986), the plasmid-like dsRNAs associated with hypovirulent strains of the chestnut blight fungus *Endothia parasitica* were among the least well characterized. To illustrate the range of structural properties exhibited by dsRNA genetic elements found in plants and plant pathogenic fungi, this chapter will summarize the results of recent structural analyses of the WTV genome and the dsRNAs of a North American hypovirulent strain of *Endothia parasitica*. (Structural properties of the dsRNA genome of the *Ustilago maydis* viruses are discussed elsewhere in this volume.)

SURVEY OF DOUBLE-STRANDED RNA GENETIC ELEMENTS ASSOCIATED WITH PLANTS AND THEIR PATHOGENIC FUNGI

It is useful to survey the dsRNA genetic elements found in plants and plant pathogenic fungi prior to a discussion of the structural properties of selected elements (Table 1). The first indication that dsRNA could serve as the genome of a plant virus came from the work of Black and Markham (1963) and Gomatos and Tamm (1963), examining the RNA isolated from WTV, now classified as a member of the Reoviridae. The number of recognized plant-infecting reoviruses now stands at 12. (For recent reviews of this virus group, see Black, 1982; Francki and Boccardo, 1983; and Nuss, 1984.) These viruses are characterized by a genome composed of 10 or 12 discrete segments of dsRNA packaged as a set within a well-defined particle. Additional properties of this virus group include the ability to replicate in insects (leafhoppers) that act as vectors and the ability to induce tumors in their plant hosts.

The other group of plant viruses that possess a dsRNA genome is the so called cryptic virus group (Boccardo et al., 1983; Boccardo et al., 1985). The genome segments of these viruses vary with respect to number (2 to 5) and size (~ 1.6 to 0.8×10^6 Da). Whether the dsRNA segments are encapsidated in the same or different particles is unclear. Cryptic viruses are also unusual in that they cause no detectable symptoms and are nontransmissible by standard means. They are, however, efficiently transmitted through seeds of infected plants.

Infection of plants by viruses containing single-stranded RNA genomes results in the accumulation of dsRNAs within infected cells. These molecules, which represent replicative intermediate forms of the viral genomes (Ralph et al., 1965; Nilsson-Tillgren, 1970; Jackson et al., 1971; Gould and Francki, 1981) and their satellite RNAs (Schneider, 1977; Murant and Mayo, 1982; Kaper and Tousignant, 1984; Francki, 1985), serve as convenient diagnostic markers for such infections (Morris and Dodds, 1979; Gould and Francki, 1981; Morris et al., 1983;

Table 1 Survey of dsRNA Genetic Elements Found in Plants and Plant Pathogenic Fungi

dsRNA genetic elements	References
Plant-infecting reovirus genome	a
Cryptic virus genome	b
Replicative intermediates of single-stranded RNA viruses and their satellite RNAs	c
Plasmidlike dsRNA in healthy plants	d
Mycovirus genome	e
Fungal plasmidlike dsRNAs	f

References listed represent review articles, articles not adequately covered by, or recent articles not included in, review articles. The list is not meant to be inclusive but to serve as an aid for readers interested in conducting their own, more-detailed literature search.
[a]Black, 1982; Francki and Boccardo, 1983; Nuss, 1984; Asamizu et al., 1985.
[b]Boccardo et al., 1983; Boccardo et al., 1985.
[c]Ralph et al., 1965; Nilsson-Tillgren, 1970; Jackson et al., 1971; Schneider, 1977; Murant and Mayo, 1982; Kaper and Tousignat, 1984; Francki, 1984; Dodds et al., 1984; Valverde et al., 1986.
[d]Grill and Garger, 1981; Wakarchuk and Hamilton, 1985; Finnegan and Brown, 1986.
[e]Lemke and Nash, 1974; Lemke, 1977; Day and Dodds, 1979; Rawlinson et al., 1973; Dunkle, 1974; Zanzinger et al., 1973; Field et al., 1983; Finkler et al., 1985.
[f]Anagnostakis, 1982; Van Alfen, 1982; Hiremath et al., 1986; L'Hostis et al., 1985; Tartaglia et al., 1986.

Dodds et al., 1984; Valverde et al., 1986). In addition to their practical importance, replicative intermediates represent useful reagents for studying the molecular details of plant viral genome replication and expression. Surprisingly, there is now only limited information regarding the structural properties of these molecules, e.g., presence of polyadenylated tails or 5' terminal caps, presence of covalently linked viral polypeptides, etc.

Double-stranded RNAs have also been detected in apparently virus-free plants. Heterogeneous and discrete forms of low-molecular-weight dsRNAs have been detected in healthy tobacco (Ikegami and Fraenkel-Conrat, 1979) and in *Phaseolus vulgaris* cv. 'Hawksburg wonder' (Gould and Francki, 1981), respectively. A high molecular weight ($\sim$10.5 × 10^6 Da) dsRNA has been implicated as the agent responsible for cytoplasmic male sterility in *Vicia faba* (Grill and Garger, 1981). More recently, autonomously replicating dsRNAs have been detected in mitochondria of maize plants with S-type male-sterile cytoplasms (Finnegan and Brown, 1986). Wakarchuk and Hamilton (1985) have described a high-molecular-weight (8 to 10 × 10^6 Da) dsRNA in *Phaseolus vulgaris* cultivars with black seed coat. Interestingly, this dsRNA was shown to hybridize to DNA isolated from the same cultivar harboring the dsRNA as well as other bean cultivars. Whether the dsRNA is homologus to the nuclear or organellar genome has not yet been reported. These authors also detected high molecular weight (greater than 5 kilobase pairs) dsRNA in apparently healthy varieties of *Rubus idaeus* (raspberry), *Prunus avium* (cherry) and *Beta vulgaris* (Swiss chard) var. *cicla*. It is likely that the future will see additional reports of the occurrence and possible function of plant associated plasmid-like dsRNA genetic elements.

Viruses have been detected in at least 40 different species of plant pathogenic fungi (Lemke, 1977; Day and Dodds, 1979). Typically, viruses associated with these fungi are isometric particles 30-50 nm in diameter that contain genomes composed of segments of dsRNA which are individually encapsidated (Day and Dodds, 1979; Ghabrial, 1980). As is the case for most fungal viruses, the common means of transmission of viruses of plant pathogenic fungi is by cytoplasmic exchange following fusion of hyphae or through spores, i.e., there appears to be no extracellular phase in their replication cycle (Lecoq et al., 1979).

The biological significance of virus infection to the fungal host is, in general, not well understood. While the presence of fungal viruses is usually symptomless, there have been a number of reports which indicate an effect of the virus on the phenotype of several plant pathogenic fungi (reviewed in Day and Dodds, 1979; Elliston, 1982). For example, some strains of the corn smut fungus, *Ustilago maydis*, produce a toxin that is lethal to other strains of *U. maydis* (Hankin and Puhalla, 1971; Day and Anagnostakis, 1972). The toxin and resistance to the toxin are encoded by viral dsRNAs (Day and Anagnostakis, 1973; Koltin and Kandel, 1978; Peery et al., 1982). In *Rhizoctonia solani* (Castanho et al., 1978; Hollings, 1978; Finkler et al., 1985) and *Helminthosporium victoriae* (Ghabrial, 1980; Ghabrial and Mernaugh, 1983), the presence of dsRNA viruses has been correlated with altered states of virulence. In the case of *R. solani*, however, the situation is somewhat confusing because the presence of dsRNA viruses has been correlated both with the attenuation (Castanho et al., 1978) and the enhancement (Finkler et al., 1985) of virulence. In addition, Zanzinger et al. (1984) reported that there was no consistent correlation between the presence of dsRNA in 49 of 50 field isolates of *R. solani* and the degree of pathogenicity. The involvement of dsRNA viruses in modulating virulence in *Gaeumannomyces graminis* and *Periconia circinata* is also inconclusive (Day and Dodds, 1979; Dunkle, 1974; Rawlinson et al., 1973).

In contrast to the results cited above, the correlative evidence that dsRNA genetic elements modulate virulence in *Endothia parasitica*, the cause of chestnut blight, is strong. Strains of *E. parasitica* that are reduced in virulence (hypovirulent) were first isolated in Europe from chestnut trees that had survived infection (Grente, 1965). Subsequent studies showed that the hypovirulent phenotype was transmissible by hyphal fusion to virulent strains setting the stage for successful use of these strains as biological control agents (reviewed by Anagnostakis, 1982; Van Alfen, 1982). Those strains able to transmit hypovirulence consistently contain dsRNA (Day et al., 1977; Anagnostakis and Day, 1979). The transmission of dsRNA from a hypovirulent strain to a virulent strain results in conversion of the virulent strain to a hypovirulent strain (Anagnostakis and Day, 1979). Furthermore, strains that lose the transmissible hypovirulent phenotype are also apparently devoid of dsRNA (stated in Van Alfen, 1982; Van Alfen and Hansen, 1983). However, direct proof that dsRNA is the causative agent of hypovirulence by cell-free transfection of virulent strains has not yet been achieved.

It was anticipated that the dsRNAs found in hypovirulent strains of *E. parasitica* represented viral genome RNAs and would be associated with virus particles typical of those described from other dsRNA-containing fungi. However, structures resembling classical mycovirus particles have not been observed in hypovirulent strains (Van Alfen, 1982; Hansen et al., 1985). Thus, hypovirulence-associated dsRNAs resemble plasmid-like dsRNAs in this respect. Additional surprising properties of these dsRNA genetic elements will be described in a subsequent section.

STRUCTURAL PROPERTIES OF THE WOUND TUMOR VIRUS GENOME

In 1962, Gomatos and his co-workers reported that human reovirus contained dsRNA. The fact that a tumor-inducing plant virus, WTV, shared a number of properties with reovirus (reviewed in Gomatos and Tamm, 1963) prompted an examination of its genetic material. As a consequence, WTV was the second virus shown to contain a dsRNA genome (Black and Markham, 1963; Gomatos and Tamm, 1963). Polyacrylamide gel electrophoretic analysis revealed the genome to be composed of 12 individual segments ranging in size from less than 1 kilobase pair to greater than 3.5 kilobase pairs (Reddy and Black, 1973) and Figure 1.

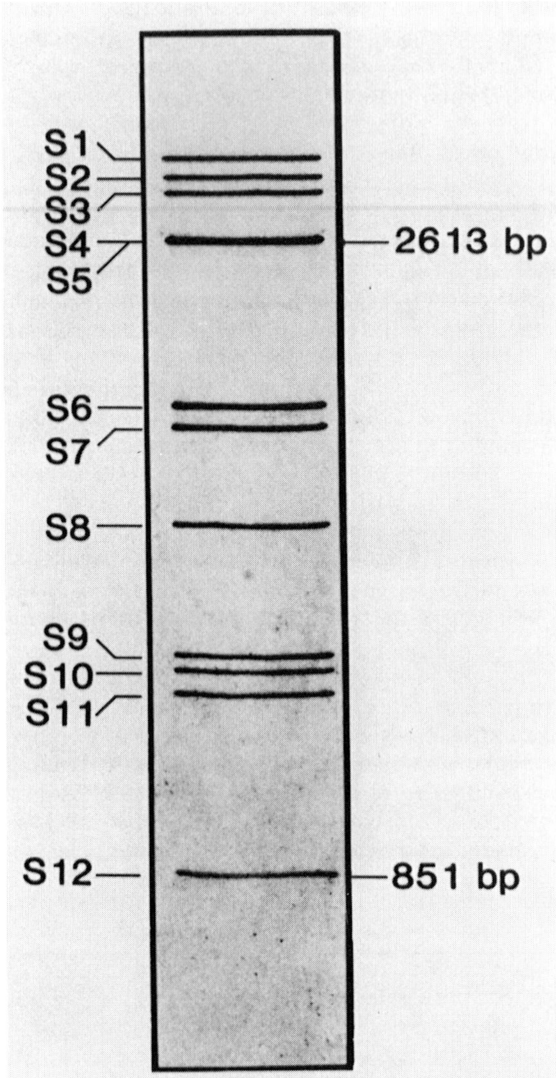

Figure 1 Relative electrophoretic mobilities of WTV genome segments in a 7.5% polyacrylamide gel. Genome segments are designated S1-S12 (Reddy and Black, 1973) based on relative mobilities: S1 is the slowest-migrating segment. The exact size (indicated at the right) of two of the genome segments, S12 (Asamizu et al., 1985) and S5 (Anzola et al., 1987) was determined by sequence analysis of cDNA clones. (*Source*: Reprinted by permission of the publisher from Nuss, 1984, copyright 1984 by Academic Press, Inc.)

The 12 WTV dsRNA segments are packaged within a particle composed of 7 of the 12 virus-specified polypeptides (Reddy and MacLeod, 1976; Nuss and Peterson, 1980). Purified virus particles exhibit RNA polymerase activity (Black and Knight, 1970) that catalyzes the synthesis of 12 single-stranded, translationally active RNA molecules that correspond to full-length copies of the plus strands of the 12 dsRNA segments (Reddy et al., 1977; Nuss and Peterson, 1980; Nuss and Peterson, 1981a) and Figure 2. Also associated with the virion are a mRNA guanylyltransferase, a guanine-7-methyltransferase, and a mRNA-2'-O-methyltransferase capable of modifying the 5'-termini of newly synthesized WTV transcripts to contain the structure $m^7G(5')$-ppp(5')A^m. . . (Rhodes et al., 1977; Nuss and Peterson, 1981b).

Analysis of the 3'-termini of partially fractionated WTV dsRNAs revealed equal amounts of uridine and cytidine (Lewandowski and Leppla, 1972). The suggestion that one strand of each WTV dsRNA segment terminates in a 3'-cytidine while the complementary strand contains a 3'-terminal uridine was confirmed by 2-dimensional (wandering spot) analyses of individual 3'-end-labeled genome segments following partial alkaline hydrolysis (Asamizu et al., 1985 and Figure 3). The exact nature of the 5'-termini of WTV genome RNA remains to be determined. We recently confirmed the presence of an adenosine residue at the 5'-end of the minus strand but we have been unable to detect a cap structure at the 5'-terminus of the plus strand (unpublished data). Closer examination of the 5'-terminus of transcripts synthesized in vitro by virus particles in the presence of the methyl donor S-adenosylmethionine (Nuss and Peterson, 1981b) has also revealed unexpected heterogeneity (D. Nuss, data). Clarification of these points will indicate whether the transcription and general replication strategy utilized by plant-infecting reoviruses differ significantly from those determined for other members of the Reoviridae.

The successful cloning of cDNA copies of the WTV genome segments was recently reported (Asamizu et al., 1985). Apparent full-length clones of 9 of the 12 genome segments (segments S4 through S12) have been identified while only partial clones of segments S1 through S3 were obtained. Sequence analysis of the terminal portions of several of the cDNA clones confirmed the results obtained by direct analysis of the 3'-termini of the dsRNA segments (Figure 3). Each WTV genome segment possesses the sequence . . . .UGAU-OH at the terminus corresponding to the 3'-end of the (+) strand and . . . .AAUACC-OH at the terminus corresponding to the 3' end of the (-) strand.

The complete nucleotide sequences of two of the WTV genome segments have now been determined (Asamizu et al., 1985; Anzola et al. 1987). Genome segment S12, the first to be sequenced, consists of 851 nucleotides. The segment possesses one long open reading frame (ORF) beginning with the first AUG triplet (residues 35-37) and extending 178 codons (Figure 4) specifying a polypeptide with a pre-

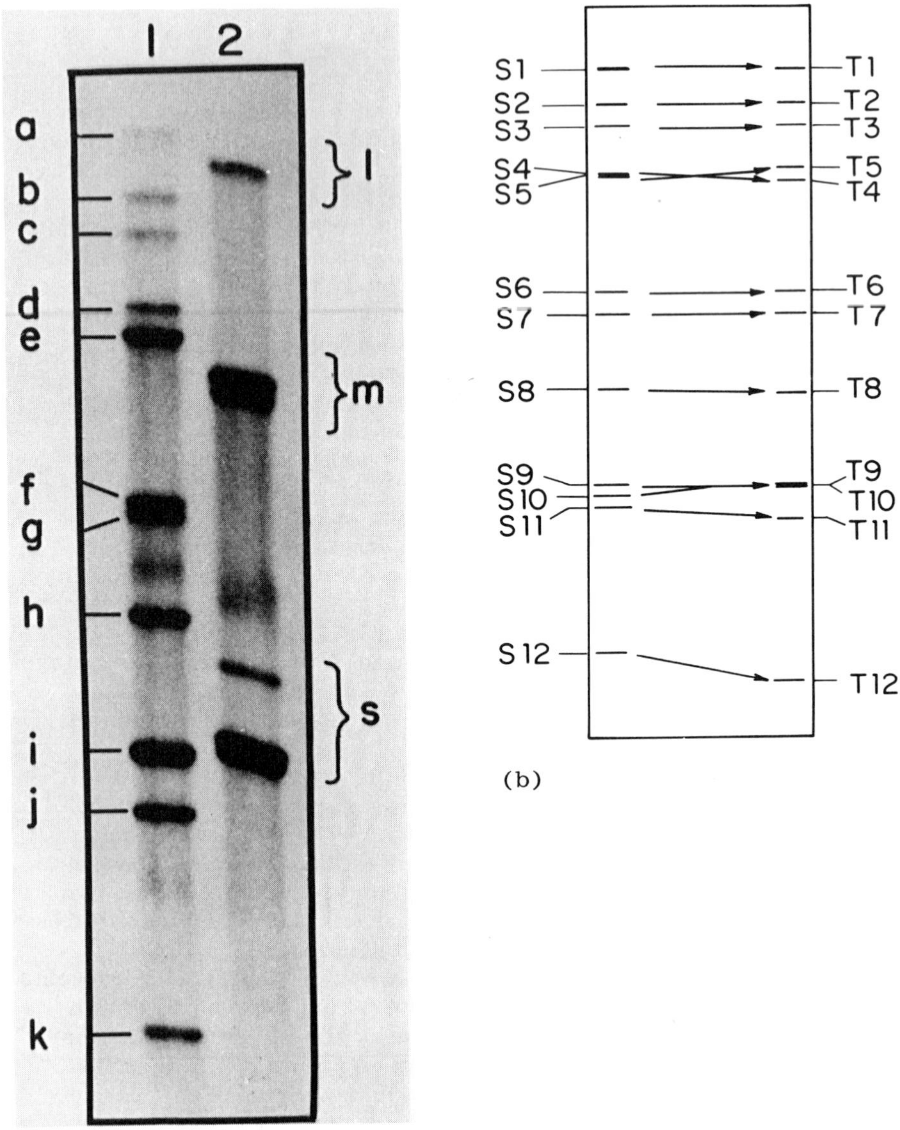

(a)

(b)

Figure 2 Resolution and genome assignment of WTV transcripts. [^{32}P]-labeled transcripts of WTV (lane 1) and reovirus (lane 2) were resolved by electrophoresis through a 2.5% to 5% polyacrylamide gel containing sodium dodecyl sulfate and 7 M urea. The letters a-k were picked arbitrarily to identify the individual WTV transcripts. The migration positions of the large (l), medium (m), and small (s) classes of the reovirus transcripts are shown at the right. The right side of the figure shows the results of assignment analysis and proposed nomenclature for WTV transcripts. Isolated ^{32}P-labeled transcripts a-k were hybridized individually to a mixture of unlabeled WTV genome RNAs. The

dicted molecular weight of 19,171 Da; a value consistent with the apparent molecular weight estimated (19,000) for the smallest of the WTV primary gene products, Pns 12 (Nuss and Peterson, 1980). A second ORF, in phase with the first and consisting of 40 codons, is present beginning at residues 674-676. While bicistronic genome segments have been reported for reoviruses (Ernst and Shatkin, 1985; Cashdollar et al., 1985; Ceruzzi and Shatkin, 1986), the functionality of the second ORF of WTV segment S12 is unclear. The sequence of genome segment S5, thought to encode the 76 kd outer coat component P5 (Nuss, 1983) was recently completed (Anzole et al., 1987). This segment contains only one large ORF which comprises 92.3% of the sequence (Figure 5), a molecular anatomy more typical of a reovirus genome segment (discussed in Asamizu et al., 1985).

The terminal nucleotide sequence domains predicted for transcripts specified by WTV segments S5 and S12 (partial information shown in Figures 4 and 5) share a number of porperties. One obvious similarity is the presence of the conserved terminal sequences 5'GGUAUU. . . . UGAU 3' found in each WTV transcript. When the nucleotide sequences comprising the 5'- and 3'-termini of the transcripts are placed in close approximation, a region of potential base pairing becomes apparent. A surprising similarity exists between the two segments in the form of the hexamer ACUACU located 7 to 8 nucleotides upstream of the first AUG codon.

An experimental approach for determining the functional role of sequence domains utilizes naturally occurring and synthetic remnants of genome segments. Remnant genome segments associated with transmission defective mutants of WTV were shown to be derived from wild-type genome segments as a result of internal deletion events (Nuss and Summers, 1984). These molecules, some of which have lost 85% of the internal portion of the parental genome sequences, retain structural domains required for transcription, replication, and packaging. The finding that terminal sequences (which include those shown in Figures 4 and 5) are retained to form a viable remnant RNA reinforces the notion that these domains play a critical role in genome function. Consequently, we have constructed a series of transcription vectors that yield transcripts with 5'- and 3'-termini identical to WTV tran-

---

resulting hybrids were analyzed by electrophoresis through a 7.5% polyacrylamide gel. The lines in this panel connect the genome segments (S1-S12) with the transcripts (T1-T12) which they encode. (*Source*: Reprinted by permission of the publisher from Nuss, 1984, copyright 1984 by Academic Press, Inc.)

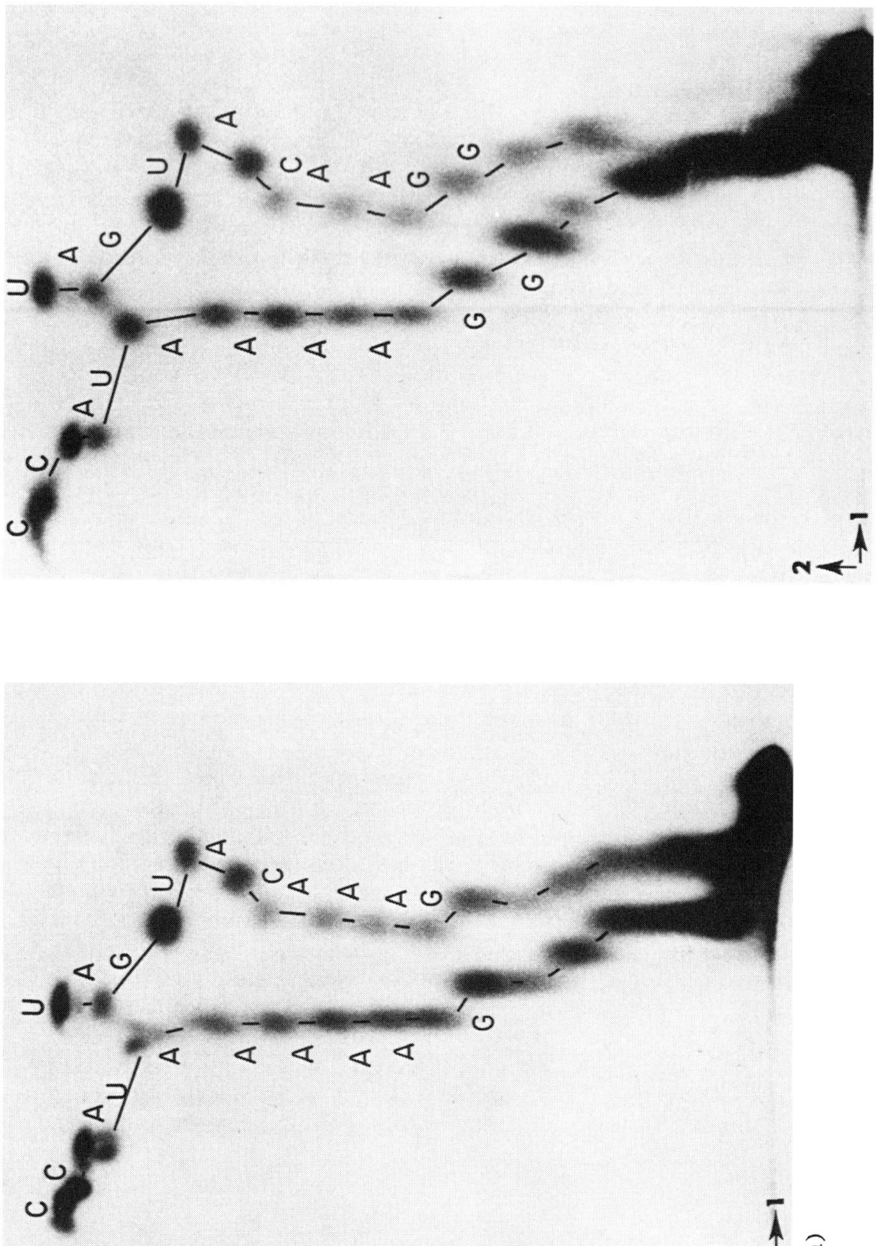

(a)

(b)

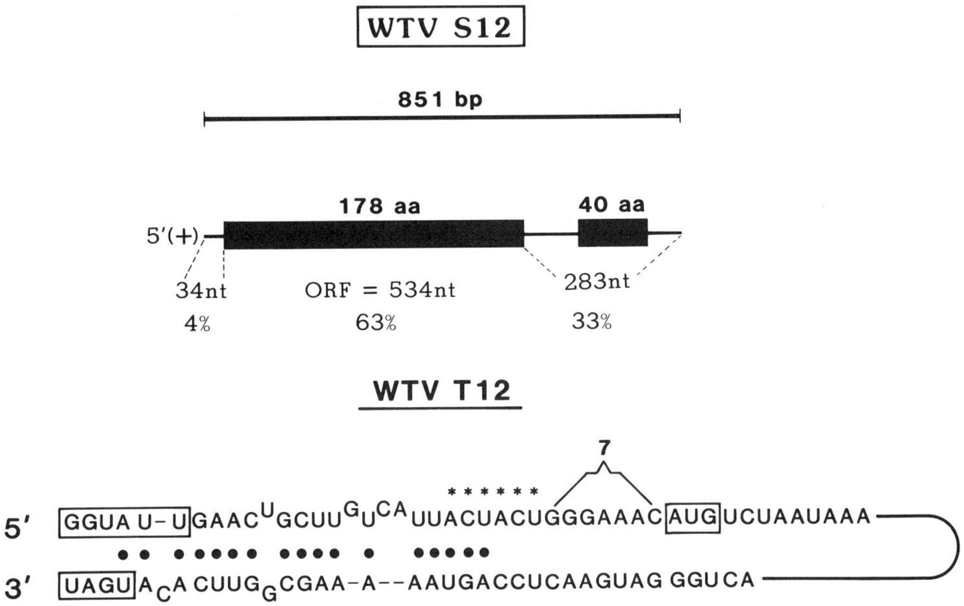

Figure 4 Schematic representation of the sequence organization of WTV segment S12. Thin bars indicate 5' and 3' non-coding regions while the thick bars represent long open reading frames. Total sequence information is presented in Asamizu et al., 1985. Also shown are the presumptive 5' and 3' nucleotide sequences of the transcript corresponding to genome segment S12. The conserved terminal sequences found in all WTV transcripts are boxed. The sequence ACUACU, found in both transcripts of segments S12 and S5 just upstream of the initiator AUG, is indicated by stars. The estimated free energy for the base-paired region of T12 is -17.7 kcal. The potential interaction between the 5' and 3' end of the transcript is purely speculative.

Figure 3 Two-dimensional oligonucleotide fingerprint analysis of partial alkaline digestion products of [^{32}P]pCp-labeled WTV genome segment S3 (a) and segment S9 (b). In each panel, the left tract represents the 3'-terminal sequence of the (-) strand while the right tract represents the 3'-terminal sequence of the (+) strand. (*Source*: Reprinted with permission of the publisher from Asamizu et al., 1985, copyright 1985 by Academic Press, Inc.)

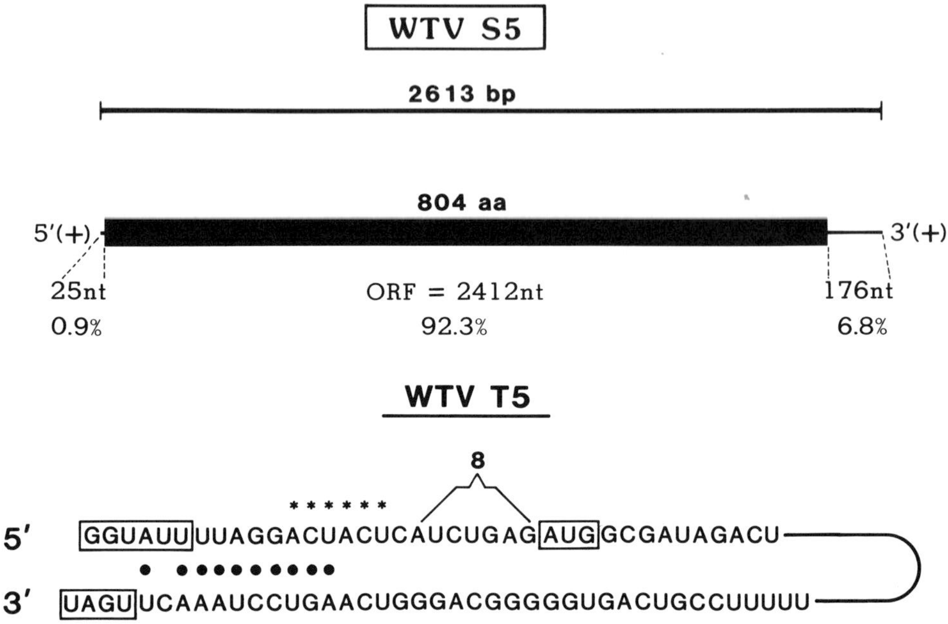

Figure 5 Schematic representation of the sequence organization of WTV genome segment S5 and presumptive terminal sequence of the corresponding transcript T5. Actual sequence information is presented (Anzola et al., 1984). The estimated free energy for the base-paired region of T5 is -18.2 kcal.

scripts and that contain all or a portion of the structural elements retained in functional remnant RNAs (see Nuss, 1987 for description). These transcripts are being tested for their ability to be replicated and packaged when introduced into cultured cells infected with helper virus. A similar approach has been successfully applied to identify regulatory sequence of the Sindbis virus genome (Levis et al., 1986). Using this strategy, it may be possible to determine the minimal sequence information required for effective transcription, replication, and packaging of this class of dsRNA genetic elements.

STRUCTURAL PROPERTIES OF THE DOUBLE-STRANDED RNAs ASSOCIATED WITH A NORTH AMERICAN HYPO-VIRULENT STRAIN OF *Endothia parasitica*

It is ironic that the premier example for the use of a natural form of biological control to manage a plant disease is for one of the most devastating plant diseases recorded: The use of hypovirulent strains of *E. parasitica* to control chestnut blight disease (Anagnostakis, 1982; Van Alfen, 1982; Garrod et al., 1985; Elliston, 1985). Hypovirulent strains capable of transmitting the hypovirulent phenotype have been isolated from surviving chestnut trees both in Europe and North America (Anagnostakis, 1982; Van Alfen, 1982; Fulbright et al., 1983; Jaynes and Elliston, 1982). Evidence that dsRNAs are the genetic elements responsible for transmissible hypovirulence is compelling (Day et al., 1977; Day and Dodds, 1979; Fulbright, 1984; Elliston, 1985). The dsRNAs associated with different hypovirulent strains vary with respect to size, concentraiton, and sequence homology (Anagnostakis and Day, 1979; Dodds, 1980; L'Hostis et al., 1985). Examples of dsRNAs isolated from a European and a North American hypovirulent strain are shown in Figure 6: dsRNA banding patterns are complex for both strains, consisting of a number of major and minor species. In this regard, banding patterns have been reported to change with time of culturing, often with the appearance of smaller dsRNA bands (Anagnostakis and Day, 1979; Garrod et al., 1985). This apparent complexity has ham-pered progress toward understanding the precise functional role of dsRNA in transmissible hypovirulence. Consequently, we have initiated a detailed characterization of specific dsRNAs associated with several hypovirulent strains of *E. parasitica*, by direct structural analysis of isolated dsRNAs and by generation of cDNA clones. The results of a recent structural analysis of the dsRNAs associated with the North American hypovirulent strain Grand Haven 2 (GH2) are presented below.

Analysis of dsRNAs purified from the GH2 strain revealed three major species of approximately 9.0 kilobase pairs (L-RNA), 3.5 kilobase pairs (M-RNA), and 1.0 kilobase pairs (S-RNA), respectively (Figure 6). Hybridization analyses of these dsRNAs indicated that L-RNA and M-RNA shared sequence homology but lacked extensive sequence homology with S-RNA (Tartaglia et al., 1986). To further characterize the relatedness of these molecules, partial and complete RNase T1 digestion patterns of the individual dsRNAs labeled at their 3'-termini with [^{32}P]pCp were compared (Figure 7). The digestion patterns of L- and M-RNA were indistinguishable (Figure 7, lanes 1, 2, 4, and 5) indicating that guanosine positions relative to the 3'-termini for each strand of L- and M-RNA are identical for at least 40 nucleotides from each end. These results confirmed the relationship between L- and M-RNA demonstrated by hybridization analyses and are consistent with the possibility that M-RNA originated as a result of an internal deletion

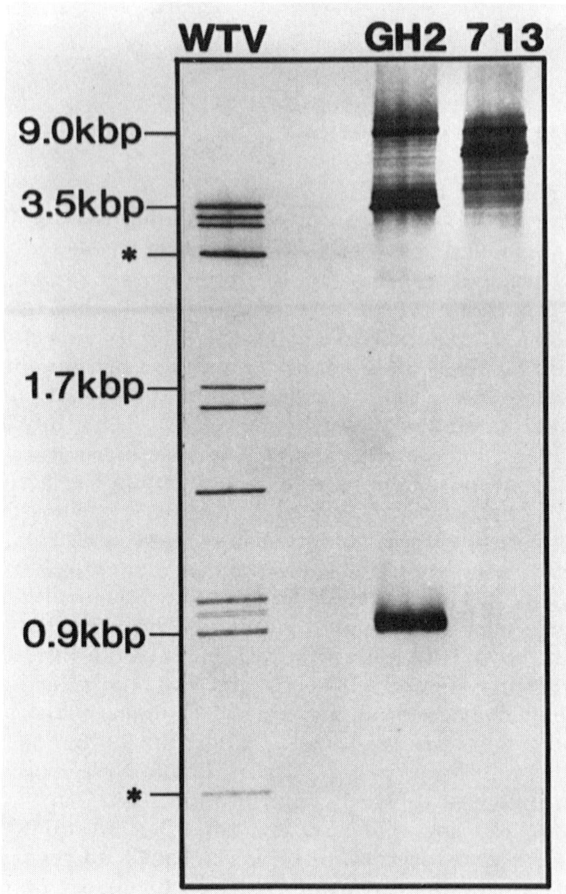

Figure 6 Electrophoretic pattern of EPGH2 and EP713 dsRNA components. Total dsRNA extracted from the GH2 strain and 713 strain of *E. parasitica* was fractionated on a 7.5% polyacrylamide gel. The dsRNA species were visualized by silver staining. Wound tumor (lane WTV) dsRNA was used as a marker. Estimated molecular sizes are shown on the left margin. The position of the EPGH2 L- and M-RNAs are indicated at the right by the 9.0 and 3.5 kilobase pair indicators (Garrod et al., 1985). The EPGH2 S-RNA is approximately 1 kilobase pair in length. The stained gels were developed to a point where the minor dsRNA species are clearly visible. The estimated sizes of WTV segments 1, 6, and 9 previously reported (Reddy and Black, 1979) are indicated. The asterisks (*) indicate WTV segments for which lengths were determined by sequence analysis of cDNA clones. Segment S5, top asterisk, is 2613 (Anzola et al., 1987) while segment S12, bottom asterisk, is 851 base pairs in length (Asamizu et al., 1985). The apparent separation of segment 10 into two distinct components represents an artifact of electrophoresis and/or gel processing.

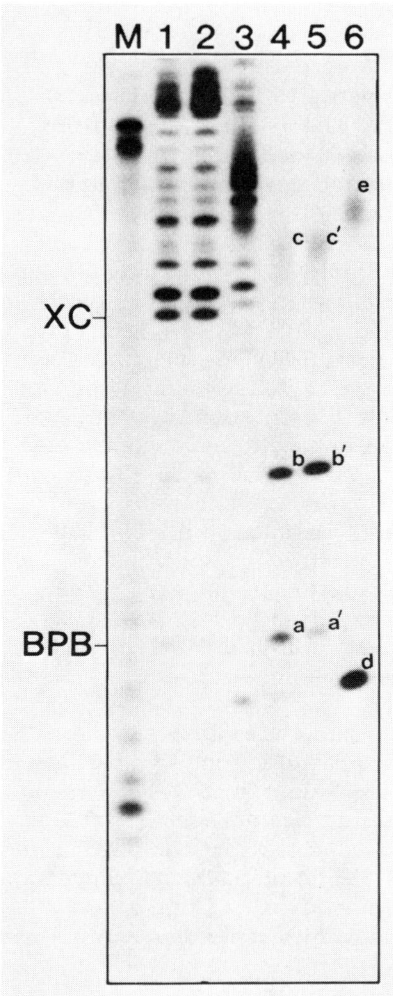

Figure 7 Partial and complete RNase T1 analysis of 3'-end labeled GH2 L-, M- and S-RNA. RNase T1 reaction products were fractionated on a 20% polyacrylamide gel containing 7 M urea and visualized by autoradiography on Kodak XAR 2 film. Lanes 1, 2, and 3 correspond to the partial RNase T1 digests of the L-, M-, and S-RNA, respectively. Lanes 4, 5, and 6 correspond to the complete digests of the L-, M- and S-RNAs, respectively. The complete RNase T1 digestion products of the L-RNA are designated a, b, and c while the complete products of the M-RNA are designated a', b', and c'. The complete digestion products of the S-RNA are designated d and e. Lane M (marker) corresponds to the partial hydrolysis of end-labeled RNA in half reactions containing 50 mM $NaHCO_3$, 1 mM EDTA (pH 9.0) at 90°C for 15 min and 30 min, respectively. The migration positions for bromophenol blue (BPB) and xylene cylanol FF (Xc) are indicated in the left margin. (*Source*: From Tartaglia et al., 1986.)

of L-RNA. S-RNA was found to be unrelated to L- and M-RNA by these criteria (Figure 7, lanes 3 and 6). The results obtained for L- and M-dsRNAs prompted us to propose (Tartaglia et al., 1986; Nuss, 1987) that the minor species of dsRNAs often associated with hypovirulent strains may also be derived from the largest dsRNA as a result of internal deletion events. The resulting remnant dsRNAs presumably replicate because they retain the structural domains recognized by the enzyme(s) responsible for the replication of the large dsRNA.

Interesting structural properties of the GH2 dsRNAs were revealed by complete RNase T1 digestion of the 3'-end labeled molecules (Figure 7; lanes 4-6). In addition to the presumptive 3'-terminal oligonucleotides a, a', b, b', and d, complete RNase T1 digestion of 3'-end labeled dsRNAs generated material that migrated in a heterogeneous manner at a position expected for oligonucleotides composed of 30-60 nucleotides (oligonucleotides c, c', and e). This heterogeneous material was resistant to pancreatic RNase under high salt conditions (330 mM KCl), sensitive to RNase U2 (an adenosine-cleavage specific nuclease), and sensitive to RNase H following hybridization to oligo-dT (Tartaglia et al., 1986). However, sensitivity to RNase U2 and RNase H required prior denaturation. These properties indicate that the RNase T1-resistant material is composed of 30-60 adenosine residues in a base-paired form.

The generation of presumptive 3'-terminal oligonucleotides a, a', b, b', as well as oligo A (oligonucleotides c and c') following RNase T1 digestion of L- and M-RNAs, indicated a degree of 3'-heterogeneity. Hybrid selection studies using M13 recombinants harboring cDNA sequences specific for the L- and M-RNAs revealed that the oligo A and oligonucleotide b/b' were located at the 3'-end of the same polarity strand (Tartaglia et al., 1986). Thus, one strand of the L- and M-dsRNAs contains a 3'-terminus composed of either oligonucleotide b/b' or oligonucleotide b/b' extended with oligo A (i.e., oligonucleotide c/c') while the other strand contains a 3'-terminus composed of oligonucleotide a/a'. The 3'-terminal heterogeneity observed with L- and M-RNAs was not evident with the S-RNA.

Complete RNase T1 digestion of 5'-end labeled L-, M-, and S-RNAs yielded heterogeneous RNase T1-resistant species similar to those observed after complete RNase T1 digestion of the 3'-end labeled dsRNAs (Tartaglia et al., 1986). The sensitivity of this material to RNase *Bacillus cereus* (a uridine and cytidine-cleavage-specific nuclease) and its insensitivity to RNase U2, coupled with the presence of 3'-terminal polyadenylic acid in a base-paired form strongly suggested that one 5'-terminus of these dsRNA molecules contains a stretch of polyuridylic acid. Digestion analyses of non-denatured dsRNAs with RNase *B. cereus* also revealed that some of the molecules have 5'-polyuridylic acid extensions.

The structural properties of the GH2 dsRNAs defined by the analyses described above are summarized in Figure 8. Double-stranded RNA molecules associated with the European strain, EP713, have also recently been shown to contain a stretch of adenosine residues at one 3'-terminus and a corresponding polyuridylic acid sequence at the 5'-terminus of the complementary strand (Hiremath et al., 1986). It is intriguing that although GH2- and EP713-dsRNAs have similar structural properties and both confer the hypovirulent phenotype, they share no detectable sequence homology (J. Tartaglia and D. L. Nuss, unpublished data). We have generated cDNA libraries of both GH2- and EP713-dsRNAs and have identified clones which encompass over

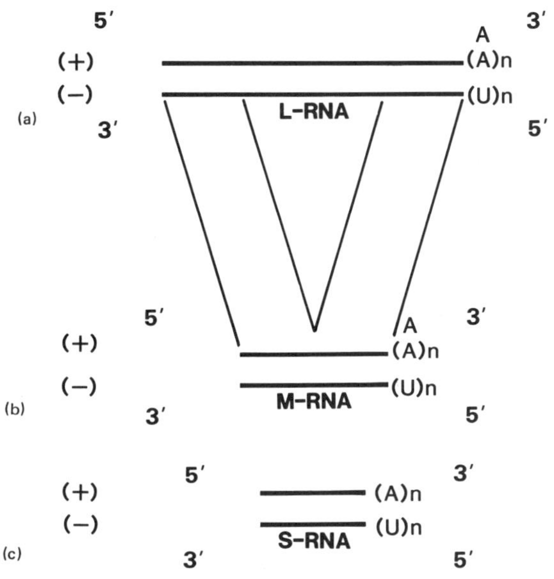

Figure 8 Schematic representation of the L-, M-, and S-RNA components associated with the GH2 hypovirulent strain of *Endothia parasitica*. (a) The 9.0 kilobase pair L-RNA. (b) The 3.5 kilobase pair M-RNA illustrated as being derived from the L-RNA as the result of an internal deletion event. (c) The 1.0 kilobase pair S-RNA which does not share any detectable sequence homology with L- and M-RNA. The RNA strand containing the 3'-terminal polyadenylic acid sequences is arbitrarily designated as the positive strand (+) while the complementary strand with the 5'-terminal polyuridylic acid sequence is designated as the minus strand (-). Also noted is the 3'-terminal heterogeneity for the L- and M-RNAs relative to polyadenylation. (*Source*: Modified from Tartaglia et al., 1986.)

80% of the largest EP713 dsRNA (J. Tartaglia and D. L. Nuss, unpublished data). The availability of such cDNA clones will allow a detailed comparison of the dsRNAs with respect to nucleotide sequence information and encoded gene products, thus yielding a clearer understanding of the relationship between dsRNAs of different hypovirulent strains.

Several structural and functional features of the *E. parasitica* dsRNAs are strikingly similar to those exhibited by relicative intermediates (RI) and replicative forms (RF) of single-stranded RNA genomes (RI and RF structures reviewed by Luria et al., 1981; Matthews, 1981). For example, RNA species of varying size classes including full-length molecules, are observed following electrophoretic analysis of the GH2 L-, M-, and S-RNAs under strand-separating conditions (J. Tartaglia and D. L. Nuss, unpublished data), a situation similar to that reported for poliovirus RI molecules (Baltimore, 1968). In addition, both *E. parasitica* dsRNAs and poliovirus RI and RF molecules contain 3'-terminal stretches of polyadenylic acid base paired to a stretch of polyuridylic acid. Finally, the dsRNAs of hypovirulent strains of *E. parasitica* are found associated with membrane structures that contain a RNA polymerase activity (Hansen et al., 1985) much like the replication complexes found within poliovirus and tobacco mosaic virus-infected cells (Girard et al., 1967; Caliquiri and Tamm, 1970a,b; Baltimore, 1964; Beachy and Zaitlin, 1975). In view of these similarities and the apparent absence of viruslike particles typical of dsRNA-containing mycoviruses (Van Alfen, 1982; Hansen et al., 1985), it is conceivable that the hypovirulence-associated dsRNAs may represent replicative forms of single-stranded genetic elements. This possibility provides an alternative perspective from which to think about the origin and function of dsRNA genetic elements found in plants and plant pathogenic fungi.

CONCLUDING REMARKS

Most of the available information concerning replication of viral dsRNA genomes comes from studies conducted with human reovirus (reviewed in Tyler and Fields, 1985). The limited results obtained from similar studies with plant-infecting reoviruses (Black, 1982; Francki and Boccardo, 1983; Nuss, 1984; Peterson and Nuss, 1986) are, so far, consistent with predictions vis-á-vis the human reovirus model. The enzymes involved in the synthesis and modification of reovirus transcripts are constituents of the virus particle. They are activated when infecting virus particles are partially uncoated in the lysosome to synthesize full-length copies of the (+) strand of each of the 10 reovirus dsRNA genome segments. The resulting transcripts can function either as messenger RNA or, once associated with viral polypeptides in nascent subviral particles, as template for the synthesis of (-) strand RNA. Available evidence suggests that viral single-stranded RNA, not dsRNA, is packaged into nascent subviral particles and that only one (-) strand is synthesized per (+) strand template. Neither (-) strand nor dsRNA is found free in the cytoplasm but is always found associated with subviral particles.

Within the context of the reovirus model, a clear distinction can be drawn between the different species of viral RNA with respect to function. The positive polarity single-stranded RNA functions as mRNA or template for (-) strand synthesis while the dsRNA functions as the repository of viral genetic information during the extracellular phase of the multiplication cycle and as the starting point for gene expression, i.e., template for (+) strand synthesis. Mycoviruses, such as the *U. maydis* viruses, have no extracellular phase. However, since the dsRNAs are found associated with discrete viral particles (Koltin et al., 1978), and these particles exhibit transcriptase activity (Ben-Tzvi et al., 1984), i.e., the virion-associated dsRNAs act as template for (+) strand synthesis, the functional roles of the different viral RNAs are still clearly defined. This distinction is less obvious in the case of the dsRNA genetic elements associated with hypovirulent strains of *E. parasitica.* No classical viruslike particles have been detected; moreover, as indicated in the previous section, these dsRNAs have a number of properties, e.g., a 3' stretch of polyadenylic acid and a complementary 5' stretch of polyuridylic acid, characteristic of replicative forms of positive-strand RNA viruses.

Reagents, in the form of cDNA clones, are now available that will allow a detailed examination of the structural and functional properties of these dsRNA genetic elements. The information obtained by direct structural analysis of the dsRNAs, as described in this chapter, provides valuable landmarks for the characterization and sequence analysis of the cDNA clones. A comparison of the structural properties of large, presumably polycistronic, dsRNA genetic elements, such as the L-dsRNA of *E. parasitica* strain GH2, with those of the presumably monocistronic genome segments of the plant-infecting reoviruses, may provide insights into the origin and evoluation of dsRNA genetic elements found in plants and plant pathogenic fungi.

ACKNOWLEDGMENT

We are grateful to Ada Hernandez for her capable assistance in preparing the manuscript.

REFERENCES

Anagnostakis, S. L. (1982). Biological control of chestnut blight. *Science 215*:466-471.

Anagnostakis, S. L., and Day, P. R. (1979). Hypovirulence conversion in *Endothia parasitica*. *Phytopathology 69*:1226-1229.

Anzola, J. V., Xu, Z., Asamizu, T., and Nuss, D. L. (1987). Segment-specific inverted repeats found adjacent to conserved terminal sequences in wound tumor virus genome and defective interfering RNAs. *Proc. Nat. Acad. Sci. USA 84*:8301-8305.

Asamizu, T. A., Summers, D., Motika, M. B., Anzola, J. V., and Nuss, D. L. (1985). Molecular cloning and characterization of the genome of wound turmor virus: A tumor-inducing plant reovirus. *Virology 144*:398-409.

Baltimore, D. (1964). *In vitro* synthesis of viral RNA by the poliovirus RNA polymerase. *Proc. Nat. Acad. Sci. USA 51*:450-456.

Baltimore, D. (1968). Structure of the poliovirus replicative intermediate. *J. Mol. Biol. 32*:359-368.

Beachy, R. N., and Zaitlin, M. (1975). Replication of tobacco mosaic virus. VI. Replicative intermediate and TMV-RNA related RNAs associated with polyribosomes. *Virology 63*:84-97.

Ben-Tzvi, B., Koltin, Y., Mevarech, M., and Tamarkin, A. (1984). RNA polymerase activity in virions from *Ustilago maydis*. *Mol. Cell. Biol. 4*:188-194.

Black, L. M. (1982). Wound tumor virus disease. In *Molecular Biology of Plant Tumors*, G. Kahl and J. Schell (Eds.). Academic, New York, pp. 69-105.

Black, D. R., and Knight, C. A. (1970). Ribonucleic acid transcriptase activity in purified wound tumor virus. *J. Virol. 6*:194-198.

Black, L. M., and Markham, R. (1963). Base pairing in the ribonucleic acid of wound tumor virus. *Neth. J. Plant Pathol. 69*:215.

Boccardo, G., Lisa, V., and Milne, R. G. (1983). Cryptic viruses of plants. In *Double-Stranded RNA Viruses*, R. W. Compans and D. H. L. Bishop (Eds.). Elsevier, New York, pp. 425-430.

Boccardo, G., Milne, R. G., Luisoni, E., Lisa, V., and Accotto, G. P. (1985). Three seed-borne cryptic viruses containing double-stranded RNA isolated form white clover. *Virology 147*:29-40.

Caliguiri, L. A., and Tamm, I. (1970a). The role of cytoplasmic membranes in poliovirus biosynthesis. *Virology 42*:100-111.

Caliguiri, L. A., and Tamm, I. (1970b). Characterization of poliovirus-specific structures associated with cytoplasmic membranes. *Virology 42*:112-122.

Cashdollar, N., Chmelo, R. A., Wiener, J. R., and Joklik, W. K.

(1985). Sequences of the S1 genes of the three serotypes of reovirus. *Proc. Nat. Acad. Sci. USA 82*:24-28.

Castanho, B., Butler, E. E., and Shepherd, R. J. (1978). The association of dsRNA with *Rhizoctonia* decline. *Phytopathology 68*:1515-1519.

Ceruzzi, M., and Shatkin, A. J. (1986). Expression of reovirus p14 in bacteria and identification in the cytoplasm of infected mouse L cells. *Virology 153*:35-45.

Day, P. R., and Anagnostakis, S. L. (1972). Heterokaryon transfer of the "Killer" factor in *Ustilago maydis*. *Phytopathology 62*:494-501.

Day, P. R., and Anagnostakis, S. L. (1973). The killer system in *Ustilago maydis*: Heterokaryon transfer and loss of determinants. *Phytopathology 63*:1017-1018.

Day, P. R., and Dodds, J. A. (1979). Virus of plant pathogenic fungi. In *Viruses and Plasmids in Fungi*, P. A. Lemke (Ed.). Marcel Dekker, New York, pp. 201-238.

Day, P. R., Dodds, J. A., Elliston, J. E., Jaynes, R. A., and Anagnostakis, S. L. (1977). Double-stranded RNA in *Endothia parasitica*. *Phytopathology 67*:1393-1396.

Dodds, J. A. (1980). Revised estimates of the molecular weights of dsRNA segments in hypovirulent strains of *Endothia parasitica*. *Phytopathology 70*:1217-1220.

Dodds, J. A., Morris, T. J., and Jordan, R. L (1984). Plant viral double-stranded RNA. *Annu. Rev. Phytopathol. 22*:151-168.

Dunkle, L. D. (1974). Double-stranded RNA mycoviruses in *Periconia circinata*. *Physiolog. Plant Pathol. 4*:107-116.

Elliston, J. E. (1982). Hypovirulence. In *Advances in Plant Pathology*, Vol. 1. D. S. Ingram and P. H. Williams (Eds.). Academic, New York, pp. 1-33.

Elliston, J. E. (1985). Characteristics of dsRNA-free and dsRNA-containing strains of *Endothia parasitica* in relation to hypovirulence. *Phytopathology 75*:151-158.

Ernst, H., and Shatking, A. J. (1985). Reovirus hemagglutinin mRNA codes for two polypeptides in overlapping reading frames. *Proc. Nat. Acad. Sci. USA 82*:48-52.

Field, L. J., Bruenn, J. A., Chang, T. H., Pinhasi, U., and Koltin, Y. (1983). Two *Ustilago maydis* viral dsRNAs of different size code for the same product. *Nucleic Acids Res. 11*:2765-2778.

Finkler, A., Koltin, Y., Barash, I., Sneh, B., and Pozniak, D. (1985). Isolation of a virus from virulent strains of *Rhizoctonia solani*. *J. Gen. Virol.* *66*:1221-1232.

Finnegan, P. M., and Brown, G. G. (1986). Autonomously replicating RNA in mitochondria of maize plants with S-type cytoplasm. *Proc. Nat. Acad. Sci. USA* *83*:5175-5179.

Francki, R. I. B. (1985). Plant virus satellites. *Annu. Rev. Microbiol.* *39*:151-174.

Francki, R. I. B., and Boccardo, G. (1983). The plant reoviridae. In *The Reoviridae*, W. K. Joklik (Ed.). Plenum, New York, pp. 505-563.

Fulbright, D. W. (1984). Effect of eliminating dsRNA in hypovirulent *Endothia parasitica*. *Phytopathology* *74*:722-724.

Fulbright, D. W., Weidlich, W. H., Haufler, K. Z., Thomas, C. S., and Paul, C. P. (1983). Chestnut blight and recovering American chestnut trees in Michigan. *Can. J. Bot.* *61*:3164-3171.

Garrod, S. W., Fulbright, D. W., and Ravenscroft, A. B. (1985). Dissemination of virulent and hypovirulent forms of marked strains of *Endothia parasitica* in Michigan. *Phytopathology* *75*:533-538.

Ghabrial, S. A. (1980). Effects of fungal viruses on their hosts. *Annu. Rev. Phytopathol.* *18*:441-461.

Ghabrial, S. A., and Mernaugh, R. L. (1983). Biology and transmission of *Helmintosporium victoriae* mycoviruses. In *Double-Stranded RNA Viruses*, R. W. Compans and D. H. L. Bishop (Eds.). Elsevier, New York, pp. 441-449.

Girard, M., Baltimore, D., and Darnell, J. E., Jr. (1967). The poliovirus replication complex: Site for synthesis of poliovirus RNA. *J. Mol. Biol.* *24*:59-74.

Gomatos, P. J., and Tamm, I. (1963). Animal and plant viruses with double-helical RNA. *Proc. Nat. Acad. Sci. USA* *50*:878-882.

Gomatos, P. J., Tamm, I., Dales, S., and Franklin, R. M. (1962). Reovirus type 3: Physical characteristics and interactions with L cells. *Virology* *17*:441-446.

Gould, A. R., and Francki, R. I. B. (1981). Immunochemical detection of dsRNA in healthy and virus-infected plants and specific detection of viral dsRNA by hybridization to labeled complementary DNA. *J. Virol. Methods* *2*:277-286.

Grente, M. J. (1965). Les formes hypovirulentes d'*Endothia parasitica*: Les espoirs de lutte contre le chancre du chataignier. *C. R. Seances Acad. Agric. Fr.* *51*:1033-1037.

Grill, L. K., and Garger, S. J. (1981). Identification and characterization of double-stranded RNA associated with cytoplasmic male sterility in *Vicia faba*. *Proc. Nat. Acad. Sci. USA 78*:7043-7046.

Hankin, L., and Puhalla, J. E. (1971). Nature of a factor causing inter-strain lethality in *Ustilago maydis*. *Phytopathology 61*:50-53.

Hansen, D. R., Van Alfen, N. K., Gillies, K., and Powell, W. A. (1985). Naked dsRNA associated with hypovirulence of *Endothia parasitica* is packaged in fungal vesicles. *J. Gen. Virol. 66*:2605-2614.

Hiremath, S. T., L'Hostis, B. L., Ghabrial, S. A., and Rhoads, R. E. (1986). Terminal structure of hypovirulence-associated dsRNAs in the chestnut blight fungus *Endothia parasitica*. *Nucleic Acids Res. 14*:9877-9896.

Hollings, M. (1978). Mycoviruses: Viruses that infect fungi. *Advan. Virus Res. 22*:3-53.

Ikegami, M., and Fraenkel-Conrat, H. (1979). Characterization of double-stranded ribonucleic acid in tobacco leaves. *Proc. Nat. Acad. Sci. USA 76*:3637-3640.

Jackson, A. O., Mitchell, D. M., and Siegel, A. (1971). Replication of tobacco mosaic virus: I. Isolation and characterization of double-stranded forms of ribonucleic acid. *Virology 45*:182-191.

Jaynes, R. A., and Elliston, J. E. (1982). Hypovirulent isolates of *Endothia parasitica* associated with large American chestnut trees. *Plant Dis. 66*:769-772.

Kaper, J. M., and Tousignat, M. E. (1984). Viral satellites: Parasitic nucleic acids capable of modulating disease expression. *Endeavour New Ser. 8*:194-200.

Koltin, Y., and Kandel, J. S. (1978). Killer phenomenon in *Ustilago*: The organization of the viral genome. *Genetics 88*:267-276.

Koltin, Y., Mayer, I., and Steinlauf, R. (1978). Killer phenomenon in *Ustilago maydis*: Mapping viral functions. *Mol. Gen. Genet. 166*:181-186.

Lecoq, H., Boissonnet-Menes, M., and Delhotal, P. (1979). Infectivity and transmission of double-stranded RNA viruses. In *Fungal Viruses*, H. P. Molitoris, M. Hollings, and H. A. Woods (Eds.). Springer-Verlag, Heidelberg and New York, pp. 34-47.

Lemke, P. A. (1977). Fungal Viruses in Agriculture. In *Beltsville Symposium in Agriculture Research. I. Virology and Agriculture*, J. A. Romberge, R. H. Foote, L. C. V. Knutson, and P. L. Lentz (Eds.). Allanheld Osmun, Montclair, New Jersey, pp. 159-175.

Lemke, P. A., and Nash, C. H. (1974). Fungal viruses. *Bacteriol. Rev.* *38*:29-56.

Levis, R., Weiss, B. G., Tsiang, M., Huang, H., and Schlesinger, S. (1986). Deletion mapping of Sindbis virus DI RNAs derived from cDNAs defines the sequences essential for replication and packaging. *Cell* *44*:137-145.

Lewandowski, L. J., and Leppla, S. H. (1972). Comparison of the 3'-termini of discrete segments of the double-stranded ribonucleic acid genomes of cytoplasmic polyhedrosis virus, wound tumor virus and reovirus. *J. Virol.* *10*:965-968.

L'Hostis, B., Hiremath, S. T., Rhoads, R. E., and Ghabrial, S. A. (1985). Lack of sequence homology between double-stranded RNA form European and American strains of *Endothia parasitica*. *J Gen. Virol.* *66*:351-355.

Luria, S. E., Darnell, J. E., Jr., Baltimore, D., and Campbell, A. (1981). *General Virology*, 3rd ed., Wiley, New York, pp. 303-342.

Matthews, R. E. F. (1981). *Plant Virology*, 2nd ed., Academic, New York, pp. 190-250.

Morris, T. J., and Dodds, J. A. (1979). Isolation and analysis of double-stranded RNA from virus-infected plant and fungal tissue. *Phytopathology* *69*:854-858.

Morris, T. J., Dodds, J. A., Hillman, B., Jordan, R. L., Lommel, S. A., Tamaki, S. J. (1983). Viral specific dsRNA: Diagnostic value for plant virus disease identification. *Plant Mol. Biol. Rep.* *1*:27-30.

Murant, A. F., and Mayo, M. A. (1982). Satellites of plant viruses. *Annu. Rev. Phytopathol.* *20*:49-70.

Nilsson-Tillgren, T. (1970). Studies on the biosynthesis of TMV. III. Isolation and characterization of the replicative form and the replicative intermediate RNA. *Mol. Gen. Genet.* *109*:246-256.

Nuss, D. L. (1983). Molecular biology of wound tumor virus transmission: Genome segment 5 and the loss of transmissibility. In *Double-Stranded RNA Viruses*, R. W. Compans and D. H. L. Bishop (Eds.). Elsevier, New York, pp. 415-428.

Nuss, D. L. (1984). Molecular biology of wound tumor virus. *Advan. Virus Res.* *29*:57-93.

Nuss, D. L. (1987). Deletion mutants of double-stranded RNA genetic elements found in plants and fungi. In *RNA Genetics, Vol. II. RNA Variability*, E. Domingo, J. Holland, and P. Ahlquist (Eds.). CRC Press, in press.

Nuss, D. L., and Peterson, A. J. (1980). Expression of wound tumor virus gene products *in vivo* and *in vitro*. *J. Virol. 34*:532-541.

Nuss, D. L., and Peterson, A. J. (1981a). Resolution and genome assignment of mRNA transcripts synthesized *in vitro* by wound tumor virus. *Virology 114*:399-404.

Nuss, D. L., and Peterson, A. J. (1981b). *In vitro* synthesis and modification of mRNA by exvectorial isolates of wound tumor virus. *J. Virol. 39*:954-957.

Nuss, D. L., and Summers, D. (1984). Variant dsRNA associated with transmission-defective isolates of wound tumor virus represent terminally conserved remnants of genome segments. *Virology 133*:276-288.

Peery, T., Koltin, Y., and Tamarkin, A. (1982). Mapping the immunity function of the *Ustilago maydis* P1 virus. *Plasmid 7*:52-58.

Peterson, A. J., and Nuss, D. L. (1986). Regulation of expression of the wound tumor virus genome in persistently infected vector cells is related to change in translational activity of viral transcripts. *J. Virol. 59*:195-202.

Ralph, R. K., Matthews, R. E. F., Matus, A. I., and Mandel, H. G. (1965). Isolation and properties of double-stranded viral RNA from virus-infected plants. *J. Mol. Biol. 11*:202-212.

Rawlinson, C. J., Hornby, C. J., Pearson, V., and Carpenter, J. M. (1973). Virus-like particles in take all fungus *Gaeumannomyces graminis*. *Ann. Appl. Biol. 74*:197-209.

Reddy, D. V. R., and Black, L. M. (1973). Electrophoretic separation of all components of the double-stranded RNA of wound tumor virus. *Virology 54*:150-159.

Reddy, D. V. R., and Black, L. M. (1977). Isolation and replication of mutant populations of wound tumor virions lacking certain genome segments. *Virology 80*:336-346.

Reddy, D. V. R., and MacLeod, R. (1976). Polypeptide components of wound tumor virus. *Virology 20*:274-282.

Reddy, D. V. R., Rhodes, D. P., Lesnaw, J. A., McLeod, R., Banerjee, A. K., and Black, L. M. (1977). *In vitro* transcription of wound tumor virus RNA by virion-associated RNA transcriptase. *Virology 80*:356-361.

Rhodes, D. P., Reddy, D. V. R., MacLeod, R., Black, L. M., and Banerjee, A. K. (1977). *In vitro* synthesis of RNA containing 5'-terminal structure 7mG(5')ppp(5')Ap^{m}. . . by purified wound tumor virus. *Virology 76*:554-559.

Schneider, I. R. (1977). Defective plant viruses. In *Beltsville Symposia in Agriculture Research. I. Virology in Agriculture*, J. A. Romberger, R. H. Foote, L. C. V. Knutson, and P. L. Lentz (Eds.). Allenheld Osmun, Montclair, New Jersey, pp. 201-219.

Tartaglia, J., Paul, C. P., Fulbright, D. W., and Nuss, D. L. (1986). Structural properties of double-stranded RNAs associated with biological control of chestnut blight fungus. *Proc. Nat. Acad. Sci. USA 83*:9109-9113.

Tyler, K. L., and Fields, B. N. (1985). Reovirus and its replication. In *Virology*, B. N. Fields, D. M. Knipe, R. M. Chanock, J. L. Melnick, B. Roizman, and R. E. Shope (Eds.). Raven, New York, pp. 823-862.

Valverde, R. A., Dodds, J. A., and Heick, J. A. (1986). Double-stranded ribonucleic acid from plants infected with viruses having elongated particles and undivided genomes. *Phytopathology 76*: 459-465.

Van Alfen, N. K. (1982). Biology and potential for disease control of hypovirulence of *Endothia parasitica*. *Annu. Rev. Phytopathol. 20*:349-362.

Van Alfen, N. K., and Hansen, D. R. (1983). Hypovirulence: A unique phenotypic expression of dsRNA in a pathogenic fungus. In *Double-Stranded RNA Viruses*, R. W. Compans and D. H. L. Bishop (Eds.). Elsevier, New York, pp. 463-467.

Wakarchuk, D. A., and Hamilton, R. J. (1985). Cellular double-stranded RNA in *Phaseolus vulgaris*. *Plant Mol. Biol.* 5:55-63.

Zanzinger, D. H., Bandy, B. P., and Tavantzis, S. M. (1984). High frequency of finding double-stranded RNA in naturally occurring isolates of *Rhizoctonia solani*. *J. Gen. Virol. 65*:1601-1605.

14

Detection and Interpretation of dsRNA

J. A. Dodds, R. A. Valverde,* and D. M. Mathews
University of California at Riverside, Riverside, California

INTRODUCTION

Franklin (1966) showed that bacteria not infected with an RNA virus contained little if any double-stranded (ds) RNA when total nucleic acid was fractionated by cellulose chromatography in the presence of buffered ethanol. Upon infection, a new class of nucleic acid, dsRNA, was present in that portion of the extract retained by cellulose in buffered 15% ethanol and eluted in the presence of ethanol-free buffer. This observation has been exploited in recent years with the intention of developing a technique for detection and diagnosis of RNA viruses in extracts from plant and fungal tissues (Morris and Dodds, 1979; Dodds et al., 1984; Morris et al., 1983; Dodds, 1986).

A cornerstone of this technique is the belief that the dsRNA in such extracts is likely to be the consequence of infection by an RNA virus, either with a single-stranded RNA genome, in which case the dsRNA is the replicative form of the viral RNA, or with a dsRNA genome. A considerable catalog of dsRNA profiles has been assembled for several groups of plant viruses, and dsRNA has been associated

*Present affiliation: Plant Pathology, Ball PanAm Plant Co., Parrish, Florida

with some diseases of uncertain etiology (Mirkov and Dodds, 1985; Reddy et al., 1985; Dale et al., 1986), suggesting that they are of viral cause. On the other hand, as a result of running many analyses of different hosts, dsRNA has also been detected in extracts from plants not thought to be diseased, and which should not have been exposed to viral infection, other than through vertical transmission from the propagule (seed, vegetative part, graft, etc.) used to establish the organism. It is unlikely that these are dsRNAs of conventional plant viruses because the dsRNA results are not those expected for the common seed-transmitted viruses of the plants in question, few of which have the rate of seed transmission (100%) attributable to the detected dsRNAs; the plants harboring these dsRNAs do not show any typical effects of virus infection.

One known source of such dsRNA is the nucleic acid found encapsidated in particles of cryptic (temperate) virus particles isolated from seedlings of several types of plants, but which are not graft-transmissible (Boccardo et al., 1983). The nature of other dsRNAs that do not resemble the genomic dsRNA of cryptic viruses is less clear. The presence of such dsRNAs makes the application of dsRNA analysis for virus detection and diagnosis in plants more complicated than hitherto imagined.

The difficulty of interpreting the nature of dsRNA in fungi is not less problematic than in plants, and this problem needs to be addressed since an increasing number of publications have used the direct analysis of dsRNA to address involvement of putative viruses in various plant pathogenic fungi.

In this chapter we will describe typical results of dsRNA analysis of plant and fungal tissues and discuss the possible origins and relationships between the classes of dsRNA that have been detected. Examples will draw heavily on our own work.

METHODS

The methods used rely on the simplicity of cellulose chromatography, designed to retain mostly dsRNA and little if any other nucleic acids in extracts from small amounts of tissue (0.5-7.0 g fresh weight). This is followed by electrophoresis in 6.0% polyacrylamide gels which maximize resolution and also further purify the dsRNA, as contaminants either do not enter or move rapidly through the gels (Morris and Dodds, 1979; Bar-Josepn et al., 1983; Jordan et al., 1983; Dodds et al., 1987). The dsRNA, present in electrophoresed gels as very thin bands, can be detected with great sensitivity by staining with ethidium bromide or by silver staining (Gildow et al., 1983). The resulting patterns can be recorded photographically and compared with standards and previous analyses (Valverde et al., 1986).

The dsRNA can be further purified by specific nuclease treatments, partially denatured into ssRNA, and used as a template either for end-labeling with ^{32}P (Garger et al., 1983; Jordan and Dodds, 1983; Rosner et al., 1983; Wakarchuk and Hamilton, 1985) or for synthesis of cDNA (Jamil et al., 1984; L'Hostis et al., 1985; Valverde and Dodds, 1986), which can then be cloned into appropriate vectors and used as a source of diagnostic probes, or for analysis of sequence and structure. In addition, the isolated dsRNAs can be melted and used as templates in cell-free franslation systems (Dalton et al., 1985). Non-denatured dsRNA can be used as an immunogen, and it is possible to obtain either a broad spectrum dsRNA specific polyclonal antibody (Zanzinger and Tavantzis, 1982; Derrick et al., 1984), or even monoclonal antibodies specific to the native dsRNA used as immunogen (Garcia-Luque et al., 1986). Several of these methods were described in greater detail in a recent review (Dodds, 1986).

When reference is made to research results from our laboratory throughout this report the assumption is made that the dsRNAs described are homogeneous, high molecular weight (0.2 to 13.3×10^6) molecules purified by CF-11 cellulose chromatography, resolved by ethidium bromide staining after electrophoresis through 6.0% polyacrylamide gels, and showing expected nuclease sensitivities.

PRESENCE OF dsRNA ASSOCIATED WITH INFECTION WITH KNOWN VIRUSES

Detection

This is best done using the isolation technique described above, and analyzing the partially purified dsRNA on 6.0% polyacrylamide gels. Several examples of expected results are illustrated in Figures 1, 2, and 3. Alternatively, the dsRNA could be detected by serological techniques such as ELISA, immunoblotting, or serologically specific electron microscopy using a broad-spectrum polyclonal antibody. In order to be sure that the dsRNAs detected are virus-specific, noninoculated plants should also be analyzed. The variety or cultivar of the infected and noninfected plants should be the same. This is not a condition that can always be satisfied when samples analyzed are of natural or field origin, and special caution should be used when interpreting results from these sources. This will also be a caution to heed when analyzing fungal samples of field origin. Not all hosts will give the same yield of dsRNA, and the relative intensity of minor compared to major dsRNAs can vary among hosts (Dodds et al., 1986; Valverde et al., 1986). Time of year, and presumably other environmental factors, can affect the yield of dsRNA, especially from field-collected samples. Citrus tissue infected with citrus tristeza virus contains little if any detectable virus-specific dsRNA in the summer months in southern California (Dodds et al., 1987). An amount of dsRNA suffi-

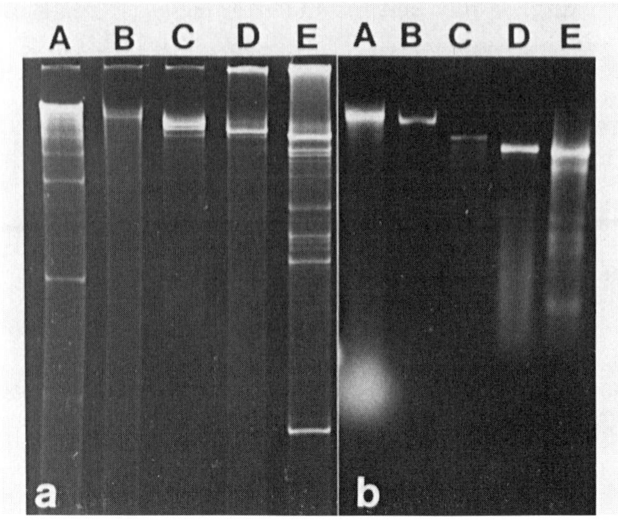

Figure 1 6.0% polyacrylamide (a) and 1.2% agarose (b) gel electrophoresis of dsRNAs (approximately 0.5 μg on each gel channel) of citrus tristeza virus (closterovirus) (A), tobacco etch virus (potyvirus) (B), dandelion latent virus (carlavirus) (C), potato virus X (potexvirus) (D), and tobacco mosaic virus (tobamovirus) (E). *Source*: Valverde et al., 1986.

cient to be readily detectable as stained bands is usually obtained from 0.5 to 2.0 g of infected plant tissue, though 7.0 g or 50.0 g is a better starting value for potyviruses and luteoviruses, respectively.

Diagnosis and Classification

When sufficient dsRNA is analyzed, results obtained on 6.0% polyacrylamide gels are of quite high resolution, and noticeably better than results on agarose gels (Figure 1). Much useful information is obtained by careful evaluation of the stained dsRNA bands. The major dsRNA bands detected in gels used to analyze dsRNAs from plants infected with known viruses are usually easy to interpret, since they are equal in number, and twice the size of the ssRNAs normally encapsidated in virions (Chu et al., 1983; Dodds and Bar-Joseph, 1983; Gildow et al., 1983; Falk and Duffus, 1984; Falk and Tsai, 1984; Mossop et al., 1985; Dodds et al., 1987; Guilford and Forster, 1986; Valverde et al., 1986). In addition to these expected dsRNAs, additional dsRNAs are frequently detected with staining intensities similar to or, more usually, less than those of the expected dsRNAs. These dsRNAs are of uncertain origin, but they are quite reproducible.

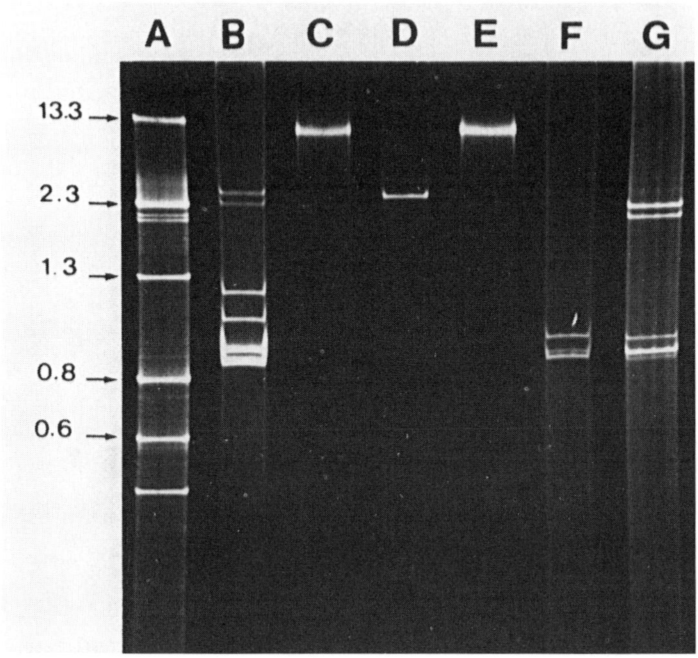

Figure 2 Polyacrylamide gel electrophoresis of dsRNAs extracted from sugar beet 'Y905' (B), bean 'Black Turtle' (C), tomato 'UC-82' (D), pepper 'Yolo Wonder' (E), and mustard 'Tendergreen' lot A (F) and lot B (G). Plant viral dsRNAs were used as molecular weight ($\times 10^6$) markers (A). *Source*: Valverde et al., 1986.

When the absolute size and number of the major and minor dsRNAs are considered, it is clear that useful information is obtained that can be used to identify viruses correctly. The results obtained when different strains of one virus (Gildow et al., 1983; Falk and Duffus, 1984; Dodds et al., 1987; Valverde et al., 1986), or different viruses of the same group (Dodds and Bar-Joseph, 1983; Valverde et al., 1986) are compared are such that it is usually possible to see a natural grouping based on similarities, and yet distinctions can usually be made based on the differences.

Mixed Infection

One of the advantages of the use of dsRNA analysis is that it is a nonspecific method. The dsRNA is purified by a single method that can be expected to give results regardless of the type of virus present

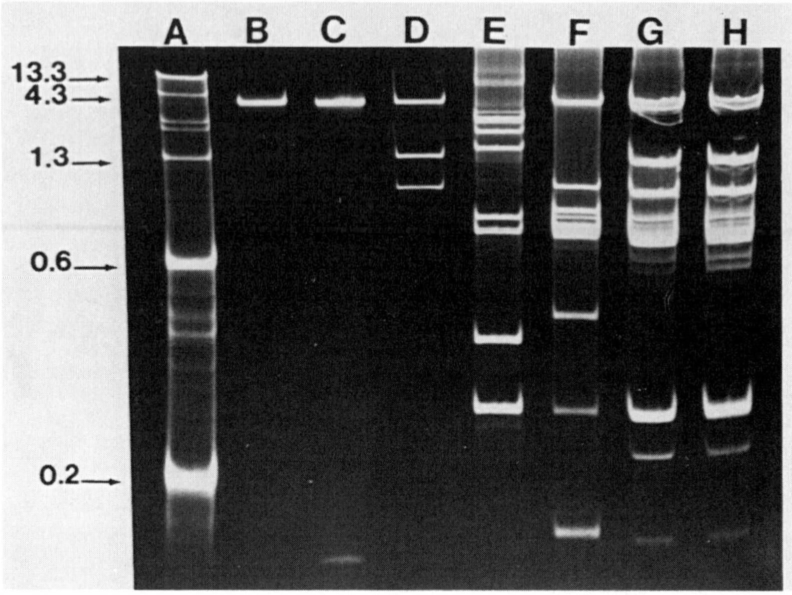

Figure 3 Polyacrylamide gel electrophoresis of dsRNAs extracted from field isolates (B-G) of *Botryodiplodia theobromae*. The dsRNA in channel H is from a single conidial subculture from the isolate illustrated in channel G. Plant viral dsRNAs were used as molecular weight ($\times 10^6$) markers (A).

so long as it is an RNA virus. This is in contrast to techniques which rely on the use of extraction procedures that will maintain the structural stability of virus particles, because in these cases different conditions are necessary for different viruses. A disadvantage is that all dsRNAs will be recovered from plants infected with more than one virus, and this may lead to initial confusion when trying to interpret dsRNA profiles in stained gels. On the other hand, once the reason for the complexity is understood to be the result of mixed infections, the benefits of using a nonspecific method are gained. The detection of a satellite RNA as a second component in mixed infections involving tobacco mosaic virus is an example of the use of dsRNA analysis to diagnose mixed infections (Valverde and Dodds, 1986).

We have observed that one of the consequences of mixed infections with strains of the same virus or viruses of the same group, including cucumoviruses (Dodds et al., 1985), tobamoviruses (Valverde and Dodds, 1986), and closteroviruses, is the likelihood that the major dsRNAs will be detected as doublets, implying small but significant differences in either the size or conformation of the genomic RNAs of the different strains. This is a useful observation, not only for diagnosis, but as a way to confirm the lack of replication of a challenge strain in cross-protection experiments.

PRESENCE OF dsRNA ASSOCIATED WITH DISEASES OF UNCERTAIN ETIOLOGY

Plants

One use of dsRNA analysis is to identify an RNA virus as the likely cause of diseases of uncertain etiology. In order to have confidence in this approach, several criteria have to be met, especially when it is not possible to transmit the disease with extracts or purified virus particles from the diseased plant. If this is possible, then Koch's postulates can be fulfilled, and infectious RNA from purified virus can also be shown to have sequence homology with the dsRNA found in diseased tissue. When this is not possible, the major condition to be met is that specific dsRNA species be detected in diseased plants and not in healthy plants, and that this is a constant association. A second condition is that those treatments that result in loss of detectable dsRNA from diseased plants also lead to a remission of disease symptoms, and vice versa. Expected results from these kinds of experiments have led to the conclusion that the presence of dsRNA is associated with lettuce big vein disease (Mirkov and Dodds, 1985), bunchy top disease of banana (Dale et al., 1986), and groundnut rosette disease (Reddy et al., 1985).

Plant Pathogenic Fungi

Transmissible hypovirulence in *Endothia parasitica* is the system in which dsRNA analysis has been used extensively to indicate a probable viral involvement in a fungal disease (cytoplasmically inherited hypovirulence is defined here as a disease). The conditions described above have been met for the most part in this system (Day et al., 1977; Fulbright, 1984), which still awaits conclusive evidence for direct transmission of the virus factors thought to be responsible for hypovirulence. This subject is the topic of Chapter 17. In the context of this chapter, the following points are made. Results have indicated that different patterns of dsRNA can be associated with hypovirulence (Dodds, 1980; Garrod et al., 1985), which is a biological property with a very wide range of values, from barely hypovirulent to nearly avirulent (Elliston, 1985a). Differences are most obvious when dsRNAs in European and American hypovirulent strains are compared (Dodds, 1980; L'Hostis et al., 1985). In addition, individual strains may contain mixed components which, when selectively cured, are shown to have separate effects on the degree of hypovirulence expressed in any given strain (Elliston, 1985b).

PRESENCE OF dsRNA ASSOCIATED WITH NO OBVIOUS DISEASE

Cryptic Viruses in Plants

One class of agents, called cryptic viruses, has been clearly identified as a source of dsRNAs in plants that are normal in appearance (Boccardo et al., 1983; Natsuaki et al., 1983; Abou-El Nasr et al., 1985; Boccardo et al., 1985; Accotto and Boccardo, 1986). These are spherical virus particles that can be purified in small amounts from some plants, and that contain segmented genomes of dsRNA. They are graft-transmitted with great difficulty or, more commonly, not at all, and, by contrast, are vertically transmitted with great efficiency. They have been detected in several plants, including sugar beet, spinach, white clover, and the fava bean, *Vicia faba*. In the hosts described below, dsRNA types are found which are multisegmented and of medium molecular weight (0.5 to 3.0 × 10^6), and which may very well be associated with the presence of particles of cryptic viruses, but to date they have not been detected.

Male Sterility in Plants

A dsRNA (M.W. = 10-13 × 10^6) that has been associated with membrane-bound particles is considered to be involved in male sterility of *V. faba* (Grill and Garger, 1981). This dsRNA can be transmitted by the use of dodder (a parasitic angiosperm), with the consequence that male sterility is conferred on the previously fertile, dsRNA-negative recipient, and loss of the dsRNA results in the recovery of fertility in previously sterile plants.

Studies of dsRNA in fertile and male-sterile corn (Sisco et al., 1984) or sunflower (Brown et al., 1986) lines did not reveal the degree of association between the presence of dsRNA and male sterility that was reported for the fava bean. Two dsRNAs (M.W. = 1.9 and 0.5 × 10^6) were detected in some S group cytoplasms of male sterile corn, and their presence was correlated with an extra degree of male sterility. The dsRNAs detected in sunflower were more variable, and resembled in complexity the examples about to be described in other plant species. Two of the dsRNAs (M.W. = 1.1 and 0.5 × 10^6) were detected consistently, but were found in fertile as well as male sterile lines and were not considered to be implicated in male sterility by the authors. The question of the involvement of dsRNA as a general phenomenon in male sterility is therefore quite an open one (Hanson and Conde, 1985; Pring and Lonsdale, 1985).

Specific Plant Cultivars

In 1983, dsRNAs that could be grouped in one of three patterns were detected in avocado plants. Nearly all possible combinations of single, double, and triple patterns were detected in individual trees. The patterns tended to be fairly stable for the cultivar/selection in which they were found. To date, no clear association between any combination of dsRNAs with any recognized disease of avocado, such as avocado black streak disease, has been established (Jordan et al., 1983).

The detection of dsRNA with M.W. ranging from 0.5 to 9.0×10^6 has now been repeated with several other plant types. A prominent high molecular weight dsRNA (M.W. = 9.0×10^6) has been detected in *Phaseolus vulgaris*, in several cultivars with black seed coats, including the cultivar 'Black Turtle Soup' (Wakarchuk and Hamilton, 1985; Valverde et al., 1986). In cucurbits, up to four dsRNA patterns, consisting of single or multiple dsRNA segments, have been detected in certain cultivars (Nameth and Dodds, 1985). In tomato, one dsRNA has been detected (Figure 2). For each of these examples, there are many cultivars of each species that lack detectable amounts of dsRNA, implying that the presence of dsRNA is not an essential part of the biochemistry of the plant.

Transmission and Therapy in Plants

Results for cryptic viruses and dsRNA associated with male sterility were described previously. Several attempts have been made to transmit dsRNAs from plant cultivars that contain specific dsRNAs to others that lack the dsRNA targeted for transmission. The recipient cultivar may or may not contain other dsRNA types. Such experiments were successful for some of the dsRNAs in avocado (Jordan et al., 1983), but were unsuccessful with dsRNAs from melon and pepper. Conventional plant viruses (zucchini yellow mosaic virus or tobacco etch virus) did move across the graft union in melon or pepper, respectively.

Relatedness between dsRNAs in Plants

We have recently paid attention to dsRNAs that can be detected in pepper and have detected several patterns, similar in diversity and complexity to those previously detected in avocado and cucurbits. One of these is a high molecular weight (M.W. = 9.0×10^6) nonsegmented dsRNA (Figure 2) and is of particular interest since a molecule of this homogeneity and size has been detected in three of the previously mentioned other examples (avocado, bean, and melon). We chose to examine the dsRNA in pepper, because it is present in higher titer than the similarly sized molecule in the other plants. This dsRNA has been used as a template from which randomly primed cDNA was transcribed, and this was used as a probe to determine sequence homology between the pepper dsRNA and other dsRNAs. Our results indicate that this dsRNA is unique, and while the cDNA hybridized to the dsRNA from which it was prepared, it did not hybridize with the similarly sized dsRNA found in bean, melon, and avocado. The probe hybridized to extracts from several cultivars of peppers that contain this size dsRNA, and it did not hybridize to extracts of several other cultivars that lack this size dsRNA but which did contain other dsRNA types that we have detected in pepper.

Relatedness to Host DNA in Plants

There is a lack of conclusive evidence to establish a relatedness between any of the dsRNAs detected in plants with either nuclear or organelle DNA. The dsRNA in male sterile corn has been associated with mitochondria (Sisco et al., 1984). There is some indication that the dsRNA in 'Black Turtle Soup' bean is homologous with DNA extracted from varieties that do not contain readily detectable amounts of the dsRNA (Wakarchuk and Hamilton, 1985).

A result that will have a bearing on future efforts to compare plant dsRNAs with host DNA is the relatedness between a dsRNA (M.W. = 4.7×10^6) in a yeast strain to mitochondrial DNA (Bielharz et al., 1982). The explanation offered for the relatedness is that large, but not complete, ssRNA transcripts in both polarities are made from mitochondrial DNA, and these hybridize to form the dsRNA present in extracts from these lines of yeast. A report of dsRNA in mice with homology to host DNA can also be included in this context (Kramerov et al., 1985). Symmetrical transcription may yield dsRNA in vivo, or the creation of this molecule may occur by association of complementary transcripts during the extraction and purification process.

Plant Pathogenic Fungi

The literature on dsRNA viruses in plant pathogenic fungi has been reviewed (Day and Dodds, 1979); attention was drawn to the lack of clear correlation between the presence of dsRNA viruses and effects on the phenotype of the fungal host. Since that time there have been new reports of dsRNA detected when fungal isolates are surveyed and compared. The fungi that have been analyzed include *Rhizoctonia solani* (Zanzinger et al., 1984; Finkler et al., 1985), *Ceratocystis ulmi* (Hoch et al., 1985), *Pyricularia oryzae* (Hunst et al., 1986), *Gaeumannomyces graminis* (Jamil et al., 1984), *Geotrichum candidum* (Mor et al., 1984), and *Puccinia striiformis* (Newton et al., 1985). Tentative conclusions that the presence of dsRNA was associated with hypovirulence have been made for some of these examples, but none have the strength of correlative evidence established for hypovirulence in *E.*

parasitica. The lack of agreement on whether dsRNA is involved in hypovirulence or virulence in *R. solani* exemplifies the problems encountered in associating the presence of dsRNA with biological effects.

A CASE HISTORY: *Botryodiplodia theobromae*

We have been interested in developing a biological control based on transmissible hypovirulence for an ascomycete disease of a perennial crop in California, in an attempt to put into practice some of the principles established by the model system, whereby chestnut blight is controlled by isolates of *E. parasitica* that contain fungal-disease specific dsRNAs.

A canker disease of grapes caused by *Botryodiplodia theobromae* is presently not well controlled. Its incidence and significance in California has been determined to be greater than previously thought. Current estimates indicate that, of the canker type disease on grapes attributable to *Eutypa armeniacae*, *B. theobromae*, or *Phomopsis viticola*, between 50% and 100% can be attributed to *B. theobromae*, depending on which part of the state is surveyed. One of the major sites for initial infection of the grapevine is at the pruning site, which suggests a method of control that has not been considered previously, and that would involve the treatment of such sites with a biological control agent that would antagonize the virulent strains of the pathogen.

The pathogen is similar in many respects to the agent that causes chestnut blight, which can be controlled by the deliberate treatment of cankers with special isolates of the causal fungus (*E. parasitica*) that are known to contain a viruslike agent that debilitates the fungus. That agent is identified by the presence of dsRNA that passes from the isolate used as a treatment, and into the target isolate in the canker. After a period during which the virus spreads through the canker, a healing reaction is observed. This occurs because virus-converted fungus in the canker is no longer able to breach the chestnut's defense reaction. Once established, the biological control agent can be self disseminating from healing cankers to other cankers by mechanisms that are not well understood. This phenomenon is known as transmissible hypovirulence.

We would like to attempt to create a similar biocontrol program for a canker disease of grapes. Initially, we wish to focus attention on *B. theobromae*, but we also believe that a similar approach could be feasible with the other canker-causing pathogens of this and other crops.

Seventy isolates of *B. theobromae* have been tested for the presence of virallike dsRNAs, and 39 (56%) of these were positive (Mathews and Dodds, 1986; and their unpublished results). Several patterns of dsRNAs have been detected. Labeled (^{32}P)cDNA made against specific dsRNAs from specific isolates failed to hybridize with other dissimilar-size dsRNAs in Northern blot analysis and hybridized with some but not all dsRNAs of similar size. These results are most easily interpreted by assuming that there are several classes of dsRNAs and that these can occur alone or in various combinations. This is a high rate of detection of dsRNA with a considerable degree of complexity. This initial success bodes well for the project.

Ten isolates have been subjected to single-spore subculturing, and the progeny were found to contain the same patterns of dsRNAs as the parent. This is quite a contrast to similar work done with *E. parasitica*, where the loss of some or all dsRNAs is common. One isolate did lose specific dsRNAs when some single-spore subcultures were tested. When two isolates have been grown together on the same culture plate, where one of the isolates was free of dsRNA and one contained dsRNA, or where each contained a markedly different set of dsRNAs, there have been no indications of transmission of dsRNA from one isolate to the other to date. This is also in contrast to work with *E. parasitica*, where transmission is possible so long as fungal isolates are of a related anastomosis group.

An effort has been made to develop a virulence assay on yams, apples, and grapes. When yam slices were inoculated with several isolates, evidence was initially obtained for a correlation between inability to rot yams and presence of dsRNA. These results were not repeated in a subsequent experiment. A consistent correlation between presence of specific dsRNAs and ability to rot apples has not been detected.

Excised grape stems have been wounded and inoculated with three isolates, one which rotted yams and was dsRNA-negative, one which rotted yams slightly and was dsRNA-positive, and one which did not rot yams and was dsRNA-positive. Results of this inoculation trial were inconclusive, since all isolates rapidly invaded the pith of the excised stem pieces, and then quickly invaded the length of the stem. An attempt is being made to develop a virulence assay on grape plants.

In summary, there are several contrasts between this system and the one that involves *E. parasitica*, not the least of which is the recognition that *B. theobromae* is a much more opportunistic pathogen, capable of attacking a much wider range of hosts, and well adapted to sustain itself in a saprophytic state. It remains to be seen if these differences are sufficient to minimize the likelihood that a mechanism of transmissible hypovirulence can be discovered or created for *B. theobromae*.

CONCLUSIONS AND DISCUSSION

The use of dsRNA analysis is quick and simple; when results are positive, considerable information is generated for use in virus diagnosis.

The dsRNA recovered is a valuable reagent that can be used for cloning, translation, and serological studies.

A major problem is that organisms may contain readily detectable dsRNAs but show no clear disease. Interpreting which of the dsRNAs detected in the host are of viral origin and which are not is difficult. A second problem is the increasing recognition that it is not sufficient to attempt to associate biological effects with presence or absence of dsRNA. It is much more important to determine the relatedness of dsRNAs, and to recognize that dsRNAs of similar size or mobility need not have sequence relatedness and may therefore be different and unrelated species. It is also important to recognize that only specific dsRNAs out of a complex set may be involved in any effects on host phenotype.

We intend to address these questions in our project on *B. theobromae*. We will determine the effects of single-spore selection, mycelial tip culturing, and chemotherapy on the detection of dsRNA. We will attempt to isolate virions and make a subsequent comparison of encapsidated dsRNAs to those in total nucleic acid extracts from mycelium, using nucleic acid homology studies involving selected dsRNAs and either cellular RNAs or nuclear and organelle DNAs. We will also perform physical examination of the dsRNA using electron microscopy and sequence analysis. It is hoped that these approaches will also be useful in attributing a biological role to dsRNAs, not likely to be of viral origin, detected in plants.

ACKNOWLEDGMENTS

The financial support of the conference organizers, BARD (grant US-700-83) and USDA (grant 83CRCR11227) is acknowledged.

REFERENCES

Abou-El Nasr, M. A., Jones, A. T., and Mayo, M. A. (1985). Detection of double-stranded RNA in particles of Vicia cryptic virus and in *Vicia faba* tissues and protoplasts. *J. Gen. Virol. 66*:2453-2460.

Accotto, G. P., and Boccardo, G. (1986). The coat proteins and nucleic acids of two beet cryptic viruses. *J. Gen. Virol. 67*:363-366.

Bar-Joseph, M., Rosner, A., Moskovitz, M., and Hull, R. (1983). A simple procedure for the extraction of double-stranded RNA from viral infected plants. *J. Virol. Methods 6*:1-8.

Bielharz, M. W., Cobon, G. S., and Nagley, P. (1982). A novel species of double-stranded RNA in mitochondria of *Saccharomyces cerevisiae*. *Nucleic Acids Res. 10*:1051-1071.

Boccardo, G., Lisa, V., and Milne, R. G. (1983). Cryptic viruses in plants. In *Double-Stranded RNA Viruses*, R. W. Compans and D. H. L. Bishop (Eds.). Elsevier, New York, pp. 425-430.

Boccardo, G., Milne, R. G., Luisoni, E., Lisa, V., and Accotto, G. P. (1985). Three seedborne cryptic viruses containing double-stranded RNA isolated from white clover. *Virology 147*:29-40.

Brown, G. G., Bussey, H., and Desrosiers, L. J. (1986). Analysis of mitochondrial DNA, chloroplast DNA and double-stranded RNA in fertile and cytoplasmic male-sterile sunflower, *Helianthus annuus*. *Can. J. Genet. Cytol. 28*:121-129.

Chu, P. W. G., Francki, R. I. B., and Randles, J. W. (1983). Detection, isolation and characterization of high molecular weight double stranded RNA in plants infected with velvet tobacco mottle virus. *Virology 126*:480-492.

Dale, J. L., Phillips, D. A., and Parry, J. N. (1986). Double-stranded RNA in banana plants with bunchy top disease. *J. Gen. Virol. 67*:371-376.

Dalton, R. E., Podila, G. K., Flurkey, W. H., and Bozarth, R. F. (1985). In-vitro translation of the major capsid polypeptide from *Ustilago maydis* virus strain. *Virus Res. 3*:153-164.

Day, P. R., and Dodds, J. A. (1979). Viruses of plant pathogenic fungi. In *Viruses and Plasmids of Fungi*, Chapter 5, P. A. Lemke (Ed.). Marcel Dekker, New York, pp. 201-238.

Day, P. R., Dodds, J. A., Elliston, J. E., Jaynes, R. A., and Anagnostakis, S. L. (1977). Double-stranded RNA in *Endothia parasitica*. *Phytopathology 67*:1393-1396.

Derrick, K. S., French, R. C., Clark, C. A., and Gabriel, C. J. (1984). Detection of double-stranded RNA by serologically specific electron microscopy. *J. Virol. Methods 9*:293-300.

Dodds, J. A. (1980). Revised estimates of the molecular weights of dsRNAs in hypovirulent strains of *Endothia parasitica*. *Phytopathology 70*:1217-1220.

Dodds, J. A. (1986). The potential for using double-stranded RNAs as diagnostic probes for plant viruses. In *Developments and Applications in Virus Testing*, R. A. C. Jones and L. Torrance (Eds.). AAB, Wellsbourne, Warwick, England, pp. 71-86.

Dodds, J. A., and Bar-Joseph, M. (1983). Double-stranded RNA from plants infected with closteroviruses. *Phytopathology 73*:419-423.

Dodds, J. A., Morris, T. J., and Jordan, R. L. (1984). Plant viral double-stranded RNA. *Annu. Rev. Phytopathology 22*:151-168.

Dodds, J. A., Lee, S. Q., and Tiffany, M. G. (1985). Cross protection between strains of cucumber mosaic virus: Effects of host and type of inoculum on accumulation of virions and double-stranded RNA of the challenge strain. *Virology 144*:301-309.

Dodds, J. A., Jarupat, T., Lee, G. J., and Roistacher, C. N. (1987). Effects of strain, host, time of harvest, and virus concentration on double-stranded RNA analysis of citrus tristeza virus. *Phytopathology 77*:422-447.

Elliston, J. E. (1985a). Characteristics of double-stranded RNA free and double-stranded RNA containing strains of *Endothia parasitica* in relation to hypovirulence. *Phytopathology 75*:151-158.

Elliston, J. E. (1985b). Further evidence for two cytoplasmic hypovirulence agents in a strain of *Endothia parasitica* from western Michigan USA. *Phytopathology 75*:1405-1413.

Falk, B. W., and Duffus, J. E. (1984). Identification of small single and double-stranded RNA species associated with severe symptoms in beet western yellows virus infected *Capsella bursa-pastoris. Phytopathology 74*:1224-1229.

Falk, B. W., and Tsai, J. H. (1984). Identification of single-stranded RNA and double-stranded RNA associated with maize stripe virus. *Phytopathology 74*:909-915.

Finkler, A., Koltin, Y., Barash, I., Sneh, B., and Pozniak, D. (1985). Isolation of a virus from virulent strains of *Rhizoctonia solani. J. Gen. Virol. 66*:1221-1232.

Franklin, R. M. (1966). Purification and properties of replicative intermediate of the RNA bacteriophase R17. *Proc. Nat. Acad. Sci. USA 55*:1504-1511.

Fulbright, D. W. (1984). Effect of eliminating dsRNA in hypovirulent *Endothia parasitica. Phytopathology 74*:722-724.

Garcia-Luque, I., Brieva, A., Diaz-Ruiz, J. R., and Rubio, N. (1986). Isolation and partial characterization of a monoclonal antibody specific for a naturally occurring double-stranded RNA. *Virology 152*:252-255.

Garger, S. J., Turpen, T., Carrington, J., Morris, T. J., Jordan, R. L., Dodds, J. A., and Grill, L. K. (1983). Rapid detection of plant viral RNA viruses by dot blot hybridization. *Plant Molec. Biol. Reporter 1*:21-25.

Garrod, S. W., Fulbright, D. W., and Ravenscroft, A. V. (1985). Dissemination of virulent and hypovirulent forms of marked strains of *Endothia parasitica* in Michigan. *Phytopathology 75*:533-538.

Gildow, F. E., Ballinger, M. E., and Rochow, W. F. (1983). Identification of double-stranded RNAs associated with barley yellow dwarf virus infection in oats. *Phytopathology 73*:1570-1572.

Grill, L. K., and Garger, S. J. (1981). Identification and characterization of double stranded RNA associated with cytoplasmic male sterility in *Vicia faba*. *Proc. Nat. Acad. Sci. USA 78*:7043-7046.

Guilford, P. J., and Forster, R. L. S. (1986). Detection of polyadenylated subgenomic RNA in leaves infected with the potexvirus Daphne virus X. *J. Gen. Virol. 67*:83-90.

Hanson, M. R., and Conde, M. F. (1985). Functioning and variation of cytoplasmic genomes: Lessons from cytoplasmic nuclear interactions affecting male fertility in plants. *Internat. Rev. Cytol. 94*:213-267.

Hoch, J. G., Tavantzis, S. M., Campana, R. J., and Anagnostakis, S. L. (1985). Evaluation of the presence of double-stranded RNA in *Ceratocystis ulmi*. *Can. J. Bot. 63*:297-300.

Hunst. P. L., Latterell, F. M., and Rossi, A. E. (1986). Variation in double-stranded RNA from isolates of *Pyricularia oryzae*. *Phytopathology 76*:674-678.

Jamil, N., Buck, K. W., and Carlile, M. J. (1984). Sequence relationships between virus double-stranded RNA from isolates of *Gaeumannomyces graminis* in different vegetative compatibility groups. *J. Gen. Virol. 65*:1741-1748.

Jordan, R. L., and Dodds, J. A. (1983). Hybridization of 5'-end-labelled RNA to plant viral RNA in agarose and acrylamide gels. *Plant Molec. Biol. Reporter 1*:33-37.

Jordan, R. L., Dodds, J. A., and Ohr, H. D. (1983). Evidence for viruslike agents in avocado. *Phytopathology 73*:1130-1135.

Kramerov, D. A., Bukrinsky, M. I., and Ryskov, A. P. (1985). DNA sequences homologous to long double-stranded RNA transcription of intracisternal A-particle genes and major long repeat of the mouse genome. *Biochim. Biophys. Acta. 826*:20-29.

L'Hostis, B., Hiremath, S. T., Rhoads, R. E., and Ghabrial, S. A. (1985). Lack of sequence homology between double-stranded RNA from European and American hypovirulent strains of *Endothia parasitica*. *J. Gen. Virol. 66*:351-356.

Mathews, D. M., and Dodds, J. A. (1986). Development of hypovirulence for the control of a grape disease caused by *Botryodiplodia theobromae*. Abstr. *Phytopathology 76*:845.

Mor, H., Steinlauf, R., and Barash, I. (1984). Viruslike particles and double-stranded RNA in *Geotrichum candidum*, the causal agent of citrus sour rot. *Phytopathology 74*:921-924.

Mirkov, T. E., and Dodds, J. A. (1985). Double-stranded ribonucleic acid is associated with lettuce big vein disease. *Phytopathology 75*:631-635.

Morris, T. J., and Dodds, J. A. (1979). Isolation and analysis of double-stranded RNA from virus-infected plant and fungal tissue. *Phytopathology 69*:854-858.

Morris, T. J., Dodds, J. A., Hillman, B., Jordan, R. L., Lommel, S. A., and Tamaki, S. J. (1983). Viral specific dsRNA: Diagnostic value for plant virus disease identification. *Plant Molec. Biol. Reporter 1*:27-30.

Mossop, D. W., Elliott, D. R., and Richards, K. D. (1985). Association of closterovirus-like particles and high molecular weight double-stranded RNA with grapevines affected by leaf-roll disease. *New Zealand J. Agric. Res. 28*:419-426.

Nameth, S. T., and Dodds, J. A. (1985). Double-stranded RNA detected in cucurbit varieties not inoculated with viruses. *Phytopathology 75*:1293 (Abstr.).

Natsuaki, T., Yamashita, S., Doi, Y., Okuda, S., and Teranaky, M. (1983). Two seed-borne double stranded RNA viruses, beet temperate virus and spinach temperate virus. *Ann. Phytopathol. Soc. Jpn. 49*:709-712.

Newton, A. C., Caten, C. E., and Johnson, R. (1985). Variation for isoenzymes and double-stranded RNA among isolates of *Puccinia striiformis* and two other cereal rusts. *Plant Pathol. 34*:235-247.

Pring, D. R., and Lonsdale, D. M. (1985). Molecular biology of higher plant mitochondrial DNA. *Internat. Rev. Cytol. 97*:1-46.

Reddy, D. V. R., Murant, A. F., Raschke, J. H., Mayo, M. A., and Ansa, O. A. (1985). Properties and partial purification of infective material from plants containing groundnut rosette virus. *Ann. Appl. Biol. 107*:65-78.

Rosner, A., Bar-Joseph, M., Moscovitz, M., and Mavarech, M. (1983). Diagnosis of specific viral RNA sequences in plant extracts by hybridization with a polynucleotide kinase labelled double-stranded RNA probe. *Phytopathology 73*:699-701.

Sisco, P. H., Garcia-Arenal, F., Zaitlin, M., Gracen, V. E., and Earle, E. D. (1984). LBN, a male sterile cytoplasm of maize contains 2 double stranded RNAs. *Plant Sci. Lett.* *34*:127-134.

Valverde, R. A., and Dodds, J. A. (1986). Evidence for a satellite RNA associated naturally with the U5 strain and experimentally with the U1 strain of tobacco mosaic virus. *J. Gen. Virol.* *67*: 1875-1884.

Valverde, R. A., Dodds, J. A., and Heick, J. A. (1986). Double-stranded ribonucleic acid from plants infected with viruses having elongated particles and undivided genomes. *Phytopathology* *76*: 459-465.

Wakarchuk, D. A., and Hamilton, R. I. (1985). Cellular double-stranded RNA in *Phaseolus vulgaris* cultivar 'Black Turtle Soup.' *Plant Molec. Biol.* *5*:55-64.

Zanzinger, D. H., and Tavantzis, S. M. (1982). Improved detection of homologous double-stranded RNA (dsRNA) antigens by the indirect method of solid-phase enzyme immunoassay. *Phytopathology* *72*:954 (Abstr.).

Zanzinger, D. H., Bandy, B. P., and Tavantzis, S. M. (1984). High frequency of finding double-stranded RNA in naturally occurring isolates of *Rhizoctonia solani*. *J. Gen. Virol.* *65*:1601-1606.

15

dsRNA and Disease Factors of the Aggressive Subgroup of *Ophiostoma ulmi*

H. J. Rogers* and K. W. Buck
Imperial College of Science and Technology, London, England

C. M. Brasier
Forest Research Station, Farnham, Surrey, England

THE AGGRESSIVE AND NONAGGRESSIVE SUBGROUPS OF *Ophiostoma ulmi*

Dutch elm disease, caused by the fungus *Ophiostoma* (*Ceratocystis*) *ulmi* (Buisman) Nannf. was first recorded in 1918 in Western Europe and quickly spread to the United States. A far more devastating new epidemic began in Britain around 1970 and was followed by the discovery that there are two major subgroups of the fungus: a highly pathogenic "aggressive strain" responsible for the recent epidemic and a less pathogenic "nonaggressive strain" believed to have been endemic in Britain and adjacent parts of Europe since the early part of the twentieth century. Studies of isolates of *O. ulmi* from many parts of Europe have shown that the current second epidemics of the disease are due to two separate races of the aggressive subgroup. One, called the North American (NAN) race, is believed to have been imported into Britain from North America during the mid-1960s and

Present affiliation: Institute of Plant Science Research, Trumpington, Cambridge, England

has subsequently spread into neighboring countries of northwest Europe and Iberia. A second race, called the Eurasian (EAN) race, appears to have migrated westward from central and southern Europe or from further east (Gibbs and Brasier, 1973; Brasier and Gibbs, 1972; Brasier, 1979,1983a,b,1986a,b).

The aggressive and nonaggressive subgroups of *O. ulmi* are readily distinguished by their different morphologies and temperature optima for growth (Brasier, 1983a), and differ in an array of other physiological properties. A fertility barrier largely prevents hybridization between them; when crossed, the behavior of their progeny indicates multiple gene differences. Moreover, the absence of isolates in nature with recombinant phenotypes suggests that they are effectively reproductively isolated in the wild. The two subgroups are probably equivalent at least to subspecies of *O. ulmi*. Their geographical origins remain unknown (Brasier, 1977,1983a,b,1986a,b; Brasier and Gibbs, 1976). The EAN and NAN races of the aggressive subgroup are more similar and are usually distinguished on the basis of the low efficiency of the EAN race to act as the female parent in a genetic cross with the NAN race (Brasier, 1979,1984,1986b) and on the occurrence of a striking mycelial dimorphism in the EAN (Brasier, 1986b). It seems likely that the EAN and NAN races arose comparatively recently from a common progenitor (Brasier, 1986d), EAN/NAN hybrids are now appearing at locations where the two races overlap (Brasier, 1987a).

In Europe and Southwest Asia the nonaggressive subgroup has been rapidly replaced by one or the other race of the aggressive subgroup and seems likely to become extinct (Brasier, 1983b,1986a,b). A similar process is occurring in those regions of North America where both the NAN aggressive and nonaggressive subgroups have been found (Gibbs et al., 1979; Houston, 1985).

Double-stranded RNA (dsRNA) was first reported in isolates of *O. ulmi* (from the United States) by Pusey and Wilson (1981,1982); however, their preliminary correlation of multiple dsRNA segments with isolates from states with a long history of Dutch elm disease (Pusey and Wilson, 1981) or with "less aggressive" isolates (Pusey and Wilson, 1982), and conversely absence of dsRNA segments in isolates from states with a shorter history of Dutch elm disease or in "more aggressive" isolates, has not been upheld by subsequent investigations. Hoch et al. (1985) found no clear association between the presence of dsRNA or between specific segments of dsRNA and the classification of American isolates as aggressive or nonaggressive. Multiple dsRNA segments were present in several isolates of the aggressive subgroup, whereas no dsRNA was detected in five out of eight nonaggressive isolates. Our group also failed to detect dsRNA in a number of European and American nonaggressive isolates. These results are consistent with the earlier genetic studies referred to above (e.g., Brasier and Gibbs, 1976; Brasier, 1977) that indicated multiple and profound genetic differences between the aggressive and nonaggressive subgroups at the nuclear, rather than cytoplasmic, level.

CYTOPLASMICALLY TRANSMITTED DISEASES OF THE AGGRESSIVE SUBGROUP OF *O. ulmi*: DISCOVERY OF d-FACTORS

A cytoplasmically transmitted disease of *O. ulmi* was first reported in a French isolate, H321, of the NAN race of the aggressive subgroup by Brasier (1983c). The diseased (d-infected) isolate was characterized by slow growth, impaired reproductive fitness, and the ability to transmit the disease by hyphal anastomosis to healthy recipient isolates. Transmission was characterized by the development of sectors of weak or abnormal growth in the young recipient culture: Either there were expanding sectors of sunken infected mycelium with a "butterfly" appearance or growth of the newly infected sector and that of much of the leading edge of the recipient culture ceased completely. Such interactions between diseased isolates and healthy recipients were termed d-reactions; the cytoplasmic determinants of the disease were called d-factors. The d-factor in isolate H321 was termed the d^1-factor, isolate H321 was renamed H321d^1, and some of the subsequently discovered d-factors have been designated d^2-, d^3-, d^4-factors, and so on (Brasier, 1986c).

Transmission of d-factors occurs very efficiently to recipient isolates of the same vegetative compatibility (v-c) group, but less efficiently to recipients of different v-c groups (Brasier, 1984,1986c). Vegetative incompatibility in *O. ulmi*, as in other ascomycetes, is heterogenic, i.e., it results form one or more v-c gene differences between two isolates; there are probably at least four different v-c genes with at least one of them (the *w* gene) being multiallelic (Brasier, 1984, 1987b). Four different types of vegetative incompatibility reactions have been observed as a result of pairings of *O. ulmi* isolates (Brasier, 1984). The most common reaction is the wide or 'w' reaction, characterized by a wide (10 to 15 mm) diffuse white mycelial barrage and synnematal lines parallel to the barrage, which probably results from differences in all or most v-c genes, including a different *w* allele. A less common reaction is the narrow or "n" reaction with a narrow (4 to 7 mm), denser, white mycelial barrage, some synnematal lines close to the barrage, most v-c genes different, and a common *w* allele. Uncommon reactions are the line or "l" reaction, with thin (1 to 3 mm) white mycelial barrage, no synnemata, and a few (>1) v-c genes different, and the line-gap or "l-g" reaction, with a gap or barrage (1 to 3 mm) between the two colonies, no synnemata, and only one v-c gene different. When a d^2-infected NAN donor was paired against 111 wild NAN recipient isolates, the percent d-factor transmission was 100%

for fully compatible interactions (all v-c genes the same), 66% for l/lg reactions, 50% for n reactions and <4% for w reactions. Broadly, transmission is reduced according to the number of v-c gene differences, but interactions between different v-c alleles of the same gene, and between different v-c genes and the background genotype, may also affect transmission (Brasier, 1984,1986b). Similar results have been reported regarding the effect of vegetative incompatibility on cytoplasmic transmission of hypovirulence in the chestnut blight fungus, *Endothia parasitica* (Anagnostakis, 1983).

Most isolates in regional or "world-wide" *O. ulmi* collections belong to different v-c groups. However, significant proportions of NAN isolates in collections from Britain and Europe (but not in collections from North America) belong to the same v-c group, designated the NAN v-c supergroup. Similarly, a significant proportion of EAN isolates were also found to belong to a single v-c group, termed the EAN v-c supergroup. When isolates from the NAN v-c supergroup were paired with those from the EAN v-c supergroup, "w" reactions occurred, indicating that the two supergroups are quite distinct. At least in older epidemic areas, within the v-c supergroups the isolates give the appearance of being otherwise genetically heterogeneous. Brasier (1984) initially suggested that the v-c supergroups originated from a founder effect resulting from the recent rapid spread into Europe from outside of the NAN and EAN races of the aggressive strain. Subsequent mutation and sexual recombination could have resulted in a gradual increase in both the number of additional v-c groups and the genetic heterogeneity within the v-c supergroups. However recent research on NAN populations at epidemic fronts in Portugal, where the NAN supergroup comprises >90% of the population, suggests that selection for a high level of fitness may be involved (Brasier and Mitchell, 1986; and cf. "episodic selection," Brasier, 1986a,1987c).

The occurrence of v-c supergroups would be expected to increase the frequency of transmission of d-factors; this is reflected in the relatively high proportion of d-infected isolates at epidemic fronts. For example, near the current epidemic front in Poland, ∿47% of a sample of the EAN population were found to belong to the EAN v-c supergroup, of which ∿31% were severely d-infected (Brasier, 1986c).

Brasier (1983c,1986c) found that ascospore progeny from genetic crosses between d-infected and non-d-infected isolates, or between two d-infected isolates, were healthy, i.e., d-factors appeared to have been eliminated during sexual reproduction. Outcrossing would also tend to reduce the spread of d-factors in populations by increasing the number of v-c groups. Transmission of d-factors occurs through conidia and the yeast phase, but with variable efficiencies. Overall the survival of d-factors in natural populations probably depends on a balance between their spread by hyphal anastomosis in the mycelial phase of the disease in the bark, and their loss through sexual, and to a lesser extent asexual, sporulation, and environmental factors selecting against d-infected isolates. If this balance were more consistently altered in favor of d-infection it might lead to natural supression of the disease. In addition, it might be possible to use d-factors in disease control programs.

A recent, important study has shown that while 1000 spores are required for infection to occur with a healthy NAN isolate, 50,000 spores are required to achieve infection with a d^2-infected isolate (Webber, 1986). Thus, d-factors probably have the potential to break the cycle of Dutch elm disease. The impact of d-factors on Dutch elm disease and their possible uses in biological control have been reviewed in some detail by Brasier (1986c).

dsRNA IN DISEASED AND HEALTHY EAN AND NAN AGGRESSIVE ISOLATES

Introduction

Cytoplasmically transmitted phenotypes of fungi can be associated with dsRNA, e.g., the killer phenotypes of *Saccharomyces cerevisiae* (Bruenn, 1986) and *Ustilago maydis* (Koltin, 1986) and hypovirulence in *Endothia parasitica* (Van Alfen, 1986); DNA plasmids, e.g., the killer phenotype of *Kluyveromyces lactis* (Gunge, 1983); or alterations in mitochondrial DNA, e.g., senescence in *Podospora anserina* and the ragged phenotype of *Aspergillus amstelodami* (reviewed by Bockelmann et al., 1986). The absence of transmission of d-factors into ascospore progeny, even when both male and female parents are d-infected (Brasier, 1986c), suggests that d-factors are not transmitted in the mitochondria, unless d-infected isolates contain mixtures of healthy and defective mitochondria, of which only the former are transmitted into ascospores. In this section we report on the dsRNA content of d-infected isolates and apparently healthy isolates of the EAN and NAN races of the aggressive subgroup, review the evidence for association of the d^2-factor with specific segments of dsRNA, and suggest that mitochondria may be the targets for dsRNA in d^2-infections.

Numbers and Sizes of dsRNA Segments in Healthy and Diseased Isolates

The molecular weights of dsRNA segments in 23 d-infected and 21 apparently healthy isolates, classified according to v-c group, are shown in Table 1. All d-infected isolates contained dsRNA, although the number of segments ranged from 1 to 14 in different isolates. Some healthy isolates appeared to be dsRNA-free, but most contained dsRNA, with segment numbers ranging from two to nine. It is clear, therefore, that d-infection is not caused generally by a generic effect of

Table 1 dsRNA Segments in Diseased and Healthy Isolates[a] of *Ophiostoma ulmi*

Isolate	Location[b]	Date of isolation	Molecular weights of dsRNA segments ($\times 10^6$)[c]
NAN v-c supergroup			
W2	GB	1972	2.40, 0.61
Log1/3-7d[2]	GB[d]	1981	2.40, 2.08, 1.84, 1.67, 1.60, 1.52, 0.65, 0.61, 0.33, 0.23
Log1/3-8d[2]	GB	1981	2.40, 2.08, 1.84, 1.60, 1.67, 1.52, 0.65, 0.61, 0.33, 0.23
Log1/3-13d[2]	GB	1981	2.40, 2.08, 1.84, 1.67, 1.60, 1.52, 0.63, 0.61, 0.33, 0.23
Log1/3-15d[2]	GB	1981	2.40, 2.08, 1.84, 1.67, 1.60, 1.52, 0.65, 0.61, 0.33, 0.23
Log1/3-63	GB	1981	2.40, 0.61
Log1/3-83	GB	1981	2.40, 0.61
T210d	GB	1983	1.62, 1.54
H106	FR	1977	1.88, 1.57
H301	IR	1978	No dsRNA detected
H351	BE	1979	No dsRNA detected
H363	IR	1979	No dsRNA detected
H625	JE	1981	2.08, 1.67, 1.52, 0.70
W4	GB	1972	1.75, 1.52, 0.33
NAN v-c group 2[e]			
H321d[1]	FR	1978	1.26, 1.09, 1.04, 0.55, 0.50, 0.46, 0.38
NAN v-c group 3			
C103d[4]	GB	1982	1.89, 1.79, 1.64, 1.52, 0.56
C113	GB	1982	1.89, 1.79, 1.64, 1.52, 0.56
NAN v-c group 4			
ES118d	SP	1984	1.96, 1.66, 1.50, 1.11, 0.90, 0.54, 0.30
ES694d	SP	1984	1.96, 1.66, 1.50, 1.11, 0.90, 0.30

Table 1 (Continued)

Isolate	Location[b]	Date of isolation	Molecular weights of dsRNA segments ($X\ 10^6$)[c]
B1/3d	SP	1984	1.96, 1.66, 1.50, 0.54, 0.30
B5/5d	SP	1984	1.96, 1.66, 1.50, 0.54, 0.30
NAN v-c group 5			
AL-E58d	GB	1981	1.70, 1.66, 1.58, 1.25
AL-E29	GB	1981	1.78, 1.70, 1.19
NAN v-c group 6			
AL-E65d	GB	1981	1.75, 1.66, 1.58, 1.52, 1.25
AL-E30d	GB	1981	1.66, 1.25
AL-E62d	GB	1981	1.61
AL-E9	GB	1981	No dsRNA detected
EAN v-c supergroup			
P111d[5]	PO	1980	2.50, 2.17, 1.86, 1.69, 1.64, 1.53, 0.63, 0.53
P104d[6]	PO	1980	2.50, 2.17, 1.86, 1.64, 0.68
R108d[7]	RO	1980	2.01, 1.86, 1.74, 1.64, 1.23, 1.18, 0.34, 0.31
P109d[8]	PO	1980	2.37, 2.17, 1.86, 1.64, 1.37, 1.22, 1.06, 0.82, 0.72, 0.65, 0.56, 0.53, 0.33, 0.25
P123d[9]	PO	1980	2.17, 1.86, 1.64, 1.51, 1.43, 1.32
P99Bd	PO	1980	2.50, 2.06, 1.85, 1.71, 1.62, 1.53, 0.59, 0.49
R114d	RO	1980	2.15, 1.87, 1.76, 1.71, 1.62, 0.52, 0.36, 0.33
R66	RO	1980	2.15, 1.98, 1.76, 1.67, 1.51, 0.91, 0.55, 0.36, 0.33
R82	RO	1980	2.06, 1.67, 1.51
R95	RO	1980	2.06, 1.98, 1.76, 1.67, 1.51, 0.36, 0.33
R132	RO	1980	1.85, 1.66, 1.51, 0.30

Table 1 (Continued)

Isolate	Location[b]	Date of isolation	Molecular weights of dsRNA segments (X 10^6)[c]
P107	PO	1980	1.85, 1.76, 1.62, 1.51
P114	PO	1980	No dsRNA detected
P121	PO	1980	1.85, 1.67, 1.54, 0.68
P134	PO	1980	1.85, 1.67, 1.54
H224	DE	1978	2.17, 1.86, 1.64, 0.53
EAN v-c group 2			
H254d^3	SU	1978	2.50, 2.18, 2.0 (do)[f], 1.75 (do), 1.62, 1.54 (do), 1.23, 1.17, 0.34

[a]All d-infected isolates have the suffix d; all other isolates are healthy.
[b]Location: BE, Belgium; DE, Denmark, FR, France; GB, Great Britain; IR, Ireland; JE, Jersey; PO, Poland; RO, Romania; SP, Spain; SU, Soviet Union.
[c]Determined by polyacrylamide gel electrophoresis. Values for most of the isolates in the NAN v-c supergroup were reported by Rogers et al. (1986a).
[d]Isolates within each of the series log 1/3-, C103d^4/C113, ES/B1/B5, and AL-, are all from the same geographic location; within each group Log 1/3-, B1/B5, and Al-, all are from the same piece of diseased elm bark.
[e]Nonsupergroup v-c groups are numbered solely to enable different groups to be distinguished in this article; they are not intended for general use.
[f]do = Doublet.

dsRNA. The finding that one d-infected isolate (C103d^4) and one apparently healthy isolate (C113), both in the same v-c group and obtained from the same locality (Chichester, England), have five dsRNA segments with identical gel mobilities, does not necessarily imply that dsRNA is not the cause of the d^4-infection, since identity of size does not imply identity of sequence and even a single base change could be sufficient to alter the genetic properties of a dsRNA segment. The d^4-factor in C113 could also have been in a latent state (see below). However, it should be borne in mind that not all d-infections may have the same genetic basis. At least ten different killer toxins are produced by different strains of yeasts, but only three of these are encoded by dsRNA segments (Young and Yagiu, 1978).

All the isolates listed in Table 1 are wild isolates with the exception of isolate series Log1/3-(7d^2, 8d^2, 13d^2, 15d^2, 63, and 83).

These isolates were obtained from an experiment in which an elm tree was inoculated with W2*toll*, a mutant of NAN v-c supergroup isolate W2 tolerant to MBC fungicide. After several months, during which the isolate had spread from the xylem to the bark, MBC-tolerant isolates were obtained from the inner bark surface (Webber and Brasier, 1984). Some of these were found to be d-infected (Brasier, 1986c) and to have acquired several dsRNA segments not present in W2*toll* (Rogers et al., 1986a), presumably as a result of transmission to W2*toll* via hyphal anastomosis from a wild *O. ulmi* genotype, brought into the bark by breeding scolytid beetles (wild *O. ulmi* isolates are generally MBC-sensitive). Other isolates were healthy and possessed the same size dsRNA segments as did W2*toll*. Because all four d-infected isolates had the same, or very similar, sizes of dsRNA segments and were isolated from regions of the bark in close proximity, it was considered likely that they had all acquired the same d-factor, designated the d^2-factor (Brasier, 1986c).

Most other d-infected isolates contain different numbers and/or sizes of dsRNA segments (Table 1). When d-infected isolates are in different v-c groups it has been assumed that they are infected with different d-factors and they have been given specific "d" numbers. Some of the phenotypes of d-infected isolates appear to be qualitatively different (Brasier, 1986c). For example, the d^1- and d^2-factors have been compared in the same genetic background and shown to cause different effects; d^2-infected isolates generally give stronger d-reactions and grow more slowly than d^1-infected isolates. It is now clear that d-infection of *O. ulmi* in the wild is relatively common and it is considered prudent to restrict further numbering of d-factors until more information on the genetic basis of d-infection and possible relationships between different d-factors has been obtained. Consequently, some of the d-infected isolates in Table 1 have not been given a d number, although all carry the "d" suffix to distinguish them from the healthy isolates.

Rogers et al. (1986a,b) have reported that ascospore progeny arising from d^1- or d^2-infected parents, as well as being healthy (Brasier, 1986b), are usually free from dsRNA. Similar elimination of dsRNA has been reported in ascospore progeny of *Gaeumannomyces graminis* var. *tritici* (McFadden et al., 1983), *Helminthosporium maydis* (Bozarth, 1977), and *Endothia parasitica* (Day et al., 1977) and may be general in the filamentous ascomycetes. If it is true for the ascospore progeny of all dsRNA-containing individuals of *O. ulmi*, the infrequency of dsRNA-free healthy isolates in the sample studied (Table 1) could indicate that ascospores play a relatively minor role in the life cycle of *O. ulmi*. However, perithecia and ascospores are usually produced abundantly during the long overwintering bark phase of the fungus and the resultant recombinant genotypes are known to recolonize the bark (Brasier, 1986c; Webber and Brasier, 1984; Webber et al., 1987). Behind epidemic fronts, ascospores sometimes constitute a significant proportion of inoculum on later-emerging vector beetles (Webber and Brasier, 1984 and unpublished data). Hence, it is more likely that the infrequency of dsRNA-free healthy isolates in the sample studied (Table 1) reflects a high frequency of dsRNA spread between isolates by hyphal anastomosis in the bark-to-bark cycle (Brasier 1986a,b), especially as the sample was heavily biassed toward EAN and NAN v-c supergroup isolates, and toward isolate series within other v-c groups from a single geographical location. Further samples are required, including a wider range of v-c groups, before firmer conclusions can be reached.

Transmission of d-Factors and dsRNA by Hyphal Anastomosis

Transmission of the d^2-factor has been studied in detail (Rogers et al., 1986a). As the d^2-infected isolate series Log1/3-($7d^2$, $8d^2$, $13d^2$, and $15d^2$) are all MBC-tolerant, transmission to MBC-sensitive recipients was easily monitored. Five recipient NAN v-c supergroup isolates from different geographical origins were found, after transmission of the d-infection from Log1/3-$8d^2$, to contain dsRNA segments of the same size as those in the donor isolate. Three of these (H301, H351, and H363) were originally dsRNA-free, but the other two (W4 and H625) contained two and three segments, respectively, of the same size as those of the donor (Table 1); therefore, transmission of segments of this size could not be monitored. Isolates W4 and H265 each contained an additional segment not present in the donor; these segments were maintained after d-infection in W4d^2 and H365d^2. Hence, the dsRNA segments of isolate Log1/3-$8d^2$ appeared to be transmitted faithfully along with the d^2-factor irrespective of host genetic background, and there was no evidence of incompatible interactions between dsRNA segments in the donor and recipient isolates.

Transmission of d-factors was accomplished from a selection of other d-infected isolates to recipients from the same v-c group, which were either wild isolates or ascospore isolates obtained from genetic crosses in the laboratory, and dsRNA was analyzed in the newly d-infected isolates, i.e., the transfer isolates. Nuclear genetic markers were not available for these transmissions, but isolations were made from areas of the recipient colonies, which in analogous experiments with MBC-tolerant donors were free from donor nuclei (Brasier, 1983, 1986c), and shown to be d-infected. When donor and recipient isolates contained dsRNA, the presence of dsRNA segments characteristic of both donor and recipient was taken as evidence of cytoplasmic mixing. In general, there was faithful transmission of dsRNA segments, along with the d-factors, from donor to recipient, but in some instances dsRNA segments characteristic of the donor isolates were lost, and/or new dsRNA segments appeared in the transfer isolates (Table 2). Loss of segments could be due to incompatibility reactions between dsRNA segments of the type described for *Ustilago maydis* by Wigderson and Koltin (1982) or to the inability of a dsRNA segment to replicate efficiently in the new host genetic background. The latter mechanism may be more likely, since no segments were lost from recipient isolates as a result of the transfer. Generation of new segments might be due to dsRNA segments present at undetectably low levels in the donor being more efficiently replicated in the new host genetic background; this is feasible because it was also noticed that some dsRNA segments present at low levels in the donor increased in amount after transfer to the recipient. The generation of deletions or duplications of sequences in some dsRNA segments as a result of transfer to the new hosts is another possibility. Analogous alterations in the number, size, and amounts of dsRNA segments after transfer to some, but not all, new host isolates have been reported for *Endothia parasitica* (anagnostalkis and Day, 1979; Anagnostakis, 1981) and the ability of dsRNA segments to undergo deletions and duplications is well documented in the S dsRNA segments of *Saccharomyces cerevisiae* (Lee et al., 1986).

Latent d-Infection

Single conidial isolates arising from isolate Log1/3-8d^2 were found to be of three types (Rogers et al., 1986a,b,c): (i) Overtly d^2-infected isolates had properties similar to the parent isolates, i.e., slow growth, degenerate "amoeboid" morphology, ability to give a d-reaction with, and transfer the d-infection to, healthy isolates of the same v-c group, and retention of the ten dsRNA segments of the parent. (ii) Latently d^2-infected isolates had growth rates and morphology similar to those of healthy wild NAN isolates, but had a tendency to revert to the overt d^2-infected phenotype either spontaneously or as a result of growth under a variety of conditions. Their conidial progeny included a proportion of overtly d^2-infected isolates and they retained all ten dsRNA segments characteristic of the parent. (iii) Healthy isolates had growth rates and morphology similar to those of wild healthy isolates and were never found to revert to the overt d^2-infected phenotype. No overtly d^2-infected isolates were recovered from their conidial progeny. They had lost three or more of the dsRNA segments of the parent.

Latently d^2-infected isolates evidently carried the d^2-factor in a form that was not overtly expressed, and several possible explanations have been put forward to account for the latent state (Rogers et al., 1986a,b,c). These include unusually high mutation rates of d-factors from wild-type to inactive mutants and vice versa; reversible integration and excision fo the d^2-factor (or a cDNA copy) into/from the nuclear or mitochondrial DNA; alterations in d^2-factor expression as a result of nuclear gene switch mechanisms, e.g., DNA transposi-

Table 2 Gains and Losses of dsRNA Segments during Transmission from Donor to Recipient Isolates

Transfer		Molecular weights of dsRNA segments ($\times 10^6$)	
Donor	Recipient	Present in donor isolate, but not in transfer isolate	Present in transfer isolate, but not in donor or recipient isolates
$H321d^1$	$13F_2P_{11}$[a]	None	0.77
T210d	Asc202[a]	None	None
$P111d^5$	H224	None	None
$P104d^6$	H224	None	None
$R108d^7$	H224[b]	2.01, 1.23, 1.18, 0.31	2.17, 1.06, 0.49, 0.44
$R108d^7$	H224[b]	2.01, 1.18, 0.31	2.17
$P109d^8$	H224	1.22, 1.06	1.28, 0.50
$P123d^9$	H224	None	None

[a] $13F_2P_{11}$ and Asc 202 are dsRNA-free, single-ascospore isolates of the same v-c group as the donor isolates.

[b] Two separate transmissions from $R108d^7$ to H224 were carried out.

tion or induction/repression, or of the levels or expression, of other cytoplasmic factors; or alterations in the levels of the d^2-factor itself. Not all of these explanations are consistent with a dsRNA d^2-factor, since the levels of all ten dsRNA segments are similar in overtly and latently d^2-infected isolates. When latently d^2-infected isolates are paired with overtly d^2-infected isolates, d-reactions ensue in which the latently d^2-infected isolates appear to act as the recipients. However, it is not possible to determine whether this is due to transfer of the d^2-factor or to induced switching from latent to overt d^2-infection.

There is some evidence that latency can occur in d-infections other than d^2. For example, isolate T210d appears healthy, but after transfer of its d-factor to Asc202, the transfer isolate was overtly d-infected. Similarly, isolates P99Bd and R114d appeared to be healthy when first isolated, but after several years' storage at -20°C were found to be overtly d-infected. Hence latency of d-infections may be a general phenomenon. Latency would probably increase the survival of d-factors in wild populations since selection against latently d-infected individuals would be expected to be lower than against overtly d-infected individuals.

The existence of latent d-infections means that great care must be taken before an isolate can be classified as healthy. Probably the only reliable test of an isolate's freedom from d-infection is its ability to produce only healthy conidial derivatives. Of the healthy isolates listed in Table 1, this has been checked rigorously only for isolate H351, which is also dsRNA-free. It is therefore possible that some of the apparently healthy isolates may be latently d-infected.

Loss of dsRNA Segments from *O. ulmi* Isolates: Association of d-factors with Specific dsRNA Segments

Loss of dsRNA segments from d^2-infected isolates of *O. ulmi* can occur through ascospore formation, through conidiogenesis, or through passage of isolates through the elm xylem, the latter presumably as a result of multiplication of the fungus in the yeast form (Rogers et al., 1986a). Numbers of segments lost ranged from ten (i.e., complete loss of all segments) in most ascospore isolates, to as few as three in many conidial isolates and reisolations from inoculated trees. All isolates that had lost dsRNA segments 4, 7, and 10 (numbering the dsRNA segments of isolate Log 1/3-8d^2 in order of decreasing molecular weight), and which did not give rise to any additional segments in transfer isolates, appeared to be healthy. Absence of possible latent d^2-infections in several of these was confirmed by the production from them of only healthy conidial progeny. No other dsRNA segments were consistently lost in all the non-d^2-infected derivatives. All derivative isolates that retained segment 4, 7, and 10 were shown to be either overtly or latently d^2-infected. It is therefore possible that the d^2-factor consists of one or more of dsRNA segments 4, 7, and 10. However, until this has been proved by dsRNA transfection experiments, the possibility must remain that d^2-infection could be caused by another cytoplasmic factor that is lost coincidentally with segments 4, 7, and 10.

Analysis of the dsRNA segments of derivatives of d^2-infected isolates after one or two cycles of conidiogenesis or passages through elm xylem has shown that the loss of segments is not random. The patterns of segments retained, and the frequency of retention of individual segments, in a collection of 52 such derivative isolates are shown in Table 3. It is noteworthy that the segments lost always included segments 4, 7, and 10. This suggests that replication of these segments is dependent on factors, produced by the host or by other dsRNA segments, that become limiting during conidiogenesis or at some stage of the yeast phase cell cycle. The situation may be analogous to the M dsRNA segment in killer strains of *Saccharomyces cerevisiae*, which are dependent for replication on the products of both host genes and a "helper" L dsRNA segment (Bruenn, 1986). The overall frequency of retention of the dsRNA segments (Table 3) followed the pattern $6 \sim 2 > 3 \sim 5 \sim 1 > 8 \sim 9 > 4 = 7 = 10$ and probably indicates increasing degrees of dependence on other dsRNA segments or host factors. The finding of dsRNA segment 2 alone in an ascospore isolate (Asc 217), derived from a genetic cross between Log1/3-8d^2 and H351 (Rogers et al., 1986a), indicates that this segment can replicate independently of the other dsRNA segments. It is also noteworthy that segment 6 was lost in only two out of the 52 isolates.

Like the d^2-factor, the d^1-factor and its associated dsRNA segments were lost in sexual reproduction. For example, ten single ascospore isolates arising from a genetic cross between H321d^1 and $13F_4P_1d^1$ (an ascospore isolate of the same v-c group of H321, to which the d^1-factor has been transmitted by hyphal anastomosis) were found to be healthy and free from dsRNA. However, following conidiogenesis, d-infection and dsRNA segments appeared to be lost more readily from d^1-infected isolates than from d^2-infected isolates, which retained two to seven dsRNA segments, all those healthy conidial isolates tested that had lost the d^1-factor were free from dsRNA. Furthermore, several conidial isolates, which still carried the d^1-factor (as evidenced by their ability to give d-reactions when paired with a healthy isolate of the same v-c group) had lost up to four of the original seven dsRNA segments of H321d^1 (Table 4). It is noteworthy that some of the conidial isolates also contained segments not detected in H321d^1. These could have been formed by one or more of the mechanisms previously discussed; one of these segments, designated 3b, is the same size as the new segment detected after transmission of the d^1-factor from H321d^1 to $13F_2P_{11}$ by hyphal anastomosis (Table 2). The dsRNA

Table 3 Retention and Loss of dsRNA Segments in Derivatives of d^2-Infected Isolates

Number of isolates with dsRNA segment pattern	dsRNA Segment number[a]										Number of dsRNA segments lost
	1	2	3	4	5	6	7	8	9	10	
16	+[b]	+	+		+	+		+	+		3
7	+	+	+		+	+		+			4
4	+	+	+		+	+		+			4
1	+	+	+			+		+	+		4
1	+	+	+		+	+					5
1	+	+	+		+			+			5
1	+	+	+			+			+		5
1		+	+		+	+			+		5
1	+	+	+			+					6
1	+				+	+		+			6
1		+	+			+			+		6
1	+				+	+					7
2		+	+			+					7
3		+			+	+					7
2			+		+	+					7
7		+				+					8
1					+	+					8
1		+									9
Number of isolates in which dsRNA segment retained	34	47	38	0	38	50	0	26	24	0	

[a]dsRNA segments of isolate Log1/3-8d^2 numbered in order of decreasing molecular weight.
[b]A + indicates dsRNA segment retained.

Table 4 dsRNA Segments in Single Conidial Isolates which had Retained the d^1-Factor

	dsRNA Segment number[a]										
Isolate	1	2	3	3b	4	5	5b	6	7	8	9
Con 1	+[b]	+	+	+	+	+		+	+		
Con 2	+	+	+	+	+	+					
Con 3	+	+		+	+	+	+				
Con 4	+	+				+					
Con 7	+	+		+		+					
Con 9	+	+	+	+	+	+	+		+	+	+

[a]Segments corresponding to those of H321d^1 (Table 1) are numbered 1 to 7 in order of decreasing molecular weight. Additional dsRNA segments, detected in some of the single conidial isolates, but not in H321d^1, are underlined and have molecular weights as follows: 3b, 0.77×10^6; 5b, 0.49×10^6; 8, 0.33×10^6; 9, 0.31×10^6.
[b]A + indicates presence of dsRNA segment.

segments 1, 2, and 5 were the only ones retained in all the d^1-factor single conidial isolates (Table 4), and therefore the d^1-factor could be comprised of one or more of these segments.

Transmission of the d^2-Factor and dsRNA, but not Chloramphenicol-Resistant Mitochondria, by Hyphal Anastomosis

Correlations of one or more specific dsRNA segments, either by transmission, loss, or retention in different derivative isolates, with a particular d-factor do not establish unequivocally that such dsRNA segments comprise that d-factor. However close the association, the possibility always remains that another cytoplasmic factor could have cosegregated with the dsRNA segments in question. Careful searches in extracts of isolate Log1/3-8d^2 for isometric virus particles of the type found in many fungi (Buck, 1986), and for club-shaped vesicles of the type found in some hypovirulent isolates of *Endothia parasitica* (Dodds, 1980; Hansen et al., 1985), proved negative, and methods for transfection of *O. ulmi* by native or denatured dsRNA have not yet been developed. It is therefore not yet possible to prove that specific dsRNA segments cause d-infections by direct infection, transfection, or transformation. Furthermore, it is not yet known if the diseased phenotype of d-infected isolates results from the synthesis of specific proteins encoded by d-factors. Hence, in vitro translation of denatured dsRNA, successfully used to prove that M dsRNA encodes the protein toxin secreted by killer strains of *Saccharomyces cerevisiae* (Bostian et al., 1980), cannot yet be used to establish possible roles of dsRNA segments in d-infections. A more indirect approach is to investigate whether or not other cytoplasmic factors are determinants of d-infections; experiments to this end with regard to mitochondria are described in this section.

To monitor the transmission of the d^2-factor and mitochondria separately, Brasier (1986c) transferred the d^2-factor by hyphal anastomosis from Log1/3-8d^2 to H106*CR1*, a chloramphenicol-tolerant mutant of H106. Chloramphenicol tolerance in *O. ulmi* shows classical maternal inheritance (Brasier and Kirk, 1986) and, like that in other fungi, is probably due to a mutation in the mitochondrial DNA. The site of the mutation conferring chloramphenicol tolerance in *Saccharomyces cerevisiae* has been sequenced; it lies within the mitochondrial 21S rRNA gene (Dujon, 1980). When the transfer isolate H106*CR1*d^2, which carried both the d^2-factor and the chloramphenicol tolerance marker, was paired with its wild-type self H106 (healthy, chloramphenicol-sensitive) to induce a further d-reaction, subcultures taken from the resulting newly d-infected sectors of the recipient were found to be d-infected but chloramphenicol-sensitive. Similar results were obtained with four other isolate pairs (W4 + H106*CR1*d^2; H363 + H106*CR1*d^2; RDT38 + RDT38*CR1*d^2; H363 + RDT38*CR1*d^2). Hence d^2-infection and chloramphenicol tolerance were probably encoded by different genetic elements.

Further information was obtained with additional pairings in which the recipient isolates were spontaneous MBC (fungicide)-tolerant mutants of wild isolates. (W2*toll* + H106*CR1*d^2; RDT38*toll* + H106*CR1*d^2; W2*toll* + RDT38*CR1*d^2; RDT38*toll* + RDT38*CR1*d^2). MBC tolerance is a stable, single-locus nuclear marker (Brasier and Gibbs, 1975), and its presence in these recipients enabled them to be distinguished from the MBC-sensitive donors. Such pairings led to the isolation of MBC-tolerant, d^2-infected, chloramphenicol-sensitive transfer isolates, confirming transmission of the d^2-factor independently of the chloramphenicol tolerance marker.

In the experiments described above, spontaneous chloramphenicol-tolerant mutants were selected from the wild isolates and then the d^2-factor was transferred to them by hyphal anastomosis; hence it remained possible that the transfer isolate contained some chloramphenicol-sensitive mitochondria transmitted along with the d^2-factor. If the chloramphenicol-sensitive mitochondria persisted alongside the chloramphenicol-tolerant mitochondria and if the chloramphenicol-sensitive mitochondria were subsequently transmitted with the d^2-factor in preference to the chloramphenicol-tolerant mitochondria, segregation of the d^2-factor from the chloramphenicol tolerance marker would not necessarily prove its segregation from (chloramphenicol-sensitive) mitochondria which might therefore carry the d^2-factor.

In an attempt to obtain a d^2-infected, chloramphenicol-tolerant isolate in which the mitochondria were more likely to be homogeneous, the d^2-factor was first introduced into wild isolate RDT38 from Log1/3-8d^2 via hyphal anastomosis to generate RDT38d^2 and then a spontaneous chloramphenicol-resistant mutant RDT38d^2*CR1* was selected directly from RDT38d^2. Examination of 295 single conidial isolates derived from RDT38d^2*CR1* revealed that some of them were d-infected while others were apparently healthy (tests for latency were not carried out). However, all were chloramphenicol-tolerant, suggesting that the mitochondria in RDT38d^2*CR1* are homogeneous, within the limits of the experiment. Rowlands and Turner (1974) reported segregation frequencies in conidia of 20 to 30% in either direction between two phenotypic cytoplasmic markers, cold sensitivity and oligomycin resistance, in heteroplasmons of *Aspergillus nidulans*. Such a segregation frequency in conidia derived from RDT38d^2*CR1* would have been detected easily in the sample of 295 conidial isolates used.

To test whether the d^2-factor and dsRNA could be transmitted independently of the chloramphenicol-tolerance marker of RDT38d^2*CR1*, isolate RDT38d^2*CR1* (MBC-sensitive) was paired against a healthy conidial isolate derived from Log1/3-8d^2, designated Log1/3-8(Con18), which was MBC-tolerant. A strong d-reaction developed, and three isolations made from the newly d-infected recipient sector were found to be MBC-tolerant, chloramphenicol-sensitive, and to contain the ten dsRNA segments characteristic of Log1/3-8d^2 and RDT38d^2*CR1*. These results therefore confirm the results of Brasier (1986c) described above and extend them to show that both the d^2-factor and at least eight dsRNA segments [Log1/3-8(Con18) contains segments 1 and 6] are transmitted by hyphal anastomosis independently of the chloramphenicol-tolerance marker.

None of the above experiments show unequivocally that the d^2-factor is not a defective mitochondrial DNA. The possibility remains that chloramphenicol-tolerant, d^2-infected isolates contain a mixture of normal, chloramphenicol-tolerant mitochondria and defective, chloramphenicol-sensitive mitochondria. If cells containing only the defective mitochondria were inviable, formation of chloramphenicol-sensitive condidia would not be possible. However transmission of the chloramphenicol-sensitive defective mitochondria without the chloramphenicol-resistant normal mitochondria by hyphal anastomosis would be possible because the recipient contained chloramphenicol-sensitive normal mitochondria which could substitute for the chloramphenicol-resistant normal mitochondria of the donor in maintaining cell viability. The stable coexistence of normal and defective mitochondria has been shown in senescent strains of *Podospora anserina* (Koll et al., 1984). Furthermore, the dominance of defective mitochondrial DNA in transfer by hyphal anastomosis has been shown for abnormal phenotypes, such as poky and stopper, in neurospora, although the reasons for this are still not clear (reviewed by Bockelmann et al., 1986).

Examination of *Eco*RI restriction fragment patterns of mitochondrial DNA from Log1/3-8d^2 and the healthy conidial derivative Log1/3-8(Con18) by gel electrophoresis revealed no differences. Hence, d^2-infection is not caused by major alterations of the mitochondrial DNA such as occur in senescent strains of *P. anserina* (Bockelmann et al., 1986), although the presence of very small deletions in the mitochondrial DNA of d^2-infected isolates, such as have been reported to be the cause of the poky phenotype in *Neurospora crassa* (Akins and Lambowitz, 1984), cannot be excluded. It is clear that if mitochondria in Log1/3-8d^2 are heterogeneous, the normal and defective mitochondria have similar restriction patterns. This suggests that restriction fragment length polymorphisms may be better markers for mitochrondria than chloramphenicol tolerance. If a recipient isolate could be found with a mitochondrial restriction pattern significantly different from that of Log1/3-8d^2, then transmission of dsRNA, mitochondrial DNA (both defective and normal), and the d^2-factor to this recipient could be monitored. The isolation of newly d^2-infected recipients that contained the ten dsRNA segments of Log1/3-8d^2 but only the mitochondrial DNA restriction pattern of the recipient would be good evidence that the genetic determinant of the d^2-phenotype was not defective mitochondrial DNA or contained within defective mitochondria.

Recently, we have shown that d^2-infected isolates have a deficiency of cytochrome oxidase and do therefore contain defective mitochondria (but not necessarily defective mitochondrial DNA); furthermore, most of the dsRNA in such isolates copurifies with isolated mitochondria, although no homology between dsRNA and mitochondrial DNA has been detected (Rogers et al., 1987). Considering all of this, it appears likely that d^2-infected isolates probably do not contain defective mitochondrial DNA, but that one or more of dsRNA segments 4, 7, and 10 interact with mitochondria, either directly or via products they encode, possibly translated within mitochondria, to cause the cytochrome oxidase deficiency. If the dsRNA replicates exclusively in the mitochondria, the results of the transmission experiments can only be explained if d^2-infected isolates contain two populations of mitochondria, defective mitochondria containing dsRNA and normal mitochondria lacking dsRNA; dsRNA would then be necessarily transmitted along with the defective mitochondria. However, it is also possible that there are two pools of dsRNA, one in a site separate from the mitochondria, and in which the dsRNA replicates, and the other in the mitochondria; this could represent a target for the dsRNA. In that case, transmission of the dsRNA might occur independently of the defective mitochondria, and targeting of the dsRNA to the mitochondria in the recipient could cause the mitochondrial defect and hence the d-phenotype. Use of donors and recipients with distinct mitochondrial DNA restriction fragment patterns could distinguish the possibilities as described above.

Sequence Relationships between dsRNA Segments in d-Infected Isolates

Since there was good evidence correlating the presence of specific dsRNA segments with the d^2-phenotype, it was of interest to determine whether there was any nucleic acid sequence homology between the dsRNA segments of Log1/3-8d^2(NAN) and those in isolates infected with other d-factors. The dsRNA from Log1/3-8d^2 was purified by cellulose chromatography, treatment with DNase I and S_1 nuclease, centrifugation through a cesium chloride density gradient to ensure that it was free from single-stranded RNA and DNA, partially hydrolyzed with alkali, and then labeled with $\gamma[^{32}P]$ATP and polynucleotide kinease. In dot hybridization assays carried out essentially as described by Thomas (1980), no hybridization could be detected with single-stranded RNA from NAN isolate H301 (dsRNA-free) but hybridization was detected with denatured dsRNA from a number of other d-infected isolates with the intensity of the dots detected by autoradiography (judged visually) decreasing in the order: C103d^4 > P104d^6 > P123d^9 ∿ T210 ∿ R108d^7 > P111d^5 > P109d^8. No hybridization was detected, with the relatively high conditions of stringency used, with denatured dsRNA from isolates $13F_2P_{11}d^1$(NAN) and H254d^3 (EAN). An interesting feature was that hybridization was obtained between dsRNA from isolates from the EAN and NAN races from several different countries.

The relationships between dsRNA segments in isolate Log1/3-8d^2 and those in isolates C103d^4 and P104d^6 were further investigated by electrophoresis of dsRNA from these three isolates alongside each other in two replicate agarose gels containing methyl mercuric hydroxide as a denaturant, followed by transfer of the RNA to GeneScreen membrane. The two Northern blots were then separately hybridized with $[^{32}P]$-labeled probes prepared to dsRNA from isolates C103d^4 and P104d^6. In this gel system, the ten dsRNA segments of Log1/3-8d^2 were not fully resolved, but it was possible to identify the segments which hybridized with C103d^4 dsRNA and P104d^6 dsRNA as belonging to dsRNA segment groups 4, 5, 6 and 7, 8. It is noteworthy that these two groups, include the dsRNA segments 4 and 7 which are closely associated with the d^2-factor. Because of the closely similar mobilities of some of the segments, even in polyacrylamide gels, further analysis would probably best be done with cloned cDNA probes specific for individual dsRNA segments.

We thank Susan Kirk (Forest Research Station) for technical assistance and the SERC for financial support of H. J. R.

REFERENCES

Akins, R. A., and Lambowitz, A. M. (1984). The [*poky*] mutant of *Neurospora* contains a 4-base-pair deletion at the 5' end of the mitochondrial small rRNA. *Proc. Nat. Acad. Sci. USA 81*:3791-3795.

Anagnostakis, S. L. (1981). Stability of double-stranded RNA components of *Endothia parasitica* through transfer and subculture. *Exp. Mycol.* 5:236-242.

Anagnostakis, S. L. (1983). Conversion to curative morphology in *Endothia parasitica* and its restriction by vegetative compatibility. *Mycologia 75*:777-780.

Anagnostakis, S. L., and Day, P. R. (1979). Hypovirulence conversation in *Endothia parasitica. Phytopathology 69*:1225-1229.

Bockelmann, B., Osiewacz, H. D., Schmidt, F. R., and Schulte, E. (1986). Extrachromosomal DNA in fungi—organization and function. In *Fungal Virology*, K. W. Buck (Ed.). CRC Press, Boca Raton, Florida, pp. 237-283.

Bostian, K. A., Hopper, J. E., Rogers, D. J., and Tipper, D. J. (1980). Translational analysis of killer-associated virus-like particle dsRNA genome of *Saccharomyces cerevisiae*: M dsRNA encodes toxin. *Cell 19*:403-414.

Bozarth, R. F. (1977). Biophysical and biochemical characterization of virus-like particles containing a high molecular weight dsRNA from *Helminthosporium maydis. Virology 80*:149-157.

Brasier, C. M. (1977). Inheritance of pathogenicity and cultural characteristics in *Ceratocystis ulmi.* Hybridisation of protoperithecial and non-aggressive strains. *Trans. Brit. Mycol. Soc. 68*:45-52.

Brasier, C. M. (1979). Dual origin of recent Dutch elm disease outbreaks in Europe. *Nature 281*:78-80.

Brasier, C. M. (1983a). Occurrence of three sub-groups within *Ceratocystis ulmi.* In *Proceedings of the Dutch Elm Disease Symposium*, Winnipeg, Manitoba, October 5-9, 1981, E. S. Kondo, Y. Hiratsuka, and W. B. G. Denyer (Eds.). Manitoba Dept. Nat. Resources, Manitoba, Canada, pp. 298-321.

Brasier, C. M. (1983b). The future of Dutch elm disease in Europe.

In *Research on Dutch Elm Disease in Europe. Forestry Commission Bulletin 60*:96-104.

Brasier, C. M. (1983c). A cytoplasmically transmitted disease of *Ceratocystis ulmi. Nature 305*:220-223.

Brasier, C. M. (1984). Inter-mycelial recognition systems in *Ceratocystis ulmi*: Their physiological properties and ecological importance. In *The Ecology and Physiology of the Fungal Mycelium*, D. H. Jennings and A. D. M. Rayner (Eds.). Cambridge Univ. Press, London, England, pp. 451-497.

Brasier, C. M. (1986a). The population biology of Dutch elm disease: The principal features and implications for other host-pathogen systems. *Advan. Plant Pathol.* 5:53-118.

Brasier, C. M. (1986b). Comparison of pathogenicity and cultural characteristics in the EAN and NAN aggressive subgroups of *Ophiostoma ulmi. Trans. Brit. Mycol. Soc. 87*:1-13.

Brasier, C. M. (1986c). The d-factor in *Ceratocystis ulmi*--its biological characteristics and implications for Dutch elm disease. In *Fungal Virology*, K. W. Buck (Ed.). CRC Press, Boca Raton, Florida, pp. 177-208.

Brasier, C. M. (1986d). Dutch elm disease--*Ophiostoma* (*Ceratocystis*) *ulmi*. The emergence of EAN and NAN hybrids on Europe. *Report on Forest Research, 1986*, HMSO, London, England.

Brasier, C. M. (1987a). Recent genetic changes in the *Ophiostoma ulmi* population: The threat to the future of the elm. In *Populations of Plant Pathogens: Their Dynamics and Genetics*, M. S. Wolfe and C. E. Caten (Eds.). Blackwell, Oxford, England, pp. 213-226.

Brasier, C. M. (1987b). Genetic systems in *Ophiostoma ulmi*. In *Advances in Plant Pathology*, Vol. 6, *Genetics of Pathogenic Fungi*, G. S. Sidhu (Ed.). Academic, London, in press.

Brasier, C. M. (1987c). The dynamics of fungal speciation. In *Evolutionary Biology of the Fungi*, A. D. M. Rayner, C. M. Brasier, and D. Moore (Eds.). Cambridge Univ. Press, London, England, pp. 231-260.

Brasier, C. M., and Gibbs, J. N. (1972). Origin of the Dutch elm disease epidemic in Britain. *Nature 242*:607-609.

Brasier, C. M., and Gibbs, J. N. (1975). MBC tolerance in aggressive and non-aggressive isolates of *Ceratocystis ulmi. Ann. Appl. Biol. 80*:231-235.

Brasier, C. M., and Gibbs, J. N. (1976). Inheritance of pathogenicity and cultural characters in *Ceratocystis ulmi*: Hybridization of aggressive and non-aggressive strains. *Ann. Appl. Biol. 83*:31-37.

Brasier, C. M., and Kirk, S. A. (1986). Maternal inheritance of chloramphenicol tolerance in *Ophiostoma ulmi*. *Trans. Brit. Mycol. Soc. 87*:460-462.

Brasier, C. M., and Mitchell, A. G. (1986). Dutch elm disease—*Ophiostoma (Ceratocystis) ulmi*. Vegetative compatibility 'supergroups' in *O. ulmi (Ceratocystis ulmi)*. *Report on Forest Research, 1986*, HMSO, London, England, p. 38.

Bruenn, J. (1986). The killer systems of *Sacharomyces cerevisiae* and other yeasts. In *Fungal Virology*, K. W. Buck (Ed.). CRC Press, Boca Raton, Florida, pp. 85-108.

Buck, K. W. (1986). Fungal virology--an overview. In *Fungal Virology*, K. W. Buck (Ed.). CRC Press, Boca Raton, Florida, pp. 1-84.

Day, P. R., Dodds, J. A., Elliston, J. E., Jaynes, R. A., and Anagnostakis, S. L. (1977). Double-stranded RNA in *Endothia parasitica*. *Phytopathology 67*:1393-1396.

Dodds, J. A. (1980). Association of type 1 viral-like dsRNA with club-shaped particles in hypovirulent strains of *Endothia parasitica*. *Virology 107*:1-12.

Dujon, B. (1980). Sequence of the intron and flanking exons of the mitochondrial 21S rRNA gene of yeast strains having different alleles at the *w* and *rib-1* loci. *Cell 20*:185-197.

Gibbs, J. N., and Brasier, C. M. (1973). Correlation between cultural characters and pathogenicity in *Ceratocystis ulmi* from Europe and North America. *Nature 241*:381-383.

Gibbs, J. N., Houston, D. R., and Smalley, E. B. (1979). Aggressive and non-aggressive strains of *Ceratocystis ulmi* in North America. *Phytopathology 69*:1215-1219.

Gunge, N. (1983). Yeast DNA plasmids. *Annu. Rev. Microbiol. 37*:253-276.

Hansen, D. R., Van Alfen, N. K., Gillies, K., and Powell, W. A. (1985). Naked dsRNA associated with hypovirulence of *Endothia parasitica* is packaged in fungal vesicles. *J. Gen. Virol. 66*:2605-2614.

Hoch, J. G., Tavantzis, S. M., Campana, R. J., and Anagnostakis, S. L. (1985). Evaluation of the presence of double-stranded RNA in *Ceratocystis ulmi*. *Canad. J. Bot. 63*:297-300.

Houston, D. R. (1985). Spread and increase of *Ceratocystis ulmi* with cultural characteristics of the aggressive strain in north eastern North America. *Plant Dis. 69*:677-680.

Koll, F., Begel, O., Keller, A.-M., Vierny, C., and Belcour, L. (1984). Ethidium bromide rejuvenation of senescent cultures of *Podospora anserina*: Loss of senescence-specific DNA and recovery of normal mitochondrial DNA. *Curr. Genet. 8*:127-134.

Koltin, Y. (1986). The killer systems of *Ustilago maydis*. In *Fungal Virology*, K. W. Buck (Ed.). CRC Press, Boca Raton, Florida, pp. 109-141.

Lee, M., Pietras, D. F., Nemeroff, M. E., Corstanje, B. J., Field, L. J., and Bruenn, J. A. (1986). Conserved regions in defective interfering viral double-stranded RNAs from a yeast virus. *J. Virol. 58*:402-407.

McFadden, J. J. P., Buck, K. W., and Rawlinson, C. J. (1983). Infrequent transmission of double-stranded RNA virus particles but absence of DNA proviruses in single ascospore cultures of *Gaeumannomyces graminis*. *J. Gen. Virol. 64*:927-937.

Pusey, P. L., and Wilson, C. L. (1981). Possible role of fungal viruses in the distribution and spread of the Dutch elm disease fungus. *J. Arboriculture 7*:230-232.

Pusey, P. L., and Wilson, C. L. (1982). Detection of double-stranded RNA in *Ceratocystis ulmi*. *Phytopathology 72*:423-428.

Rogers, H. J., Buck, K. W., and Brasier, C. M. (1986a). Transmission of double-stranded RNA and a disease-factor in *Ophiostoma ulmi*. *Plant Pathol. 35*:277-287.

Rogers, H. J., Buck, K. W., and Brasier, C. M. (1986b). The molecular nature of the d-factor in *Ceratocystis ulmi*. In *Fungal Virology*, K. W. Buck (Ed.). CRC Press, Boca Raton, Florida, pp. 209-220.

Rogers, H. J., Buck, K. W., and Brasier, C. M. (1986c). The d^2-factor in *Ophiostoma ulmi*: Expression and latency. In *Biology and Molecular Biology of Plant Pathogen Interactions*, J. A. Bailey (Ed.). Springer-Verlag, Berlin, pp. 393-400.

Rogers, H. J., Buck, K. W., and Brasier, C. M. (1987). A mitochondrial target for double-stranded RNA in diseased isolates of the fungus that causes Dutch elm disease. *Nature 329*:558-560.

Rowlands, R. T., and Turner, G. (1974). Recombination between the extranuclear genes conferring oligomycin resistance and cold sensitivity in *Aspergillus nidulans*. *Mol. Gen. Genet. 133*:151-161.

Thomas, P. S. (1980). Hybridization of denatured RNA and small DNA fragments transferred to nitrocellulose. *Proc. Nat. Acad. Sci. USA* 77:5201-5205.

Van Alfen, N. K. (1986). Hypovirulence of *Endothia* (*Cryphonectria*) *parasitica* and *Rhizoctonia solani*. In *Fungal Virology*, K. W. Buck (Ed.). CRC Press, Boca Raton, Florida, pp. 143-162.

Webber, J. F. (1986). Dutch elm disease--*Ophiostoma* (*Ceratocystis*) *ulmi*. Colonisation of beetle feeding grooves by *O. ulmi* spores carrying a d-factor. *Report on Forest Research, 1986*, HMSO, London, England.

Webber, J. F., and Brasier, C. M. (1984). The transmission of Dutch elm disease: A study of the processes involved. In *Invertebrate-Microbial Interactions*, J. M. Anderson, A. D. M. Rayner, and D. W. H. Walton (Eds.). Cambridge Univ. Press, Cambridge, England, pp. 271-306.

Webber, J. F., Brasier, C. M., and Mitchell, A. G. (1987). The role of the saprophytic phase in Dutch elm disease. In *Fungal infestation of plants: Establishment, progress and outcome of infection*, G. F. Pegg and P. G. Ayres (Eds.). Cambridge Univ. Press, London, England, in press.

Wigderson, M., and Koltin, Y. (1982). Dual toxin specificities and the exclusion relations among the *Ustilago* dsRNA viruses. *Curr. Genet.* 5:127-136.

Young, T. W. F., and Yagiu, M. (1978). A comparison of the killer character in different yeasts and its classification. *Antonie van Leeuwenhoek J. Microbiol. Serol.* 44:59-77.

16

Viruses of *Helminthosporium victoriae*

Said A. Ghabrial
University of Kentucky, Lexington, Kentucky

INTRODUCTION

The viruses of the plant pathogenic fungus *Helminthosporium victoriae* are of special interest because they are associated with a disease of their host, whereas the majority of fungal viruses are avirulent (Sanderlin and Ghabrial, 1978; Ghabrial, Sanderlin, and Calvert, 1979; Ghabrial, 1980; Ghabrial and Mernaugh, 1983). Moreover, diseased isolates of *H. victoriae* are hypovirulent; they produce little or no "victorin" (Lindberg, 1960; Ghabrial, 1986). The latter is the host-specific toxin that determines pathogenicity in this fungus (Luke and Wheeler, 1955). A further incentive for studying the *H. victoriae* system is the earlier finding that diseased isolates, but not normal ones, produce an antibiotic with a broad spectrum of antifungal and antibacterial activity (Lindberg, 1964). We have recently confirmed this report, and preliminary results suggest that the antibiotic material is of proteinaceous nature (S. A. Ghabrial, unpublished data, 1986). It will be of considerable interest to determine whether this putative protein is virus-encoded, since the killer phenomenon has not yet been found in filamentous fungi.

Two isometric dsRNA viruses, designated according to their sedimentation values, have been isolated from diseased isolates of *H. victoriae* (Sanderlin and Ghabrial, 1978). The relationship between the two viruses and their role in disease development, however, is not known. To shed some light on this, recent studies have focused on investigating in vitro transcription of virion dsRNAs and isolation and characterization of in vivo and in vitro gene products of the two viruses. This chapter will review these recent studies and discuss their relevance to other similar mycoviruses. A recent review on the disease of *H. victoriae* (Ghabrial, 1986) should be consulted for details on the biology of these viruses.

BACKGROUND

The Host

H. victoriae was first described in 1946 as the causal agent of Victoria blight of oats (Meehan and Murphy, 1946). It is a highly specialized pathogen, because it inflicts its damage on only those oat varieties with the Victoria type of resistance to crown rust (Litzenberger and Murphy, 1947; Litzenberger, 1949). The blight caused by *H. victoriae* rose to epidemic proportions in 1947 and 1948 and caused considerable yield losses in most oat-cropping areas in the United States. As a result, Victoria-derived oat cultivars were largely abandoned in the major oat growing regions except in the southern states, where they continued to be grown during the 1950s. In 1959, a disease of *H. victoriae*, transmissible by hyphal anastomosis, was reported for isolates from blighted "Victorgrain" oat plants in Louisiana (Lindberg, 1959). Colonies of diseased isolates are characterized by stunted growth, frequent sectoring, poor sporulation, collapse of aerial mycelium, and generalized lysis (Lindberg, 1959; Ghabrial et al., 1979).

Over 25 years ago, a viral etiology was suggested for the disease of *H. victoriae*, but it is only recently that evidence in support of this has been presented (Sanderlin and Ghabrial, 1978; Ghabrial et al., 1979; Ghabrial and Mernaugh, 1983; Ghabrial, 1986). Diseased isolates have been found to contain two serologically and electrophoretically distinct viruses designated according to their sedimentation values as the 190S and 145S viruses. Normal isolates, on the other hand, are either virus-free or contain low concentration of only the 190S virus. Because disease severity appeared to correlate well with the 145S titer in colonies derived from protoplasts of a diseased isolate, it has been suggested that either the 145S virus alone, or a mixed infection with the two viruses, is the cause of disease (Ghabrial et al., 1979). However, no isolate has ever been found that contained the 145S virus alone; the possibility that the 145S is a satellite of the 190S virus cannot be ruled out.

The Viruses

The 190S and 145S viruses have similar-size particles, about 40 nm in diameter (Figure 1). The 190S virus contains a single dsRNA component with a M.W. of 3.0×10^6. Four dsRNA species, with molecular weights of 2.4, 2.2, 2.1, and 2.0×10^6, are associated with the 145S virus (Sanderlin and Ghabrial, 1978). Most virus preparations from diseased isolates contain disproportionately larger amounts of the 190S than of the 145S virus. Purified 190S virus of high quality could thus be obtained in quantities adequate for characterization and for production of specific antisera. The 190S virus has recently been the subject of extensive characterization studies (Ghabrial and Havens, 1986; Ghabrial et al., 1987; and see the section on the 190S virus). On the other hand, only preliminary characterization of the 145S virus has been carried out (Sanderlin and Ghabrial, 1978). The reasons for this have been the poor yield of the 145S virus and the difficulty of completely freeing 145S preparations of 190S particles. Attempts to maintain selections of diseased isolates that support relatively high yields of the 145S virus have been unsuccessful. This may be explained by the apparent positive correlation between disease

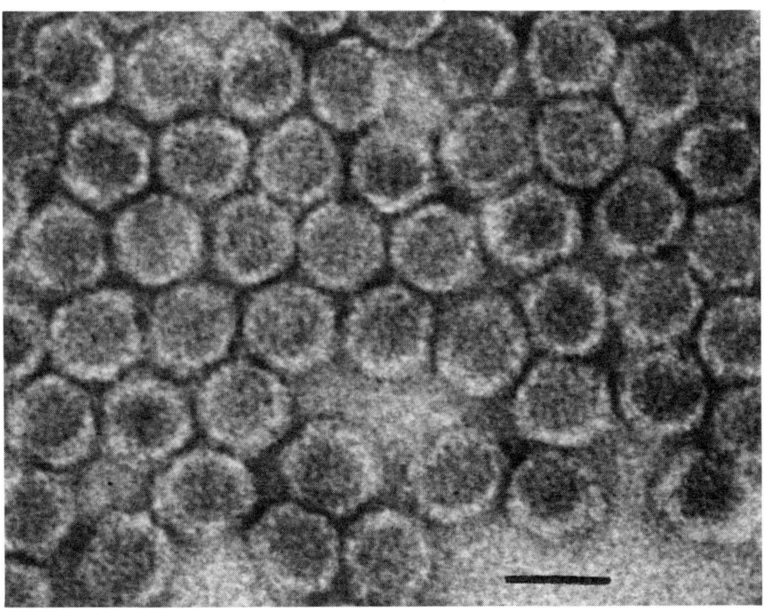

Figure 1 Electron micrograph of a purified preparation of the 190S virus stained with 1% uranyl acetate. The bar represents 50 nm.

severity and the titer of the 145S virus (Ghabrial et al., 1979). Future investigations will seek mutants of *H. victoriae* that tolerate high concentrations of the 145S virus. The possibility that other *Helminthosporium* species may serve as alternative hosts for the 190S and 145S viruses will be pursued.

THE 190S VIRUS AND COMPARISON TO VIRUSES IN THE FAMILY *TOTIVIRIDAE*

Viral Genome

The 190S virus of *H. victoriae*, like other members of the family *Totiviridae* (Brown, 1986), has an undivided dsRNA genome. This property is unique among dsRNA viruses which, as a rule, have segmented genomes (Joklik, 1983). The molecular weight of the 190S dsRNA was estimated at 3.0×10^6 by electrophoresis on polyacrylamide gels, using reovirus dsRNAs as standards (Sanderlin and Ghabrial, 1978). The value of 3.0×10^6 (4.5 kilobase pairs), which was derived by linear extrapolation, was later confirmed using dsRNA size markers covering the range from 0.46×10^6 to 5.7×10^6 (Ghabrial, 1986). However, the 190S-dsRNA is apparently slightly larger than the dsRNA from *Saccharomyces cerevisiae* virus L_A (ScV-L_A), the type species of the family *Totiviridae* (Figure 2). Several estimates for the molecular weight of ScV-L_A dsRNA have been reported that ranged from 4.5 to 4.9 kilobase pairs (Bruenn, 1980). The M.W. of the 190S-dsRNA should also fall within this range since denatured 190S-dsRNA has recently been estimated as 1.5×10^6 or 4.5 kilobases (Ghabrial et al., 1987). Unlike the L-dsRNAs associated with ScV (Tipper and Bostian, 1984), no variant or minor dsRNAs have been detected in preparations of the 190S virus. Preliminary characterization studies of the terminal sequences of the 190S-dsRNA are consistent with a single, homogeneous species of dsRNA (S. A. Ghabrial, unpublished data, 1985).

Capsid Polypeptides

SDS-polyacrylamide gel electrophoresis of the 190S virus, using the continuous buffer system of Weber and Osborn (1969), has revealed two major proteins, p88 and p83, which occurred in equimolar amounts (Sanderlin and Ghabrial, 1978; Ghabrial et al., 1987). Using the discontinuous buffer system of Laemmli (1970), however, the capsid polypeptides were resolved into three components; in addition to p88 and p83, which still occurred in approximately equimolar amounts, a third minor polypeptide, of M.W. 78,000 (p78) was revealed (Ghabrial et al., 1987; Figure 3a). Although p78 was a minor component in most preparations analyzed, in some preparations p83 was a minor component and p88 and p78 comprised the two major polypeptides (Figure 3b). In no

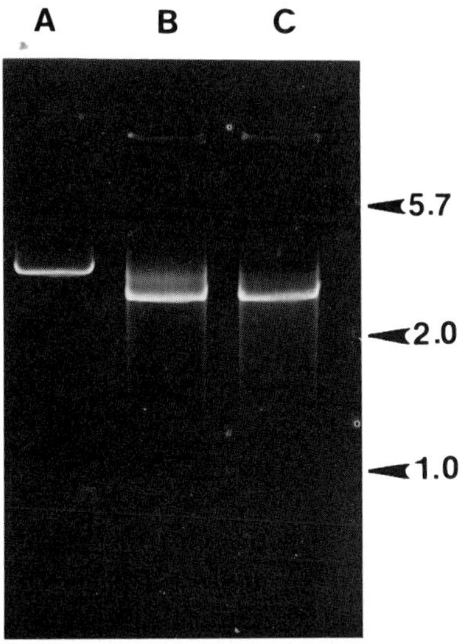

Figure 2 Polyacrylamide gel electrophoresis of dsRNA from (A) 190S virus, (B) ScV-L_A, and (C) ScV-L_{BC}. Electrophoresis was in 5% gels for 18 hr at 50 V in Tris-borate (0.089 M tris-0.089 M borate-2 mM EDTA) buffer, pH 8.0. The numbers to the right indicate M.W. ($\times 10^6$). Positions of dsRNA standards are indicated by arrows: dsRNAs from *Helminthosporium maydis* virus (5.7×10^6); *Penicillium chrysogenum* virus (2.0×10^6); *P. stoloniferum* viruses S and F (1.0×10^6).

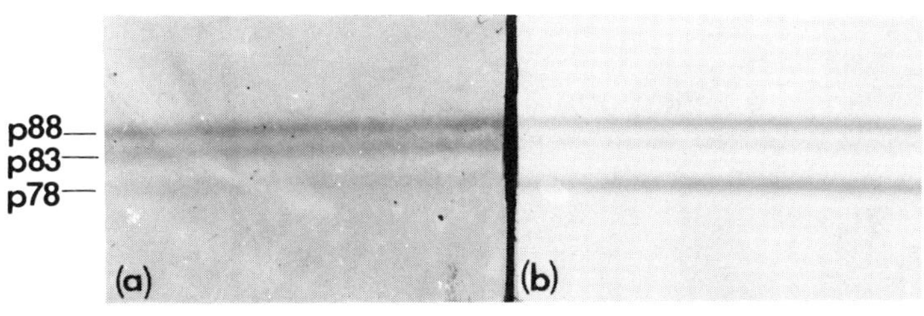

Figure 3 SDS-PAGE of the 190S virus capsid polypeptides. (a) and (b) represent two different preparations; p88 was always a major polypeptide with p83 as the second major protein in most preparations (a). In some preparations (represented in b), p78 was the second major polypeptide. Electrophoresis was carried out in preparative slab gels containing 10% polyacrylamide using Laemmli's discontinuous buffer system (Laemmli, 1970). Numbers to the left indicated M.W. ($\times 10^3$) of capsid polypeptides.

instance was p88 a minor component. Comparative peptide mapping of the capsid proteins using *Staphylococcus aureus* V8 protease and SDS-polyacrylamide gel electrophoresis (Cleveland et al., 1977), has shown p83 and p78 to be related to p88 (Ghabrial et al., 1987). The possibility that p83 and p78 may have originated from p88 by partial proteolysis during virus preparation and storage has been investigated in some detail. Such studies have concluded that neither artificial proteolysis during virus isolation nor aging of virus in vivo or in vitro could explain the occurrence of two or more related proteins in the capsids of the 190S virus.

It is possible that p83/p78 could be derived from p88 by autocatalytic or proteolytic processing prior to or during particle assembly and association with viral RNA. Alternatively, the various peptides could represent modified derivatives of a single, primary translation product, e.g., glycosylated, adenylylated, or phosphorylated forms. Reovirus polypeptides, when examined by two-dimensional gel electrophoresis, were shown to exhibit extensive charge heterogeneity (Ewing et al., 1985). This suggested that considerable posttranslational modification of reovirus proteins occurs during particle morphogenesis (Mora and Carter, 1983).

It is not clear why p78 is a minor component in some preparations of the 190S virus and a major one in others. Peptide mapping has shown p78 to be almost identical to p83 (Ghabrial et al., 1987). It is possible that p78 differs from p83 in charge rather than size and that it represents a modified form of p83. It is generally assumed, however, that charge effects are eliminated in SDS gels. On the other hand, Tung and Knight (1971) have presented evidence that modification of charge can influence electrophoretic mobility in such gels. The finding that p88 and p83 were the only capsid proteins resolved when virus preparations were analyzed using the continuous buffer system of Weber and Osborn (1969) lends some support to the "charge" hypothesis. If modified proteins are believed to play a role in virus replication, the amount of p78 detected could represent the proportion of virus particles involved in transcription and replication.

The possession of two or more closely related capsid polypeptides appears to be of common occurrence among isometric dsRNA mycoviruses with undivided genomes. In addition to the 190S virus, at least three examples can be cited: the dsRNA virus from the yeast *Yarrowia lipolytica* (El-Sherbeini et al., 1987), the 87-1-H virus from *Gaeumannomyces graminis* (Jamil and Buck, 1986), and possibly ScV-$L_{B/C}$ (Sommer and Wickner, 1982). The capsids of these four viruses are believed to be encoded by single genes (see section on in vitro translation) and, therefore, they are classified as members of the family *Totiviridae*.

Radioiodination studies of the 190S virus have been carried out to determine whether the structurally related capsid polypeptides are topologically distinct (Ghabrial et al., 1987). In these studies, intact virions were labeled with ^{125}I using Iodogen under conditions considered optimum for surface specific iodination. SDS polyacrylamide gel electrophoresis of [^{125}I]-labeled virions have shown p88 to be more heavily labeled than p83 and p78 (Figure 4). In further studies, the 190S virus was first dissociated, then iodinated. In such experiments, the intensity of label distribution was similar to the Coomassie blue staining profile of the capsid polypeptides. Radioiodination studies thus suggested that p88 is external to p83/p78 in capsid arrangement. This concept was further supported by immunological evidence. In Western blots, p88 has been found to be more reactive than p83/p78 to antibodies produced against glutaraldehyde stabilized virus (Figure 5b). However, the reactivities of the three capsid proteins to antibodies to nontreated virus were comparable to their relative abundance in the virus preparation as indicated by Coomassie blue staining pattern [Ghabrial et al., 1987; Figures 5(a) and (c)].

To explain the results of radioiodination and immunological experiments, Ghabrial et al. (1987) have proposed that the 190S virus may

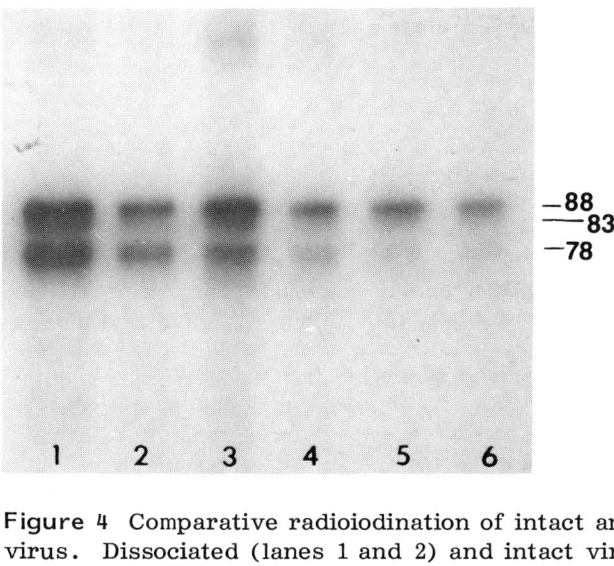

Figure 4 Comparative radioiodination of intact and dissociated 190S virus. Dissociated (lanes 1 and 2) and intact virus (lanes 3-6) were labeled with ^{125}I using Iodogen (Pierce Chemical Co., Rockford, Illinois) then analyzed by SDS-PAGE. The amount of Iodogen used in the iodination reaction and the amounts of viral proteins (in parentheses) layered per well were (1) 10 μg (20 μg); (2) 10 μg (10 μg); (3) 10 μg (20 μg); (4) 10 μg (10 μg); (5) 3 μg (20 μg); and (6) 3 μg (10 μg).

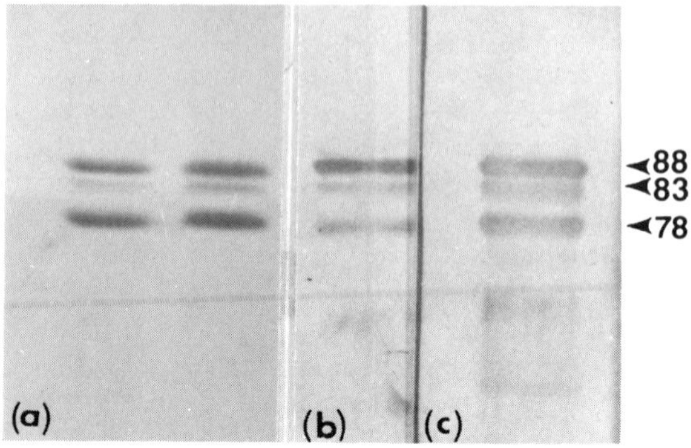

Figure 5 Comparison of the reactivities of the 190S capsid polypeptides to antibodies to glutaraldehyde-stabilized virus (190S-G) and antibodies to nontreated virus (190-NG), using Western blots. Aliquots of the same virus sample were analyzed by SDS-PAGE, transferred to nitrocellulose membranes, and reacted with antibodies to 190S-G (b) or antibodies to 190S-NG (c). The Coomassie blue staining profile for an aliquot from the same virus sample, but which was not processed for Western blotting, is shown in (a).

have double-shelled capsids with the inner core comprised of p83/p78 and the outer capsid composed of p88. This model simulates the basic structure of viruses in the Reoviridae and the distribution of related proteins between inner and outer capsids is reminiscent of the $\mu 1$ and $\mu 1C$ capsid proteins of reovirus (Joklik, 1983); $\mu 1$ is a core component, whereas its cleavage product $\mu 1C$ is the major component of the outer capsid. However, in the proposed model for the 190S virus, the outer capsid is comprised of the primary translation product. From ultrastructural studies, Ushiyama and Nakai (1982) have reported that the polyhedral particles (39 nm in dimaeter) of the dsRNA mycovirus from the Basidiomycete *Lentinus edodes* have double-shelled capsids with somewhat labile outer shells and stable inner core shells. The outer shell, but not the inner core, was digestible by chymotrypsin. The protein composition of *L. edodes* virus capsids, however, has not been determined. In preliminary experiments, we have recently found that the 190S virus particles, when incubated in Beckman B-2 barbital buffer at pH 8.6, appear to convert to smaller particles of about 33 nm in diameter (Figure 6). Clearly, more comprehensive studies are required to confirm these observations.

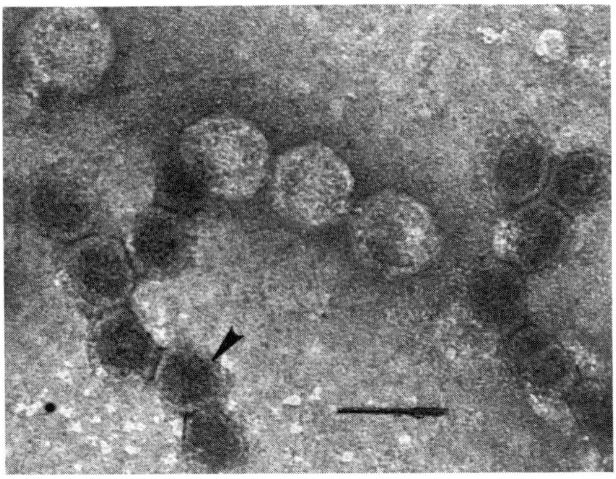

Figure 6 Electron micrograph of purified 190S virus particles (40 nm in diameter) that were dialyzed overnight against barbital buffer, pH 8.6, then stained with uranyl acetate. Note virus particles with a diameter of 33 nm (arrowhead). Bar represents 50 nm.

In Vitro Translation

In a cell-free rabbit reticulocyte translation system, 190S dsRNA, denatured by incubation with methyl mercuric hydroxide, has been found to direct the synthesis of a single major polypeptide of M.W. 88,000 (Ghabrial et al., 1987; Figure 7). The translation product has been shown to be identical to capsid polypeptide p88 by coelectrophoresis using authentic p88, specific immunoprecipitation using an antiserum to the 190S virus, and by comparative peptide mapping of native p88 and the 88K translation product (Ghabrial et al., 1987). No processing of the principal translation product p88 was detected in the in vitro system when translation products were incubated for a further 3 hr. There were no in vitro translation products of comparable size to the in vivo products p83 and p78 (Ghabrial et al., 1987; Figure 7). El-Sherbeini et al. (1987) recently reported that, when a cell-free rabbit reticulocyte lysate system was primed with denatured dsRNA from *Y. lipolytica* virus, the in vitro translation products were identi cal to the in vivo capsid polypeptides P1 (83K) and P2 (77K). These authors suggested that these two related polypeptides (P1 and P2) may represent the products of two overlapping genes which differ at the amino or carboxy terminus because of alternate translational start or stop signals. El-Sherbeini et al. (1987), however, acknowledged that the correspondence in size between in vivo and in vitro products could be fortuitous and merely represent premature termination of translation.

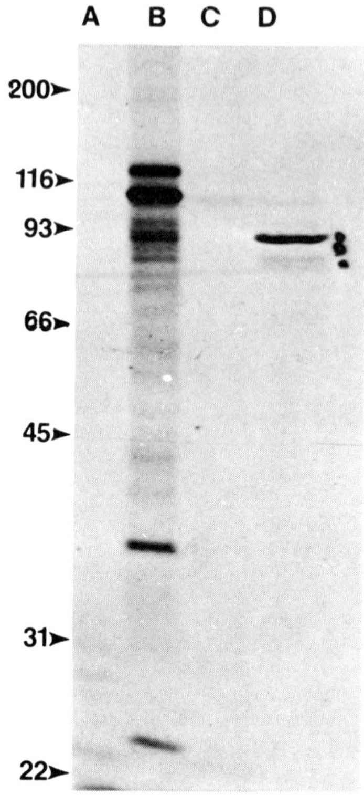

Figure 7 In vitro translation of denatured 190S-dsRNA in rabbit reticulocyte lysates. Unfractionated brome mosaic virus RNA (lane B), nondenatured (lane C) and denatured 190S-dsRNA (lane D) were translated and the products were analyzed in 10% SDS polyacrylamide gels using Laemmli's discontinuous buffer system (Laemmli, 1970). Lane (A) represents endogenous products of translation (no RNA added). Numbers to the left indicate the M.W. ($\times 10^3$) of marker proteins. The positions of 190S capsid polypeptides p88, p83, and p78, viewed by Coomassie staining, were marked with radioactive ink, and indicated to the right.

The order of cistrons in the 190S-dsRNA is proposed to be as follows: 5'-capsid polypeptide p88-RNA polymerase-3'. Similar genome organization could be deduced for ScV-L_A and possibly other members of the family *Totiviridae*. There is, however, no direct evidence that the capsid polypeptide gene for viruses in the *Totiviridae* is proximal to the 5'-terminus. The finding that the in vivo partial transcript "1_a", presumed to represent a 5' fragment of the full-length transcript "1" of ScV-L_A (Bruenn et al., 1985), had in vitro mRNA activity for capsid protein P1, supports the proposed genomic organization scheme (Bostian et al., 1983). It is not known whether the gene(s) proximal to the 3'-terminus do indeed encode for the RNA polymerase of viruses in the *Totiviridae*. The expression of these genes appear to be mediated through the generation of subgenomic mRNAs. This could be brought about by nuclease processing of the full-length transcript. The detection in vivo of "1_a", the partial transcript of ScV-L_A dsRNA, is consistent with this strategy.

In Vitro Transcription

The 190S virus of *H. victoriae*, like all dsRNA viruses, possesses virion-associated RNA polymerase activity (Ghabrial and Havens, 1986). In in vitro studies, the 190S-associated RNA polymerase has been found to catalyze the synthesis and release of full-length, single-stranded RNA transcripts. The transcriptase activity was stoichiometrically dependent on virus concentration in the range of 50-250 μg/ml. RNA synthesis was insensitive to actinomycin D and rifampin, but required all four nucleoside triphosphates and magnesium ions. The optimum concentration of Mg^{2+} has been determined as 5 mM and the pH optimum for the transcriptase activity has been shown to be 8.4.

Characterization of the size and strandedness of the products of the 190S-associated RNA polymerase activity has been performed in reactions containing either [^{3}H]- or [^{32}P-α]UTP. Over 98% of label has been shown to be incorporated in the ssRNA fraction when the reaction products were analyzed by selective precipitation of ssRNA in 2 M LiCl, CF-11 cellulose chromatography, or when sensitivity of the products to RNase in low or high salt was tested. When annealed with increasing amounts of denatured, nonlabeled 190S-dsRNA, up to 75% of the [^{3}H]-labeled ssRNA product was found to hybridize to denatured template dsRNA in increasing proportions. It could be deduced that only a single strand of the parental duplex is transcribed because negligible hybridization was detected when [^{3}H]ssRNA was incubated under annealing conditions without denatured genomic dsRNA. By analogy to the ScV-L_A system, the ssRNA transcript is expected to have mRNA activity for the capsid polypeptide. This, however, needs to be confirmed by translating the transcript in a cell-free system.

Electrophoretic analyses of the 190S-RNA polymerase-labeled products have been carried out in 4% polyacrylamide gels containing 8 M urea. A major ssRNA product of M.W. of about 1.5×10^6 has been detected by autoradiography (Figure 8). This has been considered to represent a full-length transcript of parental dsRNA (MW. 3.0×10^6) because it comigrated with denatured 190S-dsRNA (denaturation by incubation in 90% dimethyl sulfoxide for 15 min at 50°C) and was

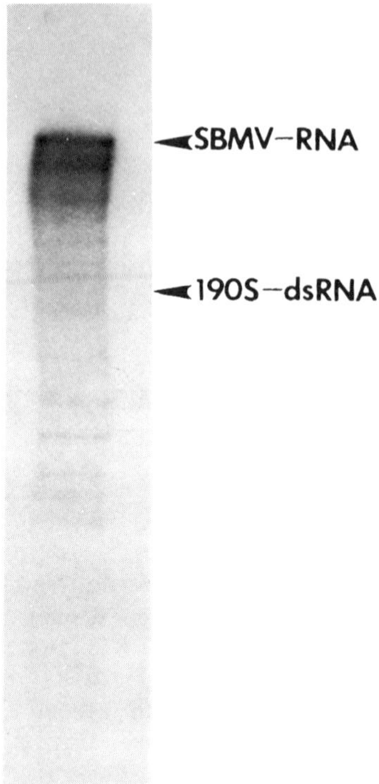

Figure 8 Polyacrylamide gel electrophoresis analysis of the products of 190S-RNA polymerase. RNA was extracted from a 4-hr reaction mixture and subjected to electrophoresis in a 4% polyacrylamide gel containing 8 M urea at 100 V for 14 hr in TBE buffer, pH 8.0 (see legend for Figure 2). The positions of southern bean mosaic virus (SBMV)-RNA and nondenatured 190S-dsRNA, viewed with ethidium bromide staining, were marked with radioactive ink and indicated with arrows.

slightly slower than the ssRNA from southern bean mosaic virus (molecular weigh about 1.4×10^6). It is not clear whether the minor bands shown in Figure 8 represent partial transcripts, degradation products, or pause products in the synthesis of full-length molecules. All labeled RNA products have been shown to be single-stranded, since no label was detected by autoradiography when the reaction products were first incubated with RNase in either high or low salt, then subjected to electrophoresis. No incorporation of label has been detected in dsRNA (Figure 8).

Evidence that the in vitro ssRNA transcripts are released from the virus particles has been based on electron microscopy of transcribing particles and on time course analysis of total reaction mixture by agarose-gel electrophoresis. In the latter experiment, ssRNA, representing full-length transcripts, was first detected after 30 min of incubation (S. A. Ghabrial, unpublished data, 1986).

It could be deduced that in vitro transcription of the 190S dsRNA occurs conservatively because, in the presence of [^{32}P-α]UTP, parental duplex remains unlabeled and the labeled full-length ssRNA transcrips are released from the particles. This is consistent with conservative transcription, although semiconservative transcription by a very small but highly active virus subpopulation cannot be ruled out on this basis alone. It may also be possible to demonstrate in vitro conservative replication for the 190S-dsRNA, entailing asynchronous synthesis of the minus-strand, as typified by reovirus (Joklik, 1983). Our inability to detect incorporation of label into dsRNA may be because the virions used in these experiments were derived from the main 190S virus peak in sucrose density gradients containing full-length dsRNA molecules, which appear to possess transcriptase rather than replicase activity. Convincing evidence for in vitro conservative transcription and replication for ScV-L_A has recently been reported (Nemeroff and Bruenn, 1986; Fujimura et al., 1986). Future investigations will probably prove conservative replication to be the modus operandi for viruses in the *Totiviridae*.

CONCLUDING REMARKS

The paucity of fungal phenotypes that can be directly and reliably related to virus infection makes the study of the *H. victoriae*/virus system attractive. It also mandates that the evidence for viral etiology of the disease of *H. victoriae* be conclusive. This evidence can best be obtained by infectivity assays using purified virus and virus-free fungal protoplasts (Ghabrial, 1980). One major problem in experiments involving extracellular transmission of *H. victoriae* viruses has been the inability of newly infected fungal colonies to support virus replication to levels that allow virus reisolation and/or unambiguous virus deteciton and characterization (Ghabrial and Mernaugh, 1983; Ghabrial, 1986). Even though there is much evidence to support a viral etiology of the disease of *H. victoriae*, it remains equivocal, since satisfaction of Koch's postulates has not yet been completed. Future investigations of experimental transmission of *H. victoriae* viruses will rely on a recently acquired virus-free fungal isolate (Hv-MN), which has been shown to be susceptible to virus infection, via hyphal anastomosis (S. A. Ghabrial, unpublished data, 1986). Future studies will take advantage of some recent technological advances, including the use of electric pulses or of lithium cations to induce the intake of nucleic acids or virus particles by fungal cells (Potter et al., 1984; Dhwale et al., 1984), and the use of cloned viral genes in developing fungal transformation systems (Bussey and Meaden, 1985). For details on new developments in methodology of extracellular transmission of the yeast viruses, see Chapter 9.

Recent studies on *H. victoriae* viruses have focused on the 190S virus (Ghabrial et al., 1987). These investigations have revealed some interesting structural features that may be common to other mycoviruses with undivided dsRNA genomes. The occurrence of two or more closely related polypeptides in the capsids of the 190S virus and some other members of the family *Totiviridae* may reflect functional heterogeneity. Shortening and/or chemical modification of the primary translation product of these viruses, i.e., the capsid polypeptide, may serve to expand and modulate its properties and functions. Maintenance of the viral RNA in the appropriate conformation for transcription and replication, switch-on of transcriptase activity, and release of RNA transcripts are some of the functions proposed for the capsid polypeptide and its cleaved or modified forms. Although no evidence has been presented for glycosylation, adenylylation, or phosphorylation of the capsid polypeptide of the 190S virus or any other member of the *Totiviridae*, oligonucleotides representing pause products of virion-associated dsRNA transcriptase of ScV-L_A have been isolated and characterized (Brennan et al., 1981). Oligoadenylates have also been reported in particles of *Allomyces arbuscula* virus (Khandjian and Turian, 1977); oligo(A) synthesis presumably represents an alternative activity of capsid-associated RNA polymerase of this virus. It is conceivable that these oligonucleotides play a role in phosphorylation or adenylylation of the capsid polypeptides. This, however, will require the occurrence of virally specified modifying enzymes. Assignment of multiple functions to the capsid polypeptide of viruses in the *Totiviridae* may be justified, considering that the gene for the coat protein accounts for as much as half of the coding capacity of the viral genome.

The results of in vitro transcription of the 190S dsRNA, namely the synthesis and release of ssRNA transcripts without displacement of the components of the parental duplex, are consistent with a conservative mode of replication analogous to that found for reovirus (Joklik, 1983). The similarity of the 190S to reovirus may extend to the basic structure of the virus particle since radioiodination and immunological studies (Ghabrial et al., 1987), have suggested that the 190S virus may have double-shelled capsids, unlike most other fungal viruses. Comprehensive ultrastructural studies, however, are needed to verify the proposed structure.

ACKNOWLEDGMENTS

I am grateful to Dr. T. P. Pirone for reading the manuscript and offering many critical suggestions and to Mrs. Wendy Havens for preparation of illustrations.

REFERENCES

Bostian, K. A., Burn, V. E., Jayachandran, S., and Tipper, T. J. (1983). Yeast killer dsRNA plasmids are transcribed *in vivo* to produce full and partial-length plus stranded RNAs. *Nucl. Acids Res. 11*:1077-1097.

Brennan, V. E., Bobek, L. A., and Bruenn, J. A. (1981). Yeast dsRNA viral transcriptase pause products: Identification of the transcript strand. *Nucl. Acids Res. 9*:5049-5059.

Brown, R. (1986). The classification and nomenclature of viruses: Summary of results of meetings of the International Committee on Taxonomy of Viruses in Sendai, September 1984, *Intervirol. 25*: 141-143.

Bruenn, J. A. (1980). Virus-like particles of yeast. *Annu Rev. Microbiol. 34*:49-68.

Bruenn, J., Madura, K., Siegel, A., Miner, Z., and Lee, M. (1985). Long internal inverted repeat in a yeast viral double-stranded RNA. *Nucl. Acids Res. 13*:1575-1591.

Bussey, H., and Meaden, P. (1985). Selection and stability of yeast transformants expressing cDNA of a M1 killer toxin-immunity gene. *Curr. Genet. 9*:285-291.

Cleveland, D. W., Fisher, S. G., Kirschner, M. W., and Laemmli, U. K. (1977). Peptide mapping by limited proteolysis in sodium dodecyl sulfate and analysis by gel electrophoresis. *J. Biol. Chem. 252*:1102-1106.

Dhwale, S. S., Paietta, J. V., and Marzluf, G. A. (1984). A new, rapid and efficient transformation procedure for Neurospora. *Curr. Genet. 8*:77-79.

El-Sherbeini, M., Bostian, K. A., LeVitre, J., and Mitchell, D. S. (1987). Gene-protein assignments within the yeast *Yarrowia lipolytica* dsRNA viral genome. *Curr. Gen. 11*:483-490.

Ewing, D. D., Sargent, M. D., and Borsa, J. (1985). Switch-on of transcriptase function in reovirus: Analysis of polypeptide changes using 2-D gels. *Virology 144*:448-456.

Fujimura, T., Esteban, R., and Wickner, R. B. (1986). *In vitro* L-A double-stranded RNA synthesis in virus-like particles from *Saccharomyces cerevisiae*. *Proc. Nat. Acad. Sci. USA 83*:4433-4437.

Ghabrial, S. A. (1980). Effects of fungal viruses on their hosts. *Annu. Rev. Phytopath. 18*:441-461.

Ghabrial, S. A. (1986). A transmissible disease of *Helminthosporium victoriae*: Evidence for a viral etiology. In *Fungal Virology*. K. W. Buck (Ed.). CRC Press, Boca Ration, Florida, pp. 163-176.

Ghabrial, S. A., and Havens, W. M. (1986). Transcription of double-stranded RNA in virions of *Helminthosporium victoriae* virus. *Phytopathology 76*:1090.

Ghabrial, S. A., and Mernaugh, R. L. (1983). Biology and transmission of *Helminthosporium victoriae* mycoviruses. In *Double-stranded RNA Viruses*, R. W. Compans and D. H. L. Bishop (Eds.). Elsevier, New York, pp. 441-449.

Ghabrial, S. A., Sanderlin, R. S., and Calvert, L. A. (1979). Morphology and virus-like particle content of *Helminthosporium victoriae* colonies regenerated from protoplasts of normal and diseased isolates. *Phytopathology 69*:312-315.

Ghabrial, S. A., Bibb, J. A., Price, K. H., Havens, W. M., and Lesnaw, J. A. (1987). The capsid polypeptides of the 190S virus of *Helminthosporium victoriae*. *J. Gen. Virol. 68*:1791-1800.

Jamil, N., and Buck, K. W. (1986). Capsid polypeptides in a group III virus from *Gaeumannomyces graminis* var. *tritici* are related. *J. Gen. Virol. 67*:1717-1720.

Joklik, W. K. (1983). *The Reoviridae*, W. K. Joklik (Ed.). Plenum, New York and London. 571 p.

Khandjian, E. W., and Turian, G. (1977). Adenine-rich RNA in the mycovirus of *Allomyces arbuscula*. *J. Gen. Virol. 36*:195-197.

Laemmli, U. K. (1970). Cleavage of structural proteins during the assembly of the head of bacteriophage T4. *Nature 227*:680-685.

Lindberg, G. D. (1959). A transmissible disease of *Helminthosporium victoriae*. *Phytopathology 49*:29-31.

Lindberg, G. D. (1960). Reduction in pathogenicity and toxin production in diseased *Helminthosporium victoriae*. *Phytopathology 50*:457-460.

Lindberg, G. D. (1964). Disease-induced antibiotic production in *Helminthosporium victoriae*. *Phytopathology 54*:898-899.

Litzenberger, S. C. (1949). Nature of susceptibility to *Helminthosporium victoriae* and resistance to *Puccinia coronata* in Victoria oats. *Phytopathology 39*:300-318.

Litzenberger, S. C., and Murphy, H. C. (1947). Methods for determining resistance of oats to *Helminthosporium victoriae*. *Phytopathology 37*:790-800.

Luke, H. H., and Wheeler, H. E. (1955). Toxin production by *Helminthosporium victoriae*. *Phytopathology 45*:453-458.

Meehan, F., and Murphy, H. C. (1946). A new *Helminthosporium* blight of oats. *Science 104*:413-414.

Mora, M., and Carter, C. A. (1983). The role of phosphoproteins in reovirus morphogenesis. In *Double-stranded RNA Viruses*, R. W. Compans and D. H. L. Bishop (Eds.). Elsevier, New York, pp. 207-211.

Nemeroff, M. E., and Bruenn, J. A. (1986). Conservative replication and transcription of *Saccharomyces cerevisiae* viral double-stranded RNA *in vitro*. *J. Virol. 57*:754-758.

Potter, H., Weir, L., and Leder, P. (1984). Enhancer-dependent expression of human K immunoglobulin genes introduced into mouse pre-B lymphocytes by electroporation. *Proc. Nat. Acad. Sci. USA 81*:7161-7165.

Sanderlin, R. S., and Ghabrial, S. A. (1978). Physicochemical properties of two distinct types of virus-like particles from *Helminthosporium victoriae*. *Virology 87*:142-151.

Summer, S. S., and Wickner, R. B. (1982). Yeast L dsRNA consists of at least three distinct RNAs: Evidence that the non-Mendelian genes [HOK], [NEX] and [EXL] are on one of these dsRNAs. *Cell 31*:429-441.

Tipper, D. J., and Bostian, K. A. (1984). Double-stranded ribonucleic acid killer systems in yeasts. *Microbiol. Rev. 48*:125-156.

Tung, J. S., and Knight, C. A. (1971). Effect of charge on the determination of molecular weight of proteins by gel electrophoresis in SDS. *Biochem. Biophys. Res. Comm. 42*:1117-1121.

Ushiyama, R., and Nakai, Y. (1982). Ultrastructural features of fungal virus-like particles from *Lentinus edodes*. *Virology 123*: 93-101.

Weber, K., and Osborn, M. (1969). The reliability of molecular weight determination by dodecyl sulfate-polyacrylamide gel electorphoresis. *J. Biol. Chem. 244*:4406-4412.

17

Viruses of *Endothia parasitica*

Neal K. Van Alfen
Utah State University, Logan, Utah

INTRODUCTION

Viruses in the filamentous fungus *Endothia* (*Cryphonectria*) *parasitica* cause a reduction in virulence (hypovirulence) of this fungal plant pathogen. The discovery of hypovirulence of *E. parasitica* was the first known example of a transmissible agent responsible for reducing the virulence of a pathogen. Virulence control by transmissible agents has been reported in a number of bacterial systems (Kosuge and Nester, 1984). In all cases, the agent responsible for controlling virulence causes an increase in virulence expression by the pathogen. One can advance arguments as to how transmissible agents which increase virulence could provide a selective advantage to populations of pathogenic organisms. It is more difficult to offer agruments concerning the advantage to the pathogen of transmissible hypovirulence. In this regard, transmissible hypovirulence caused by viral dsRNA is unique. At our current level of understanding, we must consider this dsRNA as a disease of the fungus that is detrimental to its host, unlike the bacterial systems of virulence control and the killer systems of yeast, each of which can benefit their hosts.

The question of whether a fungal virus benefits or damages its host is not trivial. In the case of transmissible hypovirulence it is a critical question. The success of the dsRNA as a biological control agent, over periods of time measured in generations of trees, may be determined by whether the dsRNA is detrimental enough to act as a selective pressure in the fungal population. Field observations suggest that hypovirulence may be stable within the fungal population (Grente and Berthelay-Sauret, 1978; Mittembergher, 1978). An understanding of how the dsRNA affects *E. parasitica* may provide clues as to whether the relationship is unique and highly evolved or one that is relatively primitive, i.e., the dsRNA makes the fungus "sick" merely by its presence. The former relationship would likely be more stable.

One of the fascinating aspects of transmissible hypovirulence is that it appears to be caused by a number of different, nonhomologous viruses. Each of the viruses cause specific symptoms which indicate its presence in the fungus. These viruses produce a plethora of morphological symptoms, but in almost all cases the viruses reduce virulence expression and sporulation by the fungus. The investigation of how these viruses interact with the fungus to affect virulence expression may lead to a better understanding of how viruses are able to affect their hosts adversely, as well as to help us better understand the nature of virulence expression by fungal pathogens.

THE FUNGUS *Endothia parasitica*

This fungus causes a canker disease of chestnut and oak trees. In some species of these trees the canker is limited in size by host defense reactions, but in the American chestnut (*Castanea dentata*) the fungus is not stopped in its penetration of host periderm, phloem, and cambium (Bramble, 1936). Eventually the pathogen girdles the tree. In the case of the American chestnut, total lack of resistance to this disease resulted in the elimination of the species from its former dominant position in the deciduous forests of the eastern United States (Hepting, 1974). The relative resistance of Oriental species of *Castanea* suggests that the pathogen evolved in the Orient (Galloway, 1926). The cankers caused on Oriental chestnuts tend to be small annual, or slowly spreading perennial, cankers. In its relationship with Oriental chestnuts, the fungus is able to invade sufficient host tissue to allow sporulation (Hebard et al., 1984), but the spread of canker diameters is slow enough that it is not a serious disease under normal conditions. The destruction caused by *E. parasitica* after being introduced from the Orient to North America is a classic example of the defenselessness of hosts that have not evolved with their pathogens.

E. parasitica is an ascomycete that forms two spore stages. Pycnidia produce conidia soon after a canker is established. The orange-colored pycnidia are produced in a stroma visible on the bark of infected trees. The conidia produced are single-celled and generally contain a single nucleus. They are contained in a gelatinous matrix that oozes from the ostiole of the pycnidium (Anderson, 1913; Shear and Stevens, 1917). These spores are readily dispersed in water and may also be spread by insects, mites, and animals that visit the cankers (Anagnostakis, 1982; DePalma, 1981; Wendt et al., 1982). The pycnidia form readily in culture on most solid media. They appear to require light for formation, but this has not been confirmed by careful study.

Perithecia are formed following mating. There are two mating types known: A and a (Anagnostakis, 1982). Ascogenous nuclei may at times come from fusion of one female and at least two male nuclei (Anagnostakis, 1982). Sexual crosses are not yet possible on artificial media, but sterilized chestnut wood can be used (Anagnostakis, 1979). Perithecia form beneath pycnidia about six weeks after crosses are made on sterile wood in culture. A long black perithecial neck protrudes above the surface of the stroma amid the pycnidia, indicating the presence and location of the embedded perithecia (Figure 1). Although quite small, the perithecia can be dissected out of the stroma for isolation of single ascospores (Puhalla and Anagnostakis, 1971). The perithecia can be recognized by their spherical shape and gray walls.

The anastomosis of hyphae occurs readily and frequently in culture. This is particularly true of self-anastomosis, but it also occurs between different strains in some cases. The control of anastomosis has been found to be typical of many filamentous fungi (Fincham et al., 1979). Vegetative compatibility genes determine whether such anastomosis occurs in a stable manner.

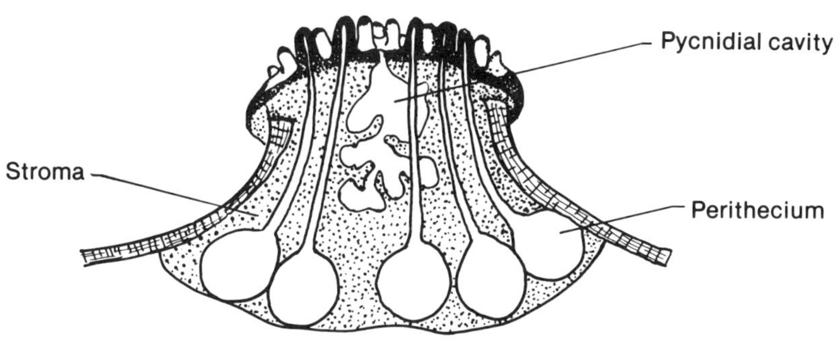

Figure 1 Diagrammatic illustration of a stroma, containing perithecia and pycnidial cavity. Redrawn from Anderson (1913).

Vegetative compatibility is detected in *E. parasitica* by the lack of a visible reaction or barrage line formation when two strains meet on solid media (Anagnostakis, 1977). Such compatible strains are considered to be in the same vegetative compatibility (v-c) group. When two strains of different v-c groups meet, growth stops where they meet. At the contact point between the strains, two parallel lines of dense hyphae form with a clear space between (barrage). This visible reaction occurs because incompatible strains do not form a stable anastomosis. In *E. parasitica* there have been no studies of the cytology of anatomosis, but if it is similar to that of other examples of fungal incompatibility studied (Alymore and Todd, 1986), anastomosis occurs, followed by death of the interconnecting hyphal segments. The dsRNA of *E. parasitica* can move at detectable frequencies between incompatible strains (Anagnostakis and Day, 1979; Anagnostakis and Waggoner, 1981). This movement suggests that intact cytoplasmic bridges remain for at least a short time following anastomosis between strains of different v-c groups in *E. parasitica*.

VIRULENCE EXPRESSION BY *E. parasitica*

Virulence is a quantitative assessment of a pathogen's ability to cause disease. In the case of a canker pathogen such as *E. parasitica*, virulence can be assessed simply on its relative ability to grow in host tissue. However, if the entire disease cycle of the pathogen is considered, virulence assessments can involve quantitative measurements of functions such as sporulation, dormancy of spores, and spore germination. Quantitative assessment of virulence of *E. parasitica* is usually based upon canker diameter or area measured after about three months of growth in a chestnut tree. This method has been extended by Elliston (1985) to include canker area and sporulation measurements after a second season of growth. Such long-term virulence studies, however are not practical for routine virulence tests. In addressing this problem, J. E. Elliston (unpublished data) also showed that apple fruits can be used as a substrate to measure relative virulence of *E. parasitica*. The advantage of this method is that, unlike chestnut trees, apple fruits are universally available, and differences in virulence are evident in a few weeks rather than months. Fulbright (1984) confirmed that this method of virulence assessment can be used routinely in place of that utilizing chestnut trees. The apple method is most useful if strains to be compared are at the extremes of virulence expression.

MECHANISMS OF VIRULENCE EXPRESSION

Little is known of how *E. parasitica* is able to attack and colonize chestnut and oak trees. Plant pathologists have described two essentially

different approaches used by pathogens in extracting nutrients from their hosts. One approach, generally felt to be the more primitive, is for the pathogen to secrete substances that kill plant cells in advance of mycelial growth. This type of pathogen has been termed a necrotroph. The other approach, extracting nutrients from living organisms, is exhibited by the biotrophs. These organisms generally are highly specialized pathogens that do not immediately kill the cells upon which they feed. Rather, they either penetrate the cell by means of haustoria, or they may make the host membranes leaky so that nutrients become available to the intercellular hyphae. *Endothia parasitica* is a necrotroph.

The host range of *E. parasitica* is limited, but it is not completely specific to a single host (Griffin et al., 1986). When chestnut blight was spreading through the forests of Eastern United States, there must have been an enormous amount of inoculum being produced on the large number of cankered trees. If the fungus had hosts other than chestnuts, chinquapins, and some species of oaks, it would have been evident under the very high inoculum pressure present in the forests at that time. Thus, even though the fungus is able to cause cankers on unrelated trees such as chinquapins and oaks, it is quite limited in its host range.

The success of *E. parasitica* in killing American chestnut trees so quickly is a reflection of the lack of effective defenses in the host: *E. parasitica* is not a "super" pathogen. On Oriental chestnuts and on oak it would be classed as a "weak" pathogen. It is thus important in considering pathogen virulence to be clear about what is occurring. In the case of the interaction of *E. parasitica* with *C. dentata*, once a wound has been invaded by the fungus, the resistance expressed by the host is ineffective in controlling further spread by the pathogen. The difference in the outcome between *E. parasitica* on Oriental and American chestnuts is due to unidentified resistance factors in Oriental chestnuts. Hypovirulence of *E. parasitica* is thus best detected and studied in a susceptible host such as American chestnut.

E. parasitica lacks the ability to penetrate bark tissue directly. This requirement for wounding is not unusual in canker pathogens and probably reflects the inability of germ tubes or individual hyphae to grow through bark. Colonization of the wound is the first step in the establishment of the pathogen on a new host. Once colonization of the wound has occurred, the pathogen is able to attack the bark (Bramble, 1936; Hebard et al., 1984).

Although much less is known of the ecology of wounds than that of leaf or root surfaces (Foster, 1986; Windels and Lindow, 1985), it is probably reasonable to assume that pathogens are not the only, nor the first, colonizers of wounded tissue. There are nutrients from the broken or otherwise injured cells that are available to various microorganisms. Wounding initiates a plant response (Shigo, 1984). Various names have been given to the plant tissue that forms in response to wounding. Necrophylactic periderm (NP) is the most commonly accepted term to describe the tissue formed to protect living tissue from water loss and other adverse effects of wounding (Biggs, Merrill and Davis, 1984). *E. parasitica* does not appear to invade quickly through the NP into living tissue. Its ability to invade is thus not based upon overwhelming host defenses. Rather, it first colonizes the wound. Between two and three weeks after being inoculated into wounds (Hebard et al., 1984) specific hyphae become organized into structures called mycelial fans. It is these juggernaut-like invasion structures that are able to invade the living tissues of the host successfully (Bramble, 1936). By the time that the mycelial fans form, the wound is enclosed by NP. The pathogen must therefore be able to break through this lignified tissue prior to general invasion of other bark tissues.

Hebard et al. (1984) compared the ability of virulent and hypovirulent strains to colonize and invade the susceptible *C. dentata* and the highly resistant *Castanea mollissima*. They found that the primary difference between virulent and hypovirulent strains was that virulent strains produced more and larger mycelial fans than did hypovirulent strains. Figure 2 shows a typical mycelial fan that forms under the bark of a chestnut tree. Figure 3(a) is a micrograph showing how these fans are organized. Little is known of how mycelial fans promote the pathogen's attack. The structures are composed of a mass of interconnected hyphae that grow from the tip of the fan in a synchronous manner (Figure 3). The suppression of branching, the coordinated growth, and the common direction of growth indicate that mycelial fans result from specific morphogenic regulation. Figure 3(b) indicates that the mycelial fan is not just a physical wedge being driven by growth through host tissue. The host cells die before the fans reach them. Death of the host cells may be the result of a host defensive response, but may also indicate that the fan secretes substances that are toxic to the host cells in advance of its growth. This is a question that will require careful study to resolve.

Likely candidates for secretion products have been proposed by a number of workers. As with other plant pathogens, toxins and cell wall-degrading enzymes have been detected in culture fluids of *E. parasitica*. The first toxic substances to be detected were described by Gäumann and Naef-Roth (1957). One of these nonhost specific toxins was given the trivial name diaporthin. It was reported that a wide variety of organisms, including bacteria, are affected by low concentrations of this toxin. Another compound from culture fluids found to be toxic toward plant cells is the pigment skyrin (Gäumann et al., 1959).

Studies of the role of diaporthin in host-pathogen relations were reported by Grente and Berthelay-Sauret (1978). They found that diaporthin is produced in much lower amounts in vitro by hypovirulent

Figure 2 Photograph of mycelial fan of *Endothia parasitica* growing under the bark of a large chestnut tree (Anderson, 1913).

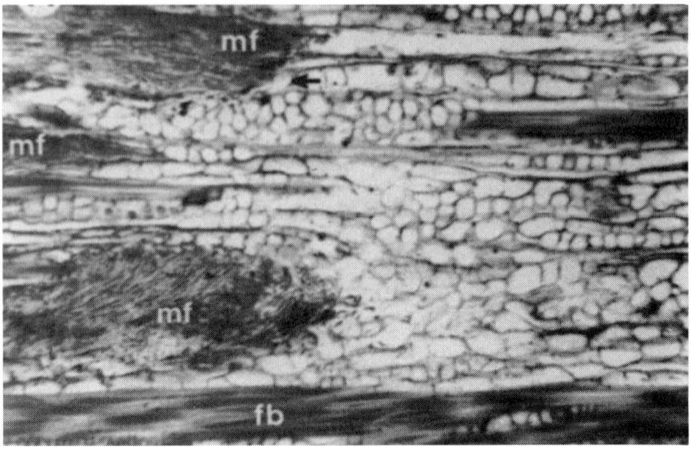

(a)

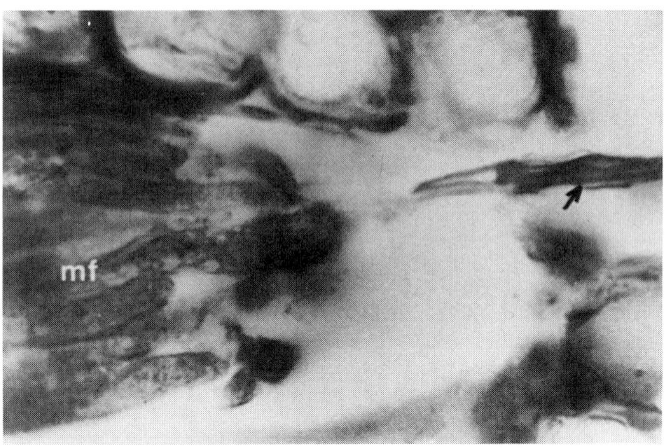

(b)

Figure 3 (a) Mycelial fans (mf) of *Endothia parasitica* 23 days after inoculation into *Castanea dentata*. Fiber bundles (fb) are also shown. (b) Tip of a mycelial fan 21 days after inoculation into *C. dentata*. The arrow points to apparent swelling of plant cell middle lamella in advance of mycelial fan growth (Hebard et al., 1984).

strains than by virulent ones. By exogenously adding diaporthin to cankers formed by hypovirulent strains, they were able to prolong the time that the fungus is able to colonize the wounds. They reported that diaporthin interferes with wound healing. McCarroll and Thor (1985b) questioned whether diaporthin and skyrin are important virulence factors in this disease. They were unable to detect these compounds except in necrotic or sporulating fungal tissues. If they are important in virulence they should be present at the edge of mycelial fans. Neither diaporthin, nor the pigments skyrin and ruglosin, were detected at the leading edges of the hyphae in infected tissue (McCarroll and Thor, 1986b).

Oxalate, an organic acid that has been implicated as a virulence factor in a number of host-pathogen systems, is produced in much lower quantities by hypovirulent strains than by virulent ones (Havir and Anagnostakis, 1983,1985). Oxalate may be locally toxic by reducing the pH to a level that is injurious. It may also play a role in chelation of calcium, which in turn may make plant cell walls more susceptible to degradation by fungal-produced extracellular pectinases (McCarroll and Thor, 1985a).

The effect on the fungus of infection of the hypovirulence-causing viruses is pleiomorphic: virulence, sporulation, and colony morphology are all clearly affected. The physiological basis for hypovirulence is probably not simply the lack of production of a single toxin or enzyme by hypovirulent strains. It is not likely that all of these effects result from lack of production of a single molecule by hypovirulent strains. It is much more likely that a multitude of physiological differences between virulent and hypovirulent strains will be found.

NATURE OF THE VIRUSES

The first suggestion that there was a virus in hypovirulent strains was the report by Moffitt and Lister (1975) that antiserum to dsRNA indicated the presence of dsRNA in two hypovirulent but not in two virulent strains of the fungus. They attempted to confirm this observation by isolation of dsRNA or viral particles, but were unsuccessful. Day et al. (1977) succeeded in isolating dsRNA from 13 hypovirulent strains but not from 15 virulent strains of *E. parasitica*. They found that multiple segments of dsRNA were present in hypovirulent strains and that there was variation in the dsRNA segment sizes and numbers among the different hypovirulent strains. Since these reports, evidence has continued to accumulate that dsRNA is the cause of hypovirulence. The correlation between dsRNA and hypovirulence has been demonstrated by transfer of dsRNA strain-to-strain (Anagnostakis and Day, 1979) as well as by curing strains of dsRNA (Fulbright, 1984). In each type of experiment, hypovirulence is expressed only when dsRNA is detectable in the strain. To date, purified dsRNA, or cell-free fractions containing dsRNA, have not been found to be infective, so direct evidence of the role of dsRNA in hypovirulence is lacking. The correlation between dsRNA and hypovirulence, however, is so clear that, as with killer of yeast, the question of whether dsRNA causes hypovirulence has become moot. In the yeast system, infection by pure virus has recently been achieved (Sturley et al., this volume).

Recent work suggests that more than one virus is able to cause hypovirulence. This conclusion was reached by several groups using genome hybridization data (L'Hostis et al., 1985). A similar conclusion was reached by Elliston (1985) from experiments in which he was able to segregate two different hypovirulent phenotypes from a single North American strain by culturing single conidial isolates. These two phenotypes were each associated with distinctive sets of dsRNA segments. Studies such as those described by Tartaglia and Nuss in this volume clearly demonstrated the differences between dsRNA strains. Although a number of strains lacking sequence homology have been studied, it appears that the genome organization of these viruses is similar.

To date, no viral capsid has been detected in any hypovirulent strain. In strain EP 113, naked (not encapsidated) dsRNA has been found to be associated with membrane vesicles (Hansen et al., 1985). Vesicles that are similar in size and composition have been isolated from virulent strains, but these lack dsRNA. This suggests that the vesicles have a function in normal cells. Significant amounts of polysaccharide are found in vesicles from both virulent and hypovirulent strains. These polysaccharides are composed of the same sugars as the fungal cell wall, suggesting that these vesicles may function in cell wall synthesis (Hansen et al., 1985).

The vesicles from hypovirulent strains contain an RNA polymerase that is not present in the vesicles of the virulent strain. The products of this polymerase are both single-stranded and double-stranded RNA, with greater than 70% of the product being single-stranded. The dsRNA product is of the same size as the segments serving as templates (D. R. Hansen and N. K. Van Alfen, unpublished data).

HOW DOES THE VIRUS CAUSE HYPOVIRULENCE?

The phenotypes or symptoms caused by the viruses are not simple ones, in this context meaning that the fungus is not obviously sick. As indicated before, significant viral-induced changes occur in colony morphology, sporulation, and virulence. Death is not one of the symptoms. It is very unusual for hypovirulent isolates to die in culture. The effects of the virus on the fungus are specific and are a function of which virus strain is present. Such specificity of symptoms suggests that the virus affects the fungus in limited and unique ways rather than merely making the fungus sick. This view is supported by the facts that (1) mutants of the fungus have been isolated which mimic much of the phenotype of the virus, and (2) the virus has been found to affect accumulation of specific polyA RNAs produced by virulent strains of the fungus.

MUTANTS THAT MIMIC HYPOVIRULENCE

If the virus affects virulence regulation in a specific manner it is reasonable to assume that mutants can be isolated that have the same effect. This would assume that nonspecific effects of viral replication on fungal health would be minimal. Seeking such mutants is in fact a powerful tool to analyze how the virus affects fungal virulence.

In working with the fungus we have frequently observed mutants that lack pigment and otherwise look in culture like strains containing dsRNA. Such mutant strains show the symptoms that Grente and Sauret (1969) described as type B hypovirulence. These mutants do not contain dsRNA and were derived from strains which, to our knowledge, have never contained dsRNA. This phenotype is due to a pleiotrophic mutation causing a substantial reduction in formation of pycnidia, reduced production of the fungal pigments, and greatly reduced virulence. The mutation has been mapped both by sexual crosses and complementation studies to a single nuclear locus. Mutants at this locus (*Pig*) are indistinguishable from type B hypovirulence in every way, except they do not contain dsRNA (N. K. Van Alfen, unpublished data).

Grente and Sauret (1969) described another hypovirulent strain type which they called JR. This strain was different from type B in that it was stable (type B could yield on subculture either type B, JR, or a virulent revertant). JR was described as being pigmented and producing conidia from conidiophores on hyphae rather than in pycnidia. Anagnostakis (1984) showed that JR is caused by a single-locus nuclear mutation and not by dsRNA as previously thought. The mutation (*flat*) is hypovirulent, with the JR type phenotype, and its morphological phenotype is dominant to dsRNA-caused type B hypovirulence when dsRNA is present. The JR type of hypovirulence was thus shown to be caused by a nuclear mutation that occurred very frequently, particularly when dsRNA is present in the cell (Anagnostakis, 1984).

Both of these mutants show that pleiotropic phenotypes such as reduced sporulation and virulence can be affected by single locus mutations. The differences in the mutant phenotypes, as well as genetic evidence, suggest that these mutations are at different loci (N. K.

Van Alfen, unpublished data), yet each apparently is involved in regulating sporulation and virulence. Perhaps the virus affects the same pathways. The mutants are important tools for addressing this question, as well as being good evidence that such regulatory mechanisms exist.

EVIDENCE OF REGULATION OF FUNGAL POLYA RNA BY dsRNA

We have addressed the question of how the dsRNA affects the fungus by seeking polyA RNA specific to either virulent or hypovirulent strains. We used the technique of differential hybridization to detect such polyA RNAs. These studies were done with strains that were specifically constructed using nuclear and mitochondrial DNA markers to assure that they were isogenic except for the presence of dsRNA in the hypovirulent strain (E. Gobbi, W. Powell, and N. K. Van Alfen, unpublished data). We have detected two high copy number polyA RNAs which are present in the virulent but not the hypovirulent strain. Using genomic clones of these polyA RNAs as probes, we have detected these polyA RNAs in all virulent strains of the fungus that have been tested to date, but not in hypovirulent strains (Powell and Van Alfen, 1987a). It is interesting that dsRNA from both a European strain (EP 113) and an American strain (EP 915) affect accumulation of these same two polyA RNAs. The polyA RNAs do not accumulate in a *pig* mutant strain either. Thus, ds RNA of two viral strains that have no homology to one another by hybridization experiments (L'Hostis et al., 1985) and a mutant that mimics the phenotype of dsRNA-caused-hypovirulence each greatly reduce accumulation of the same polyA RNAs. These results strongly suggest that the dsRNA viruses are all affecting the fungus in a similar way and that the effect is a specific perturbation of regulatory pathways rather than debilitation of the fungus (Powell and Van Alfen, 1987b).

SUMMARY

The factors responsible for virulence expression by *E. parasitica* are not clearly understood but probably include: organization of a mycelial fan for penetration of host tissue, production of fungal metabolites toxic to the host, and secretion of extracellular enzymes that hydrolyze complex molecules of host origin. Both mycelial fan formation and secretions of some of these fungal metabolites are affected by the presence of dsRNA viruses. There are a number of different viruses that lack homology with one another yet are able to cause hypovirulence. Recent studies suggest that these viruses affect virulence expression by the fungus in specific ways. The best evidence for this is that the different types of dsRNA viruses significantly reduce the accumulation of at least two specific polyA RNAs that are very abundant in virulent strains. Hypovirulence may be the result of viral effects on an easily perturbed fungal virulence regulatory system rather than the result of a debilitating disease of the fungus.

ACKNOWLEDGMENTS

Studies in the author's laboratory were supported by grants from the National Science Foundation (PCM-840245) and the Science and Education Administration of the United States Department of Agriculture under grant No. 85-CRCR-1-1895 from the Competitive Research Grants Office.

REFERENCES

Alymore, R. C., and Todd, N. R. (1985). Cytology of non-self hyphae fusions and somatic incompatibility in *Phanerochaete velutina*. *J. Gen. Microbiol.* *132*:581-591.

Anagnostakis, S. L. (1977). Vegetative incompatibility in *Endothia parasitica*. *Exp. Mycol.* *1*:306-316.

Anagnostakis, S. L. (1979). Sexual reproduction of *Endothia parasitica* in the laboratory. *Mycologia* *121*:213-215.

Anagnostakis, S. L. (1982a). Genetic analyses of *Endothia parasitica*: Linkage data for four single genes and three vegetative compatibility types. *Genetics* *102*:25-28.

Anagnostakis, S. L. (1982b). The origin of ascogenous nuclei in *Endothia parasitica*. *Genetics* *100*:413-416.

Anagnostakis, S. L. (1982c). Carpenter ants as carriers of *Endothia parasitica*. *Proceedings of the USDA Forest Service American Chestnut Cooperators' Meeting*, pp. 111-113.

Anagnostakis, S. L. (1984). Nuclear gene mutations in *Endothia (Cryphonectria) parasitica* that affect morphology and virulence. *Phytopathology* *74*:561-565.

Anagnostakis, S. L., and Waggoner, P. E. (1981). Hypovirulence, vegetative incompatibility, and the growth of cankers of chestnut blight. *Phytopathology* *71*:1198-1202.

Anderson, P. J. (1913). *The Morphology and Life History of the Chestnut Blight Fungus*. Commission for Investigation and Control of the Chestnut Tree Blight Disease in Pennsylvania, Bulletin 7.

Biggs, A. R., Merrill, W., and Davis, D. D. (1984). Discussion: Response of bark tissue to injury and infection. *Can. J. For. Res. 14*:351-356.

Bramble, W. C. (1936). Reaction of chestnut bark to invasion by *Endothia parasitica. Am. J. Botan. 23*:89-94.

Day, P. R., Dodds, J. A., Elliston, J. E., Jaynes, R. A., and Anagnostakis, S. L. (1977). Double-stranded RNA in *Endothia parasitica. Phytopathology 67*:1393-1396.

DePalma, N. (1981). Dissemination of *Endothia parasitica* by birds and small mammals. *USFS American Chestnut Cooperators' Meeting General Technical Report NE-64*, 20 p.

Elliston, J. E. (1985). Further evidence for two cytoplasmic hypovirulence agents in a strain of *Endothia parasitica* from Western Michigan. *Phytopathology 75*:1405-1413.

Fincham, J. R. S., Day, P. R., and Radford, A. (1979). *Fungal Genetics.* University of California, California.

Foster, R. C. (1986). The ultrastructure of the rhizoplane and rhizosphere. *Annu. Rev. Phytopathol. 24*:211-234.

Fulbright, D. W. (1984). Effect of eliminating dsRNA in hypovirulent *Endothia parasitica. Phytopathology 74*:722-724.

Galloway, B. T. (1926). *The Search in Foreign Countries for Blight-Resistant Chestnuts and Related Tree Crops.* USDA, Circ. 383.

Gäumann, E., Kern, H., and Obrist, W. (1959). Der Einfluss einigen Welketoxine auf den Wasserhaushalt abgeschnittener Tomatensprosse. *Phytopathologische Z. 36*:111-121.

Gäumann, E., and Naef-Roth, S. (1957). Uber die Toxine der *Endothia parasitica* (Murr.) And. *Pflanzensdrutz-Berichte 15*:9-16.

Grente, J., and Berthelay-Sauret, S. (1978). Biological control of chestnut blight in France. *Procedings of the American Chestnut Sympoisum,* West Virginia Univ., Morgantown, West Virginia, pp. 30-34.

Grente, J., and Berthelay-Sauret, S. (1978). Research carried out in France into diseases of the chestnut tree. *Proceedings of the American Chestnut Symposium*, West Virginia Univ., Morgantown, West Virginia, pp. 88-92.

Grente, J., and Sauret, S. (1969). L'"hypovirulence exclusive," Est elle controlée par des determinants cytoplasmiques? *C. R. Acad. Sci. Ser. 268*:3173-3176.

Griffin, G. J., Elkins, J. R., and Roane, M. K. (1986). *Chestnut Blight, Other Endothia Diseases, and the Genus Endothia.* APS Books, St. Paul, Minnesota, 53 pp.

Hansen, D. R., Van Alfen, N. K., Gillies, K., and Powell, W. (1985). Naked dsRNA associated with hypovirulence of *Endothia parasitica* is packaged in fungal visicles. *J. Gen. Virol. 66*:2605-2614.

Havir, E. A., and Anagnostakis, S. L. (1983). Oxalate production by virulent but not by hypovirulent strains of *Endothia parasitica. Physiol. Plant Pathol. 23*:369-376.

Havir, E. A., and Anagnostakis, S. L. (1985). Oxaloacetate acetylhydrolase activity in virulent and hypovirulent strains of *Endothia (Cryphonectria) parasitica. Physiol. Plant Pathol. 26*:1-9.

Hebard, F. V., Griffing, G. J., and Elkins, J. R. (1984). Developmental histopathology of cankers incited by hypovirulent and virulent isolates of *Endothia parasitica* on susceptible and resistant chestnut trees. *Phytopathology 74*:140-149.

Hepting, G. H. (1974). Death of the American Chestnut. *J. For. Hist. 18*:60-67.

Kosuge, T., and Nester, E. W. (Eds.) (1984). *Plant-Microbe Interactions. Molecular and Genetic Perspectives,* Vol. 1. MacMillan, New York, 429 pp.

L'Hostis, B., Hiremath, S. T., Rhoads, R. E., and Ghabrial, S. A. (1985). Lack of sequence homology between double-stranded RNA from European and American hypovirulent strains of *Endothia parasitica. J. Gen. Virol. 66*:351-355.

McCarroll, D. R., and Thor, E. (1985a). Do "toxins" affect pathogenesis by *Endothia parasitica? Physiol. Plant Pathol. 26*:357-366.

McCarroll, D. R., and Thor, E. (1985b). Pectolytic, cellultic activities expressed by cultures of *Endothia parasitica*, and inhibition of these activities by components extracted from Chinese and American chestnut inner bank. *Physiol. Plant Pathol. 26*:367-378.

Mittembergher, L. (1978). The present status of chestnut blight in Italy. *Proceedings of the American Chestnut Symposium,* West Virginia Univ., Morgantown, West Virginia, pp. 34-37.

Moffitt, E. M., and Lister, R. M. (1975). Application of a seriological screening test for detecting double-stranded RNA mycoviruses. *Phytopathology 65*:851-859.

Paul, C. P., and Fulbright, D. W. (1985). Homology relationships with Michigan hypovirulent strains of *Endothia parasitica*. *Phytopathology 61*:1325 (Abstr.)

Powell, W. A., and Van Alfen, N. K. (1987a). Differential accumulation of poly(A) RNA between virulent and double-stranded RNA-induced hypovirulent strains of *Cryphonectria (Endothia) parasitica*. *Mol. Cell. Biol. 7*:3688-3693.

Powell, W. A., and Van Alfen, N. K. (1987b). Two nonhomologus viruses of *Cryphonectria (Endothia) parasitica* reduce accumulation of specific virulence-associated polypeptides. *J. Bacteriol. 169*: 5324-5326.

Puhalla, J. E., and Anagnostakis, S. L. (1971). Genetics and nutritional requirements of *Endothia parasitica*. *Phytopathology 61*:169-173.

Shear, C. L., and Stevens, N. E. (1971). *Endothia parasitica and related species*. USDA Bulletin 380.

Shigo, A. L. (1984). Compartmentalization: A conceptual framework for understanding how trees grow and defend themselves. *Annu. Rev. Phytopathol. 22*:189-214.

Wendt, R. J., Weidhaas, J., Griffing, G. J., and Elkins, J. R. (1982). Association of *Endothia parasitica* with mites isolated from cankers of American chestnut trees. *Proceedings of the USDA Forest Service American Chestnut Cooperators' Meeting*. West Virginia Univ., Morgantown, West Virginia, pp. 114-116.

Windels, C. E., and Lindow, S. E. (1985). *Biological Control of the Phylloplane*. The American Phytopathological Society, St. Paul, Minnesota, 169 pp.

18

dsRNA Virus of *Rhizoctonia solani*

Aliza Finkler, Bat-Sheva Ben-Zvi, and Yigal Koltin
Tel Aviv University, Ramat Aviv, Israel

INTRODUCTION

Rhizoctonia solani, a soil-borne fungal plant pathogen, causes damping-off in some 130 plant species, and is found in most agricultural areas throughout the world (Baker, 1970). The characteristic features of *R. solani* Kuhn [*Thanatephorus cucumeris* (Frank) Donk] include multinucleate hyphal cells that do not generally sporulate, and the formation of sclerotia by vegetative cells. Vegetative incompatibility is determined by six anastomosis groups, and different strains can form hyphal fusions and exchange cytoplasm only if both mates are of the same anastomosis group. Sexuality is very rare and had been reviewed by Sherwood (1970). Strains investigated displayed either homothallism or bipolar incompatibility (Whitney and Parmeter, 1963).

While dsRNA viruses of fungi are abundant and have been under investigation for some time (reviewed by Lemke, 1979; Buck, 1986), the role of these viruses in fungal ecology has not yet been elucidated. Transmissible hypovirulence in *Endothia parasitica* is the only case in which an effect on the etiology of a disease can be ascribed to dsRNA molecules. Attempts to correlate the presence of viruses with unique phenotypes or biological properties of fungal carriers is feasible in few cases and is quite confusing in most others.

The majority of fungal infections with viruses are symptomless (Hollings, 1978), and in only few cases do the carriers exhibit a unique phenotype that can be ascribed to viral infection: toxin production in *Saccharomyces cerevisiae* and *Ustilago maydis* is related to viral infection (Bevan et al., 1973; Wood and Bozarth, 1973); conditional lysis of certain strains of the genus *Penicillium* (Borre et al., 1971); the disease of the cultivated mushroom *Agaricus bisporus* (Marino et al., 1976), and diseased mycelium in *Helminthosporium victoriae* (Ghabrial and Mernaugh, 1983; see Chapter 16) are among the few such cases. Other effects, such as reduced growth rate and poor sporulation, have been reported, but these correlations with viral infections are complex. A disease of *H. victoriae*, the causative agent of Victoria blight, appeared to be transmitted to normal strains by cytoplasmic exchange (Ghabrial and Mernaugh, 1983). Diseased strains were characterized by reduced growth, poor sporulation, and lysis. Both normal and diseased strains contain 190S virus particles, but only the diseased strain contains an additional group of 145S virus particles. It is suggested that the 145S virus is the causal agent of the disease of *H. victoriae*.

The correlation between virus-infected isolates of the wheat take-all fungus *Gaeumannomyces graminis*, characterized by poor sporulation and instability in culture, is quite confusing due to the complexity of viruses reported to infect individual isolates of this species (Buck et al., 1981). It is possible that take-all decline depends upon an unstable balance of dsRNA components in the fungal population in soil adjacent to host roots (Day and Dodds, 1979).

The relation between virulence and dsRNA virus particles in *R. solani* was initially examined by Castanho et al. (1978) following a study of *Endothia parasitica*, the causative agent of the chestnut blight (Van Alfen et al., 1975), in which the attenuation of virulence by cytoplasmic factors was suggested. It appeared that dsRNA molecules act as attenuators of virulence in *E. parasitica* and that transmission of dsRNA by cytoduction from hypovirulent to virulent strains suppresses virulence (see Chapters 13 and 17, this volume). The early studies with *R. solani* revealed the occurrence of a similar phenomenon. The attenuation of virulence was transmissible but the hypovirulent strains were debilitated and it was this debilitated state that was transmitted. In these studies, dsRNA, but not virus particles, was detected in the mycelia of hypovirulent strains. Virulent strains were reported devoid of dsRNA. The transmission of the debilitated growth to a virulent strain of *R. solani* through cytoplasmic exchange along with dsRNA suggested that the phenomena of hypovirulence in *E. parasitica* and *R. solani* were similar. However, since hypovirulence and debilitation of growth are two separable phenotypes, it is clear that rigorous criteria must be set in defining hypovirulence to avoid confusion between growth retardation and hypovirulence. Therefore, in studies by Finkler et al. (1985) to reexamine the relation of dsRNA viruses to virulence in *R. solani*, hypovirulent strains were defined as those strains that infect a wide variety of hosts but incur almost no discernible damage to the host. Such strains should have no growth disadvantage, and their growth rate should be similar to that of the virulent strains. Virulent strains were those that affect all the tested hosts and incur damage of high severity. Furthermore, since *R. solani* has a very wide host range, a distinction between virulent and hypovirulent strains must represent this potential. An examination of the response of a variety of hosts to each strain is a minimal requirement in the definition of such a complex parameter as virulence. Therefore, results reported by Zanzinger et al. (1984), in which dsRNA occurred in both virulent and hypovirulent strains, may be misleading by the use of nonstringent criteria for the distinction between virulent and hypovirulent strains. It appears that the data reported thus far in a number of systems is subject to various interpretations due to the inaccurate definition of hypovirulence. We have, therefore, undertaken the study of the relationship of dsRNA viruses to the virulence of *R. solani*, by examination of a local population and determining those elements that appear as most critical in the phenomenon of hypovirulence. Furthermore, in an effort to resolve whether the dsRNA viruses play a role in the regulation of virulence, we addressed the major question of whether virulence or hypovirulence is cytoplasmically transmitted.

VIRULENCE AND OCCURRENCE OF dsRNA VIRUSES IN *Rhizoctonia solani*

Isolation

Strains of *R. solani* were isolated from soil samples in Israel. The isolates were from diverse ecological niches (Ichielevich-Auster et al., 1985a) and varied in host range and disease severity when tested on as many as 11 different host plants (Table 1). Approximately 30% of the isolates were nonpathogenic to all the host plants tested, and these are considered to be hypovirulent. Strains selected for the study displayed normal growth rate and with no significant difference in growth rate between the virulent and hypovirulent strains. Similarly, no other morphological differences were distinguished, except that hypovirulent strains produced fewer sclerotia and their mycelia appeared to contain less melanin. Formation of sclerotia and melanin may be useful markers for identification of virulent strains, but the number of strains examined was insufficient to establish a correlation. In a sample of 109 strains tested, most anastomosis groups were represented and both virulent and hypovirulent strains were found within each group. In addition, no special geographic distribution was noticed among the virulent and the hypovirulent strains or for the anastomosis groups.

Table 1 Host Range and Disease Severity Index of Strains of *Rhizoctonia solani*

Strain No.	AG[b]	Disease severity index[a]					
		R	T	C	B	Co	O
Virulent							
82	4	5.00 ± 0.00[c]	1.00 ± 0.00	5.00 ± 0.00	5.00 ± 0.00	2.00 ± 0.00	4.80 ± 0.00
13	4	5.00 ± 0.00	5.00 ± 0.00	5.00 ± 0.00	3.00 ± 0.00	5.00 ± 0.00	5.00± 0.00
56	4	5.00 ± 0.00	5.00 ± 0.00	4.00 ± 0.00	3.00 ± 0.00	5.00 ± 0.00	5.00 ± 0.00
53	4	5.00 ± 0.00	5.00 ± 0.00	5.00 ± 0.00	5.00 ± 0.00	5.00 ± 0.00	0.00
189HT[c]	1	5.00 ± 0.00	5.00 ± 0.00	4.01 ± 0.09	2.00 ± 0.00	1.00 ± 0.00	0.00
Hypovirulent							
160	2	1.00 ± 0.00	0.00	0.00	0.00	0.00	0.00
521	4	0.70 ± 0.03	0.00	0.00	0.00	0.00	0.00
558	4	1.00 1 0.00	0.00	2.00 1 0.00	1.00 ± 0.00	0.00	1.00 ± 0.00
751	1	2.00 ± 0.00	1.09 ± 0.03	1.00 ± 0.00	1.00 ± 0.00	0.00	0.00
189a[c]	1	0.38 ± 0.01	0.00	0.00	0.00	0.00	0.00

[a]Virulence ratings were based on the infected area of hypocotyl of seedlings. The results with six hosts are presented (R, radish; T, tomato; C, carrot; B, bean; Co, cotton; O, onion). Tests conducted with five additional hosts gave similar results without any exception. The scale of 1 to 5 refers to the area of infection: 0, no detectable infection; 1, 1 to 10%; 2, 11 to 30%; 3, 31 to 50%; 4, 51 to 80%; 5, over 80%.
[b]Anastomosis group.
[c]Strains kindly sent by Dr. E. E. Butler, Department of Plant Pathology, University of California, Davis, California.
Source: Finkler et al. (1985).

Detection of dsRNA Viruses

Of the nine strains selected for the study, five were virulent and four hypovirulent as determined by pathogenicity tests. A unique nucleic acid was extractable only from the virulent strains. The unique nucleic acids extracted from the mycelia of each of the virulent strains were shown to be dsRNA by resistance to RNase at high ionic strength and to DNase treatments, and sensitivity to RNase at low ionic strength. Also, binding properties to a cellulose column (CF11) routinely used to separate between single-stranded and double-stranded RNA (Franklin, 1966) was indicative of the dsRNA nature of this material. The high-molecular-weight material seen in gels of extracts consisted of a heterogenous population of DNA molecules, as determined by digestion with DNase and with restriction enzymes as well as by electron microscopy.

The results indicated that only virulent strains contained dsRNA (Figure 1); hypovirulent strains were devoid of dsRNA (Finkler et al., 1985). By electrophoresis of the dsRNA, a number of segments were identified in polyacrylamide gel and their pattern of segmentation varied slightly. Three distinct patterns were identified: class I included strains 13 and 56, which consisted of two segments with M.W. of 1.45 and 1.32×10^6; class II included strains 82 and 53 which contained an additional segment with a M.W. of 1.6×10^6, and class III, represented by strain W, contained two dsRNA segments with M.W.s of 1.6 and 1.32×10^6, but lacked the 1.45×10^6 species. Additional segments with M.W.s of 1.7, 1.25, and 1.225×10^6 were identified only in strain 82. Repeated extractions of dsRNA from the mycelia of *R. solani* maintained on potato-dextrose agar plates and in soil for over a year revealed some differences in the pattern of dsRNA segments. For example, strain W displayed an additional segment with M.W. of 1.45×10^6 in later extractions; strain 13 was found lately to contain two additional segments with M.W.s of 1.6 and 1.25×10^6. The variation did not yield totally unknown segments, and looking at the population of isolates, it appears that four segments are predominant: 1.6, 1.45, 1.32, and 1.25×10^6. These fluctuations in dsRNA segments are not uncommon among dsRNA fungal viruses and a change in titer was also shown in *G. graminis* by Buck et al. (1981) and noticed in *U. maydis* (Koltin, 1986).

Virus particles were sought in virulent and hypovirulent strains but were detected only in virulent strains (Figure 2). Repeated extractions of dsRNA from the viruses indicated only two segments, M.W. 1.45 and 1.32×10^6, are recovered from the particles. Thus, it appears that the additional segments detected in the mycelia of a number of isolates are probably not encapsidated.

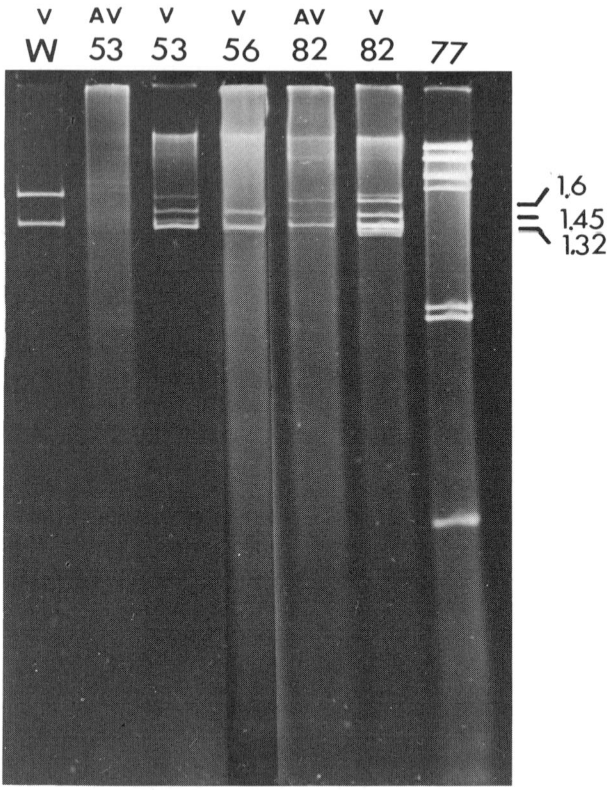

Figure 1 Polyacrylamide gel electrophoresis of dsRNA isolated from virulent strains of *R. solani* (V) and from virulent strains transformed into hypovirulence (AV). *Ustilago maydis* P4 viral (77) dsRNA was used as the molecular weight marker. *Source*: Finkler et al., 1985.

Characterization of *R. solani* Viruses

The viruses isolated from virulent strains of *R. solani*, as described by Finkler et al. (1985), are isometric, with a diameter of 33 nm (Figure 2). The virus particles were purified on a 10 to 40% sucrose gradient and identified in specific fractions by electron microscopy. Virus-associated RNA-dependent RNA polymerase has been detected in most fungal viruses (Buck, 1979; Ratti and Buck, 1979; Ben-Zvi et al., 1984). Virus-associated RNA-dependent RNA polymerase activity was tested in each fraction from the gradient and maximum activity of this enzyme coincided with those fractions containing dsRNA (Figure 3).

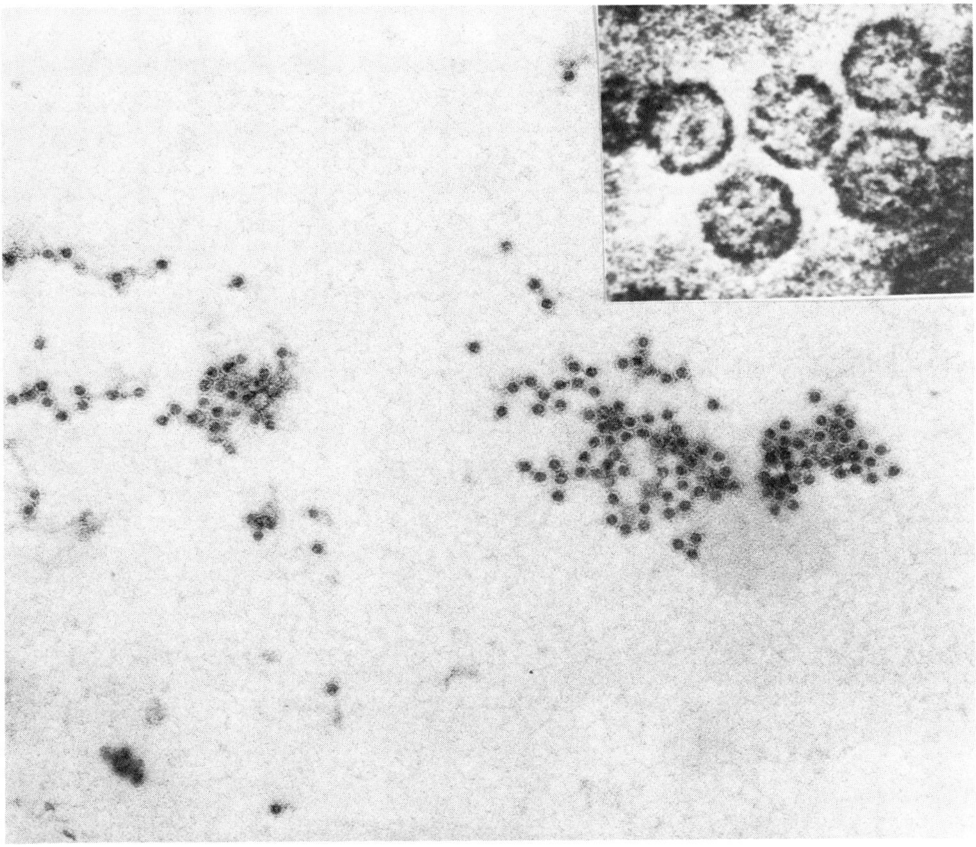

Figure 2 Electron micrograph of negatively stained virus particles purified from a virulent strain of *R. solani*.

Examination of the viral structural proteins on SDS-polyacrylamide gels revealed a single major capsid protein, 54 kDa (Figure 4). The results of in vitro translation of viral transcripts suggested that the additional minor proteins observed in the gel were not encoded by the viral nucleic acids.

Analysis of Viral Transcripts

Full-size transcripts were synthesized by the viral-associated RNA polymerase in reaction mixtures containing [^{32}P]UTP as described by Ben-Zvi et al. (1984). The reaction products were analyzed on a denaturing polyacrylamide gel (Figure 5). The transcripts were 0.74

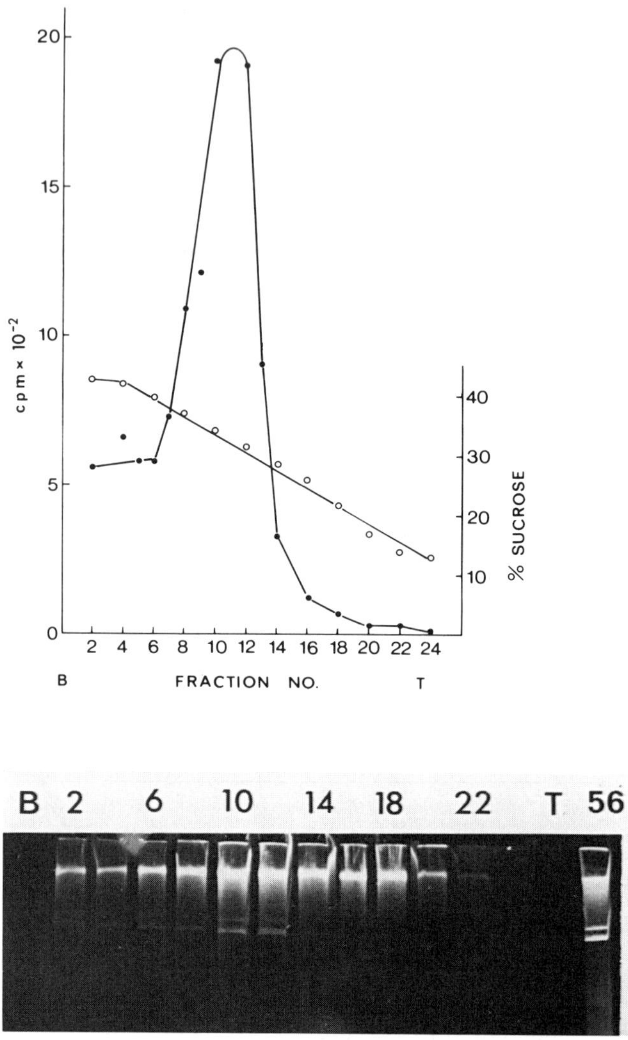

Figure 3 Distribution of RNA polymerase activity of *R. solani* virus and dsRNA content in fractions of a 10 to 40% sucrose gradient. The graph indicates RNA polymerase activity (●) and sucrose density (○) of the fractions. The photograph is an ethidium bromide-stained polyacrylamide gel of dsRNA extracted from each gradient fraction. B and T indicated bottom and top of gradient. The dsRNA extracted from virus particles isolated from strain 56 are shown in the right lane. *Source*: Finkler et al., 1985.

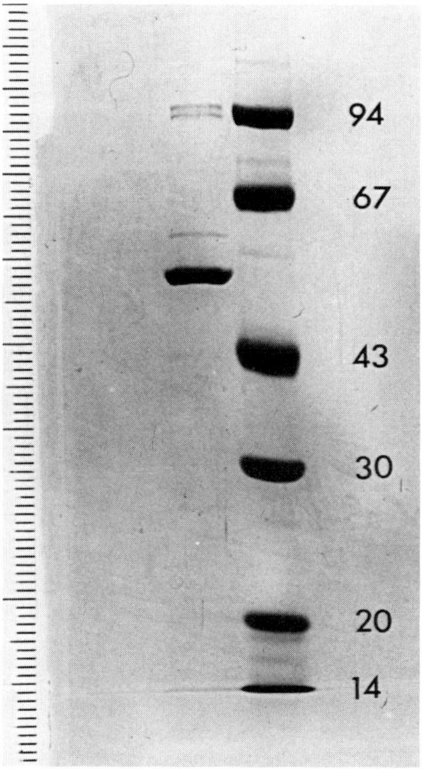

Figure 4 SDS-PAGE of viral proteins.

and 0.66 × 10^6 M.W., corresponding to the encapsidated dsRNA templates with M.W.s of 1.45 and 1.32 × 10^6, respectively. While dsRNA extracted from virus particles contained two segments of 1.45 and 1.32 × 10^6 M.W., the dsRNA segment, 1.6 × 10^6 M.W., detected in the mycelia of a number of virulent strains, was never extracted from the viruses. The relatedness of the 1.6 × 10^6 segment to those encapsidated was examined by hybridization experiments. Labeled transcripts obtained from the viruses were used as probes in hybridization experiments. The results presented in Figure 6 indicate that the transcripts synthesized from strain 82 did not hybridize with the 1.6 × 10^6 segment, thus providing further support to the contention that this segment is not encapsidated.

To determine the relatedness between the different viral isolates, the viruses of one isolate were used for hybridization with other dsRNA

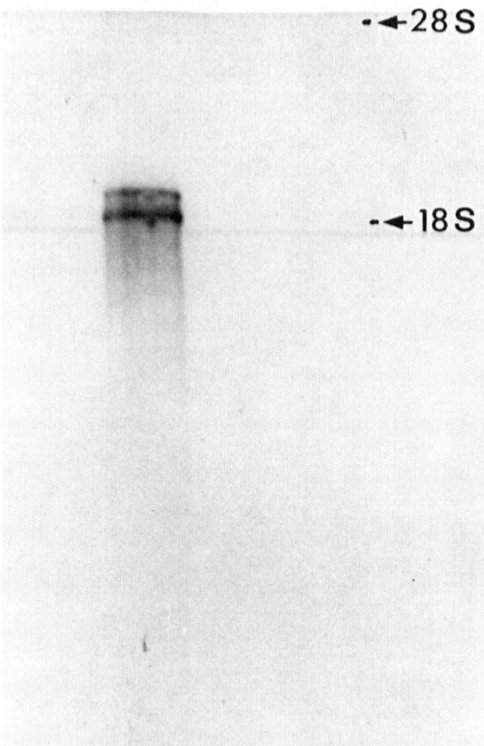

Figure 5 Autoradiography of ^{32}P-labeled virus transcripts subjected to electrophoresis in a denaturing polyacrylamide gel. Transcripts were denatured at 65°C with 5 M urea and electrophoresed in 3% polyacrylamide gel containing 7 M urea at 110 V for 24 hr. *Source*: Finkler et al., 1988.

segments obtained from different isolates. The results indicate that the viruses found among different virulent isolates of *R. solani* that are collected in different geographical locations are all interrelated since they cross-hybridize. In addition, these transcripts were used for hybridization studies with fungal host DNA. The results obtained indicated that no homology exists between host DNA and the viral genome. Therefore, similarly to other fungal dsRNA viruses, no "provirus" state occurs in *R. solani* (Bruenn, 1980; McFadden et al., 1983; Ben-Zvi et al., 1984).

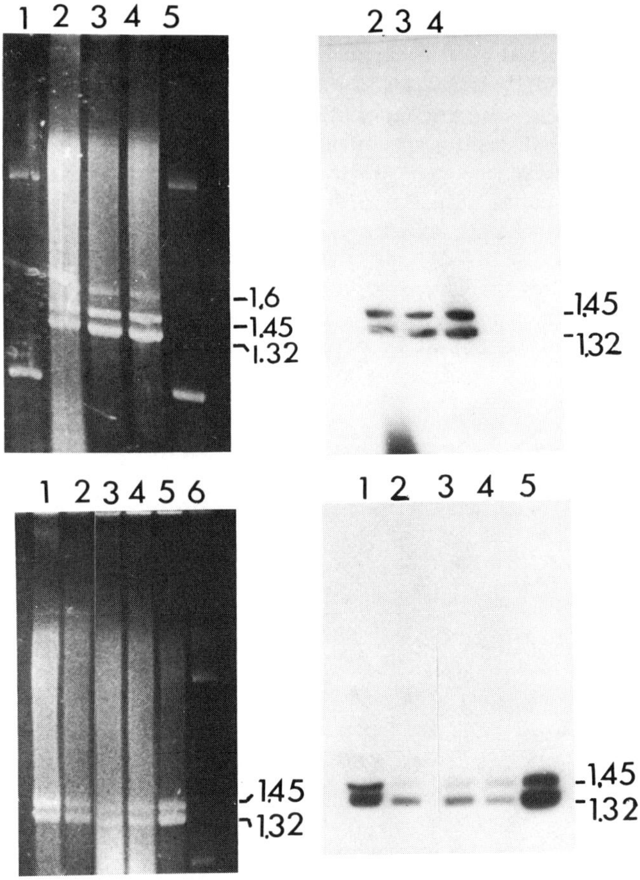

Figure 6 Hybridization experiments: Top, left panel: ethidium bromide staining of dsRNA extracted from strains 82 (lane 2) and 53 (lanes 3, 4) and *U. maydis* P6 (lanes 1, 5). Right panel: autoradiography of ^{32}P-labeled transcript isolated from strain 82 and used as a probe in hybridization with the dsRNA shown in the left panel. Bottom, left panel: ethidium bromide staining of dsRNA isolated from *R. solani* 13 (lane 1), 82 (lane 2), 56 (lanes 3, 4), W (lane 5) and *U. maydis* P6 (lane 6). Right panel: autoradiography of ^{32}P-labeled transcript synthesized from strain 82 used as a probe in hybridization with the dsRNAs shown in the left panel.

Translation Studies of the Viral Genome

To study the coding potential of the viral genome, purified viral transcripts were used for in vitro translation experiments using the rabbit reticulocyte lysate system. The incorporation of [^{35}S]methionine into proteins was analyzed on SDS-polyacrylamide gels. Identical patterns of in vitro translation products were obtained from the transcripts of two virus preparations from strains 13 and 82. Maximum incorporation of label was obtained with 100 μg/ml of purified ssRNA. A major protein, about 55 kDA and a close submajor product of 53 kDa were synthesized in vitro in addition to a number of other "minor" proteins: 47, 45, 43, 40, 38, and 35 kDa (Figure 7). It is not known whether these smaller proteins are the result of premature termination. Extension of the incubation period of translation fourfold did not alter the pattern of these minor proteins. Some of the additional bands de-

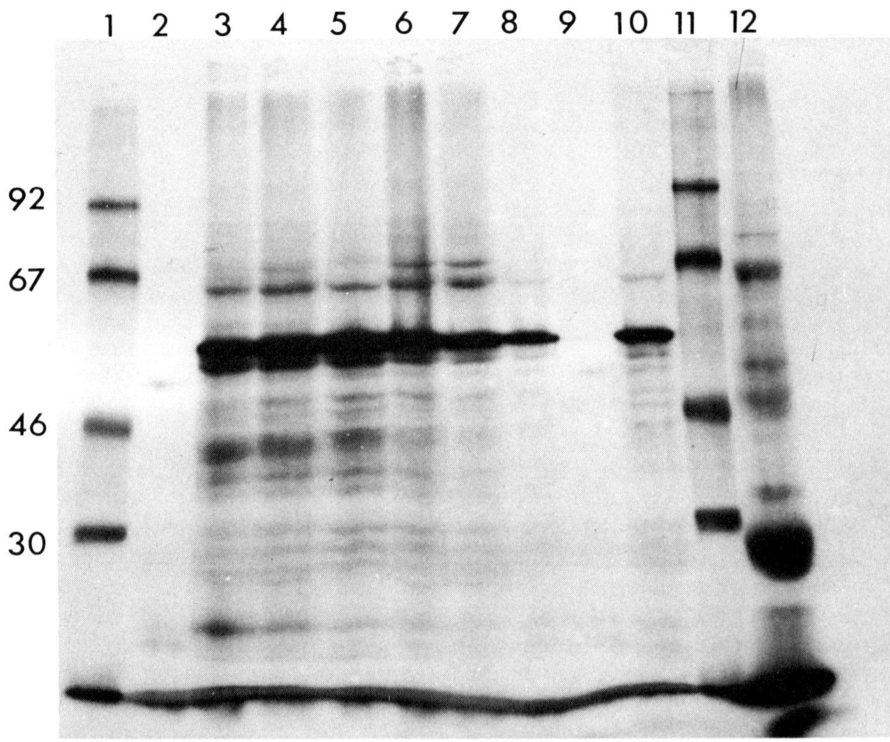

Figure 7 Translation of *R. solani* viral transcript in vitro. Lanes: (1 and 11) [^{14}C]methyl protein molecular weight standards; (2 and 9) in vitro translation products with no added RNA; (3-8) reticulocyte lysate reaction products of transcript RNA (50, 100, 150, 200, 250, 300 μg/ml, respectively); synthesized from strain 13; (10) translation products of transcript synthesized from strain 82 (100 μg/ml); (12) translation of globin mRNA. *Source*: Finkler et al., 1988.

tected among the products of in vitro translation were also found in control assays lacking RNA and many represent a low level of endogenous translation, nonspecific binding of [^{35}S]methionine to lysate proteins, or short peptides bound to tRNA.

Viral structural proteins were identified by immunoprecipitating the in vitro translation products with rabbit antiserum and with partially purified γ-globulins prepared against purified virus particles. The major protein, 55 kDa, was precipitated in all experiments along with a submajor product, 53 kDa (Figure 8). These results suggest that at least two proteins (55 kDa and 53 kDa) are associated with viral capsids.

The identity of the major product synthesized by in vitro translation of viral RNA was further elucidated by peptide mapping. The patterns of the peptide map of the authentic viral capsid and the major in vitro protein product, analyzed by digestion with *Staphylococcus aureus* V8 protease and examined by electrophoresis on 15% SDS-polyacrylamide gel, were identical (Figure 9). This result indicates that the 55 kDa major protein obtained by in vitro translation is indeed the major capsid protein.

REGULATION OF VIRULENCE

dsRNA Viruses and Virulence

If pathogenicity of *R. solani* is controlled by dsRNA viruses, it should be possible to convert virulent strains to hypovirulence. This conversion should occur simultaneously with the loss of dsRNA. Indeed, hypovirulent strains were recovered by hyphal tip isolation from virulent strains. While the frequency of recovery of hypovirulent strains was about 1%, virulence was never recovered from hypovirulent strains. Loss of total or specific dsRNA segments was associated with the recovery of hypovirulent strains from virulent strains (Figure 1). No direct correlation was made between a specific segment of dsRNA and virulence. Since elimination of the dsRNA segments with M.W.s of 1.25 and 1.225 × 10^6 in strain 82 was also associated with diminishing levels of the 1.45 × 10^6 segment, it may be speculated that a balanced interaction between the various segments is required to regulate virulence.

Further evidence supporting the relation between dsRNA viruses and virulence was provided by cytoplasmic transmission of virulence (Finker et al., 1985). A heterokaryon was formed between compatible strains, one of which (82) was virulent and the other (521) hypovirulent. The two strains differed in their dsRNA content and in their resistance to the fungicides Benomyl and Bay NTN. The virulent strain was Benomyl-resistant but sensitive to Bay NTN, while the hypovirulent strain was resistant to Bay NTN but sensitive to Benomyl. While

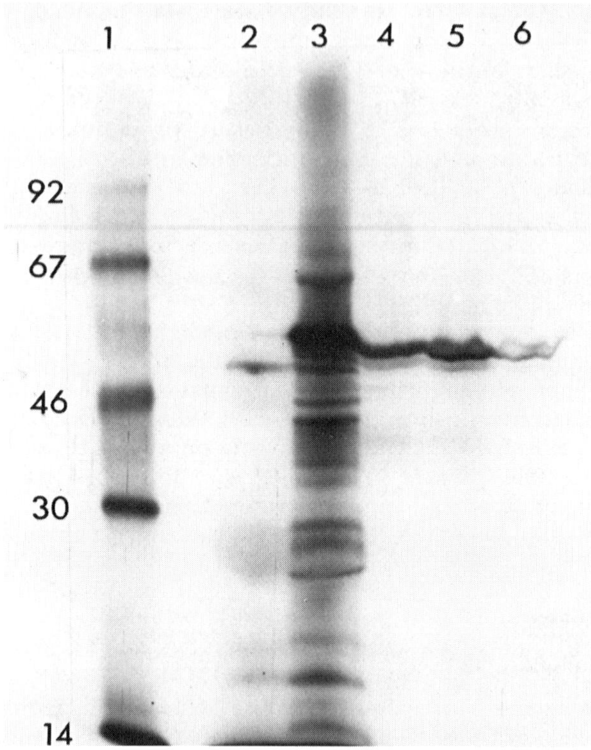

Figure 8 Immunoprecipitation of in vitro translation products with anti-virus antibodies. Lanes: 1: [^{14}C]methyl protein standards; 2: reticulocyte lysate reaction without added RNA; 3: in vitro translation products; 4-6: translation products immunoprecipitated with varying amounts of γ-globulins prepared against virus particles. *Source*: Finkler et al., 1988.

strain 82 contained a number of dsRNA segments, the hypovirulent 521 strain totally lacked dsRNA molecules. Using the differential sensitivities as genetic markers, selection of hypovirulent strains transformed to virulence was performed (Figure 10). The new virulent cultures, Benomyl-sensitive, Bay NTN-resistant, had the morphological characters of the hypovirulent strain 521 (less melanin and producing fewer sclerotia). To determine whether the transformation of the hypovirulent strain to virulence occurred simultaneously with the transmission of the dsRNA viruses, the dsRNA of the transformed strains was extracted. Two segments of M.W.s of 1.45 and 1.32×10^6 were recovered from the transformed strains (Figure 11). This pattern of segmentation was identical to the viral dsRNA content of the donor

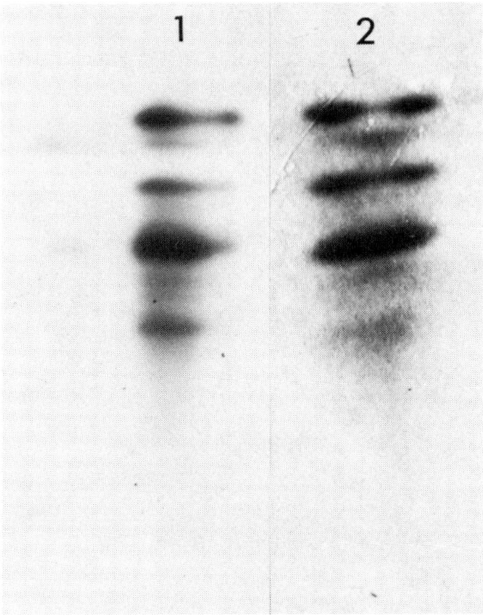

Figure 9 Comparison of peptide map patterns of the major 55 kDa in vitro translation product (1) and authentic capsid protein (2). *Source*: Finkler et al., 1988.

virulent strain. The additional segments, present in the mycelium of the donor virulent strain but not encapsidated, were not detected in the recipient strain. Therefore, unlike the transmissible hypovirulence of *E. parasitica*, in which dsRNA molecules suppress virulence, the dsRNA viruses in *R. solani* appear to be associated with induction of virulence.

Pectolytic Enzymes of *R. solani* and Virulence

Enzymatic activities associated with virulence of *R. solani* are unknown. It is therefore difficult to relate viral proteins to a specific enzymatic activity or to the induction of specific enzymes in the host. However, due to lack of sufficient biological characterization of virulent and hypovirulent strains, a study of pectolytic enzymes in a number of isolates of *R. solani*, as possible markers for virulence, was pursued. The role of pectolytic enzymes in the pathogenesis of *R. solani* has been well established (Weinhold and Motta, 1973). Four enzymes, known as endopolygalacturonase I (endo PG-I) and II, pectinesterase (PE) and endopectinolyase (endo-PL), were purified to homogeneity from a number of *R. solani* isolates (Marcus et al., 1986). The endo-PL was found in all the virulent strains used in this study, but was not detected in any of the five hypovirulent strains examined. Although

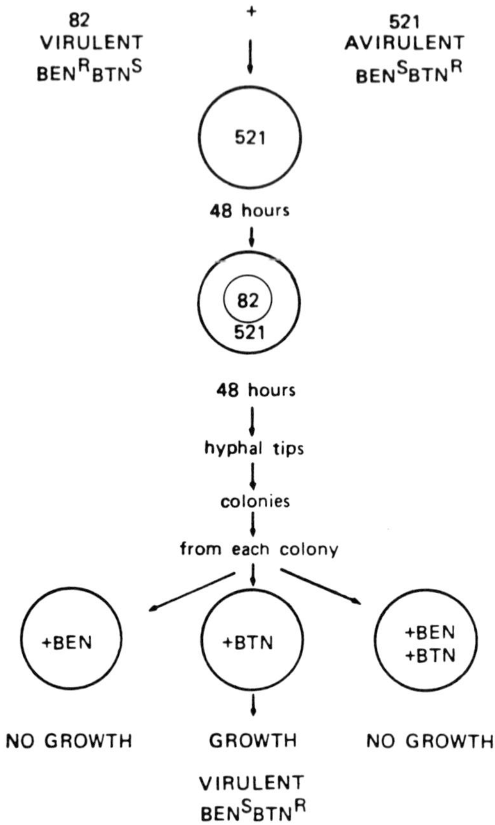

BEN = Benomyl BTN = Bay NTN 19701

Figure 10 Transmission of virulence by heterokaryon from a virulent to a hypovirulent strain.

the role of endo-PL in the virulence of *R. solani* is supported by other reports in which detection of the enzyme was associated with tissue degradation (Sherwood, 1966), the late production of endo-PL in culture suggested that this activity is essential only in the later stages of infection. The association between endo-PL and transmission of virulence will be studied by cytoplasmic exchange between virulent and hypovirulent isolates. The occurrence of endo-PL activity after transformation of the hypovirulent strains into virulence may indicate a viral-encoded or -induced activity.

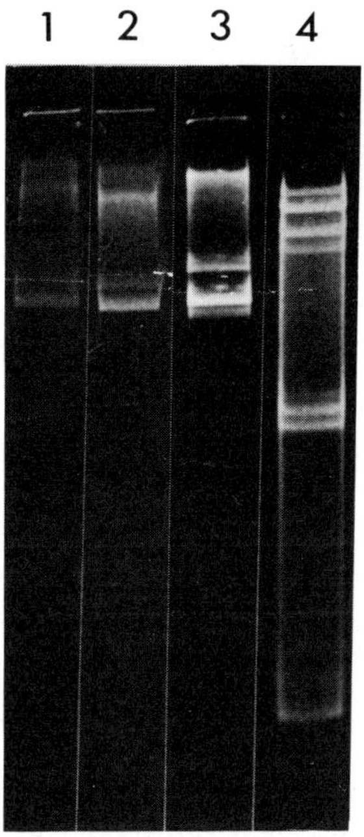

Figure 11 Polyacrylamide gel containing dsRNA isolated from the virulent donor strain (82) and the recipient strain (521) transformed to virulence. Lanes 1 and 2: dsRNA from the two strains "transformed" to virulence. Lane 3: dsRNA from the virulent donor strains. Lane 4: dsRNA molecular weight marker (*U. maydis* P4 virus). *Source*: Finkler et al., 1985.

Biological Control of *R. solani*

Bernard (1909) reported the protection of orchid bulbs against a virulent strain of *Rhizoctonia repens* following inoculation of orchid roots with a weak isolate of the same organism. Similar studies were conducted by Ichielevich-Auster et al. (1985b) to verify whether a similar phenomenon existed in *Rhizoctonia solani* isolates. Indeed, a high degree of protection against damping-off caused by virulent strains of *R. solani* was obtained using hypovirulent isolates. The protection of seedlings of diverse plant species such as cotton, wheat, radish, carrot, and lettuce was effective when the seedlings were infected with a hypovirulent strain prior to a challenge with a virulent strain (Figure 12). Apparently, in an experiment with one hypovirulent strain (521), the damage was reduced by 75-94%. The protection phenomenon was also observed in plants grown in soil infested by the hypovirulent strain and then transferred into soil infested with the virulent isolates. This suggested that the hypovirulent strain was acting

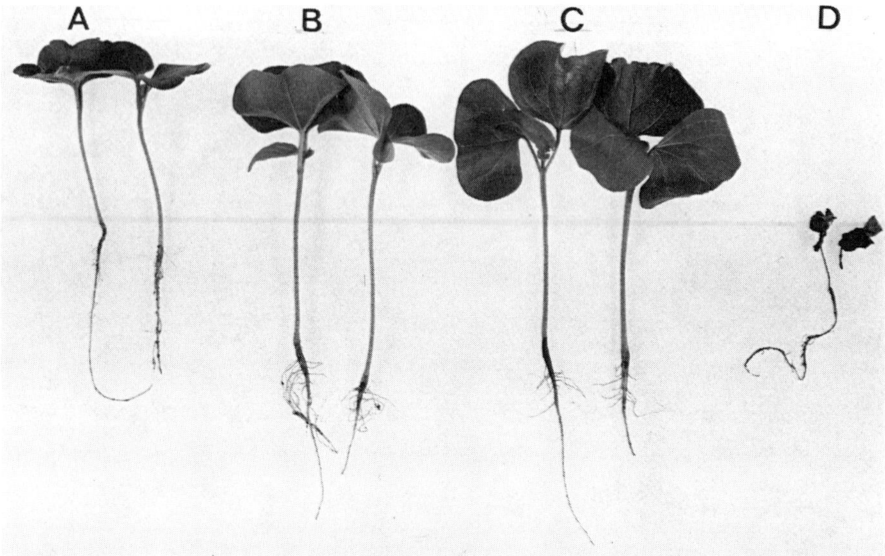

Figure 12 Protection of cotton seedlings by a hypovirulent strain of *R. solani*. A, control, uninoculated; B, inoculated with the hypovirulent strain; C, as B but challenged 3 days later with the virulent isolate, and D, inoculated with the virulent isolate.

in or on the plant. Moreover, cytoplasmic compatibility was probably not essential for protection since a similar effect could be obtained between strains of *R. solani* and *Rhizoctonia zeae*. Therefore, biological control of *R. solani* does not depend only on cytoplasmic exchange, since it can be induced by an isolate from one anastomosis group against virulent strains of different anastomosis groups. Biological control of *E. parasitica* and *R. solani* by hypovirulent strains are probably achieved via different mechanisms. While the cytoplasmic transmissible dsRNA molecules contain some virulence-suppressive information in *E. parasitica*, the dsRNA viruses in *R. solani* may be associated with induction of virulence. The protection of plants by hypovirulent strains of *Rhizoctonia* appears to be in part by induction of a specific defense mechanism. Thus, hypovirulence is not directly related to the dsRNA viruses.

It should be pointed out that the hypovirulent strain used in this study was not diseased and exhibited a normal growth rate which did not differ significantly from virulent isolates, unlike the diseased hypovirulent isolates used by Castanho and Butler (1978). Biological control of *R. solani* by hypovirulent strains was attributed by these authors to transmission of dsRNA molecules, presumed to contain some suppressive factors. However, our results are different, and the mechanism by which protection is obtained is as yet unresolved. Protection, however, is not induced by competition for nutrients, since repeated addition of nutrients to the host's hypocotyl did not reduce the protective effect. It is unclear whether the hypovirulent strain elicits a natural resistance response, or it creates a barrier against superinfection merely by efficient superficial invasion of the plant tissue. Since hypovirulent strains are known to lack dsRNA viruses, protection could be conferred in addition by prevention of viral transmission. Various mechanisms are known from other systems that prevent this type of transmission. As many as 30 nuclear genes in *Saccharomyces cerevisiae* are involved in the maintenance and expression of dsRNA viruses in yeast (reviewed by Tipper and Bostian, 1984). A single gene mutation in one of the nuclear genes affecting maintenance could be sufficient to prevent viral transmission (Wickner, 1986). Other mechanisms may involve vegetative incompatibility (Burnett, 1975), viral exclusion mechanisms by cytoplasmic factors (Wigderson and Koltin, 1982), defective viruses and specific dsRNA plasmids (Sommer and Wickner, 1982; Field et al., 1982).

Another phenomenon related to the hypovirulent strain was its capability to promote plant growth, which was also expressed in increased yields (Sneh et al., 1986). There is no evidence for stimulation of growth via hormonal regulation by the hypovirulent strain, and enhancement may result from increased area for absorption of nutrients and minerals.

CONCLUDING REMARKS

The occurrence of dsRNA viruses in virulent strains of *R. solani* and their absence from hypovirulent strains are described in this chapter. The results presented are contradictory to those published by Castanho et al. (1978). These authors reported the occurrence of dsRNA molecules in the mycelia of hypovirulent strains only. However, the hypovirulent strain used by Castanho et al. (1978) was debilitated, and therefore the degree of virulence of this strain may have been masked by its diseased state. Indeed, while virulent strains were derived from Castanho's hypovirulent strains in their studies, virulence was never detected among hyphal tip cultures from a number of hypovirulent strains used in our study. Moreover, reexamination of Castanho's strains in our laboratory indicated the occurrence of dsRNA molecules in both hypovirulent and virulent cultures. The differences between these results and those reported in our study may therefore stem from differences in definition of hypovirulence. Hypovirulence as defined in our studies refers to a low level of virulence resulting from the regulation of virulence and not from secondary effects as reduced growth rate of the debilitated state of such strains. Evidence is also presented correlating induction or regulation of virulence with dsRNA viruses in *R. solani*. With the present state of the art one cannot exclude transmission of other cytoplasmic organelles or plasmids. More direct support should be provided by infection of protoplasts derived from hypovirulent strains with purified virus particles and the conversion of hypovirulent strains to virulent. Infection experiments will exclude other cytoplasmic factors that may confuse the issue since clearly *R. solani* contains a number of cytoplasmic elements in addition to the dsRNA viruses: unencapsidated dsRNA and a DNA plasmid (Hashiba et al., 1984), and perhaps others. The identity of the viral-induced virulence-promoting activity has not yet been elucidated. Indeed, activity of pectinolyase is absent from hypovirulent strains, but it is most probably not the primary and only virulence factor. Further studies will be directed toward resolution of the mechanism that regulates or induces virulence by dsRNA viruses in *R. solani*. Identification of virulence-specific mRNA from the host may permit us to determine the alteration of transcription and translation in hypovirulent strains transformed to virulence.

It appears that fungal pathogens display an array of interactions with the host, ranging from a high degree of virulence to hypovirulence. The former may be a virus-mediated effect and the latter as a form of suppression of viral replication, as in *R. solani*, or an independent fungal effect attenuated by a virus as in *E. parasitica*. Equilibrium in the fungal-plant interaction is maintained by the fungal genome in *R. solani* and by the viral genome in *E. parasitica*.

REFERENCES

Baker, K. F. (1970). Types of *Rhizoctonia* diseases and their occurrence. In *Rhizoctonia solani*: *Biology and Pathology*, J. R. Parmeter (Ed.). University of California Press, Los Angeles, pp. 125-148.

Ben-Zvi, B., Koltin, Y., Mevarech, M., and Tamarkin, A. (1984). RNA polymerase activity in viruses from *U. maydis*. *Mol. Cell. Biol. 4*:188-194.

Bernard, N. (1909). L'évolution dans la symbiose des orchidées et leur champignon commenseaux. *Ann. Sci. Nat. Bot. 9*:1-196.

Bevan, E. A., Herring, J. A., and Mitchell, D. J. (1973). Preliminary characterization of two species of dsRNA in yeast and their relationship to the killer character. *Nature 245*:81-86.

Borre, E., Morgantini, L. E., Ortali, V., and Tonolo, A. (1971). Production of lytic plaques of viral origin in *Penicillium*. *Nature 229*:568-569.

Bruenn, J. A. (1980). Virus-like particles of yeast. *Annu. Rev. Microbiol. 34*:49-68.

Buck, K. W. (1979). Replication of double-stranded RNA mycoviruses. In *Viruses and Plasmids in Fungi*, P. A. Lemke (Ed.). Marcel Dekker, New York, pp. 94-162.

Buck, K. W. (1986). Fungal virology—an overview. In *Fungal Virology*, K. W. Buck (Ed.). CRC Press, Boca Raton, Florida, pp. 1-84.

Buck, K. W., Almond, M. R., McFadden, J. J. P., Romanos, M. A., and Rawlinson, C. J. (1981). Properties of thirteen viruses and virus variants obtained from eight isolates of the wheat take-all fungus, *Gaeumannomyces graminis. J. Gen. Virol. 53*:235-245.

Burnett, J. H. (1975). *Mycogenetics.* Wiley, New York.

Castanho, B., and Butler, E. E. (1978). *Rhizoctonia* decline: Studies on hypovirulence and potential use in biological control. *Phytopathology 68*:1511-1514.

Castanho, B., Butler, E. E., and Shepherd, R. J. (1978). The association of dsRNA with *Rhizoctonia* decline. *Phytopathology 68*: 1515-1519.

Day, P. R., and Dodds, A. (1979). Viruses of plant pathogenic fungi. In *Viruses and Plasmids in Fungi*, P. A. Lemke (Ed.). Marcel Dekker, New York, pp. 202-239.

Field, L. J., Bobek, L., Brennan, V., Reilly, J. D., and Bruenn, J. (1982). There are at least two yeast double-stranded RNAs of the same size: An explanation for viral exclusion. *Cell 31*:193-200.

Finkler, A., Koltin, Y., Barash, I., Sneh, B., and Pozniak, D. (1985). Isolation of a virus from virulent strains of *Rhizoctonia solani. J. Gen. Virol. 66*:1221-1232.

Finkler, A., Ben-Zvi, B. S., Koltin, Y., and Barash, I. (1988). Transcription and in vitro translation of the dsRNA virus isolated from *Rhizoctonia solani. Virus Genes*, in press.

Franklin, R. M. (1966). Purification and properties of the replicative intermediate of the RNA bacteriophage R17. *Proc. Nat. Acad. Sci. USA 55*:1504-1511.

Ghabrial, S. A., and Mernaugh, R. L. (1983). Biology and transmission of *Helminthosporium victoriae* mycoviruses. In *Double-Stranded RNA Viruses*, R. W. Compans and D. H. L. Bishop (Eds.). Elsevier, New York, pp. 441-449.

Hashiba, T., Homma, Y., Hyalumadi, M., and Matsudo, I. (1984). Isolation of a DNA plasmid in the fungus *Rhizoctonia solani*. *J. Gen. Microbiol.* *130*:2067-2070.

Hollings, M. (1978). Mycoviruses: Viruses that infect fungi. *Advan. Virus Res.* *22*:3-53.

Ichielevich-Auster, M., Sneh, B., Koltin, Y., and Barash, I. (1985a). Pathogenicity, host specificity and anastomosis groups of *Rhizoctonia* spp. isolated from soils in Israel. *Phytoparasitica 13*:103-112.

Ichielevich-Auster, M., Sneh, B., Koltin, Y., and Barash, I. (1985b). Suppression of damping-off caused by *Rhizoctonia* species by a non-pathogenic isolate of *R. solani*. *Phytopathology 75*:1080-1084.

Koltin, Y. (1986). *Ustilago maydis* virus-encoded killer system. In *Extrachromosomal Elements in Lower Eukaryotes*, R. B. Wickner, A. Hinnebusch, A. Lambowitz, I. C. Gunsalus, and A. Hollaender (Eds.). Plenum, New York, pp. 239-257.

Lemke, P. A. (Ed.). (1979). *Viruses and Plasmids in Fungi*. Marcel Dekker, New York.

Marcus, L., Barash, I., Sneh, B., Koltin, Y., and Finkler, A. (1986). Purification and characterization of pectolytic enzymes produced by virulent and hypovirulent isolates of *Rhizoctonia solani*. *Physiol. Mol. Plant Pathol.* *29*:325-336.

McFadden, J. J. P., Buck, K. W., and Rawlinson, C. J. (1983). Infrequent transmission of double-stranded RNA virus particles but absence of DNA proviruses in single ascospore cultures of *Gaeumannomyces graminis*. *J. Gen. Virol.* *64*:927-937.

Marino, R., Saskena, K. N., Schuler, M., Mayfield, J. E., and Lemke, P. A. (1976). Double-stranded RNA from *Agaricus bisporus*. *Appl. Environ. Microbiol.* *32*:433-438.

Ratti, G., and Buck, K. W. (1979). Transcription of double-stranded RNA in viruses of *Aspergillus foetidus* virus S. *J. Gen. Virol.* *42*:59-72.

Sherwood, R. T. (1966). Pectic lyase and polygalacturonase production by *Rhizoctonia solani* and other fungi. *Phytopathology 56*:279-286.

Sherwood, R. T. (1970). Physiology of *Rhizoctonia solani*. In *Rhizoctonia solani, Biology and Pathology*, J. R. Parmeter (Ed.). University of California Press, Los Angeles, pp. 69-92.

Sneh, B., Zeiden, M., Ichielevich-Auster, M., Barash, I., and Koltin, Y. (1986). Increased growth responses induced by a non-pathogenic *Rhizoctonia solani*. *Can. J. Bot.* *64*:2372-2378.

Sommer, S. S., and Wickner, R. B. (1982). Yeast L dsRNA consists of at least three distinct RNAs: Evidence that the non-Mendelian genes [HOK], [NEX] and [EXL] are on one of these dsRNAs. *Cell 31*:429-441.

Tipper, D. J., and Bostian, K. A. (1984). Double-stranded RNA killer systems in yeast. *Microbiol. Rev. 48*:125-156.

Van Alfen, N. K., Jaynes, R. A., Anagnostakis, S. L., and Day, P. R. (1975). Chestnut blight: Biological control by transmissible hypovirulence in *Endothia parasitica*. *Science 189*:890-891.

Weinhold, A. R., and Motta, J. (1973). Initial host responses in cotton infection by *Rhizoctonia solani*. *Phytopathology 63*:157-162.

Whitney, H. S., and Parmeter, J. R. (1963). Synthesis of heterokaryon in *Rhizoctonia solani*. *Canad. J. Bot. 41*:879-886.

Wickner, R. B. (1986). Double-stranded RNA replication in yeast: The killer system. *Annu. Rev. Biochem. 55*:373-395.

Wigderson, M., and Koltin, Y. (1982). Dual toxin specifities and the exclusion relations among the *Ustilago* dsRNA viruses. *Curr. Genet. 5*:127-131.

Wood, H. A., and Bozarth, R. F. (1973). Heterokaryon transfer of virus-like particles associated with a cytoplasmically inherited determinant in *Ustilago maydis*. *Phytopathology 63*:1019-1021.

Zanzinger, D. H., Bandy, B. P., and Tavanzis, S. M. (1984). High frequency of finding double-stranded RNA in naturally occurring isolates of *Rhizoctonia solani*. *J. Gen. Virol. 65*:1601-1605.

19

Viruses of Eukaryotic Chlorella-like Algae

James L. Van Etten, Anne M. Schuster, and Russel H. Meints
University of Nebraska, Lincoln, Nebraska

INTRODUCTION

As reported in this volume, a lot of information has accumulated in the past 20 years on the physicochemical and biological properties of the viruslike particles (VLP) found in many fungi. In contrast, there is less information on viruses or VLPs in eukaryotic algae. About 35 reports have appeared since 1971 showing that certain eukaryotic algae harbor particles, usually polyhedral in shape, which resemble viruses (see reviews in Lemke, 1976; Sherman and Brown, 1978; Dodds, 1983; Van Etten et al., 1987). With a few exceptions (Gibbs et al., 1975; Skotnicki et al., 1976; Waters and Chan, 1982; Cole et al., 1980; Dodds and Cole, 1980; Stanker et al., 1981) these reports are primarily ultrastructural studies. The paucity of information on algal viruses or VLPs is due primarily to the difficulty in obtaining sufficient quantities of particles. Several factors have contributed to this lack of material: (1) usually only a few cells in a population contained particles, (2) usually the particles are contained only within one stage of the algal life cycle, (3) in many instances the particles did not lyse the cells, and (4) in most cases, the particles were not infectious. These factors, plus the fact that some particles were present in multicellular filamentous algae, hindered the development of a biological assay for them.

These difficulties do not exist, however, for a family of large polyhedral, dsDNA-containing viruses that infect and replicate in certain strains of unicellular, exsymbiotic chlorella-like green algae. These viruses, commonly found in freshwater ponds, lakes, and streams, can be produced in milligram quantities in culture. Furthermore, the viruses can be assayed by plaque formation on lawns of the host chlorella (Van Etten et al., 1983a,1985a,c; Schuster et al., 1986a). One-step growth studies, life cycle studies, and genetic studies (Tessman, 1985) are underway for some of these viruses. This report will describe some of the pertinent properties of these chlorella viruses focusing primarily on the most studied one, PBCV-1. Additional reviews of these chlorella viruses have recently appeared (Meints et al., 1985; Van Etten et al., 1986b, 1987).

HOST CHLORELLA

The hosts for the viruses are three exsymbiotic chlorella isolates (designated strains NC64A, N1a, and ATCC-30562). These isolates originally existed as hereditary endosymbionts in the protozoan *Paramecium bursaria* (Van Etten et al., 1983a; Meints et al., 1986). At the ultrastructural level, the three strains appear identical, and virus PBCV-1 replicates at the same rate in each of them. Chlorellas NC64A and N1a, isolated 25 years apart, contain identical chloroplast genomes, as shown by DNA restriction polymorphisms (R. H. Meints and S. Amberg, unpublished data). However, chlorella strain N1a grows slightly faster in culture (doubling time of 20 hr versus 27 hr for NC64A) and reaches a higher cell density (5×10^8 cells/ml versus 5×10^7 cells/ml) than does NC64A. The majority of the experiments on the chlorella viruses have used strain NC64A as the host.

The virus PBCV-1 attaches rapidly and specifically to cell walls of host strains but not to cell walls of nonhost, free-living chlorellas (Meints et al., 1984). Thus the host range for PBCV-1 and presumably the other viruses is limited to chlorellas with the appropriate viral receptor.

The ease with which plaque forming viruses are found on chlorella NC64A suggests that viruses infecting other eukaryotic, free-living algae may be common in nature. However, attempts to find plaque forming viruses for other chlorella or chlamydomonas species have been unsuccessful (See Van Etten et al., 1985c,1987 for more comments on the natural history of the viruses.)

It should be noted that the genus *Chlorella* encompasses a heterogeneous collection of asexually reproducing, unicellular green algae (Shihra and Krauss, 1965; DaSilva and Gyllenberg, 1972). This diversity is reflected in the guanosine plus cytosine content of their nuclear DNAs, ranging from 43 to 79% (Hellmann and Kessler, 1974). In addition, the chemical composition and morphology of chlorella cell walls vary widely (Northcote et al., 1958; Atkinson et al., 1972; Takeda and Hirokawa, 1978,1979; Yamada and Sakaguchi, 1982). As a consequence of this diversity, properties discovered in one species or isolate of chlorella may or may not apply to another.

PROPERTIES OF VIRUSES

PBCV-1

PBCV-1 is easy to purify and about 2×10^{12} plaque forming units (PFUs) of purified virus are obtained per liter of lysate (Van Etten et al., 1983b). Purified virus can be stored at 4°C for at least 2 years without appreciable loss of infectivity; however, freezing and thawing rapidly destroy virus infectivity. PBCV-1 has a M.W. of about 1×10^9 (Yonker et al., 1985) and is composed of 64% protein, 21 to 25% DNA, and 5 to 10% lipid (Van Etten et al., 1983b; Skrdla et al., 1984). PBCV-1 contains at least 50 structural proteins ranging from 10 to 135 kDa. The major protein of 54 kDa comprises about 40% of the total virus protein. This protein and a 135 kDa protein are probably glycoproteins. These two glycoproteins, plus two additional proteins (14.5 and 71 kDa), are labeled when intact virus is treated with ^{125}I, indicating their surface location (Skrdla et al., 1984).

Phosphatidylcholine, phosphatidylethanolamine, and one unidentified component are the major virus lipids (Skrdla et al., 1984). Treatment with chloroform, ethyl ether, or toluene destroys PBCV-1 infectivity; thus the lipid component, which we believe is internal to a protein shell, is required for infectivity.

The large genome of $\sim$335 kilobase pairs, based on restriction fragment summation (Van Etten et al., 1982; Girton and J. L. Van Etten, 1987), has the potential to code for at least 300 proteins. Discrete 5' and 3' ends of the DNA were not labeled by end-labeling techniques (L. E. Girton and A. M. Schuster, unpublished data). The structure of the DNA is unknown.

Additional Viruses

Twenty-nine additional plaque forming viruses, which infect the same chlorellas as PBCV-1, have been characterized sufficiently to group them into 11 classes (25 of these viruses are listed in Table 1). Like PBCV-1, each of these viruses contains many structural proteins, a large dsDNA genome (at least 300 kilobase pairs) with a 40% guanosine plus cytosine content, and the infectivity of each of them is reduced

Table 1 Separation of 25 Chlorella Viruses into 11 Classes[a]

Virus	Class	Plaque size (mm)	Reacts with antibody to[b] PBCV-1	NY-2A	Restriction group[c]	Level of methylation 5mC[d]	6mA[e]
NE-8D	1	3	Yes	No	A	0.44	nd[f]
NYb-1		3	Yes	No	A	1.60	nd
CA-4B		3	Yes	No	A	0.12	nd
AL-1A	2	3	No	Partial	A	0.45	nd
NY-2C		3	No	Partial	A	0.39	nd
NC-1D		3	No	Partial	A	0.33	nd
PBCV-1	3	3	Yes	No	C	1.86	1.45
NC-1C		3	Yes	No	C	1.72	1.59
CA-1A	4	3	Yes	No	B	10.0	nd
IL-2A		3	Yes	No	B	9.4	nd
IL-3A		3	Yes	No	B	9.7	nd
IL-3D		3	Yes	No	B	12.6	nd
SC-1A	5	1	Yes	No	D	1.94	7.34
SC-1B		1	Yes	No	D	2.04	7.45
NC-1A	6	3	Yes	No	D	7.1	7.3
NE-8A	7	3	Yes	No	E	14.3	8.1
MA-1E		3	Yes	No	E	14.9	8.1
NY-2F		3	Yes	No	E	14.6	8.1
NYs-1	8	1	No	Yes	F	47.5	11.3
IL-5-2s1	9	1	No	Yes	G	45.0	16.1
AL-2A		1	No	Yes	G	35.8	14.6
MA-1D		1	No	Yes	G	47.0	16.7
NY-2B		1	No	Yes	G	36.5	16.2
CA-4A	10	1	No	Yes	H	39.8	19.6
NY-2A	11	0.5	No	Yes	I	44.9	37.0

[a]Based on at least one of the following criteria: plaque size, reaction with virus antisera, sensitivity of virus DNAs to 13 restriction endonucleases, and percentage of 5mC and 6mA in their genomic DNAs.
[b]5 μg of purified virus was mixed with twofold serial dilutions of antiserum and the precipitant monitored 2 hr later (Ball, 1974); "yes" means a reaction was obtained at an antibody dilution of 1:64 or greater; "partial" a dilution of 1:8 to 1:32, and "no" a dilution of 0 to 1:4.

several orders of magnitude by treatment with chloroform. Negative staining indicates that the viruses are polyhedrons, with diameters ranging from 150 to 190 nm. The DNAs of some of the viruses hybridize extensively with PBCV-1 DNA, while others hybridize poorly with PBCV-1 DNA (Schuster et al., 1986b).

The viruses can be distinguished from PBCV-1 and from each other plaque size, reaction with viral antisera, DNA restriction patterns, sensitivity of the DNAs to 13 bacterial restriction endonucleases, and (most important), the nature and abundance of methylated bases in their genomes. DNA from each of the viruses contained 5-methyldeoxycytidine (5mC); the concentration of 5mC, as a percentage of deoxycytidine, varied from 0.12% to 47.5% (Table 1). In addition, 15 of the 25 viral DNAs contained N^6-methyldeoxyadenosine (6mA); the concentration of 6mA, as a percentage of deoxyadenosine, varied from 1.45% to 37%.

Methylation of the cytidines in the virus DNAs is required for virus replication since 5-azacytidine (5-AzaC), an inhibitor of cytidine

Table 1 (Continued)

[c]Viral DNAs were tested for susceptibility to the 13 restriction endonucleases: *Bgl*II, *Eco*RI, *Bam*HI, *Sal*I, *Xho*I, *Bcl*I, *Pst*I, *Hind*III, *Sst*I, *Mbo*I, *Dpn*I, *Msp*I, and *Hpa*II. The DNAs were grouped according to the resistance of the DNAs to these enzymes.

Group	Resistant to
A	*Dpn*I
B	*Hind*III, *Sst*I, and *Dpn*I
C	*Bcl*I and *Mbo*I
D	*Sal*I, *Xho*I, *Bcl*I, *Pst*I, and *Mbo*I
E	*Sal*I, *Xho*I, *Bcl*I, *Pst*I, *Hind*III, *Sst*I, and *Mbo*I
F	*Bam*HI, *Bcl*I, *Pst*I, *Hind*III, *Sst*I, *Mbo*I, *Msp*I, and *Hpa*II
G	*Bam*HI, *Sal*I, *Xho*I, *Bcl*I, *Pst*I, *Hind*III, *Sst*I, *Mbo*I, *Msp*I, and *Hpa*II
H	*Eco*RI, *Bam*HI, *Sal*I, *Xho*I, *Bcl*I, *Pst*I, *Hind*III, *Sst*I, *Mbo*I, *Msp*I, and *Hpa*II
I	*Bgl*II, *Eco*RI, *Bam*HI, *Sal*I, *Xho*I, *Bcl*I, *Pst*I, *Hind*III, *Sst*I, *Mbo*I, *Msp*I, and *Hpa*II

[d]Percentage 5mC per dC + 5mC + deoxyuridine.
[e]Percentage 6mA per dA + 6mA + deoxyinosine.
[f]nd, not detected.
Source: Modified from Schuster et al. (1986b).

methylation (Jones, 1984), inhibits virus replication. As reported in Table 2, there is a direct correlation between increasing concentrations of 5mC in the virus genome and sensitivity of virus replication to 5-AzaC. It will be interesting to isolate 5-AzaC-resistant mutants of some of these viruses to see if their DNAs have reduced levels of 5mC and to determine if these changes effect DNA replication and gene expression.

The finding of methylated bases in the genomes of the chlorella viruses was unexpected since DNA containing viruses that infect eukaryotes typically lack methylated bases (Doerfler, 1981,1983). Besides these chlorella viruses, the other exceptions are the two iridoviruses, frog virus 3 (FV3) (Willis and Granoff, 1980), and fish lymphocystis disease virus (Wagner et al., 1985), in which 20 to 22% of the deoxycytidines are methylated, Epstein-Barr HR-1 virus with 15% 5mC (Diala and Hoffman, 1983), and human papilloma virus, which has small amounts of 5mC (Danos et al., 1980).

VIRUS REPLICATION

Infection

PBCV-1 attaches rapidly and specifically to the host chlorella cell wall. The host receptor, to which the virus attaches irreversibly, has not been positively identified, although our evidence suggests that it is a carbohydrate (R. H. Meints, unpublished data). Ultrastructural studies indicate that PBCV-1 attaches to the cell wall by one of its hexagonal vertices and enzymatically digests the wall at the attachment point. Virus DNA is then released into the cell interior, leaving an empty capsid on the surface (Meints et al., 1984). PBCV-1 DNA is not exposed to the environment during infection, since exogenous DNase has no effect on virus infectivity (J. L. Van Etten and D. E. Burbank, unpublished data).

Each host chlorella cell can bind 5000 PBCV-1 virus particles as measured by Scatchard analysis (R. H. Meints and J. L. Van Etten, unpublished data). We calculate that 3000 to 5000 virus particles can fit on the surface of a single cell, assuming the host cell has a mean diameter of 5 μm. The similarity of these values indicates that the entire cell surface contains PBCV-1 receptors.

Life Cycle

The destination of PBCV-1 DNA and the intracellular site of virus DNA replication and transcription are unknown, although inhibitor studies indicate that virus protein synthesis occurs on cytoplasmic ribosomes (Skrdla et al., 1984). PBCV-1 capsid assembly is first observed 1 to 2 hr after infection. Capsid assembly followed by DNA packaging

Table 2 Effect of 5-Azacytidine on the Replication of Four Chlorella Viruses[a]

Concentration of 5-azacytidine (μm)	Infective centers			
	CA-4B (0.12%)[b]	PBCV-1 (1.9%)	IL-3A (9.7%)	NY-2A (45%)
0	6.3×10^6	4.4×10^6	16×10^6	3.7×10^6
1	5.5×10^6	5.0×10^6	18×10^6	3.4×10^6
5	5.7×10^6	3.8×10^6	11×10^6	4.8×10^5
10	5.5×10^6	4.2×10^6	5.7×10^6	3.2×10^5
25	5.1×10^6	2.4×10^6	5.1×10^5	1×10^1
50	4.2×10^6	3.0×10^5	0	0
100	1.5×10^6	1.1×10^4	0	0

[a]Chlorella NC64A cells were infected with each of the viruses in the presence of 5-azacytidine. After 60 min the cells were centrifuged and plaqued for infective centers in lawns of chlorella NC64A; the lawns also contained 5-azacytidine.
[b]The number in brackets is the percentage of 5mC in the viral DNA. (See Table 1.)

occurs in virus assembly centers located in the cytoplasm (Meints et al., 1986). Progeny virus are released by localized lysis of the host cell wall and first appear 3 to 4 hr after infection; by 6 to 8 hr, release is complete (Figure 1). The PBCV-1 burst size is 200 to 350 plaque forming units (PFU). Photosynthesis is not required for PBCV-1 replication, since the virus replicates in cells grown in the dark or in cells previously treated with the photosynthetic inhibitor 3-(3,4-dichlorophenyl)-1,1 dimethylurea. Apparently, a functional host nucleus is not required for virus replication, since PBCV-1 also replicates, although more slowly and with a smaller burst size, in UV-irradiated chlorella (Van Etten et al., 1986a).

Growth curves for five other viruses are shown in Figure 1. Viruses that form plaques similar in size to PBCV-1 replicate at about the same rate as PBCV-1 (e.g., NC-1A and IL-3A), whereas viruses that form small plaques (e.g., SC-1A, CA-4A, and NY-2A) take about 2 to 3 times longer to replicate. The burst size for each of the viruses is typically 200 to 350 PFUs.

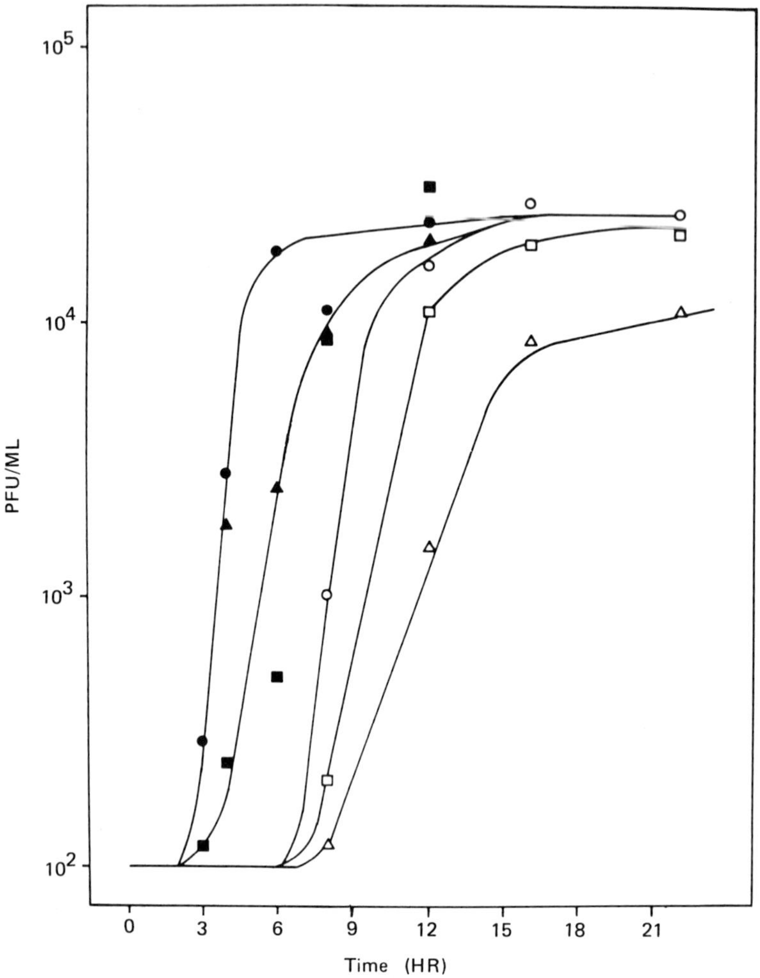

Figure 1 One-step growth curves for chlorella viruses: PBCV-1 (●) [230], IL-3A (■) [310], NC-1A (▲) [200], SC-1A (○) [250], CA-4A (□) [290], and NY-2A (△) [110]. The number in brackets is the burst size. Only one line is shown for IL-3A and NC-1A.

DNA Synthesis

PBCV-1 infection affects DNA metabolism in several ways (Van Etten et al., 1984). (1) Immediately following virus infection host DNA synthesis stops. (2) Virus DNA synthesis begins about 45 min after infection and requires de novo synthesis of a protein or proteins. This implies that PBCV-1 either directs the synthesis of its own DNA polymerase(s) or an essential component required for DNA synthesis. (3) Both host nuclear and chloroplast DNAs are degraded beginning about 1 hr after infection. A virus-encoded protein(s) is, at least partially, responsible for this process since addition of cycloheximide at the time of PBCV-1 infection prevents host DNA degradation and host DNA is degraded if UV-inactivated chlorella are infected with PBCV-1 (Xia et al., 1986a). (4) PBCV-1 infection increases the total DNA in the chlorella cell by at least three- to fourfold 4 hr after infection. Consequently, PBCV-1 infection must place a tremendous demand on the host for energy and deoxynucleotide intermediates. Some of these intermediates presumably come from recycled host DNA, but these DNA-derived intermediates can supply only 25% of the total required.

RNA Synthesis

Studies on RNA synthesis after PBCV-1 infection have led to the following observations and conclusions (Schuster et al., 1986a). (1) Virus infection results in a rapid, but possibly not complete, inhibition of host RNA synthesis. (2) Chloroplast ribosomal RNAs, but not cytoplasmic ribosomal RNAs, are degraded beginning about 30 min after virus infection. Presumably a virus encoded protein(s) is at least partially involved in this process since addition of cycloheximide at the time of PBCV-1 infection prevents degradation. (3) Virus transcription is programmed and some PBCV-1 transcripts are detected as early as 5 min after infection. (4) A few, but not all, early viral transcripts are synthesized in the absence of de novo protein synthesis. Additional synthesis of later transcripts depends on the translation of an early gene product(s). (5) The transition from early to late virus transcription occurs between 40 and 60 min after infection and correlates with the start of virus DNA synthesis. As reported in Table 3 the incorporation of [^{3}H]adenine into polyadenylate containing [poly A(+)] RNA also changes abruptly at about 60 min after PBCV-1 infection. Polyadenylate segments are virtually absent from late transcripts while at least some early virus transcripts are probably polyadenylated. (6) Early and late genes are not clustered within the virus genome, since several PBCV-1 DNA clones hybridized to both early and late transcripts. (7) Three of four PBCV-1 DNA clones tested hybridized to multiple transcripts which additively were larger than the corresponding DNA probe. This finding suggests that either virus transcripts undergo posttranscriptional modification, the virus

Table 3 Distribution of RNA, from PBCV-1 Infected Chlorella NC64A, into PolyA(-) and PolyA(+) Fractions[a]

Cells	Pulse time	PolyA(-) (cpm × 10^4)	PolyA(+) (cpm × 10^4)	Percentage polyA(+)
Uninfected	0-3 hr	370	46	12.4
Infected	0-30 min	24	5.4	22.5
Infected	30-60 min	19	4.9	25.8
Infected	60-90 min	56	3.4	6.1
Infected	90-120 min	42	1.9	4.5
Infected	120-240 min	170	3.7	2.2

[a]Chlorella cells infected with PBCV-1 at a m.o.i. of 5, were labeled with [^{3}H]adenine for the indicated times; total RNA was isolated and separated by chromatography on an oligo dT-cellulose column.

has overlapping genes, or transcription occurs from both DNA strands. Either one of these latter possibilities are especially intriguing since a virus with such a large genome would seem to be under little selective pressure to optimize its coding capacity.

PBCV-1 INFECTION INDUCES A MODIFICATION AND RESTRICTION SYSTEM

Both PBCV-1 and host chlorella nuclear DNAs contain methylated bases. Twenty-one percent of the deoxycytidines in the nuclear DNA are 5mC and 0.6% of the deoxyadenosines are 6mA; PBCV-1 DNA contains 1.9% 5mC and 1.5% 6mA (Van Etten et al., 1985b). However, at least some, if not all of the methylated bases in PBCV-1 DNA are located in different base sequences than those in the host DNA. For example, PBCV-1 DNA contains 6mA in the sequence GATC, whereas the host DNA does not. This finding suggests that PBCV-1 infected chlorella cells contain a DNA methyltransferase which specifically methylates deoxyadenosines in virus GATC sequences. We have found this methyltransferase activity in PBCV-1 infected chlorella cells but not in uninfected cells (Xia and Van Etten, 1986). The methyltransferase (named M.*Cvi*AI) activity is first detected 30 to 60 min after infection, which is the time that PBCV-1 DNA synthesis begins. M.*Cvi*AI activity increases until 3 to 4 hr after infection.

PBCV-1 infection also results in the appearance of a type II DNA restriction endonuclease that recognizes the same GATC sequence recognized by M.*Cvi*AI (Xia et al., 1986a). The restriction endonuclease (named *Cvi*AI) cleaves DNA 5' to the G and does not cleave DNAs containing $G^{m}ATC$ sequences. In vitro, *Cvi*AI cleaves host chlorella DNA but not PBCV-1 DNA, because host DNA contains GATC sequences while PBCV-1 DNA contains $G^{m}ATC$ sequences. *Cvi*AI activity, which is not present in uninfected cells, is first detected 30 to 60 min after virus infection. The appearance of *Cvi*AI activity coincides with the onset of host DNA degradation following PBCV-1 infection.

The following observations suggest that both *Cvi*AI and M.*Cvi*AI are encoded by PBCV-1. (1) Both enzyme activities first appear 30 to 60 min after PBCV-1 infection. (2) The appearance of both enzyme activities is dependent on de novo protein synthesis since neither activity appears if cycloheximide and PBCV-1 are added to the cells at the same time. (3) Both enzyme activities appear in UV-irradiated chlorella inoculated with PBCV-1. Presumably only virus-encoded genes are expressed in these irradiated cells since endogenous host nucleic acid and protein syntheses are reduced to nearly background levels (Van Etten et al., 1986a). (4) Infection of the same chlorella species with viruses other than PBCV-1 induces the synthesis of DNA methyltransferases and restriction endonucleases that recognize different DNA sequences than the sequence recognized by the PBCV-1-induced enzymes. However, conclusive evidence that *Cvi*AI and M.*Cvi*AI are encoded by PBCV-1 will require isolation and characterization of their genes.

It will be interesting to compare the genes for M.*Cvi*AI and *Cvi*AI with genes from bacteria that code for methyltransferases (e.g., *Escherichia coli dam* methylase) and restriction endonucleases (e.g., *Mbo*I) which recognize the same GATC sequence. This will allow us to determine their degree of homology and perhaps their evolutionary relationships. We have probed PBCV-1 DNA with the *E. coli dam* methylase gene (kindly provided by Martin Marinus) at low stringency; no hybridiczation was detected (K. E. Narva and J. L. Van Etten, unpublished data).

VIRUS INFECTION INDUCES OTHER RESTRICTION MODIFICATION SYSTEMS

DNA from each of the viruses listed in Table 1 contain 5mC and 15 of the 25 viruses also contain 6mA. In the majority of the viruses it is not known which DNA sequences contain methylated bases; however, we assume that a specific DNA sequence contains either 5mC or 6mA if a DNA is resistant to a restriction endonuclease known to be inhibited by a 5mC or 6mA in its recognition site. This conclusion implies that infection with these viruses, e.g., PBCV-1, induces the appearance of new DNA methyltransferase(s). Enzyme extracts from virus

NC-1A-infected chlorellas contain at least three distinct adenosine DNA methyltransferase activities; separate enzymes methylate adenosines in TCGA, GATC, and GANTC sequences of the NC-1A genome (Y. Xia, K. E. Narva, and J. L. Van Etten, unpublished data).

We have also assayed extracts prepared from chlorella cells infected with different viruses for DNA restriction endonuclease activity. NC-1A infected cells contain a restriction endonuclease that recognizes the sequence GANTC. This enzyme (named *Cvi*BI), like the bacterial restriction endonuclease *Hin*fI, cleaves DNA between G and A; the presence of 6mA in the recognition sequence probably inhibits its activity (Xia et al., 1986b). The restriction endonucleases induced by infection with each of the three viruses included in class 7 (Table 1) also recognize the sequence GANTC and cleave within this sequence; however, the precise cleavage site has not been determined for these three enzymes.

When chlorella cells are infected with yet another virus, NY-2A, at least two DNA restriction endonuclease activities are separated on a phosphocellulose column. One of these enzymes, named *Cvi*II, cleaves unmethylated pBR322 DNA into the same size fragments as the bacterial restriction endonuclease *Rsa*I; thus it apparently recognizes the sequence GTAC (Xia et al., 1987). The other NY-2A restriction endonuclease(s) cleaves unmethylated pBR322 DNA into small fragments but does not cleave NY-2A DNA. The recognition site of this enzyme(s) is not known. (See Van Etten et al., 1986b for more comments on the virus-induced methyltransferases and restriction endonucleases.)

We believe the virus-induced DNA methyltransferase and restriction endonuclease enzymes constitute virus-encoded modification and restriction systems. The restriction endonuclease(s) probably cleaves host nuclear and chloroplast DNA into fragments that are digested to deoxynucleotides by nonspecific DNases; after phosphorylation the deoxynucleotides are recycled into virus DNA. The newly synthesized virus DNA is resistant to the restriction endonuclease because its DNA is methylated by the corresponding methyltransferase. Consistent with this hypothesis is the finding that in every system we have examined the virus-induced restriction endonuclease can cleave the host nuclear DNA but not the parent virus DNA in vitro; thus the virus DNA is presumably modified in the DNA sequence recognized by the virus induced-restriction endonuclease.

SIMILARITY OF THE CHLORELLA VIRUSES TO OTHER VIRUSES

The chlorella viruses have several properties in common with the Iridoviridae, also called icosahedral cytoplasmic deoxyriboviruses (Goorha and Granoff, 1979; Matthews, 1982; Willis, 1985) and probably will be placed in this group. Iridoviruses are widely distributed in nature and infect invertebrates and some vertebrates. Common properties between the Iridoviridae and the chlorella viruses include icosahedral morphology, a large dsDNA genome, a lipid component comprising 5 to 10% of the virus weight, and a cytoplasmic site of virus assembly. The most studied iridovirus is FV3. FV3, like the chlorella viruses, is one of a few known viruses infecting a eukaryotic organism that contains 5mC in its genome (Willis and Granoff, 1980); FV3 also encodes a methyltransferase enzyme (Willis et al., 1984). Another similarity is that FV3 transcripts (Willis and Granoff, 1976) and, at least late PBCV-1 transcripts, lack polyadenylated segments. However, FV3 and PBCV-1 also differ in several ways. First, the major structural protein in the chlorella viruses is probably a glycoprotein; FV3 lacks glycoproteins (Elliott and Kelly, 1980). Second, FV3 typically is uncoated inside the host cell (Hauts et al., 1974; Kelley, 1975) whereas PBCV-1 is uncoated at the cell surface. Third, FV3 cannot initiate replication in UV-inactivated host cells (Goorha et al., 1977) while PBCV-1 initiates and completes viral replication at a slow rate in UV-inactivated chlorella.

The replication of FV3 DNA is unusual for viruses that infect eukaryotic organisms because DNA replication occurs at two intracellular sites (Goorha et al., 1978; Goorha, 1982). Initially, genomic lengths of FV3 DNA are synthesized in the nucleus. These progeny DNAs move to the cytoplasm where they participate in a second stage of DNA replication leading to DNA concatemers. The DNA concatemers are apparently processed during FV3 DNA packaging. The finding that FV3 genomes are circularly permuted and terminally redundant (Goorha and Murti, 1981) is consistent with this model. While we have concluded tentatively that PBCV-1 DNA maps as a circle, it remains to be determined if PBCV-1 DNA is circularly permuted and terminally redundant and if its replication involves two intracellular sites.

FINAL COMMENTS

There are several reasons why we believe these chlorella viruses are interesting and deserve investigation. (1) Virus-infected cells are excellent model systems for studying gene regulation and expression in a photosynthetic eukaryote. (2) The viruses differ from most viruses infecting higher plants since they contain a dsDNA genome. We are attempting to construct vectors for transferring genes into algae using portions of the viral genomes such as the origin(s) of replication and promoters. (3) Virus-infected chlorella cells may serve as an important new source of DNA methyltransferase and DNA restriction endonuclease enzymes.

ACKNOWLEDGMENTS

This manuscript is dedicated to the late Dr. William Muir, a dedicated teacher at Carleton College. His encouragement was responsible for one of us (J.L.V.E.) pursuing a career as a research scientist.

Research from our laboratories has been supported by grants from the Department of Energy (J.L.V.E.), the National Institutes of Health (J.L.V.E.), the National Science Foundation (R.H.M.), the United States Department of Agriculture (R.H.M.), and the Solar Energy Research Institute (R.H.M.).

REFERENCES

Atkinson, A. W., Gunning, B. E. S., and John, P. C. L. (1972). Sporopollenin in the cell wall of chlorella and other algae: Ultrastructure, chemistry, and incorporation of [^{14}C]acetate, studied in synchronous cultures. *Planta 107*:1-32.

Ball, E. (1974). *Serological Tests for the Identification of Plant viruses.* American Phytopathological Society, St. Paul, Minnesota, p. 31.

Cole, A., Dodds, J. A., and Hamilton, R. I. (1980). Purification and some properties of a double-stranded DNA containing virus-like particle from *Uronema gigas*, a filamentous eukayotic green alga. *Virology 100*:166-174.

Danos, O., Katinka, M., and Yaniv, M. (1980). Molecular cloning, refined physical map and heterogeneity of methylation sites of papilloma virus type 1a DNA. *Eur. J. Biochem. 109*:457-461.

DaSilva, E. J., and Gyllenberg, H. G. (1972). A taxonomic treatment of the genus chlorella by the technique of continuous classification. *Arch. Mikrobiol. 87*:99-117.

Diala, E. S., and Hoffman, R. M. (1983). Epstein-Barr HR-1 virion DNA is very highly methylated. *J. Virol. 45*:481-483.

Dodds, J. A. (1983). New viruses of eukaryotic algae and protozoa. In *A Critical Appraisal of Viral Taxonomy*, R. E. F. Matthews (Ed.). CRC Press, Boca Raton, Florida, pp. 177-188.

Dodds, J. A., and Cole, A. (1980). Microscopy and biology of *Uronema gigas*, a filamentous eukaryotic green alga, and its associated tailed virus-like particle. *Virology 100*:156-165.

Doerfler, W. (1981). DNA Methylation—a regulatory signal in eukaryotic gene expression. *J. Gen. Virol. 57*:1-20.

Doerfler, W. (1983). DNA methylation and gene activity. *Annu. Rev. Biochem.* *52*:93-124.

Elliott, R. M., and Kelly, D. C. (1980). Frog virus 3 replication: Induction and intracellular distribution of polypeptides in infected cells. *J. Virol.* *33*:28-51.

Gibbs, A., Skotricki, A. H., Gardiner, J. E., Walker, E. S., and Hollings, M. (1975). A tobamovirus of a green alga. *Virology* *64*:571-574.

Girton, L. E., and Van Etten, J. L. (1987). Restriction site map of the chlorella virus PBCU-1 genome. *Plant Mol. Biol.* *9*:247-257.

Goorha, R. (1982). Frog virus 3 DNA replication occurs in two stages. *J. Virol.* *43*:519-528.

Goorha, R., and Granoff, A. (1979). Icosahedral cytoplasmic deoxyriboviruses. In *Comprehensive Virology*, Vol. 14, H. Fraenkel-Conrat and R. R. Wagner (Eds.). Plenum, New York, pp. 347-399.

Goorha, R., and Murti, K. G. (1982). The genome of frog virus 3, an animal DNA virus, is circularly permuted and terminally redundant. *Proc. Nat. Acad. Sci. USA* *79*:248-252.

Goorha, R., Willis, D. B., and Granoff, A. (1977). Macromolecular synthesis in cells infected by frog virus 3. VI. Frog virus 3 replication is dependent on the cell nucleus. *J. Virol.* *21*:802-805.

Goorha, R., Murti, G., Granoff, A., and Tirey, R. (1978). Macromolecular synthesis in cells infected by frog virus 3. VIII. The nucleus is a site of frog virus 3 DNA and RNA synthesis. *Virology* *84*:32-50.

Hauts, G. E., Gravell, M., and Granoff, A. (1974). Electron microscopic observation of early events of frog virus 3 replication. *Virology* *58*:589-594.

Hellmann, V., and Kessler, E. (1974). Physiologische und biochemische Beitrag zur Taxonomie der Gattung chlorella. VIII Die Basenzusammensetzung der DNS. *Arch. Microbiol.* *95*:311-318.

Jones, P. A. (1984). Gene activation by 5-azacytidine. In *DNA Methylation: Biochemistry and Biological Significance*, A. Razin, H. Cedar and A. D. Riggs (Eds.). Springer-Verlag, New York, pp. 165-187.

Kelly, D. C. (1975). Frog virus 3 replication: Electron microscope observations on the sequence of infection in chick embryo fibroblasts. *J. Gen. Virol.* *26*:71-86.

Lemke, P. A. (1976). Viruses of eukaryotic microorganisms. *Annu. Rev. Microbiol. 30*:105-145.

Matthews, R. E. F. (1982). Classification and nomenclature of viruses. *Intervirology 17*:1-199.

Meints, R. H., Lee, K., Burbank, D. E., and Van Etten, J. L. (1984). Infection of a chlorella-like alga with the virus, PBCV-1: Ultrastructural studies. *Virology 138*:341-346.

Meints, R. H., Schuster, A. M., and Van Etten, J. L. (1985). Chlorella viruses. *Plant Mol. Biol. Rept. 3*:180-187.

Meints, R. H., Lee, K., and Van Etten, J. L. (1986). Assembly site of the virus PBCV-1 in a chlorella-like green alga: Ultrastructural studies. *Virology 154*:240-245.

Northcote, D. H., Goulding, K. J., and Horne, R. W. (1958). The chemical composition and structure of the cell wall of chlorella pyrenoidosa. *Biochem. J. 70*:391-397.

Schuster, A. M., Girton, L., Burbank, D. E., and Van Etten, J. L. (1986a). Infection of a chlorella-like alga with the virus PBCV-1: Transcriptional studies. *Virology 148*:181-189.

Schuster, A. M., Burbank, D. E., Meister, B., Skrdla, M. P., Meints, R. H., Hattman, S., Swinton, D., and Van Etten, J. L. (1986b). Characterization of viruses infecting a eukaryotic chlorella-like green alga. *Virology 150*:170-177.

Sherman, L. A., and Brown, R. M. (1978). Cyanophages and viruses of eukaryotic algae. In *Comprehensive Virology*, Vol. 12, H. Fraenkel-Conrat and R. R. Wagner (Eds.). Plenum, New York, pp. 145-234.

Shihra, I., and Krauss, R. W. (1965). Chlorella: Physiology and taxonomy of forty-one isolates. University of Maryland, College Park, Maryland.

Skotnicki, A., Gibbs, A., and Wrigley, N. G. (1976). Further studies on *Chara corrallina* virus. *Virology 75*:457-468.

Skrdla, M. P., Burbank, D. E., Xia, Y., Meints, R. H., and Van Etten, J. L. (1984). Structural proteins and lipids in a virus, PBCV-1, which replicates in a chlorella-like alga. *Virology 135*: 308-315.

Stanker, L. H., Hoffman, L. R., and MacLeod, R. (1981). Isolation and partial chemical characterization of a virus-like particle from a eukaryotic alga. *Virology 114*:357-369.

Takeda, H., and Hirokawa, T. (1978). Studies on the cell wall of chlorella. I. Quantitative changes in cell wall polysaccharides during the cell cycle of *Chlorella ellipsoidea*. *Plant Cell Physiol.* *19*:591-598.

Takeda, H., and Hirokawa, T. (1979). Studies on the cell wall of chlorella. II. Mode of increase of glucosamine in the cell wall during synchronous growth of *Chlorella ellipsoidea*. *Plant Cell Physiol.* *20*:489-491.

Tessman, I. (1985). Genetic recombination of the DNA plant virus PBCV-1 in a chlorella-like alga. *Virology* *145*:319-322.

Van Etten, J. L., Meints, R. H., Kuczmarski, D., Burbank, D. E., and Lee, K. (1982). Viruses of symbiotic chlorella-like algae isolated from *Paramecium bursaria* and *Hydra viridis*. *Proc. Nat. Acad. Sci. USA* *79*:3867-3871.

Van Etten, J. L., Burbank, D. E., Kuczmarski, D., and Meints, R. H. (1983a). Virus infection of culturable chlorella-like algae and development of a plaque assay. *Science* *219*:994-996.

Van Etten, J. L., Burbank, D. E., Xia, Y., and Meints, R. H. (1983b). Growth cycle of a virus, PBCV-1, that infects chlorella-like algae. *Virology* *126*:117-125.

Van Etten, J. L., Burbank, D. E., Joshi, J., and Meints, R. H. (1984). DNA synthesis in a chlorella-like alga following infection with the virus, PBCV-1. *Virology* *134*:443-449.

Van Etten, J. L., Burbank, D. E., Schuster, A. M., and Meints, R. H. (1985a). Lytic viruses infecting a chlorella-like alga. *Virology* *140*:135-143.

Van Etten, J. L., Schuster, A. M., Girton, L., Burbank, D. E., Swinton, D., and Hattman, S. (1985b). DNA methylation of viruses infecting a eukaryotic chlorella-like green alga. *Nucleic Acids Res.* *13*:3471-3478.

Van Etten, J. L., Van Etten, C. H., Johnson, J. K., and Burbank, D. E. (1985c). A survey for viruses from fresh water that infect a eucaryotic chlorella-like green alga. *Appl. Environ. Microbiol.* *49*:1326-1328.

Van Etten, J. L., Burbank, D. E., and Meints, R. H. (1986a). Replication of the algal virus PBCV-1 in ultraviolet irradiated chlorella. *Intervirology* *26*:115-120.

Van Etten, J. L., Xia, Y., Narva, K. E., and Meints, R. H. (1986b). Chlorella algal viruses. In *Extrachromosomal Elements in Lower Eukaryotes*, R. B. Wickner, A. Hinnebusch, A. M. Lambowitz, I. C. Gunsalau, and A. Hollaender (Eds.). Plenum, New York, pp. 337-347.

Van Etten, J. L., Xia, Y., and Meints, R. H. (1987). Viruses of a chlorella-like green alga. In *Plant-Microbe Interactions*, Vol. 2, T. Kosuge and E. W. Nester (Eds.). MacMillan, New York, pp. 307-325.

Wagner, H., Simon, D., Werner, E., Gelderblom, H., Darai, C., and Flugel, R. M. (1985). Methylation pattern of fish lymphocystis disease virus DNA. *J. Virol. 53*:1005-1007.

Waters, R. E., and Chan, A. T. (1982). *Micromonas pusilla* virus: The virus growth cycle and associated physiological events within the host cells; host range mutation. *J. Gen. Virol. 63*:199-206.

Willis, D. B. (Ed.). (1985). Iridoviridae. *Curr. Top. Microbiol. Immunol. 116*:1-170.

Willis, D. B., and Granoff, A. (1976). Macromolecular synthesis in cells infected with frog virus 3. V. The absence of polyadenylic acid in the majority of frog virus 3-specific mRNA species. *Virology 73*:543-547.

Willis, D. B., and Granoff, A. (1980). Frog virus 3 DNA is heavily methylated at CpG sequences. *Virology 107*:250-257.

Willis, D. B., Goorha, R., and Granoff, A. (1984). DNA methyltransferase induced by frog virus 3. *J. Virol. 49*:86-91.

Xia, Y., and Van Etten, J. L. (1986). A DNA methyltransferase induced by PBCV-1 virus infection of a chlorella-like green alga. *Mol. Cell. Biol. 6*:1440-1445.

Xia, Y., Burbank, D. E., Uher, L., Rabussay, D., and Van Etten, J. L. (1986a). Restriction endonuclease activity induced by PBCV-1 virus infection of a chlorella-like green alga. *Mol. Cell. Biol. 6*:1430-1439.

Xia, Y., Burbank, D. E., and Van Etten, J. L. (1986b). Restriction endonuclease activity induced by NC-1A virus infection of a chlorella-like green alga. *Nucleic Acids Res. 14*:6017-2030.

Xia, Y., Narva, K. E., and Van Etten, J. L. (1987). The cleavage site of the *Rsa*I isoschizomer, *C*viII, is G↓TAC. *Nucleic Acids Res. 15*:10063.

Yamada, T., and Sakaguchi, K. (1982). Electron microscopic studies of *Chlorella ellipsoidea* protoplast formation. *J. Gen. Microbiol. 128*:1319-1327.

Yonker, C. R., Caldwell, K. D., Giddings, J. C., and Van Etten, J. L. (1985). Physical characterization of PBCV virus by sedimentation field flow fractionation. *J. Virol. Methods 11*:145-160.

Index

A

ADH2, 76
Algal viruses, 411-423
Anastomosis groups, 387,389, 391,392,399,401
Antiviral activity, 1
Aspergillus, 200,344
Attenuation, 388
Autonomously replicating sequences, 273-275,277
Avocado black streak disease, 317

B

Bacteriophage M13, 124
Basidiomycete, 209
Biological control, 295,372,403-406
Blastomyces dermatiditis, 257
Botryodiplodia theobromae, 319-321
Bunchy top disease, 315

C

Candida albicans, 220,248,250-256
Candida tropicalis, 256
Candida, killing activity, 248
Capsid polypeptides, 118,356-363, 366,399
Carboxypeptidase Y, 168-169
Castanea dentata, 372-383
Castanea mollissima, 376
Cellulose chromatography, 309-310
Cell wall receptor, 169
Ceratocystis ulmi, 318
Chestnut blight, 295,371-383
Chloramphenicol resistance, 342-344
Chlorella viruses, 411-423
Chromosomal rearrangements, 63-83
CLO, 94,97
Clonal variation, 225
Closteroviruses, 314
Colicins, 217,235-236
Compatibility, 209-210,373-374, 387,399,405
Complementation, 213
Concerted events, 80-83
Condidia, 330
Conidial isolates, 344
Conidiogenesis, 340
Cryphonectria parasitica (see Endothia parasitica)
Cryptic viruses, 284,285,316
Cryptococcus neoformans, 251
killer activity, 251
Cucumoviruses, 314

Cucurbits, 317
Curing, 225
CYC1, 71-75
Cytochrome oxidase, 345
Cytoduction, 220-221,388
Cytoplasmic exchange, 402
Cytoplasmic inheritance, 213
Cytoplasmic male sterility, 285, 286
Cytoplasmic mixing, 214
Cytoplasmic transmission of virulence, 399

D

D-factors, 329
DEL1, 71-75
Deletions, 63-83
Delta element, 16-25,43-58,63-83
Density components, 226
Deoxyribonucleic acid
  circular, 21
  direct repeats, 68
  methyltransferase, 420
  mitochondrial, 345
  plasmids, 331
  synthesis, 419
  strong stop, 17,19
  viruses, 265-277,411-423
Diaporthin, 376,379
Dibasic endopeptidase, 166
Drift, 231
Duplications, 63-83
Dutch elm disease, 327-351
Dyad symmetry, 127

E

EDR1, 78
Elm xylem, 339-340
Endothia parasitica, 284,287,295-300,315-316,318-320,330, 342,371-383,387,388,401, 406
Enhancer, 17,19,23
Epidemic fronts, *Ophiostoma ulmi*, 336
Epidemiological typing, 252-254
Episodic selection, 330
Epsilon element, 17
Epstein-Barr virus, 416
Eutypa armeniacae, 319
Evolution, 32-34
Exclusion, 134-219,235
EXL, 92
Extracellular pectinase, 379
Extracellular transmission, 198

F

Fish lymphocystis disease virus, 416
Frog virus, 416
Fungi, plant pathogenic, 283

G

Gaeumannomyces graminis, 286, 318,335,358,388
Gag gene, 28-30
Gene conversion, 63-83
Genetic map, 215
Geotrichum candidum, 318
(1,6)-β-D-glucan, 169
Glycosylation, 164-165
Graft transmission, 317
Groundnut rosette, 315
Growth retardation, 388

H

Helminthosporium maydis, 335
Helminthosporium victoriae, 286, 353-368,388
  145S virus, 354-356
  190S virus, 354-368
  disease, 354-368
Helper function, double-stranded RNA, 340

Heteroduplex analysis, 231
Heterokaryon, 211-214
HIS4, 75
Histoplasma capsulatum, 257
HOK, 92
Host range, 412
Host receptor, 416
Human papilloma virus, 416
Hybridization, 215
Hybrids, *EAN/NAN*, 328
Hyphal anastomosis, 330,335,336, 340,344
Hypovirulence, 287,331,342,371-383,389,391,399,401,403
  type B, 381
  type JR, 381

I

Immunity, 181,184,187-189,229
  viral, 162-174
Incompatibility, 96,217,268,269, 271,273,329-331
Infection, 180,198-204
  cross-species, 200-204
Infectivity, 24
Influenza virus, 145,153
Insertion mutations, 42
Interallelic interactions, 213
Interferon, 2,5-11
Intergenic interactions, 213
Intermolecular bonds, 235
Internal deletion events, 291
Interstrain inhibition, 217
Intron, 23,24
Inversions, 63-83
Inverted terminal repeats, 266, 274,275,277
Iridoviridae, 416,422-423
Irregular meiosis, 210

K

KEX genes, 95,179,184,187-188, 193
[*KEX* genes]
  KEX1, 167-169,184,187-188,193
  KEX2, 166-167,184,188
Killer phenotype
  Aspergillus amselodami, 331
  Kluyveromyces lactis, 265-267, 273-275
  Saccharomyces cerevisiae, 27, 179,184,266-269,331,337, 340,343
  Ustilago maydis, 209-237
Killer toxins, 117,161-174,233-235,244-255,
  structure, 170-172
  receptors, 236
KRE genes, 95
  KRE1, 187,189-190
  KRE2, 187,189-190
  KRE3, 187,189-190

L

L-sarcin, 236
Latent d-infection, 337
Lentinus edodes, 360
Lettuce big vein disease, 315
Long terminal repeats, 16-25

M

MAK genes, 94,179
Male sterility in plants, 316
Mating type, 17,213
MBC fungicide, 335
Meiotic recombination, 214
Messenger RNA, 140-156
5-methyldeoxycytidine, 415-416
Mitochondria, 342-346
  DNA, 268,269,271
  mutants, 269
Mitotic recombination, 213
Mixed infection, dsRNA, 313
MKS genes, 95
MKT genes, 95

Modification and restriction, viral, 420-422
Mouse encephalomyocarditis virus, 2,11
Mutants, 162-165
Mutations, site-directed, 173
Mutator, 71-75
Mycelial fan, 376-379

N

N6-Methyldeoxyadenosine, 415
Necrophylactic periderm, 376
Necrotroph, 375
Neoplasms, 10
Neoplastic growth, 210
Neurospora crassa, 345
NEX, 92
Northern blots, 346
Nuclear genes, 221
Nuclease, 215
Nucleoside triphosphate phosphotransferase, 129,136-139
Nucleus, 27

O

Ophiostoma ulmi
- aggressive and unaggressive subgroups, 327-351
- ascospore progeny, 320,321, 335,339,340
- EAN v/c supergroup, 330
- NAN v/c supergroup, 330
- races, 327
- reactions, L, L-g, N, and W, 329-330
- yeast phase, 330,339,340

Overlap, gag/pol, 21
Oxalate, 379

P

Paramecium bursaria, 412
Pathogenicity, 399
Pathogenicity tests, 391
Pectolytic enzymes, 401
Penicillium chrysogenum, 6,195,197
Penicillium funiculosum, 4
Penicillium stoloniferum, 1,3,4
Penicillium variabile, 4
Penicillium viruses, 200,388
Pepper, 317
Peptide mapping, 228,399
Periconia circinata, 286,287
Perithecia, 335,373
Phaseolus vulgaris, 317
PHO5, 180,181,189
Phomopsis viticola, 319
Phosphorylation, 29
Pichia kluyveri, 170
Plant pathogenic fungi, 318
Plasma membrane receptor, 173
- instability, 268,269,271
- pGKL, 265-277
- linear, 265,266,276,277
- transfer, 267
- vectors, 273
- yeast, 265,266

Poky, 345
Pol gene, 32
Polyadenylic acid, 134,298-301
Poly rI:rC, 4
Polyuridylic acid, 298-301
Premature termination, 398
Preprotoxin, yeast killer virus, 133,136,162,181,184-186, 189,191,193
- alpha domain, 181,185,189,191, 193,194
- beta domain, 181,185,189,191, 193,194
- delta domain, 181,185,191
- gamma domain, 181,189,191,193

Primates, 6
Processing proteases, 165-169
Protease, 18
Protein
- modification, 29
- M1-P1, 185
- p14, 185
- p22, 185-187,189,193

Protoplasts, 117,215
  fusion, 267
Protoxin, yeast killer virus, 184
Provirus, 396
Pseudogenes, 26
PTX1, 181-185,191,192
Puccinia striiformis, 318
Pycnidia, 373
Pyricularia oryzae, 318

R

RAD7, 71
RAD52, 79
Radioiodination, 358
Recombination, 63-83
  site-specific, 74
  unequal sister-strand, 63-83
Reoviridae, 117,142,144,284,287-294,301
Repair, 214
Resistance, 221
Restriction endonuclease, 414, 415,420-422
Restriction fragment length polymorphisms, 345
Retrotransposons, 15-39,41-56, 63-83
Retroviruses, 15-39,64
Reverse transcriptase, 22,24,27, 32,68
  cellular gene, 33
REX genes, 95,185,187-188,193
  REX1, 185,187,193
Rhizoctonia repens, 403
Rhizoctonia solani, 286,318,387-406
  anastomosis groups, 387,389, 391,392,399,401
Rhizoctonia zeae, 404
Ribonuclease, 6-10
Ribonuclease H, 32
Ribonucleic acid
  antisense, 141
  capped, 151-153
[Ribonucleic acid]
  double-stranded, 5-8,91-110, 133-153,222-236,243,255-258,266,268-270,272,273, 283-301,309-321,331-351, 371-383,386-406
    deletion mutants, 337
    L, 179-194
    L1, 179,194
    Ly, 197
    M, 180,181,186,194
    M1, 179-181,184,194
  initiator methionine tRNA, 17,26
  pause products, 121-123
  polyA, 381-383
  polymerase, 103-111,135,228-231,380,392-397,419-420
  processing, 230
  single-stranded, 309,312,318
  variation, 391
Ribosomes, 140-148
Ruglosin, 379

S

S-adenosyl-methionine, 121
Saccharomyces cerevisiae, 15-208, 243-245,249-250,265-271, 273-275,388,405
Saccharomyces kluyveri, 265-266
Satellite tobacco mosaic virus, 314
Schizosaccharomyces pombe, 200, 201,273-275
Scolytid beetles, 335
ScV-M1, 198,199
ScV-M2, 198
ScV-P1, 194
SEC genes, 95,184
Secretion, 161-174
Sedimentation coefficients, 226
Senescence, *Podospora anserina*, 345
Sensitivity to toxins, 251-255

Serological detection of RNA, 311-312
Sexual reproduction, 330
Sigma element, 70-71
Signal sequence, 164-166
SKI genes, 95,201
Skyrin, 376
Solo delta insertion mutation, 44
Solopathogenic cells, 212
SPT genes, 45
  SPT2, 79
  SPT3, 46-56,80
  SPT7, 53-56
  SPT8, 54
Statolon, 1,2-4
Stopper, 345
SUF23, 77
SUP4, 65-83
Suppression, 45,98
Suppressive factors, 405
Suppressive mutants, 118-129

T

Target site, 66
TATA region, 55
Tau element, 70-71
Teliospore, 213
Tobamoviruses, 314
Tobramycin, 6-10
Tomato, 317
Torulopsis glabrata, 244-248, 255-256
Torulopsis toxin, 249
Totiviridae, 356,362,363,365,366
Transcription, 118-129
  conservative, 135-140
  initiation, 48-52,136-139,223
  in vitro, 363-366
  vectors, 291
Transcripts, viral, 393
Transformation, 214-215,267, 273,275
Translation, 129,140-156,226-229, 361,393,398-399
[Translation]
  hybrid-stimulated, 143,145, 148-151
  initiation, 140-156
Translocation, reciprocal, 63-83
Transmissible hypovirulence, 295, 315-316,318-320,371-383, 387-389,401
Transposable elements, 15-83
Transposition, 63-83
Trichoderma harzianum, 215
tRNA, 66
Ty elements, 15-83
  frameshift, 28
  protease, 31,32
  read-through, 28,31
  transposition mechanism, 17-26
  Tya, 17,28-30
  Tyb, 17,31-32
  Tyb-90 protein, 32
  Ty-VLP, 18,19,26-27

U,V

Ustilago maydis, 200,209-237,286, 388
Vesicles, 380
Vicia faba, 316
Victorin, 353
Viral exclusion, 405
Viral structural proteins, 194,197, 199,399
Viral transmission, transformation, 400-401
Virulence, 374-379,839-391
  transmissible, 399,401-405
Virus assembly centers, 417
Viruses in pathogenic yeast, 255-258
Virus lipids, 413

W,Y

Wheat takeall fungus, 388
Wound tumor virus, 283,287-294
Yarrowia lipolytica, 195,197,200, 201,358,361